Lecture Notes in Computer Science   4059

Commenced Publication in 1973
Founding and Former Series Editors:
Gerhard Goos, Juris Hartmanis, and Jan van Leeuwe

T0216286

Lars Arge   Rusins Freivalds (Eds.)

Algorithm Theory – SWAT 2006

10th Scandinavian Workshop on Algorithm Theory
Riga, Latvia, July 6-8, 2006
Proceedings

 Springer

Volume Editors

Lars Arge
University of Aarhus
Department of Computer Science
IT-Parken, Aabogade 34, 8200 Aarhus N, Denmark
E-mail: large@daimi.au.dk

Rusins Freivalds
University of Latvia
Institute of Mathematics and Computer Science
Raina bulvaris 29, Riga, LV-1459, Latvia
E-mail: rusins@cclu.lv

Library of Congress Control Number: 2006927810

CR Subject Classification (1998): F.2, E.1, G.2, I.3.5, C.2

LNCS Sublibrary: SL 1 – Theoretical Computer Science and General Issues

ISSN	0302-9743
ISBN-10	3-540-35753-X Springer Berlin Heidelberg New York
ISBN-13	978-3-540-35753-7 Springer Berlin Heidelberg New York

Springer is a part of Springer Science+Business Media

springer.com

© Springer-Verlag Berlin Heidelberg 2006
Printed in Germany

Typesetting: Camera-ready by author, data conversion by Scientific Publishing Services, Chennai, India
Printed on acid-free paper   SPIN: 11785293   06/3142   5 4 3 2 1 0

Preface

This volume contains the papers presented at SWAT 2006, the 10th Scandinavian Workshop on Algorithm Theory. The workshop, which is really a full-fledged international conference, is intended as a forum for researchers in the area of design and analysis of algorithms and data structures. Since 1988 SWAT has been held biennially in the five Nordic countries; it has a loose association with WADS (Workshop on Algorithms and Data Structures) that is held in odd-numbered years in North America. This 10th SWAT was held in the neighboring Baltic region. More precisely, it was held on July 6-8, 2006, at the Institute of Mathematics and Computer Science in the University of Latvia in Riga.

The call for papers invited contributions in all areas of algorithms and data structures, including approximation algorithms, computational biology, computational geometry, distributed algorithms, external-memory algorithms, graph algorithms, online algorithms, optimization algorithms, parallel algorithms, randomized algorithms, string algorithms and algorithmic game theory. A total of 154 papers were submitted, out of which the Program Committee selected 36 for presentation at the workshop. In addition, invited lectures were given by Kazuo Iwama (Kyoto University), Raimund Seidel (Universität des Saarlandes) and Robert E. Tarjan (Princeton University).

We would like to thank all the people who contributed to making SWAT 2006 a success. In particular, we thank the Program Committee and all of our many colleagues who helped the committee evaluate the submissions. We also thank Gerth S. Brodal for his invaluable help with the submission process and the Program Committee software.

May 2006   Lars Arge and Rusins Freivalds

Organization

SWAT 2006 Program Committee

Pankaj K. Agarwal, Duke University
Lars Arge, University of Aarhus (Co-chair)
Gerth S. Brodal, University of Aarhus
Adam Buchsbaum, AT&T Labs Research
Karlis Cerans, University of Latvia
Erik Demaine, Massachusetts Institute of Technology
Lars Engebretsen, Google Switzerland
Jeff Erickson, University of Illinois at Urbana-Champaign
Rusins Freivalds, University of Latvia (Co-chair)
Pinar Heggernes, University of Bergen
J. Ian Munro, University of Waterloo
S. Muthukrishnan, Rutgers University
Rasmus Pagh, IT University of Copenhagen
Hadas Shachnai, The Technion
Gerhard Woeginger, Eindhoven University of Technology

SWAT 2006 Organizing Committee

Rusins Freivalds, University of Latvia
Lelde Lace, University of Latvia
Andrzej Lingas, Lund University
Inara Opmane, University of Latvia (Chair)
Juris Smotrovs, University of Latvia

SWAT Steering Committee

Lars Arge, University of Aarhus
Magnús M. Halldórsson, University of Iceland
Rolf Karlsson, Lund University
Andrzej Lingas, Lund University
Jan Arne Telle, University of Bergen
Esko Ukkonen, University of Helsinki

SWAT 2006 Referees

Andre Allavena
Andris Ambainis
Spyros Angelopoulos
Mihai Bădoiu
Jeremy Barbay
Anne Berry
Philip Bille
Markus Bläser
Andrej Brodnik
Edith Cohen
Jose Correa
Ovidiu Daescu
Tamel Dey
Frederic Dorn
Alon Efrat
Leah Epstein
Rolf Fagerberg
Lene Favrholdt
Sandor Fekete
Jon Feldman
Jiri Fiala
Fedor Fomin
Dimitris Fotakis
Gudmund S. Frandsen
Karlis Freivalds
Martin Fuhrer
Rajiv Gandhi
Emden Gansner
Jie Gao
Serge Gaspers
Rezza Dorri Giv
Alex Golynski
Fabrizio Grandoni
Inge Li Gørtz
M. Hajiaghayi
Bjarni Halldorsson
Magnus M. Halldorsson
Angele Hamel
Kristoffer A. Hansen
Sariel Har-Peled
Stefan Hougardy
Alex Hudek
Thore Husfeldt

Peter Høyer
Riko Jacob
David Johnson
Tibor Jordan
Irit Katriel
Paulis Kikusts
Yoo-Ah Kim
Bettina Klinz
Joachim Kneis
Stephen Kobourov
Spyros Kontogiannis
Amos Korman
Guy Kortsarz
Annamaria Kovacs
Lukasz Kowalik
Daniel Kral
Dieter Kratsch
Lars M. Kristensen
Asaf Levin
Moshe Lewenstein
Anna Lubiw
Rune Bang Lyngsø
Thomas Mailund
David Manlove
Fredrik Manne
Conrado Martinez
Daniel Marx
Adam Mayerson
Taneli Mielikäinen
Peter Bro Miltersen
Kamesh Munagala
Gabriel Moruz
Christian W. Mortensen
Hannes Moser
Rolf Niedermeier
Bengt Nilsson
Johan Nilsson
Martin Olsen
Martin Pal
Charis Papadopoulos
Christian N.S. Pedersen
Derek Phillips
Karlis Podnieks

Kirk Pruhs
Artem Pyatkin
J. Radhakrishnan
Harald Raecke
Prabhakar Ragde
S. Srinivasa Rao
Dieter Rautenbach
Dror Rawitz
Oded Regev
Peter Rossmanith
Tim Roughgarden
Louis Salvail
Saket Saurabh
Gabriel Scalosub
Baruch Schieber
Jiri Sgall
Shakhar Smorodinsky
Robert Spalek
Michael Spriggs
Rob van Stee
Maxim Sviridenko
Mario Szegedy
Troels Bjerre Sørensen
Tami Tamir
Jan Arne Telle
Dimitrios Thilikos
Ioan Todinca
Kasturi Varadarajan
S. Venkatasubramanian
Juris Viksna
Tomas Vinar
Berthold Vöcking
Yusu Wang
Emo Welzl
Sebastian Wernicke
Stephan Westphal
Qin Xin
Anders Yeo
Ke Yi
Neal Young
Hai Yu
Martin Zachariasen
Lisa Zhang

Table of Contents

Invited Papers

Contributed Papers

Top-Down Analysis of Path Compression: Deriving the Inverse-Ackermann Bound Naturally (and Easily)

Raimund Seidel

Universität des Saarlandes, Fachrichtung Informatik, Im Stadtwald, D-66123
Saarbrücken, Germany
rseidel@cs.uni-sb.de

Path compression is used in a number of algorithms, most notably in various very natural solutions to the so-called Union-Find problem. This problem is basic and important enough to be covered in most introductory courses and textbooks on algorithms and data structures. However the performance analysis of the solutions is more often than not at best incomplete if not omitted altogether. Already the definition of the function α, the interesting constituent of the time bound, as a quasi inverse of the Ackermann function is complicated and not easy to understand.

All the previous analyses of path compression proceed in a bottom-up fashion, employing rather intricate charging schemes, sometimes cloaked in the language of potential functions for amortized analysis, and they need to introduce the Ackermann function beforehand in order to be properly formulated.

I will present a new [1], rather easy way of analyzing the running times of union-find algorithms. It is based on a relatively simple top-down approach and naturally leads by itself to this famous "Inverse Ackermann" function without ever having to talk about the Ackermann function itself.

I will discuss how this top-down approach can also be made to work for related procedures such as path compaction. Finally I will consider the case of moderately sized instances and will derive some explicit, rather small upper bounds on the number of pointer changes.

Reference

1. Seidel, R., Sharir, M.: Top-Down Analysis of Path Compression. SIAM J. Comput.
   34 (2005) 515–525.

L. Arge and R. Freivalds (Eds.): SWAT 2006, LNCS 4059, p. 1, 2006.
© Springer-Verlag Berlin Heidelberg 2006

Results and Problems on Self-adjusting Search Trees and Related Data Structures

Robert E. Tarjan

Department of Computer Science, Princeton University, Princeton, NJ
and
Hewlett Packard, Palo Alto, CA
ret@cs.princeton.edu

The splay tree is a form of self-adjusting search tree invented almost 25 years ago. Splay trees are remarkably efficient in both theory and practice, but many questions concerning splay trees and related data structures remain open. Foremost among these is the dynamic optimality conjecture, which states that the amortized efficiency of splay trees is optimum to within a constant factor among all kinds of binary search trees. That is, are splay trees constant-competitive? A broader question is whether there is any form of binary search tree that is constant-competitive. Recently, three different groups of researchers have devised kinds of search trees that are loglog-competitive, improving on the log-competitiveness of balanced trees. At least one of these data structures, the multisplay tree, has many if not all of the nice asymptotic properties of splay trees (even though it is more complicated than splay trees). We review this recent work and look at remaining open problems, of which there are many, including resolving the question of whether splay trees themselves are loglog-competitive.

We also look at a more complicated class of data structures that maintain information about a dynamic collection of disjoint trees. We review various versions of the dynamic trees problem, describe efficient solutions (both worst-case and amortized), and list open problems.

L. Arge and R. Freivalds (Eds.): SWAT 2006, LNCS 4059, p. 2, 2006.

Classic and Quantum Network Coding[*]

Kazuo Iwama

School of Informatics, Kyoto University, Kyoto 606-8501, Japan
iwama@kuis.kyoto-u.ac.jp

Ahlswede, Cai, Li, and Yeung (IEEE Trans. Inform. Theory, 2000) showed that the fundamental law for network flow, the max-flow min-cut theorem, no longer applies for "digital information flow." The simple, nice example they gave is called the Butterfly network illustrated in Fig. 1. The capacity of each directed link is all one and there are two source-sink pairs s_1 to t_1 and s_2 to t_2. Notice that both paths have to use the single link from s_0 to t_0 and hence the total amount of (conventional commodity) flow in both paths is bounded by one, say, $1/2$ for each. In the case of digital information flow, however, the protocol shown in Fig. 2 allows us to transmit two bits, x and y, simultaneously. Thus, we can effectively achieve larger channel capacity than can be achieved by simple routing. This is known as *network coding* since this seminal paper and has been quite popular as a mutual interest of theoretical computer science and information theory.

The natural question is whether such a capacity enhancement is also possible for *quantum* information, more specifically, whether we can transmit two qubits from s_1 to t_1 and s_2 to t_2 simultaneously, as with classical network coding. Note that there are (at least) two tricks in the classical case. One is the EX-OR (Exclusive-OR) operation at node s_0; one can see that the bit y is encoded by using x as a key which is sent directly from s_1 to t_2, and vise versa. The other is the exact copy of one-bit information at node t_0. Our answer to the question obviously depends on if we can find quantum counterparts for these key operations.

Neither seems easy: For the copy operation, there is the famous no-cloning theorem. Also, there is no obvious way of encoding a quantum state by a quantum state at s_0. Consider, for example, a simple extension of the classical operation at node s_0, i.e., a controlled unitary transform U as illustrated in Fig. 3. (Note that classical EX-OR is realized by setting $U = X$ "bit-flip.") Then, for any U, there is a quantum state $|\phi\rangle$ (actually an eigenvector of U) such that $|\phi\rangle$ and $U|\phi\rangle$ are identical (up to a global phase). Namely, if $|\psi_2\rangle = |\phi\rangle$, then the quantum state at the output of U is exactly the same for $|\psi_1\rangle = |0\rangle$ and $|\psi_1\rangle = |1\rangle$. This means their difference is completely lost at that position and hence is completely lost at t_1 also.

Nevertheless, we show that quantum network coding is possible if approximation is allowed. Our results for the Butterfly network include: (i) We can send any quantum state $|\psi_1\rangle$ from s_1 to t_1 and $|\psi_2\rangle$ from s_2 to t_2 simultaneously with a fidelity strictly greater than $1/2$. (ii) If one of $|\psi_1\rangle$ and $|\psi_2\rangle$ is classical, then the

---

[*] Supported in part by Scientific Research Grant, Ministry of Japan, 1609211 and 16300003.

L. Arge and R. Freivalds (Eds.): SWAT 2006, LNCS 4059, pp. 3–4, 2006.

fidelity can be improved to 2/3. (iii) Similar improvement is also possible if $|\psi_1\rangle$ and $|\psi_2\rangle$ are restricted to only a finite number of (previously known) states. (iv) Several impossibility results including the general upper bound of the fidelity are also given.

This is a joint work with Masahito Hayashi, Harumichi Nishimura, Rudy Raymond, and Shigeru Yamashita.

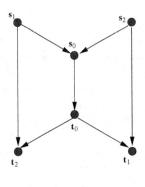

Fig. 1. Betterfly Network

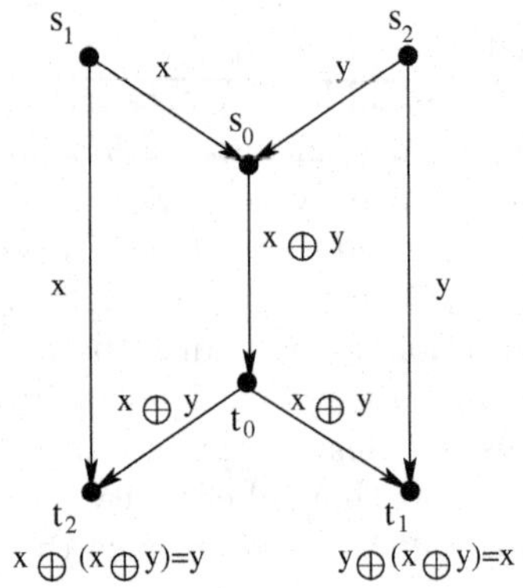

Fig. 2. Coding scheme

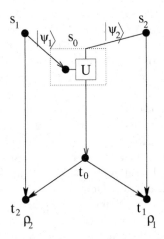

Fig. 3. Network using a controlled unitary operation

Multiplexing Packets with Arbitrary Deadlines in Bounded Buffers

Yossi Azar[1,*] and Nir Levy

School of Computer Science, Tel Aviv University, Tel Aviv, 69978, Israel
azar@tau.ac.il, levynir@tau.ac.il

Abstract. We study the online problem of multiplexing packets with arbitrary deadlines in bounded multi-buffer switch. In this model, a switch consists of m input buffers each with bounded capacity B and one output port. Each arriving packet is associated with a value and a deadline that specifies the time limit till the packet can be transmitted. At each time step the switch can select any non-empty buffer and transmit one packet from that buffer. In the preemptive model, stored packets may be preempted from their buffers due to lack of buffer space or discarded due to the violation of the deadline constraints. If preemption is not allowed, every packet accepted and stored in the buffer must be transmitted before its deadline has expired. The goal is to maximize the benefit of the packets transmitted by their deadlines. To date, most models for packets with deadlines assumed a single buffer. To the best of our knowledge this is the first time a bounded multi-buffer switch is used with arbitrary deadline constraints.

Our main result is a 9.82-competitive deterministic algorithm for packets with arbitrary values and deadlines. Note that the greedy algorithm is not competitive. For the non-preemptive model we present a 2-competitive deterministic algorithm for the unit value packets. For arbitrary values we present a randomized algorithm whose competitiveness is logarithmic in the ratio between the largest and the smallest value of the packets in the sequence.

1 Introduction

In recent years, the growth of the IP-Based networks, especially the Internet, and the growing number of services and applications that make use of those networks, such as VoD and VoIP applications, have led to an explosively growth in the number of end-users and appliances which make use of those services. As a result, various institutions such as offices, schools and factories, cellular providers and HotSpots (formatting private wireless cells), are providing access points to several end-users at the same time. In general, those service providers are equipped with switches in order to manage the incoming traffic, arriving from different users and applications (sessions). Due to bandwidth limitation

---

* Research supported in part by the Israeli Ministry of Industry and Trade and by the Israel Science Foundation.

L. Arge and R. Freivalds (Eds.): SWAT 2006, LNCS 4059, pp. 5–16, 2006.

of the outgoing link and due to the tendency of the incoming traffic to arrive in bursts, data packets from various sessions are repeatedly multiplexed in the switch with packets from other sessions that share the same outgoing link. This process of multiplexing packets requires a temporary storage of the arriving packets until the outgoing link is free and as a result, might cause delay in the transmission of the packets. For many applications this delay reveals several problems such as fluctuation in the arrival time of the packets at the target session which may cause for example high-definition media to jitter and in severe cases, when the incoming traffic is heavy, packet loss becomes unavoidable. These problems become an even bigger issue when the IP-Based networks are equipped with Quality of Service (QoS) capabilities. If this is the case and a benefit or deadline attached to each of the packets in the incoming traffic or if a guaranteed Quality of Service is given to the end-users, then, the management of the switch needs to decide effectively which packet to store and which to discard. Hence, during the past few years extensive research has been made in the area of packet multiplexing (also called multi-buffer switch).

Within this paper, we study the problems of packet multiplexing assuming the basic model of bounded multi-buffer switch. To date, most models for packets with deadlines assumed a single buffer. To the best of our knowledge this is the first time a bounded multi-buffer switch is used with arbitrary deadline constraints. We found this model worthwhile to study for its simplicity and for the reason that the simple greedy algorithm that achieves a good competitive ratio in similar models fails to be competitive in our model.

Our results: Our main result applies to the preemptive model of the multi-buffer switch. In this model we present a 9.82-competitive deterministic algorithm for packets with values and arbitrary deadlines. Next, we consider the non-preemptive model and present a 2-competitive deterministic algorithm for the unit value packets with arbitrary deadlines. For arbitrary values we present a $O(logF)$-competitive randomized algorithm where F is the ratio between the largest and the smallest value of the packets in the sequence. This result is tight up to a constant factor. We also present a lower bound of F for every deterministic algorithm in this model.

Since there are no results for multi-buffer switches with arbitrary deadlines we view the results of multi-buffer switches without deadlines and the results for a single buffer switch with deadlines separately.

Related results for throughput maximization for multi-buffer switches: In the preemptive model for packets with values, a 2-competitive algorithm for the multi-buffer model and a 4-competitive algorithm for the multi-queue model, which consists of FIFO queues, are shown in [4]. For the multi-queue model the ratio was improved to 3 by [5]. For the general non-preemptive model, using the result of Andelman *et al.* [2], a $(2e\lceil \ln F \rceil)$-competitive algorithm is given by [4]. For the multi-buffer switch with unit value packets, the competitive ratio can get below 2. Albers and Schmidt [1] show a $\frac{17}{9} \approx 1.89$-competitive deterministic algorithm, for $B \geq 2$. A 1.58-competitive for switches with large buffers is shown in [3]. For randomized online algorithms a 1.58-competitive algorithm is shown in [4] and a 1.5-competitive is shown in [12] . A lower bound of 1.58 for the competitive ratio of any deterministic online algorithm and a lower bound of 1.4659 on the performance of every randomized online algorithm, for any B and large enough m is given by Albers and Schmidt [1].

Related results for throughput maximization for single buffer switches with deadlines: When deadlines become an integral part of the model, all models presented so far assume a single buffer. Note that when the buffers are unbounded there is no difference between the single and the multi-buffer models. For packets with general deadlines, Kesselman *et al.* [10] and Hajek [9] showed that a simple greedy algorithm is 2-competitive. Later, Chrobak *et al.* [8] presented an algorithm with competitive ratio of $\frac{64}{63} \approx 1.939$. For the general case a lower bound of ϕ is known [7, 2, 9] where $\phi \approx 1.618$ is the Golden Ratio. Bartal *et al.* [6] presented a 1.58-competitive randomized algorithm. For the special case of δ-bounded delay i.e., the difference between the deadline and the arrival time is at most δ, Bartal *et al.* [6] showed a $2 - 2/\delta + o(1/\delta)$-competitive ratio algorithm. Note that for 3-bounded instances this ratio is ϕ and it matches the lower bound. For the agreeable-deadlines model i.e., earlier arriving packets have earlier deadlines, Chrobak *et al.* [8] showed a competitive ratio algorithm of 1.838. Recently, this ratio was improved by Fei *et al.* [11] presenting a competitive ratio of ϕ.

1.1   Problem Definition and Notations

In this model, we are given a switch consists of m input buffers and one output port. Each buffer has a limited capacity and can store up to B packets at a time. Time is divided into discrete time steps. At each time step packets may arrive to some of the buffers. All packets are of equal size. Each arriving packet is associated with a value and a deadline. A packet can be stored in the buffer if an empty space is available and arriving packets that are not being stored must be discarded. At each time step the switch can select any non-empty buffer and transmit one packet from that buffer. In the preemptive model, stored packets may be preempted from their buffers due to lack of buffer space or discarded due to the violation of the deadline constraints. If preemption is not allowed then, every packet accepted and stored in the buffer must be transmitted before its deadline has expired. The goal of this model is to maximize the benefit of the packets transmitted by their deadlines.

We view the sequence σ as arrival and transmission events. For each packet p we denote by $d(p)$ the deadline of p and by $v(p)$ the value of p. We let $v(\emptyset) = 0$. For any online algorithm $\mathcal{A}$ (respectively, the optimum algorithm $\mathcal{OPT}$), we denote by B_t (respectively, B_t^*), the set of all packets residing in $\mathcal{A}$ (respectively, $\mathcal{OPT}$) buffers at time step t. We denote by $B_{i,t}$ (respectively, $B_{i,t}^*$) the packets residing in buffer i of $\mathcal{A}$ (respectively, $\mathcal{OPT}$) at time step t. Let TR_t (respectivly TR_t^*) denote the set of packets transmitted by $\mathcal{A}$ (respectivly, $\mathcal{OPT}$) until and including time step t. When it is clear from the context we may abuse the notation and omit the subscript t. In addition, we denote by $\mathcal{A}(\sigma)$ (respectively, $\mathcal{OPT}(\sigma)$) the total gain achieved by algorithm $\mathcal{A}$ (respectively, $\mathcal{OPT}$) on the incoming sequence σ. We say that a deterministic online algorithm $\mathcal{A}$ is c-competitive if for every incoming sequence of packets σ we have: $OPT(\sigma) \leq c \cdot \mathcal{A}(\sigma)$ (for randomized online algorithms we change $\mathcal{A}(\sigma)$ by $E[\mathcal{A}(\sigma)]$).

2 Preemptive Algorithm for Packets with Values and Deadlines

In this section, we present a deterministic online algorithm for the preemptive multi-buffer switch for packets with values and arbitrary deadlines. The outline of the algorithm is as follows. During the arrival phase each arriving packet p that is accepted to one of the buffers is assigned to a **tentative time slot** t so that t will be the last time step at which p will reside in the buffers. That is, after time step t if p was not transmitted or preempted it will be discarded from the buffers. In order to accept a packet we need to find an available place in the buffer and a free time slot. If this requires preemption then the total value of all preempted packets should be smaller by a constant factor $\alpha > 1$ than the accepted packet. In the transmission phase the packet in the current tentative time slot (if exists) or the current heaviest packet in the buffers, is transmitted. The decision is made by comparing their ratio to a constant factor $\beta > 1$.

To describe the algorithm we begin by introducing some additional notations. A time slot t is called **empty** if it is not assigned to any of the packets in B_t. For each packet p in $\mathcal{ON}$ buffers let $S(p)$ denote the tentative time slot assigned to p. We refer to $S(\emptyset)$ as the earliest empty tentative time slot. We denote by $S^{-1}(t)$ the inverse function of S that maps for each tentative time slot t the packet that is assigned to. Similary, we denote by $S^*(p)$ the time slot to which p is assigned for transmission by $\mathcal{OPT}$. Let $SC_t^d = \{p : p \in B_t \text{ and } S(p) \leq d\}$ be the set of packets residing in $\mathcal{ON}$ buffers at time step t and are scheduled for transmission before or at time step d. Clearly $|SC_t^{d(p)}| \leq d(p)$. A detailed description of the algorithm appears in figure 1.

Theorem 1. *For an optimal choice of the parameters α and β algorithm $\mathcal{ON}$ is 9.82-competitive.*

In order to prove the theorem we present a load assignment L_M that maintains, at each time step t, an assignment of the value $v(p)$ of every packet $p \in B_t^* \cup TR_t^*$ to the packets in $B_t \cup TR_t$ such that each packet $p \in B_t$ is assigned a load $w(p)$ of at most c' times $v(p)$ and each packet $p \in TR_t$ is assigned a load $w(p)$ of at most c times $v(p)$ where $w(p)$ is the total load assigned to packet $p \in B_t \cup TR_t$. Moreover, in this assignment, no packet in TR_t^* assign its value to a packet in B_t. This implies that every packet in TR_t^* assign its value only to packets in TR_t. This implies that

$$\sum_{p \in TR^*} v(p) \leq \sum_{p \in TR} w(p) \leq \sum_{p \in TR} c \cdot v(p)$$

which then yields a c-competitive algorithm.

Algorithm [$\mathcal{ON}$]

Arrival: (a new packet p arrives to buffer i at time step t)

Let p_{min} be the packet with minimum value in $B_{i,t}$ if $B_{i,t}$ is full or $\emptyset$ if $B_{i,t}$ is not full.
Let p_{sc} be the packet with minimum value in $SC_t^{d(p)}$ if $|SC_t^{d(p)}| = d(p)$ or $\emptyset$ if $|SC_t^{d(p)}| < d(p)$.

- **If** $S(p_{min}) \leq d(p)$ *and* $v(p_{min}) \leq v(p)$ discard p_{min}, insert p and assign $S(p_{min})$ to p. **Else,**
- **If** $\alpha \cdot (v(p_{min}) + v(p_{sc})) \leq v(p)$ discard p_{min} and p_{sc}, insert p and assign $S(p_{sc})$ to p. **Else,** discard p.

Transmission: (at time step t)

Let p_e denote the packet assigned to the tentative time slot t, or $\emptyset$ if no such packet exists.
Let p_h denote a packet with maximum value in B_t.

- **If** $v(p_e) \geq \beta \cdot v(p_h)$ transmit p_e. **Else,** transmit p_h and discard p_e from its buffer.

Fig. 1. Algorithm $\mathcal{ON}$

The proof is partitioned into several steps. In the first step we introduce the load assignment L_M prove its validity and show that the complete value of every packet $p \in TR_t^*$ is assigned to the packets in TR_t. In the second step we bound the load assigned to each of the packets $p \in B_t$ during any time step t. In the third step we bound the load assigned to each transmitted packet $p \in TR_t$ by showing that for every transmitted packet $p \in TR$, $w(p) \leq c \cdot v(p)$. That will complete the proof.

Step 1: The load assignment [L_M]
The load assignment follows the arrival and transmission phases of Algorithm $\mathcal{ON}$. We partition the arrival phase into two steps. In the first step the value of a newly accepted packet p by $\mathcal{OPT}$ is divided between at most two packets in B_t. In the second step the load, assigned to packets that are preempted due to the acceptance of a new packet p by $\mathcal{ON}$, is re-assigned to the packets in B_t after the acceptance of p. In the transmission phase, we re-assign the value of the packet transmitted by $\mathcal{OPT}$ which is assigned to packets in $\mathcal{ON}$ buffers to the packet transmitted by $\mathcal{ON}$.

We present the assignment of the value of the packets accepted by $\mathcal{OPT}$ to the packets accepted and transmitted by $\mathcal{ON}$ as a directed bipartite graph $G = (V^* \cup V, E)$ from $V^* = B_t^* \cup TR_t^*$ to $V = B_t \cup TR_t$. Packets arrive on-line and each packet accepted by $\mathcal{OPT}$ or $\mathcal{ON}$ (or both) is added to B_t or B_t^* respectivly. For every packet accepted by $\mathcal{OPT}$ the load assignment creates at most two emanating edges such that each emanating edge e is assigned a positive value $v(e)$. Throughout the algorithm the load assignment is described by the edges of the bipartite graph G. Note that every packet causes the creation of up to two vertices in G, $p^{v^*} \in V^*$ and $p^v \in V$. When the context is clear we slightly abuse the notation and refer to the vertex p^{v^*} as the packet p and to the vertex p^v as the packet p.

We say that a set of edges F implies a matching in a bipartite graph $G = (V^* \cup V, F)$ if the in-degree of every vertex $v \in V$ and the out-degree of every vertex $v \in V^*$ are at most one. We say that a packet $p \in B_t$ is **unmatched** in a matching X if its in-degree is 0 with respect to X.

The load assignment consists of three types of edges $\mathcal{M}, \mathcal{Q}$ and $\mathcal{L}$ and every edge is immediately assigned a type at its creation time. The three types of edges $\mathcal{M}, \mathcal{Q}$ and $\mathcal{L}$ form the three subgraphs $G^{\mathcal{M}} = (V^* \cup V, \mathcal{M})$, $G^{\mathcal{Q}} = (V^* \cup V, \mathcal{Q})$ and $G^{\mathcal{L}} = (V^* \cup V, \mathcal{L})$ respectivly, such that $G = G^{\mathcal{M}} \cup G^{\mathcal{Q}} \cup G^{\mathcal{L}}$. Throughout the algorithm the load assignment maintains a matching in the subgraphs $G^{\mathcal{M}}$ and $G^{\mathcal{Q}}$ such that every edge $(p, p') \in \mathcal{M}$ induced by $B_t^* \cup B_t$ is always between packets of the same buffer i.e., $p \in B_{i,t}^*$ and $p' \in B_{i,t}$ for some buffer i. In addition, every edge $(p, p') \in \mathcal{Q}$ induced by $B_t^* \cup B_t$ is always between packets such that $S^*(p) = S(p')$. For the subgraph $G^{\mathcal{L}}$ the load assignment maintains an out-degree of at most two (the in-degree can be arbitrary).

For simplicity of notation we refer to $(\emptyset, p), (p', \emptyset)$ and $(\emptyset, \emptyset)$ as $\emptyset$. We denote by

$$(p, p') \in X \to (p, p'') \in Y$$

the transition of the edge emanating from p and entering p' to entering the packet p'' and the transition in the edge-type from type X to type Y. We denote by

$$\forall p \ (p, p') \in X \to (p, p'') \in Y$$

the set transition

$$\forall p \ s.t. \ (p, p') \in X \ set \ (p, p') \in X \to (p, p'') \in Y.$$

We denote by $w_m(p'), w_q(p')$ and $w_l(p')$ the load assigned to a packet p' as the total value of the edges entering to p' of type $\mathcal{M}, \mathcal{Q}$ and $\mathcal{L}$ respectively. Note that $w(p') = w_m(p') + w_q(p') + w_l(p')$. The load assignment is described in details in figure 2.

Note that the packets p_{min} and p_{sc} as defined in algorithm $\mathcal{ON}$ and the packets p_b and p_s as defined in the load assignment are functions of the arriving packets. We omit the argument of the arriving packet when it is clear from the context. We start with the following observations.

Observation 1. *Let e_m and e_q denote the edges emanating from the newly accepted packet p at time step t and were set an edge type $\mathcal{M}$ and $\mathcal{Q}$ respectively. Then, for every time step t', $t' \geq t$ if e_m and e_q are entering to packets in $B_{t'}$ we have that:*

Load Assignment:L_M

1. Let packet p arrive at time step t to buffer i.
   (a) **If p was accepted by $\mathcal{OPT}$:**
       Let $p_b \in B_{i,t}$ be any unmatched packet in $\mathcal{M}$ if $B_{i,t}$ is full or $\emptyset$ if $B_{i,t}$ is not full.
       Let $p_s = S^{-1}(S^*(p))$ be the packet assigned to the same time slot as p.
       - **If p was accepted by $\mathcal{ON}$:** Add $\mathcal{M} \leftarrow \mathcal{M}\cup(p^{v^*}, p^v)$ and assign $v(p^{v^*}, p^v) = v(p)$.
       - **If p was rejected by $\mathcal{ON}$:**
         - Add $\mathcal{M} \leftarrow \mathcal{M}\cup(p, p_b)$ and assign $v(p, p_b) = v(p) \cdot \frac{v(p_b)}{v(p_b)+v(p_s)}$.
         - Add $\mathcal{Q} \leftarrow \mathcal{Q}\cup(p, p_s)$ and assign $v(p, p_s) = v(p) \cdot \frac{v(p_s)}{v(p_b)+v(p_s)}$.
   (b) **If p was accepted by $\mathcal{ON}$:**
       Let p_{min} *and* p_{sc} be the packets as defined in the arrival phase.
       i. Let $p' \in B_{i,t} \cup \{p\}$ be any unmatched packet in $\mathcal{M}$. Set $\forall q \ (q, p_{min}) \in \mathcal{M} \rightarrow (q, p') \in \mathcal{M}$.
       ii. **If** $S(p_{min}) \leq d(p)$ set $\forall q \ (q, p_{min}) \in \mathcal{Q} \rightarrow (q, p) \in \mathcal{Q}$.
       iii. **If** $S(p_{min}) > d(p)$ set $\forall q \ (q, p_{min}) \in \mathcal{Q} \rightarrow (q, p) \in \mathcal{L}$.
       iv. Set $\forall q \ (q, p_{min}) \in \mathcal{L} \rightarrow (q, p) \in \mathcal{L}$
       v. Set $\forall q \ (q, p_{sc}) \in \{\mathcal{M} \cup \mathcal{L}\} \rightarrow (q, p) \in \mathcal{L}$
       vi. Set $(q, p_{sc}) \in \mathcal{Q} \rightarrow (q, p) \in \mathcal{Q}$
2. Let p denote the packet transmitted by $\mathcal{OPT}$ at time step t. Let $E_R \subseteq \{\mathcal{M}\cup\mathcal{L}\}$ denote a subset of edges (p, p') such that $S(p') > t$.
   - **If $\mathcal{ON}$ transmitted p_e** set $E_R \rightarrow (p, p_e) \in \mathcal{L}$.
   - **If $\mathcal{ON}$ transmitted p_h** set $\forall q \ (q, p_e) \in \{\mathcal{M} \cup \mathcal{Q} \cup \mathcal{L}\} \rightarrow (q, p_h) \in \mathcal{L}$ and $E_R \rightarrow (p, p_h) \in \mathcal{L}$.

Fig. 2. Load Assignment :L_M

1. e_m *is never set a type* $\mathcal{Q}$.
2. e_q *is never set a type* $\mathcal{M}$.

Observation 2. *The set of edges of type $\mathcal{M}$ or $\mathcal{Q}$ implies a matching on the bipartite subgraphs $G^{\mathcal{M}}$ and $G^{\mathcal{Q}}$ respectively.*

Observation 3. *Let p' be any packet preempted from $\mathcal{ON}$ buffers due to the acceptance of a new packet p. Then the following properties holds:*

1. *$\forall q$ the edge $(q, p') \in \mathcal{Q}$ and the edges $(q, p') \in \mathcal{L}$ are always re-assigned to the preempting packet p.*
2. *The edges entering to p' are always re-assigned to a packet p'' such that $v(p'') \geq v(p')$.*

In the next lemma we show that the load assignment is feasible. Specifically, the only place where it is not trivial is in step $1(a)$ where we assume an unmatched packet in $\mathcal{M}$ if the buffer B_i is not full.

Lemma 1. *The definition of the load assignment L_M is feasible. That is, whenever a packet is accepted by $\mathcal{OPT}$ to buffer i there is at least one packet $p' \in B_i$ that is unmatched in $\mathcal{M}$ or B_i is not full.*

Lemma 2. *For every time step t, every edge emanating from a packet $p \in TR_t^*$ is entering to a packet in TR_t.*

Step 2: Bounding the load on the packets in B_t

In this step we bound the load assigned to the packet in $\mathcal{ON}$ buffers as the following lemma states.

Lemma 3. *For every time t and every packet $p \in B_t$ the load $w(p)$, satisfies $w(p) \leq (2\alpha + \frac{\alpha}{\alpha-1})v(p)$.*

Proof. We first show that the value assigned to an edge is bounded by α times the value of the packet it was first assigned to. Next, we bound separately each of the loads $w_q(p')$, $w_m(p')$, and $w_l(p')$ assigned to a packet $p' \in B_t$ as a factor of $v(p')$ and conclude the bound in the lemma.

Lemma 4. *Let p denote a packet accepted by $\mathcal{OPT}$ at time step t. Let $p' \in B_t$ denote any packet assigned an edge emanating from p. Then the value assigned to the edge (p, p') at the arrival of p satisfies $v(p, p') \leq \alpha v(p')$.*

Lemma 5. *For every time step t and every packet $p \in B_t$, $w_m(p) \leq \alpha \cdot v(p)$ and $w_q(p) \leq \alpha \cdot v(p)$.*

Lemma 6. *During any time step t and every packet $p \in B_t$, $w_l(p) \leq \frac{\alpha}{\alpha-1} \cdot v(p)$.*

Proof. Let p' be any packet in B_t. According to the load assignment the transmission or the acceptance of packets by $\mathcal{OPT}$ cannot change the load assigned to $w_l(p')$. In addition, no load is assigned to $w_l(p')$ after the arrival of p'. Therefore, we prove the lemma by induction on the changes which occur to the load assigned to $w_l(p')$ only at the arrival time of p'.

First, observe that if there is no packet preempted due to the acceptance of p' then, no edges are re-assigned to p. For the initial state, the lemma clearly holds. Next, let p' denote a new packet accepted by $\mathcal{ON}$ to buffer i at time step t and consider the edges of type $\mathcal{L}$ assigned to p' on its arrival. We will consider all possible cases in which a packet might be preempted from B_t, due to the acceptance of p', and change the load $w_l(p')$. We consider the following cases:

1. **p_{min} is preempted and $S(p_{min}) \leq d(p')$:**

   If that is the case, then only p_{min} is preempted due to the acceptance of p'. Therefore, according to the load assignment $w_l(p')$ is assigned only the load from the edges assigned to $w_l(p_{min})$. Therefore,

   $$w_l(p') = w_l(p_{min}) \leq \frac{\alpha}{\alpha - 1} \cdot v(p_{min}) \leq \frac{\alpha}{\alpha - 1} \cdot v(p')$$

   Where the first inequality is due to the induction hypothesis and the second inequality is due to the first condition in the arrival phase.

2. p_{min} **is preempted and** $S(p_{min}) > d(p')$, **and/or** p_{sc} **is preempted:**

If that is the case, p_{min} and/or p_{sc} assign a load to $w_l(p')$. Therefore,

$$
\begin{aligned}
w_l(p') &\leq w_b(p_{sc}) + w_l(p_{sc}) + w_s(p_{min}) + w_l(p_{min}) \\
&\leq w_b(p_{sc}) + \frac{\alpha}{\alpha - 1} \cdot v(p_{sc}) + w_s(p_{min}) + \frac{\alpha}{\alpha - 1} \cdot v(p_{min}) \\
&\leq \alpha \cdot v(p_{sc}) + \frac{\alpha}{\alpha - 1} \cdot v(p_{sc}) + \alpha \cdot v(p_{min}) + \frac{\alpha}{\alpha - 1} \cdot v(p_{min}) \\
&= \alpha \cdot (v(p_{sc}) + v(p_{min})) \cdot (1 + \frac{1}{\alpha - 1}) \leq v(p') \cdot (1 + \frac{1}{\alpha - 1}) \\
&= \frac{\alpha}{\alpha - 1} \cdot v(p')
\end{aligned}
$$

where the first inequality is due to the maximum load that can be assigned, by the load assignment, to $w_l(p')$ from the preempted packets p_{sc} and p_{min}. The second inequality is due to the induction hypothesis and the third inequality is due to lemma 5. The fourth inequality is due to the second condition of the arrival phase. □

We are now ready to complete the proof of lemma 3.

Since $w(p) = w_q(p) + w_m(p) + w_l(p)$ we conclude by lemma 5 and lemma 6 that $w(p) \leq (2\alpha + \frac{\alpha}{\alpha - 1})v(p)$. That completes the proof of lemma 3. □

Step 3: Bounding the load assigned to the transmitted packets

Before we can bound the load w assigned to the packets transmitted by $\mathcal{ON}$ we need to bound the value of the edges emanating from the packet transmitted by $\mathcal{OPT}$ and are re-assigned to the packet transmitted by $\mathcal{ON}$. We first show that there can be at most one such edge.

Lemma 7. *Whenever a packet p is transmitted by $\mathcal{OPT}$ there is at most one edge emanating from p such that $(p, p') \in \{\mathcal{M} \cup \mathcal{L}\}$ and $S(p') > t$ for some $p' \in B_t$.*

Corollary 1. *Let p denote a packet transmitted by $\mathcal{OPT}$ at time step t and let $p' \in B_t$ denote a packet such that $e = (p, p') \in \{\mathcal{M} \cup \mathcal{L}\}$ and $S(p') > t$. Then, only the edge $(p, p_b) \in \mathcal{M}$ created at the arrival of p can meet the conditions of e.*

Lemma 8. *Let p denote the packet transmitted by $\mathcal{OPT}$ at time step t. And let p_h be the heaviest packet in B_t. Let e_R denote any edge in E_R as defined in the load assignment. Then, $v(e_R) \leq \alpha v(p_h)$.*

We are now ready to bound the load w assigned to the packet transmitted by $\mathcal{ON}$.

Lemma 9. *For every packet p transmitted by $\mathcal{ON}$ $w(p) \leq 9.82v(p)$ for $\alpha = 1.55$ and $\beta = 2.51$.*

Proof. Let p_e *and* p_h be the packets as defined in the transmission phase at time step t. Let e_R denote the edge in E_R as defined in the load assignment. We consider the following two cases:

1. **If $\mathcal{ON}$ transmitted p_e:**
   Then according to the load assignment, the load assigned to p_e on its transmission is:

$$w(p_e) = w(p_e) + v(e_R) \leq (2\alpha + \frac{\alpha}{\alpha - 1})v(p_e) + v(e_R)$$

$$\leq (2\alpha + \frac{\alpha}{\alpha - 1})v(p_e) + \alpha v(p_h)$$

$$\leq (2\alpha + \frac{\alpha}{\alpha - 1})v(p_e) + \frac{\alpha}{\beta}v(p_e)$$

$$= ((2\alpha + \frac{\alpha}{\alpha - 1}) + \frac{\alpha}{\beta})v(p_e) \leq 9.82 v(p_e)$$

   where the first inequality is due to lemma 3. The second inequality is due to lemma 8. The third inequality follows the transmission criteria of algorithm $\mathcal{ON}$ such that $v(p_e) \geq \beta v(p_h)$. The forth inequality is by setting $\alpha = 1.55$ and $\beta = 2.51$.

2. **If $\mathcal{ON}$ transmitted p_h:**
   Then according to the load assignment, the load assigned to p_h on its transmission is:

$$w(p_h) = w(p_e) + w(p_h) + v(e_R) \leq (2\alpha + \frac{\alpha}{\alpha - 1})(v(p_e) + v(p_h)) + v(e_R)$$

$$\leq (2\alpha + \frac{\alpha}{\alpha - 1})(v(p_e) + v(p_h)) + \alpha v(p_h)$$

$$\leq (2\alpha + \frac{\alpha}{\alpha - 1})(\beta v(p_h) + v(p_h)) + \alpha v(p_h)$$

$$= ((2\alpha + \frac{\alpha}{\alpha - 1})(\beta + 1) + \alpha)v(p_h) \leq 9.82 v(p_h)$$

   where the first inequality is due to lemma 3 . The second inequality is due to lemma 8. The third inequality follows the transmission criteria of algorithm $\mathcal{ON}$ such that $v(p_e) \geq \beta v(p_h)$. The forth inequality is by setting $\alpha = 1.55$ and $\beta = 2.51$.   □

That completes the proof of theorem 1.

3   Non-preemptive Algorithms

In this section, we introduce a non-preemptive deterministic online algorithm for the multi-buffer switch with unit value packets and a non-preemptive randomized online algorithm for the multi-buffer switch with packets with values. Both models assume arbitrary deadlines on the packets.

3.1   Algorithm for Unit Value Packets-Algorithm $\mathcal{JF}$

In this section, we presents a non-preemptive deterministic algorithm for unit value packets, denoted as $\mathcal{JF}$. Algorithm $\mathcal{JF}$ transmits all packets accepted in FIFO order and store newly arrived packets only if at the originated transmission time they are not expired. A description of the algorithm appears in figure 3.

Algorithm [$\mathcal{JF}$]

- **Arrival:**(a new packet p arrives to buffer j at time step t)
  If buffer j is full or $d(p) \leq |B_t|$ discard p else, insert p to buffer j.
- **Transmission:** Transmit the earliest packet that arrived and was stored in the buffers.

Fig. 3. Algorithm $\mathcal{JF}$

Theorem 2. *Algorithm $\mathcal{JF}$ is 2-competitive.*

3.2   Algorithm $\mathcal{RJF}$

In this section, we present a non-preemptive randomized algorithm, denoted as $\mathcal{RJF}$ for packets with values. We assume the ratio F between the largest and the smallest value of the packets in the sequence is known in advance. The $\mathcal{RJF}$ algorithm is stated according to the "classify and randomly select" paradigm and it makes use of the $\mathcal{JF}$ algorithm that was stated in the previous section. Our result is $O(logF)$ and it is optimal up to a constant factor since, even for a single buffer and without deadlines Andelman *et al.* [2] show a lower bound of $O(logF)$.

Algorithm $\mathcal{RJF}$ classifies the input sequence into $\lfloor logF \rfloor + 1$ classes. Then, it randomly selects a class with probability $\frac{1}{\lfloor logF \rfloor + 1}$. After a class is selected $\mathcal{RJF}$ accepts only packets associated to the chosen class and exercise algorithm $\mathcal{JF}$ on the accepted packets while ignoring their values .

A description of the $\mathcal{RJF}$ algorithm appears in figure 4.

Algorithm [$\mathcal{RJF}$]

- Randomly select class j uniformly from the $\lfloor logF \rfloor + 1$ available classes.

  On the arrival of a new packet p to buffer i at time step t:
- **If** $2^{j-1} \leq v(p) < 2^j$: Exercise algorithm $\mathcal{JF}$ on p ignoring its value. **Else,** discard p.

Fig. 4. Algorithm $\mathcal{RJF}$

Theorem 3. *Algorithm $\mathcal{RJF}$ is $O(logF)$ competitive.*

It is straightforward to show that randomization is necessary in order to achieve a $O(logF)$ bound. Specifically, a lower bound of F for every deterministic non-preemptive algorithm is shown.

Lemma 10. *Every deterministic algorithm for the non-preemptive multi-buffer switch is at least F-competitive where F is the ratio between the largest and the smallest value of the packets in the sequence.*

References

1. S. Albers and M. Schmidt. On the performance of greedy algorithms in packet buffering. In Proc. 36th ACM Symp. on Theory of Computing, 35-44, 2004.
2. N. Andelman, Y. Mansour, and A. Zhu. Competitive queueing policies for QoS switches. In Proc. 14th ACM-SIAM Symp. on Discrete Algorithms, 761-770, 2003.
3. Y. Azar and A. Litichevskey. Maximizing throughput in multi-queue switches. In Proc. 12th Annual European Symposium on Algorithms, 53-64, 2004.
4. Y. Azar and Y. Richter. Management of multi-queue switches in QoS networks. In Proc. 35th ACM Symp. on Theory of Computing, 82-89, 2003.
5. Y. Azar and Y. Richter. The zero-one principle for switching networks. In Proc. 36th ACM Symp. on Theory of Computing, 2004. 64-71.
6. Y. Bartal, F. Y. L. Chin, M. Chrobak, S. P. Y. Fung, W. Jawor, R. Lavi, J. Sgall, and T. Tichy. Online competitive algorithms for maximizing weighted throughput of unit jobs. In Proc. 21st Symp. on Theoretical Aspects of Computer Science (STACS), volume 2996 of LNCS, 187-198. Springer, 2004.
7. Francis Y. L Chin and Stanley P. Y. Fung. Online scheduling for partial job values: Does timesharing or randomization help? Algorithmica, 37:149-164, 2003.
8. M. Chrobak, W. Jawor, J. Sgall, and T. Tichy. Improved online algorithms for buffer management in QoS switches. In Proc. 12th European Symp. on Algorithms (ESA), volume 3221 of LNCS, 204-215. Springer, 2004.
9. B. Hajek. On the competitiveness of online scheduling of unit-length packets with hard deadlines in slotted time. In Conference in Information Sciences and Systems, 434-438, 2001.
10. A. Kesselman, Z. Lotker, Y. Mansour, B. Patt-Shamir, B. Schieber, and M. Sviridenko. Buffer overflow management in QoS switches. In Proc. 33rd ACM Symp. on Theory of Computing, 520-529, 2001.
11. F.Li, J. Sethuraman, and C.Stein. An optimal online algorithm for packet scheduling with agreeable deadlines. In Proc. 16th Symp. on Discrete Algorithms (SODA), 801-802, 2005.
12. M. Schmidt. Packet buffering: Randomization beats deterministic algorithms. In Proc. 22st Symp. on Theoretical Aspects of Computer Science (STACS), volume 3404 of LNCS, 293-304. Springer, 2005.

Scheduling Jobs on Grid Processors[*]

Joan Boyar and Lene M. Favrholdt

Department of Mathematics and Computer Science
University of Southern Denmark
{joan, lenem}@imada.sdu.dk

Abstract. We study a new kind of on-line bin packing motivated by a problem arising when scheduling jobs on the Grid. In this bin packing problem, the set of items is given at the beginning, and variable-sized bins arrive one by one. A closely related problem was introduced by Zhang in 1997. Our main result answers a question posed in that paper in the affirmative: we give an algorithm with a competitive ratio strictly better than 2, for our problem as well as Zhang's problem.

1 Introduction

We introduce a new on-line scheduling problem based on Grid computing, a network model where access to computing resources is sold. The problem is to schedule a given set of jobs with memory requirements on the Grid, where processors will "arrive" over time, reporting that they are available and specifying how much memory they have. In the specific bioinformatics application (blasting genomes against each other) originally motivating this work, extremely large jobs are divided into subtasks of varying size in a logical manner. These subtasks are then the jobs with memory requirements.

Motivation. It is well-known that paging slows down computation drastically: parallelizing jobs can result in better than linear speed-up by eliminating unnecessary paging [6]. Therefore, the problem becomes to schedule jobs on each processor such that the total memory requirement of jobs scheduled on the same processor does not exceed the capacity of the given processor, thus eliminating the need for paging after loading a job. On the other hand, the total memory requirement of the jobs should not be too much smaller than the capacity of the processor, or we would be wasting resources.

We view this as a bin packing problem, with the items and their sizes given initially and variable sized bins arriving one by one. The items must be packed in the bins, such that the total size of the items packed in a bin is no more than the size of the bin. We say that a bin is used, if the algorithm packs at least one item in it. Unused bins correspond to unused processors, which would not be paid for. However, to avoid unnecessary waiting time, if a bin is large enough to hold some as yet unpacked item, that bin must be used. For the same reason, an

---

[*] Supported in part by the Danish Natural Science Research Council (SNF).

algorithm is not allowed to close a bin as long as there are still unpacked items that fit in it. The goal is to pack all items, minimizing the total size of the bins used. In the scheduling scenario, this is assuming that the charge for using the processor is proportional to the size of its available memory (which is generally almost proportional to its speed). Since memory sizes are discrete, we let bins and items have integer sizes between 1 and some maximum size M. The problem is different from most on-line scheduling and (variable-sized) bin packing problems in that the jobs/items are given initially and the processors/bins arrive one by one, instead of the other way around.

The problem. To summarize, the *Grid Scheduling Problem* is defined in the following way. It is a variant of bin packing where:

- The items and their sizes are given initially and variable sized bins arrive one by one.
- The items must be packed in the bins, such that the total size of the items packed in a bin is no more than the size of the bin.
- Whenever a bin arrives, the algorithm must keep packing items in it until no as yet unpacked item fits in the bin.
- When packing a bin, the algorithm has no knowledge about possible future bins, it knows only the current and earlier bins.
- The goal is to pack all items, minimizing the total size of bins used.
- Bins and items have integer sizes between 1 and some maximum size M.

Furthermore, we assume that sufficiently many sufficiently large bins arrive that eventually any algorithm will have packed all items.

A similar problem. A closely related problem was previously studied by Zhang [7]. However, he considered the continuous version, where items can have any size in the range $(0, 1]$. Furthermore, Zhang used the restriction that the largest item is no larger than the smallest bin. Thus, as long as there are still unpacked items, the algorithm is "charged" for any bin arriving, whether the bin is used or not. We call this similar problem *Zhang's Bin Packing Problem*.

Results. We study the two problems using the competitive ratio [5, 4], which is the worst-case ratio of the on-line performance to the optimal off-line performance, up to an additive constant. More precisely, for a set S of items, a sequence I of bins, and an algorithm $\mathbb{A}$ for the Grid Scheduling Problem, let $\mathbb{A}(S, I)$ denote the total size of the bins used by $\mathbb{A}$ when packing S in the sequence I of bins. Then, the *competitive ratio* $\mathrm{CR}_{\mathbb{A}}$ of $\mathbb{A}$ is

$$\mathrm{CR}_{\mathbb{A}} = \inf_{S,I} \left\{ c \mid \exists b \colon \forall S, I \colon \mathbb{A}(S, I) \leq c \cdot \mathrm{OPT}(S, I) + b \right\},$$

where OPT denotes an optimal off-line algorithm.

We propose a new algorithm, FFD$_{2/3}$, with a competitive ratio of at most $\frac{13}{7}$, answering in the affirmative Zhang's open problem [7] concerning the existence of an algorithm with competitive ratio less than 2. For Grid Scheduling, the competitive ratio of this algorithm, $\mathrm{FFD}_{2/3}$, is at least 1.8. The algorithm is a member of a family of algorithms, FFD_α, with $\frac{1}{2} < \alpha \leq 1$. The lower and upper bounds for FFD_α are summarized in Figure 1 (p. 21).

Most results in this paper apply to both the Grid Scheduling Problem and Zhang's Bin Packing Problem. The upper bound proof for FFD_α applies to Zhang's bin packing problem, since it does not use the fact that there are only a fixed number of possible item and bin sizes. The general upper bound uses the requirement that a bin cannot be closed if there are still items remaining which fit and thus does not apply to Zhang's bin packing problem in general, but all specific algorithms considered have this property. Lower bound proofs for Grid Scheduling apply to Zhang's bin packing problem, unless bins smaller than the largest items appear. Note that lower bound proofs that apply to both problems often result in slightly stronger bounds for Zhang's bin packing problem, because item/bin sizes are not discrete. Thus, whenever a lower bound on the competitive ratio includes a term which is a function of M only, this term can be ignored for Zhang's bin packing problem.

When a result only applies to Grid Scheduling, we specify this in the statement of the theorem. The proof that a result also applies to Zhang's Bin Packing Problem is usually a trivial modification of the proof for the Grid Scheduling Problem (translating item sizes between 1 and M to item sizes between 0 and 1), so most of these proofs are not included. For some of the lower bound proofs, we only give the adversarial instance. The full proofs are given in the full paper [1].

2   General Results

Zhang [7] proved that the algorithms First-Fit, First-Fit-Decreasing, Next-Fit, and Next-Fit-Decreasing all have a competitive ratio of 2. The upper bound proof holds for any algorithm for the Grid Scheduling Problem, since it uses only the fact that no item placed in the ith bin fits in the $i - 1$st bin.

Theorem 1. *Any Grid Scheduling algorithm, $\mathbb{A}$, has $CR_\mathbb{A} \leq 2$.*

Proof. For completeness, we give the proof from [7]. For any instance of the problem, let m be the number of bins used by $\mathbb{A}$, let b_i denote the size of the ith of these bins to arrive, and let s_i denote the total size of items packed in the ith bin by $\mathbb{A}$, $i = 1, \ldots, m$. Since no item placed by $\mathbb{A}$ in the $i + 1$st bin fits in the ith bin, $s_i + s_{i+1} > b_i$, $i = 1, \ldots, m - 1$. Thus,

$$\mathbb{A}(I) = \sum_{i=1}^{m} b_i < \sum_{i=1}^{m-1}(s_i + s_{i+1}) + b_m < 2\sum_{i=1}^{m} s_i + b_m \leq 2 \cdot \mathrm{OPT}(I) + M. \qquad \square$$

The proof of the following lower bound does not hold for Zhang's bin packing problem, since the adversary may use bins that are smaller than the largest items.

Theorem 2. *Any Grid Scheduling algorithm, $\mathbb{A}$, has $CR_\mathbb{A} \geq \frac{5}{4}$.*

Proof. Let $M' = \lfloor \frac{M}{4} \rfloor$. Consider the following instance of the problem.
Items: n items of size $2M' + 1$ and $2n$ items of size $M' + 1$, for some even n.
Bins: First, n bins of size $2M' + 2$. Let q, $0 \le q \le n$, be the number of items of size $2M' + 1$ that the algorithm $\mathbb{A}$ packs in these bins. If $q \ge \frac{n}{2}$, the sequence continues with $2n$ bins of size $2M' + 1$. If $q < \frac{n}{2}$, the sequence continues with $2n$ bins of size $M' + 1$ and then $n - q$ bins of size $4M'$. $\qquad\square$

3   First-Fit-Decreasing and a Better Variant

Since the item sizes are known in advance, one can use an algorithm based on First-Fit-Decreasing for the standard off-line bin packing problem. The First-Fit-Decreasing (FFD) algorithm for this problem is as follows: for each bin b_i, repeatedly take the largest item still unpacked which fits in the remaining space in b_i and pack it there, until no more unpacked items fit in b_i.

This simple algorithm will intuitively do well, since it will tend to save the smaller, easier to place, items for later bins. However, for Zhang's bin packing problem, FFD has a competitive ratio of 2 [7]. The proof of this is very similar to our proof below in that the item sizes are decreasing, while the numbers of items of each size is doubling. The differences between consecutive item sizes decreases inverse exponentially in our case, and even faster in [7]. Thus, when the item sizes in [7] are converted to integers, the largest bin size, M, must be even larger than in ours to obtain the same lower bound.

Theorem 3. *For the Grid Scheduling Problem, $CR_{FFD} \ge 2 - \Theta\left(\frac{\log M}{M}\right)$.*

Proof. Let k be the largest integer such that $2^{k+1} \le M$, let $M' = 2^k$, and let n be a large integer. Consider the following instance of the problem:
Items: For $0 \le i \le k$, $2^i n$ items of size $x_i = (\frac{1}{2} + 2^{-i})M'$.
Bins: For $0 \le i \le k - 1$, $2^i n$ bins of size $b_i = (1 + 2^{-i})M'$, then n bins of size $\frac{3}{2}M'$, and finally $(M' - 2)n$ bins of size $M' + 1$.

FFD packs the items of size x_i in the bins of size b_i, $0 \le i \le k - 1$. The last $2^k n = M'n$ items are packed, two by two, in the n bins of size $\frac{3}{2}M'$ and, one by one, in the $(M' - 2)n$ bins of size $M' + 1$. The empty space in each bin, except the n bins of size $\frac{3}{2}M'$, will be exactly $\frac{1}{2}M'$.

The optimal strategy is to pack the $2^i n$ items of size x_i, two by two, in the $2^{i-1}n$ bins of size b_{i-1}, $1 \le i \le k$, and the n items of size $\frac{3}{2}M'$ in the n bins of that same size, not using any of the last $(M' - 2)n$ bins.

Using $M' = 2^k$, we get that the total size of the bins used by an optimal off-line algorithm is

$$\text{OPT} = \sum_{i=0}^{k-1} 2^i n(1 + 2^{-i})M' + \frac{3}{2}nM' = \left(M' + k + \frac{1}{2}\right)nM'.$$

Furthermore, FFD $>$ OPT $+ (M' - 2)nM'$. Thus,

$$\frac{\text{FFD}}{\text{OPT}} > 1 + \frac{M' - 2}{M' + k + \frac{1}{2}} = 2 - \frac{k + \frac{5}{2}}{M' + k + \frac{1}{2}} = 2 - \Theta\left(\frac{\log M}{M}\right). \qquad\square$$

3.1   A Variant of First-Fit-Decreasing

The family of algorithms, FFD_α, uses the algorithm, FFD, described above. FFD_α is defined for $\frac{1}{2} < \alpha \leq 1$. When packing a bin, it considers the remaining item sizes in decreasing order as starting sizes for applying FFD. More specifically, let L_1 be the set of items currently unpacked. Assume that there are items of k different sizes in L_1 and let $\{s_1, \ldots, s_k\}$ be the set of these sizes, sorted such that $s_i > s_{i+1}$. FFD_α tries using FFD on L_1, recording how full the bin gets. If the bin gets filled to less than α full, FFD is used on the set L_2 obtained from L_1 by deleting all items of size s_1. If, again, the bin gets filled to less than α full, the items of size s_2 are deleted from L_2 to obtain L_3. This contiunes until either a set L_i is obtained for which FFD fills the bin to more than α full or the empty set L_{k+1} is reached. In the first case, FFD's packing of L_i is used. In the latter case, among the k packings, the one that fills the bin the most is chosen. In case of ties, the first of the candidate packings is chosen.

The idea is that, choosing α greater than $\frac{1}{2}$, the algorithm may consider more possibilities than FFD when choosing the items to pack in a given bin. On the other hand, α should probably not be too close to 1, since then we lose the advantage of getting rid of large items early. Furthermore, with $\alpha < 1$, the algorithm considers fewer possibilities and thus is more time efficient.

Lower bounds. We give lower bounds that are valid for both problems (Theorems 4, 5, and 7) and a tighter lower bound for the Grid Scheduling Problem (Theorem 6). The bounds are summarized in Figure 1. The figure also shows an upper bound (Theorem 8).

Theorem 4. *For* $\frac{1}{2} < \alpha \leq 1 - \frac{3}{M}$, $CR_{FFD_\alpha} \geq \frac{2+\alpha}{1+\alpha} - \Theta\left(\frac{1}{M}\right)$.

Proof. Consider the following input to the Grid Scheduling Problem:
Items: n items of size $\lceil \alpha M \rceil$ and $2n$ items of size $\lfloor M/2 \rfloor$.
Bins: n bins of size M. Then n bins of size $\lceil \alpha M \rceil$. Finally n bins of size $M - 2$.
$\square$

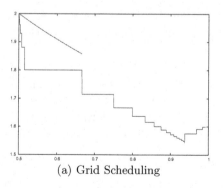
(a) Grid Scheduling

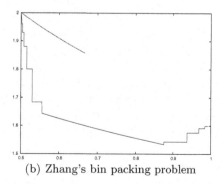
(b) Zhang's bin packing problem

Fig. 1. Lower and upper bounds on the competitive ratio of FFD_α as a fct. of α

For α close to $\frac{1}{2}$, we use a slight modification of the proof of Theorem 3 to arrive at the following stronger lower bound.

Theorem 5. *For $\frac{1}{2} < \alpha \leq \frac{m}{2m-1}$, $CR_{FFD_\alpha} \geq 2 - \Theta\left(\max\left\{\frac{\log m}{m}, \frac{\log M}{M}\right\}\right)$.*

The following result is stronger than that in Theorem 4, but since the adversary presents bins which are smaller than the largest item size, it only holds for the Grid Scheduling Problem, not for Zhang's Bin Packing Problem.

Theorem 6. *For Grid Scheduling, $CR_{FFD_\alpha} \geq \frac{3r}{2r-1} - \Theta\left(\frac{r}{M}\right)$, $\frac{1}{2} < \alpha \leq \frac{r-1}{r}$, $r \geq 3$.*

Proof. Let $M' = \lfloor M/r \rfloor$. Consider the following instance of the problem:
Items: n items of size $(r-1)M'$, rn items of size $M' - 1$.
Bins: n bins of size $rM' - 2$, rn bins of size $2M' - 3$, n bins of size $(r-1)M'$. $\square$

Corollary 1. *For Grid Scheduling, $CR_{FFD_\alpha} \geq 1.8 - \Theta\left(\frac{1}{M}\right)$, for $\frac{1}{2} < \alpha \leq \frac{2}{3}$.*

Theorem 7. *Even for identical bins of size M, $CR_{FFD_\alpha} \geq \sum_{i=1}^{\ell} \frac{1}{2^i-1}$, where $\ell = \max\{k \in \mathbb{N} \mid \alpha > 1 - 2^{-k} + \frac{2^{k-1}}{M}, \ 2^{2k} \leq M\}$. As ℓ tends to infinity, this ratio tends to approximately 1.607.*

Proof. Assume that 2^ℓ divides M. We use the following well-known instance.
Items: For $i = 1, \ldots, \ell$, n items of size $2^{-i}M + 1$.
Bins: $\sum_{i=1}^{\ell} \frac{n}{2^i-1}$ bins of size M.
    Since it is possible to pack one item of each size in one bin, OPT will use n bins.

    Using FFD starting with the items of size $2^{-i}M + 1$, the algorithm will first pack $2^i - 1$ items of size $2^{-i}M + 1$. This leaves an empty space of size $2^{-i}M - (2^i - 1)$. Since $M \geq 2^{2k}$, this will leave space for exactly one item of each size $2^{-j}M + 1$, $j = i+1, \ldots, \ell$. Thus, the bin gets filled to $(1 - 2^{-\ell})M + \varepsilon$, where $\varepsilon = (2^i - 1) + (\ell - i)$ is the number of items packed in the bin. This number is increasing with i, and starting with the second smallest items, the bin gets filled to $(1 - 2^{-\ell})M + 2^{\ell-1}$. Thus, since $1 - 2^{-\ell} + \frac{2^{\ell-1}}{M} < \alpha$, FFD_α will start with the smallest items, using $\sum_{i=1}^{\ell} \frac{n}{2^i-1}$ bins. $\square$

The (nonpolynomial) algorithm, KNAPSACK, that simply fills each bin as much as possible (considering all possible combinations of items) will behave as FFD_α on the sequences of the proof of Theorem 7. Thus, for $M \geq 2^{16}$, $CR_{\text{KNAPSACK}} > 1.6$. More generally (as mentioned in several papers, e.g., [2], [3]):

Corollary 2. *Even for identical bins of size M, $CR_{\text{KNAPSACK}} \geq \sum_{i=1}^{\ell} \frac{1}{2^i-1}$, where $\ell \in \mathbb{N}$, $2^{2\ell} \leq M$. This ratio tends to approximately 1.607 as M tends to infinity.*

3.2   An Upper Bound

In this section we prove an upper bound on the competitive ratio of FFD_α for $\frac{1}{2} < \alpha \leq \frac{2}{3}$. The bound is smaller than 2 throughout the interval $(\frac{1}{2}, \frac{2}{3}]$ and, at $\alpha = \frac{2}{3}$, the bound attains its minimum value of $\frac{13}{7} \approx 1.857$.

The size of a bin, b, will be denoted by $|b|$, and the size of an item, x, by $|x|$. For any algorithm, $\mathbb{A}$, and any set B of bins, $s_\mathbb{A}(B)$ denotes the total size of the items packed by $\mathbb{A}$ in B. Whenever the algorithm considered is FFD_α, we omit the subscript for convenience. Furthermore, $e(B)$ ($E(B)$) denotes the total empty space left by FFD_α (OPT) in a bin or set of bins, B.

We first prove four small technical lemmas.

Lemma 1. *Suppose an algorithm, $\mathbb{A}$, fills k of its bins, $b_{i_1}, b_{i_2}, \ldots, b_{i_k}$, such that $s_\mathbb{A}(b_{i_{j+1}}) > |b_{i_j}|$, $1 \leq j \leq k-1$. Then, if none of these bins is filled to more than a fraction β,*

$$\sum_{j=1}^{k} |b_{i_j}| < \frac{M}{1-\beta} \quad \text{and, for Grid Scheduling, } k \leq \log_{1/\beta} M.$$

Proof. By assumption, $\beta|b_{i_{j+1}}| \geq s_\mathbb{A}(b_{i_{j+1}}) > |b_{i_j}|$. Thus, $\beta^{k-j}|b_{i_k}| > |b_{i_j}|$, so $\beta^{k-1}|b_{i_k}| > |b_{i_1}|$ giving the bound on k. The total size of these k bins is thus

$$\sum_{j=1}^{k} |b_{i_j}| < \sum_{j=1}^{k} \beta^{k-j}|b_{i_k}| = |b_{i_k}| \sum_{l=0}^{k-1} \beta^l < |b_{i_k}| \sum_{l=0}^{\infty} \beta^l = M\frac{1}{1-\beta}. \qquad \square$$

Lemma 2. *For any set B of nonempty bins in FFD_α's packing, $e(B) < s(B)+M$.*

Proof. The result clearly holds if at most one bin in B is less than half full. Let $B^< = \{b_1, b_2, \ldots, b_m\}$ denote the set of bins in B that are less than half full, in the order they appear in the input sequence, and assume that $m \geq 2$. By the definition of the algorithm, $s(b_{i+1}) > e(b_i)$, $1 \leq i \leq m-1$. Thus,

$$\begin{aligned}
e(B) - s(B) &\leq e(B^<) - s(B^<) = \sum_{i=1}^{m} \big(e(b_i) - s(b_i)\big) \\
&< \big(e(b_m) - s(b_m)\big) + \sum_{i=1}^{m-1} \big(s(b_{i+1}) - s(b_i)\big) \\
&= \big(e(b_m) - s(b_m)\big) + \big(s(b_m) - s(b_1)\big) = e(b_m) - s(b_1) < M
\end{aligned}$$

$\square$

Lemma 3. *Let $\frac{1}{2} < \alpha \leq \frac{2}{3}$. If FFD_α fills a bin b to less than $\alpha|b|$, then no two items which FFD_α places in later bins would fit together in bin b.*

Proof. Assume b is filled to $\ell < \alpha|b|$. Then, when FFD_α was filling bin b, FFD was applied starting with every item size remaining, and the best packing was chosen. Consider any two items x and y packed later than b, with $|x| \geq |y|$.

If $|x| + |y| \leq \ell$, then $|y| \leq \frac{\ell}{2}$. But since $\ell + \frac{\ell}{2} < \frac{3}{2}\alpha|b| \leq |b|$, y would fit in b together with the items packed there, which contradicts the definition of FFD_α.

If $\ell < |x| + |y| \leq |b|$, we also get a contradiction with the definition of FFD_α: when using $|x|$ as the starting size, FFD will pack x together with y or some item z, $|x| \geq |z| \geq |y|$, filling b to more than ℓ.

Thus, we are left with the case $|x| + |y| > |b|$, proving the lemma.   □

Lemma 4. *Let $\frac{1}{2} < \alpha \leq \frac{2}{3}$. Consider an instance (S, I) and let $F^<$ be the set of bins used by FFD_α, but filled to less than α full, and not used by OPT. If there are bins in $F^<$ with more than one item, there is another instance (S, I') with $\text{FFD}_\alpha(S, I') = \text{FFD}_\alpha(S, I)$ and $\text{OPT}(S, I') \leq \text{OPT}(S, I)$, such that all bins used only by FFD_α and filled to less than α contain only one item each.*

Proof. Consider a bin $b \in F^<$ with items $y_1, y_2, \ldots, y_n$, $n \geq 2$, such that $|y_1| \leq |y_2| \leq \ldots \leq |y_n|$. Consider the sequence I' obtained from I by replacing b by n bins with sizes $|b_i| = |y_i|$, $1 \leq i \leq n - 1$, and $|b_n| = |y_n| + e(b)$. Note that the sizes of the new bins sum up to exactly $|b|$. The n bins arrive in the order $b_1, b_2, \ldots, b_n$. Since OPT did not use b, $\text{OPT}(S, I') \leq \text{OPT}(S, I)$. Clearly, FFD_α will pack y_i in b_i, $1 \leq i \leq n - 1$, so to complete the proof we just need to prove that y_n will be packed in b_n.

Let U denote the set of items that are still unpacked after FFD_α has packed the prefix of I ending with b. By Lemma 3, no two items in U could be combined in b. Since b_n is smaller than b the same is clearly true of b_n. Furthermore, each item in U is larger than $e(b)$, so no item in U can be combined with y_n in b_n. Thus, to prove that FFD_α will place y_n in b_n, we just need to prove that there is no item $x \in U$ with $|y_n| < |x| \leq |y_n| + e(b)$. Assume for the sake of contradiction that such an item, x, exists. If $|y_n| \geq \frac{1}{3}|b|$, $|x| + |y_n| > \frac{2}{3}|b|$, and this combination would be tried when packing b, unless an even better combination were found. But this is a contradiction, since b is filled to less than α full. If $|y_n| < \frac{1}{3}|b|$, all items packed in b are smaller than $\frac{1}{3}|b|$, and hence when FFD_α tries out FFD starting with size $|x|$, it will stop only when all items $y_n, y_{n-1}, \ldots, y_1$ have been used, in which case a better packing of b has been found, or b will be filled to more than $\frac{2}{3}$ full. Both cases are contradictions.   □

Theorem 8. *FFD_α has a competitive ratio of $CR_{\text{FFD}_\alpha} < \frac{2\alpha+3}{2\alpha+1}$, for $\frac{1}{2} < \alpha \leq \frac{2}{3}$. This is minimum at $\alpha = \frac{2}{3}$, where it gives $CR_{\text{FFD}_{2/3}} < \frac{13}{7} \approx 1.857$.*

Proof. Consider a worst case instance (S, I). For a bin b (or a set of bins B), the notation $x \in b$ (or $x \in B$) will refer to an item FFD_α places in b (or B). Furthermore, for an item, $x \in S$, we use the notation

$b_o(x)$: the bin where OPT places x
$b(x)$: bin where FFD_α places x.

We consider the following sets of bins, A and $\mathcal{F}^< \subseteq F^< \subseteq F$.

A: the set of bins which both FFD_α and OPT use.
F: the set of bins used by FFD_α only.

$F^<$: the bins in F which FFD_α fills to less than α full. By Lemma 4, we can assume that FFD_α packs exactly one item in each bin in $F^<$.

$\mathcal{F}^<$: a subset of $F^<$ with the following property. When FFD_α packs a bin $b \in \mathcal{F}^<$ with some item, x, there is at least one other item, x', still available that could have fit in b. Clearly, $|x'| \leq |x|$, since FFD_α does not pack x' in b.

For each $x \in \mathcal{F}^<$, define a *chain* $C(x)$ as follows: Let $x_0 = x$. If, in FFD_α's packing, the bin $b_o(x_j)$ contains only one item and is filled to less than α full, let x_{j+1} be the item FFD_α places in $b_o(x_j)$. Let $x_0, \ldots, x_\ell$ denote the sequence of items defined in this way. We let $C(x)$ denote $x_0, \ldots, x_\ell$ as well as the sequence $b(x_0), \ldots, b(x_\ell)$ of bins. For any chain of items, $x_0, x_1, \ldots, x_\ell$, we let $b_\ell(x_0) = b_o(x_\ell)$ denote the bin where OPT places the last item of the chain. Note that $b_\ell(x_0)$ is not part of the chain. We now define one more set of bins:

L: the set of bins which are $b_\ell(x)$ for some $x \in \mathcal{F}^<$.

The main ideas of the proof are that all chains are well defined, the total size of bins in $F^< \setminus \mathcal{F}^<$ is small, the items packed in $\mathcal{F}^<$ all fit in L, and most bins in L are filled to at least α full. For this purpose it is useful to show that, for any chain $C(x)$, $b(x)$ is given after all other bins in $C(x)$ and no item in $C(x)$ is smaller than x. The bins in $F \setminus F^<$ can essentially be ignored, since the competitive ratio we are trying to prove is so large.

For any chain $C(x_0)$, $b(x_0)$ arrives later than any other bin in $C(x_0)$: Note that for any chain, $b_o(x_0) = b(x_1)$ is an earlier bin than $b(x_0)$, since x_0 fits in both bins and OPT did not use $b(x_0)$.

Recall that when FFD_α packs x, there is still an unpacked item x' no larger than x. Suppose for the sake of contradiction that there exists an x_i, $i \geq 1$, in $C(x_0)$, where $b_o(x_i)$ is after $b(x_0)$. Let i be the first index where this occurs, i.e., $b(x_i) = b_o(x_{i-1})$ occurs before $b(x_0)$. We find that $|x_0| > \alpha |b(x_0)|$, contradicting that $x_0 \in F^<$: By Lemma 3, $2|x_0| \geq |x_0| + |x'| > |b(x_i)|$. Furthermore, $|x_i| < \alpha |b(x_i)|$, by the definition of a chain. Since $b_o(x_i)$ is after $b(x_0)$ and OPT did not use $b(x_0)$, $|x_i| > |b(x_0)|$. Combining these inequalities, we get $|x_0| > \frac{1}{2}|b(x_i)| > \frac{1}{2\alpha}|x_i| > \frac{1}{2\alpha}|b(x_0)| > \alpha |b(x_0)|$, where the last inequality follows from $\frac{1}{2} < \alpha \leq \frac{2}{3}$.

Using Lemma 3 and the fact that $b(x_0)$ arrives last, we get that x_0 and x' cannot be packed together in any bin $b(x_i)$, $i \geq 1$, in the chain. Combining with the fact that x_0 is packed alone in $b(x_0)$, we arrive at

Fact 1. *For any chain $C(x_0)$, $2|x_0| > |b|$, for any bin $b \in C(x_0)$.*

The first item in a chain C is no larger than any other item in C: Let x_i be an item in C. We use induction on i. The base case, $i = 0$, is trivial. For $i \geq 1$, x_{i-1} fits in $b(x_i)$ (OPT put it there). By the induction hypothesis, $|x_0| \leq |x_{i-1}|$, so x_0 also fits there. Thus, since FFD_α did not place x_0 in $b(x_i)$ when that bin was requested, $|x_0| \leq |x_i|$. This also shows that $b_o(x_i)$ is never a bin which FFD_α left empty, i.e., $L \subseteq A$. Furthermore, combining with Fact 1, we get

Fact 2. *For any chain C and any item $x_i \in C$, $2|x_i| > |b|$, for any bin $b \in C$.*

The total size of bins in $F^< \setminus \mathcal{F}^<$ is small: Note that the set $F^< \setminus \mathcal{F}^<$ of bins satisfies the conditions of Lemma 1, with $\beta = \alpha$. Thus,

$$\sum_{b \in F^< \setminus \mathcal{F}^<} |b| < 3M. \tag{1}$$

Chains do not contain cycles: Assume for the sake of contradiction that a chain $C(x)$ contains two distinct items x_i and x_j that are packed in the same bin, b, by OPT, i.e., $b_o(x_i) = b_o(x_j) = b$. Then clearly, $|x_i| + |x_j| \leq |b|$, contradicting Fact 2.

Chains do not intersect: Assume for the sake of contradiction that there is a bin b contained in two chains $C(x)$ and $C(y)$, $x \neq y$. Then b contains two items $x_i \in C(x)$ and $y_j \in C(y)$. Assume $|x_i| \leq |y_j|$. Then $2|x_i| \leq |b|$, contradicting Fact 2.

The items packed in $\mathcal{F}^<$ all fit in the bins in L: Consider a bin, $b_\ell \in L$. Obviously, all items y, such that $b_o(y) = b_\ell$, fit in the bin b_ℓ. For any two chains, $x_0, \ldots, x_\ell$ and $y_0, \ldots, y_m$, $x_\ell \neq y_m$, since no two chains intersect. In addition, all items on a chain from some x are at least as large as x. Hence, for any $b \in L$, $|b| \geq \sum_{x \mid b_\ell(x) = b} |x|$ and thus,

$$e(L) + s(L) \geq s(\mathcal{F}^<). \tag{2}$$

Most bins in L are filled to at least α: Consider two bins, $b, b' \in L$, where b occurs before b' and FFD_α fills both with at least two items, but to less than α full. Let x and y be two items in b'. Neither one fit in b after b was packed, so each is larger $\frac{1}{3}|b|$. Hence $|x| + |y| > \frac{2}{3}|b|$, so together they are larger than the total contents of b. Thus, FFD_α would have put them there (or some others which filled b to even more than $\frac{2}{3}$ full) if they fit there. This means that the contents of b' are too large to fit in b, so the set of bins in L which are filled to less than α satisfy the conditions of Lemma 1, and their total size is less than $\frac{1}{1-\alpha}M \leq 3M$. Thus,

$$3M + \frac{1}{\alpha}s(L) > e(L) + s(L) \overset{(2)}{\geq} s(\mathcal{F}^<). \tag{3}$$

Rearranging, we get

$$s(L) - e(L) > \left(2 - \frac{1}{\alpha}\right)s(L) - 3M. \tag{4}$$

Finally, we derive a few more useful inequalities. First, since $e(A) - E(A) \leq s(F)$ and $e(A \setminus L) = e(A) - e(L)$, we get

$$e(A \setminus L) \leq s(F) + E(A) - e(L). \tag{5}$$

Since, by (1), all bins in $F \setminus \mathcal{F}^<$, except possibly some with total size less than $3M$, are filled to at least α,

$$e(F \setminus \mathcal{F}^<) < \frac{1-\alpha}{\alpha}s(F \setminus \mathcal{F}^<) + 3M. \tag{6}$$

The total size of the items in S is $s(A) + s(F)$, the total size of empty space in FFD_α's packing is $e(A) + e(F)$, and in OPT's packing it is at least $E(A)$. Thus, $\text{OPT}(S, I) \geq s(A) + s(F) + E(A)$ and

$$
\begin{aligned}
\text{FFD}_\alpha(S, I) \quad &= \quad s(A) + e(A) + s(F) + e(F) \\
&\overset{\text{Lemma 2, (6)}}{<} \quad s(A) + e(A) + 2s(\mathcal{F}^<) + \frac{1}{\alpha}s(F \setminus \mathcal{F}^<) + 4M.
\end{aligned}
$$

Hence,

$$
\begin{aligned}
\frac{\text{FFD}_\alpha(S, I)}{\text{OPT}(S, I)} \quad &< \quad \frac{s(A) + e(A) + 2s(\mathcal{F}^<) + \frac{1}{\alpha}s(F \setminus \mathcal{F}^<) + 4M}{s(A) + s(F) + E(A)} \\
&\overset{\text{Lemma 2}}{<} \quad \frac{s(L) + e(L) + 2e(A \setminus L) + 2s(\mathcal{F}^<) + \frac{1}{\alpha}s(F \setminus \mathcal{F}^<) + 3M}{s(L) + e(A \setminus L) - M + s(F) + E(A)}
\end{aligned}
$$

Thus, since $\frac{\text{FFD}_\alpha(S,I)}{\text{OPT}(S,I)} \leq 2 + \frac{M}{\text{OPT}}$ (by the proof of Theorem 1),

$$
\begin{aligned}
\frac{\text{FFD}_\alpha(S, I)}{\text{OPT}(S, I)} \quad &\overset{(5)}{<} \quad \frac{s(L) + 4s(\mathcal{F}^<) + (2 + \frac{1}{\alpha})s(F \setminus \mathcal{F}^<) - e(L) + 2E(A) + 4M}{s(L) + 2s(F) - e(L) + 2E(A) - M} \\
&\leq \quad \frac{s(L) + 4s(\mathcal{F}^<) + (2 + \frac{1}{\alpha})s(F \setminus \mathcal{F}^<) - e(L) + 4M}{s(L) + 2s(F) - e(L) - M}
\end{aligned}
$$

If this ratio is smaller than $1 + \frac{1}{2\alpha}$, we are done, since $1 + \frac{1}{2\alpha} < \frac{2\alpha+3}{2\alpha+1}$. Otherwise, we can subtract $(1 + \frac{1}{2\alpha})2s(F \setminus \mathcal{F}^<)$ from the numerator and $2s(F \setminus \mathcal{F}^<)$ from the denominator, arriving at

$$
\begin{aligned}
\frac{\text{FFD}_\alpha(S, I)}{\text{OPT}(S, I)} \quad &< \quad \frac{s(L) + 4s(\mathcal{F}^<) - e(L) + 4M}{s(L) + 2s(\mathcal{F}^<) - e(L) - M} \\
&\overset{(3)}{<} \quad \frac{s(L) + 4(\frac{1}{\alpha}s(L) + 3M) - e(L) + 4M}{s(L) + 2(\frac{1}{\alpha}s(L) + 3M) - e(L) - M} \\
&\overset{(4)}{<} \quad \frac{(2 - \frac{1}{\alpha})s(L) - 3M + \frac{4}{\alpha}s(L) + 16M}{(2 - \frac{1}{\alpha})s(L) - 3M + \frac{2}{\alpha}s(L) + 5M} \\
&= \quad \frac{(2\alpha + 3)s(L) + 13\alpha M}{(2\alpha + 1)s(L) + 2\alpha M}
\end{aligned}
$$

As $s(L)$ tends to infinity, this ratio tends to $\frac{2\alpha+3}{2\alpha+1}$. If $s(L)$ does not tend to infinity, we can essentially ignore the bins in F which are filled to less than α, since by Equations (1) and (3), $s(L) > \alpha\big(s(\mathcal{F}^<) - 3M\big) > \alpha(s(\mathcal{F}^<) - 6M)$. Thus, as the total size of the items in the input tends to infinity, $\text{FFD}_\alpha(S, I)/\text{OPT}(S, I)$ tends to at most

$$
\frac{s_b(A) + \frac{1}{\alpha}s(F)}{s_b(A) + s_b(O)} \leq \frac{s_b(A) + \frac{1}{\alpha}(e(A) + s_b(O))}{s_b(A) + s_b(O)},
$$

where $s_b(A)$ and $s_b(O)$ denote the total size of the bins in A and in the bins used only by OPT, respectively. If this ratio is at most $\frac{1}{\alpha}$, we are done. Otherwise, it is bounded by

$$\frac{s_b(A) + \frac{1}{\alpha}e(A)}{s_b(A)} \overset{\text{Lemma 2}}{\leq} \frac{s_b(A) + \frac{1}{2\alpha}(s_b(A) + M)}{s_b(A)} = 1 + \frac{1}{2\alpha} + \frac{M}{2\alpha s_b(A)}. \quad \square$$

Thus, for $\frac{1}{2} < \alpha \leq \frac{2}{3}$, the algorithm FFD_α has a competitive ratio better than 2, with the minimum value being $\frac{13}{7}$ for $\alpha = \frac{2}{3}$, and this also applies to Zhang's Bin Packing Problem. It leaves open the problem of how much better than $\frac{13}{7}$ can be achieved. Note that for $\alpha > \frac{2}{3}$, the above proof fails in more than one place.

4   Conclusion

We have introduced a new scheduling problem, the Grid Scheduling Problem, that can be formulated as a bin packing problem. The main result is a new algorithm with a competitive ratio better than 2, answering an open question in [7]. There are many open problems remaining, the most interesting being finding the optimal competitive ratio for the Grid Scheduling Problem. Determining the best value of α for FFD_α and finding the exact competitive ratio of FFD_α for all α would be very interesting progress on this problem.

Acknowledgment

We would like to thank Brian Vinter for suggesting the problem, sharing his knowledge of Grid computing, and discussing the model with us.

References

1. J. Boyar and L. M. Favrholdt. Scheduling jobs on grid processors. Technical Report IMADA preprint PP-2006-6, University of Southern Denmark, 2006.
2. M. R. Garey, R. L. Graham, and J. D. Ullman. An analysis of some packing algorithms. In *Combinatorial algorithms (Courant Computer Science Symposium 9)*, pages 39–47, 1972.
3. R. L. Graham. Bounds on multiprocessing anomalies and related packing algorithms. In *Proc. 1972 Spring Joint Computer Conference*, pages 205–217, 1972.
4. A. R. Karlin, M. S. Manasse, L. Rudolph, and D. D. Sleator. Competitive snoopy caching. *Algorithmica*, 3(1):79–119, 1988.
5. D. D. Sleator and R. E. Tarjan. Amortized efficiency of list update and paging rules. *Comm. of the ACM*, 28(2):202–208, 1985.
6. B. Vinter. Personal communication. http://mig-2.imada.sdu.dk:8092/MiG/MiG/About_MiG.html, 2006.
7. G. Zhang. A new version of on-line variable-sized bin packing. *Discrete Applied Mathematics*, 72:193–197, 1997.

Variable Sized Online Interval Coloring
with Bandwidth

Leah Epstein[1], Thomas Erlebach[2], and Asaf Levin[3]

[1] Department of Mathematics, University of Haifa, 31905 Haifa, Israel
lea@math.haifa.ac.il
[2] Department of Computer Science, University of Leicester, University Road,
Leicester LE1 7RH, United Kingdom
te17@mcs.le.ac.uk
[3] Department of Statistics, The Hebrew University, Jerusalem, Israel
levinas@mscc.huji.ac.il

Abstract. We consider online coloring of intervals with bandwidth in a setting where colors have variable capacities. Whenever the algorithm opens a new color, it must choose the capacity for that color and cannot change it later. The goal is to minimize the total capacity of all the colors used. We consider the bounded model, where all capacities must be chosen in the range $(0, 1]$, and the unbounded model, where the algorithm may use colors of any positive capacity. For the absolute competitive ratio, we give an upper bound of 14 and a lower bound of 4.59 for the bounded model, and an upper bound of 4 and a matching lower bound of 4 for the unbounded model. We also consider the offline version of these problems and show that the unbounded model is polynomially solvable, while the bounded model is NP-hard in the strong sense and admits a 3.6-approximation algorithm.

1 Introduction

Online interval coloring has received much attention recently. In the basic problem, the nodes of an interval graph arrive online, one by one, together with the interval representation. The goal is to find a proper vertex coloring (i.e., each pair of adjacent vertices, i.e. intersecting intervals, are assigned distinct colors) with a minimum number of colors. The coloring has to be determined online, i.e., each new interval must be assigned a color upon arrival.

This standard problem was studied by Kierstead and Trotter [14]. They constructed an online algorithm that uses at most $3\omega - 2$ colors where ω is the maximum clique size of the interval graph. They also presented a matching lower bound of $3\omega - 2$ on the number of colors in a coloring of an arbitrary online algorithm. Note that the chromatic number of interval graphs equals the size of a maximum clique, which is equivalent in the case of interval graphs to the largest number of intervals that intersect any point (see [11]). Many papers studied the competitive ratio of First-Fit for this problem [12, 13, 17, 4]. The latter reference shows that the competitive ratio of First-Fit is strictly worse than the competitive ratio of the algorithm from [14].

L. Arge and R. Freivalds (Eds.): SWAT 2006, LNCS 4059, pp. 29–40, 2006.

Adamy and Erlebach [1] introduced the interval coloring with bandwidth problem and presented a 195-competitive algorithm. In this problem each interval has a bandwidth requirement in $(0, 1]$. The intervals are to be colored so that at each point, the sum of bandwidths of intervals colored by a certain color does not exceed 1. This problem was also studied in [16], giving an improved competitive ratio of 10, and in [8], showing a lower bound of 3.2609.

We study a variant of this problem, where colors are not necessarily of capacity 1 as in [1]. The input arrives as in this model, but an algorithm may use colors of arbitrary capacity. In an online environment, the capacity of a color is determined when the color is first used. The coloring is valid if for every color a that is used with capacity C_a, at each point the sum of bandwidths of intervals colored by a does not exceed C_a. The cost of a coloring is the sum of the capacities of the colors used. We study the *unbounded model*, with no restriction on the capacities of colors, and the *bounded model*, where capacities cannot exceed 1.

The interval coloring problem with bandwidth of [1] is a generalization of the well known bin packing problem (see e.g. [7, 5, 19]). Our problem is related to variable sized bin packing (see [15, 10, 6, 18, 20]), but does not generalize it. In the bin packing problem, allowing the usage of bins of any size (even if the sizes are bounded by 1) leads to a simple 1-competitive algorithm, which assigns every item a bin of the same size. In the variable sized bin packing problem, a set of allowed bin sizes is set in advance.

As mentioned in [1], the interval coloring problem with bandwidth arises in many applications, often from the field of communication networks. Consider a network with a line topology that consists of links, where each link has channels of constant capacity. A connection request is from one network node a to another node b and has a bandwidth associated with it. The set of requests assigned to a channel must not exceed the capacity of the channel on any of the links on the path $[a, b]$. The goal is to minimize the number of channels (colors) used. In our problem, we can choose the capacity of the channel, and therefore we pay a cost proportional to the capacity of the channel, rather than a fixed cost as in the case of unit capacity channels. A connection request from a to b corresponds to an interval $[a, b]$ with the respective bandwidth requirement and the goal is to minimize the sum of capacities of the channels used to serve all requests. We allow different capacities since not all channels are necessarily identical.

Another important application comes from scheduling. A requested job has a starting time, a duration, and a resource requirement during its execution. Jobs (intervals) arrive online and must be assigned to a machine (color) immediately. It is possible to pick a machine of any capability, which is fixed when the machine is ordered. The cost of the machine is proportional to its resource capacity. The objective is to minimize the sum of the costs of the machines used.

For an algorithm $\mathcal{A}$, we denote its cost by $\mathcal{A}$ as well. The cost of an optimal offline algorithm that knows the complete sequence of intervals is denoted by OPT. We consider the absolute competitive ratio and the absolute approximation ratio criteria. For an online algorithm we use the term competitive ratio whereas for an offline algorithm we use the term approximation ratio.

The competitive ratio of $\mathcal{A}$ is the infimum $\mathcal{R}$ such that for any input, $\mathcal{A} \leq \mathcal{R} \cdot \text{OPT}$. If the competitive ratio of an online algorithm is at most $\mathcal{C}$ we say that it is $\mathcal{C}$-competitive. The approximation ratio of a polynomial time offline algorithm is defined similarly to be the infimum $\mathcal{R}$ such that for any input, $\mathcal{A} \leq \mathcal{R} \cdot \text{OPT}$. If the approximation ratio of a polynomial time offline algorithm is at most $\mathcal{R}$ we say that it is an $\mathcal{R}$-approximation algorithm.

We first consider the online problem. We give tight bounds for the unbounded model, showing that the competitive ratio achieved by applying doubling is 4, and this is best possible. For the bounded model, we show that an adaptation of the algorithm in [16] combined with doubling is 14-competitive. We prove that no algorithm has competitive ratio better than 4.59.

We further show that the offline unbounded problem can be solved optimally using a simple polynomial algorithm, while the bounded problem is NP-hard in the strong sense. For that problem we design an approximation algorithm with ratio $\frac{18}{5} = 3.6$. Some proofs are omitted due to space restrictions.

2 Preliminaries

The following $KT_{\ell b}$ algorithm for the online interval coloring with bandwidth problem was studied by Epstein and Levy [8, 9] (see also [16]). We are given an upper bound b on the maximum bandwidth request and a parameter ℓ. The algorithm partitions the requests into classes and then colors each class using the First-Fit algorithm. The partition of the requests is performed online so that a request j is allocated to class m, where m is the minimum value so that the maximum load of the requests that were allocated to classes $1, 2, \ldots, m$ with the additional new request is at most $m\ell$. For an interval v_i that was allocated to class m a *critical point of v_i* is a point q in v_i so that the set of all the intervals that were allocated to classes $1, 2, \ldots, m-1$ prior to the arrival of v_i, together with v_i, has total load strictly larger than $(m-1)\ell$ in q (i.e., q prevents the allocation of v_i to class $m-1$). They proved the following lemmas.

Lemma 1. *Given an interval v_i that was allocated class m. For the set A_m of intervals that were allocated to class m, and for every critical point q of v_i the total load of A_m in q is at most $b + \ell$. If all intervals have the same bandwidth b, and ℓ is divisible by b, this total load is at most ℓ.*

Lemma 2. *For every m, the set A_m of intervals that were allocated to class m has a maximum load of at most $2(b+\ell)$. If all intervals have the same bandwidth, b, and ℓ is divisible by b, the set A_m of intervals that were allocated to class m has a maximum load of at most 2ℓ.*

Lemma 3. *The number of classes used by the algorithm is at most $\lceil \frac{\omega^*}{\ell} \rceil$, where ω^* is the maximum load.*

3   Online Algorithms

3.1   The Unbounded Model

Our algorithm for the unbounded model simply uses standard doubling (see [3, 2]). I.e., we keep a current "guess" of the maximum load of the complete sequence, which is actually a lower bound on the load, and a single active color. On the arrival of the first interval, we initialize the guess to be the smallest (negative) power of 2 that is not larger than the bandwidth requirement of the interval. We open the first color with capacity which is twice the guess. Each time an interval arrives we color it with the active (i.e., last opened) color if possible. If a new interval arrives that cannot be colored with the active color, this means that the maximum load is at least twice larger than the current guess. We therefore update the guess to equal twice the current guess, and open a new color with its capacity equal to twice the new value of the guess. Repeat this process until the interval can be colored with the most recently opened color. This color becomes active.

Theorem 1. *The competitive ratio of the above algorithm is 4.*

Proof. If there is a single color used by the algorithm, then its capacity is at most twice the largest load, and the competitive ratio is bounded by 2. Otherwise, consider the last time a new color was opened by the algorithm. The value L that is the current guess of the maximum load at this time is a lower bound on OPT. The new color has capacity $2L$, and since each time a new color is opened its capacity is at least twice the previous capacity, we conclude that the total cost of the algorithm is at most $2L + L + \frac{L}{2} + \cdots + \frac{L}{2^i} + \cdots \leq 4L \leq 4\text{OPT}$. $\square$

Given a non-negative small value $0 < \varepsilon < \frac{1}{6}$, we next describe a modified procedure whose asymptotic competitive ratio is $2 + \varepsilon$. The algorithm runs the $KT_{\ell b}$ algorithm with "unit" capacity that is set to $\frac{1}{\varepsilon}$. In order to use the algorithm with unit capacities, we multiply the bandwidth of all input intervals by ε. In this way we get $b = \varepsilon$ and therefore we can use $\ell = \frac{1}{2} - \varepsilon$, so that each class of the algorithm can be packed using one color. Using Lemma 3, we can show that the algorithm has the following performance guarantee:

Theorem 2. *There is an online algorithm that for each input sequence provides a solution with cost at most $(2 + \varepsilon)\text{OPT} + O(\frac{1}{\varepsilon})$.*

3.2   The Bounded Model

Our algorithm for this case is the following adaptation of the algorithm of Narayanaswamy [16] for the online interval coloring problem with bandwidth. We partition the requests into three groups. *Large requests* are requests with bandwidth in the interval $\left(\frac{1}{2}, 1\right]$, *medium requests* are requests with bandwidth in the interval $\left(\frac{1}{4}, \frac{1}{2}\right]$, and *small requests* are requests with bandwidth at most $\frac{1}{4}$. We use disjoint colors for coloring requests of distinct groups. Our algorithm is different from the algorithm of [16] mainly in the procedure for coloring the small requests.

For packing large requests we use unit capacity colors, and pack these requests using Kierstead and Trotter's algorithm [14] for online interval coloring (without bandwidth). This is equivalent to using the algorithm in Section 2 with $\ell = 1$ and ignoring bandwidth requirements. In this case the total load of a class is at most two requests at each point, and as explained in [14], each class requires at most three colors.

Lemma 4. *The total cost of the colors used for large requests is at most* $6 \cdot \text{OPT}$.

For packing medium requests we again use unit capacity colors, and pack these requests using the algorithm in Section 2, giving each interval bandwidth of $\frac{1}{2}$. This is similar to using Kierstead and Trotter's algorithm for online interval coloring (without bandwidth). Each class is packed using one color (and not three colors). This packing of each class is feasible by Lemma 2, since we use $b = \ell = \frac{1}{2}$.

Lemma 5. *The total cost of the colors for medium requests is at most* $4 \cdot \text{OPT}$.

It remains to describe the packing of the small requests. We partition the small requests into type 1 requests and type 2 requests. A *type 1* request is a request such that upon its arrival, for each point within the request the total load of previously presented type 1 requests, plus the load of the new request, is at most $\frac{1}{2}$. A *type 2* (small) request is a small request that is not a type 1 request.

We use separate sets of colors for type 1 small requests and type 2 small requests. The type 1 small requests are packed using the doubling procedure (described in the unbounded model). Recall that in that procedure, the capacity of each color that we use is an integer power of 2. Therefore, the last opened color that we use for small requests of type 1 has a capacity of at most $\frac{1}{2}$.

The packing of type 2 small requests uses only colors with unit capacity and is carried out by applying algorithm $KT_{\ell b}$ for $\ell = \frac{1}{4}$ and $b = \frac{1}{4}$. More precisely, we apply algorithm $KT_{\ell b}$ to all small requests, but the requests that are assigned to the first two classes by $KT_{\ell b}$ are actually the type 1 small requests that are handled as explained above.

The purpose of this partition into types is that if the load caused by the small intervals is very low, then opening a color of capacity 1 right away might be an overkill for the small intervals. Specifically, we want to show an absolute competitive ratio of 4, which would be impossible if a unit capacity color was opened immediately.

Lemma 6. *The total cost of the colors used for small requests is at most* $4 \cdot \text{OPT}$.

Proof. If there is no type 2 small request, then the claim holds since the doubling procedure is a 4-competitive algorithm. Thus, we can assume that there is at least one type 2 small request. Note that in this case all colors that we use to color type 1 small requests have a total cost that is at most 1. Consider the execution of the algorithm $KT_{\ell b}$ for $\ell = \frac{1}{4}$ and $b = \frac{1}{4}$ on the complete input (i.e., already starting at the first interval). All intervals of the first two classes that would have been opened by $KT_{\ell b}$ are colored in our algorithm by the set of colors which are given capacities smaller than 1. To see this last property note that by the definition of $KT_{\ell b}$, the first two classes of the algorithm contain only intervals whose total load is at most $2\ell = \frac{1}{2}$. All these intervals are by definition type 1 small requests. Therefore, if we denote by ω^* the maximum total load of the small requests, then OPT $\geq \omega^*$ and the number of unit capacity colors that the algorithm uses in order to pack the type 2 small requests is at most $\left\lceil \frac{\omega^*}{\ell} \right\rceil - 2 < 4\omega^* + 1 - 2 = 4\omega^* - 1$. Since the total cost of the type 1 small requests is 1, we conclude that the algorithm packs the small requests using colors with total cost that is at most $4\omega^* \leq 4 \cdot$ OPT. □

Using Lemmas 4, 5 and 6, we establish the following theorem.

Theorem 3. *There is a 14-competitive online algorithm for the bounded model.*

4   Lower Bounds

4.1   The Unbounded Model

We next show that the competitive ratio of our algorithm for the unbounded model is best possible. In the proof we again apply methods similar to [3].

Theorem 4. *Any online algorithm for the unbounded model has a competitive ratio of at least 4.*

Proof. Before we construct the lower bound we note that we assume for ease of presentation that bandwidth requirements can be numbers larger than 1. Clearly, the unbounded model is equivalent to any model where the bandwidths are bounded by some constant (not necessarily 1). Before presenting the sequence, we can compute a bound on the largest bandwidth needed for the proof, and thus our lower bound satisfies the model.

Our construction of the lower bound for the unbounded model is based on instances in which OPT equals the maximum load, whereas the algorithm tries to guess an upper bound on the maximum load, and pays the sum of all its guesses. We consider input sequences with the following structure. The first interval is $[0, 1]$ with a unit bandwidth request. Given an arbitrary prefix of intervals for which the algorithm opened the set of colors with capacities $c_1 \leq c_2 \leq \cdots \leq c_k$ the next interval is disjoint to all the previous intervals with bandwidth request $c_k + \varepsilon$ for a sufficiently small value of ε. Then, the algorithm needs to open another color with capacity at least $c_{k+1} \geq c_k + \varepsilon$. Note that at this step OPT $= c_k + \varepsilon$ as all the intervals are disjoint and therefore they all fit into a common color with capacity $c_k + \varepsilon$, whereas the algorithm pays $\sum_{j=1}^{k+1} c_j$.

Given a fixed value of ρ that is strictly smaller than 4, we will show that if our input sequence is long enough an online algorithm cannot pay at each step k at most ρ times the cost of OPT at this step (the sequence can be stopped at any point, preventing all future intervals from arriving). Assume that this does not hold, and that there is a ρ-competitive online algorithm with $\rho = 4 - \delta$ for some $\delta > 0$. Denote this algorithm by $\mathcal{A}$. Assume that given the above input sequence for the value of ε that satisfies $\frac{1}{1+\varepsilon} = 1 - \delta^2$, $\mathcal{A}$ opens colors with capacities $c_1 < c_2 < \cdots < c_k < \cdots$. Then, since $\mathcal{A}$ is ρ-competitive, the inequalities $c_1 \leq \rho$ and $\sum_{j=1}^{k+1} c_j \leq \rho(c_k + \varepsilon)$ must hold. Let $r_{k+1} = 4 - \delta - \frac{\sum_{j=1}^{k} c_j}{c_k + \varepsilon}$, for $k \geq 1$. The inequality above implies $\frac{c_{k+1}}{c_k + \varepsilon} \leq r_{k+1}$. Note that if $r_{k+1} < 1$, $\mathcal{A}$ cannot open a color of sufficient capacity in step $k + 1$ without violating the assumption that its competitive ratio is ρ. We will show that the values r_{k+1} for $k = 1, 2, \ldots$ form a decreasing sequence so that r_{k+1} must be strictly less than 1 for some large enough value of k (depending only on ε). This is a contradiction to $r_{k+1} \geq 1$ and shows that such a sequence of c_k's cannot exist, hence no algorithm can achieve competitive ratio $4 - \delta$ for any $\delta > 0$.

First, we observe that $r_2 = 4 - \delta - \frac{c_1}{c_1 + \varepsilon} \leq 4 - \delta$. Next, we will show that $r_{k+2} \leq r_{k+1}/(1+\gamma)$ for all $k \geq 1$ (as long as $r_{k+1} \geq 1$), where $\gamma > 0$ is a constant chosen in such a way that $\frac{4}{1+\gamma} \geq 4 - \delta^2$ is satisfied. Assuming that $r_{k+1} \leq 4 - \delta$ was shown by induction, we can use elementary calculations to bound r_{k+2} as follows:

$$r_{k+2} = 4 - \delta - \frac{\sum_{j=1}^{k+1} c_j}{c_{k+1} + \varepsilon} \leq 5 - \delta - \frac{1}{1+\varepsilon} - \frac{4-\delta}{r_{k+1}(1+\varepsilon)}$$

This expression can be shown to be bounded by $r_{k+1}/(1+\gamma)$. $\qquad\square$

4.2 The Bounded Model

In order to construct the lower bound, we use as a black box the lower bound of Kierstead and Trotter [14] given originally for the standard online interval coloring problem. They designed for any integer k a lower bound sequence where the clique size is at most k, whereas any online algorithm is forced to use $3k - 2$ colors. In [8] it was shown that this construction can be adapted to the case where the value k or bounds on it are known in advance to the algorithm.

Theorem 5. *Any online algorithm for the bounded model has a competitive ratio of at least* 4.5.

Proof. Let k be a large enough integer. We are going to have at most two such constructions, where there is no overlap between the intervals of the two constructions. Let $\varepsilon > 0$ be a small value, such that $P = \frac{1}{2\varepsilon}$ is an integer. We start with such a construction where all intervals have bandwidth $\frac{1}{2} + \varepsilon$. Since the largest possible capacity of a color is 1, no two overlapping intervals can receive the same color, and therefore the algorithm is forced to use $3k - 2$ colors, whereas an optimal offline algorithm can use at most k colors, each of capacity $\frac{1}{2} + \varepsilon$.

The second construction will use intervals of bandwidth $\frac{1}{2} + j\varepsilon$ for some $2 \leq j \leq P$. In this construction as well the algorithm is forced to use $3k - 2$ colors of capacity at least $\frac{1}{2} + j\varepsilon$, whereas the construction is k-colorable. An optimal offline algorithm uses k colors of capacity $\frac{1}{2} + j\varepsilon$ each, and these colors are used to color all intervals of the first construction as well. Consider the $3k - 2$ colors with largest capacity opened by the algorithm for the first construction. Let s be the number of colors out of these colors whose capacity is strictly smaller than $\frac{1}{2} + j\varepsilon$. The algorithm has to open at least s new colors of capacity $\frac{1}{2} + j\varepsilon$.

Already in the first construction, the algorithm only needs to open colors whose capacities are in the set $\{\frac{1}{2} + \varepsilon, \frac{1}{2} + 2\varepsilon, \ldots, \frac{1}{2} + P\varepsilon = 1\}$. Consider only the $3k - 2$ colors of largest capacities that are opened for the first construction. Let X_j for $1 \leq j \leq P$ be the number of colors of capacity $\frac{1}{2} + j\varepsilon$.

Let C be the competitive ratio. The cost of the algorithm for the first construction is at least $\sum_{j=1}^{P}(\frac{1}{2} + j\varepsilon)X_j$. Note that according to the definition of the values X_j, $\sum_{j=1}^{P} X_j = 3k - 2$, therefore we can write this lower bound on the cost as $\frac{3k}{2} - 1 + \varepsilon\sum_{j=1}^{P} jX_j$. Since the optimal cost is $(\frac{1}{2} + \varepsilon)k$, we get $\frac{3k}{2} - 1 + \varepsilon\sum_{j=1}^{P} jX_j \leq C(\frac{1}{2} + \varepsilon)k$. This is equivalent to

$$\sum_{j=1}^{P} jX_j \leq CP(1 + 2\varepsilon)k - 3kP + 2P. \tag{1}$$

For every $2 \leq j \leq P$ we get a lower bound on the cost of the algorithm for the second construction of $\sum_{i=1}^{P}(\frac{1}{2} + i\varepsilon)X_i + (3k - 2 - \sum_{i=j}^{P} X_i)(\frac{1}{2} + j\varepsilon) = (3k - 2)(1 + j\varepsilon) + \varepsilon\sum_{i=1}^{j-1} iX_i + \varepsilon\sum_{i=j}^{P}(i - j)X_i - \frac{1}{2}\sum_{i=j}^{P} X_i$. As this value must be at most $Ck(\frac{1}{2} + j\varepsilon)$, we get

$$\sum_{i=1}^{j-1} iX_i + \sum_{i=j}^{P}(i - j)X_i - P\sum_{i=j}^{P} X_i \leq P(C - 6)k + 4P + j(C - 3)k + 2j. \tag{2}$$

For each $1 \leq j \leq P$, we multiply the inequality for j by a_j, and add up the resulting inequalities. The coefficients are $a_1 = \frac{P+1}{2}$ (for equation (1)), and for $j > 1$, $a_j = 1$. Next, we compute the coefficient of each value X_i, $1 \leq i \leq P$, in the resulting inequality. Given a value X_i, its coefficient in the inequality (1) is i. Its coefficient in the inequality (2) for $j > i$ is i and for $j \leq i$ is $i - j - P$. Thus, we get

$$\frac{P+1}{2}i + \sum_{j=2}^{i}(i - j - P) + \sum_{j=i+1}^{P} i = i\left(\frac{3P - 1}{2}\right) - P(i - 1) - \left(\frac{i(i+1)}{2} - 1\right)$$

$$= \frac{iP}{2} + \frac{i}{2} + iP - i - Pi + P - \frac{i^2}{2} - \frac{i}{2} + 1 \geq (P - i) \cdot \frac{i+2}{2} \geq 0.$$

Therefore the left hand side of the resulting inequality is non-negative. Next, consider the right hand side. It is equal to $\frac{P+1}{2}(CP(1 + 2\varepsilon)k - 3kP + 2P) + \sum_{j=2}^{P}(P(C - 6)k + 4P + j(C - 3)k + 2j) = \frac{P+1}{2}(CP(1 + 2\varepsilon)k - 3kP + 2P) + (P - 1)(P(C - 6)k + 4P) + ((C - 3)k + 2)(\frac{P(P+1)}{2} - 1)$. Letting k tend to infinity, we get the following inequality on C.

$$0 \leq (P^2 + P)(C(\frac{1}{2} + \varepsilon) - \frac{3}{2}) + (P^2 - P)(C - 6) + (C - 3)\frac{P^2 + P - 2}{2}$$

Next, we let P tend to infinity and get $0 \leq \frac{C-3}{2} + (C-6) + \frac{C-3}{2} = 2C - 9$. This gives a lower bound of 4.5 on C. □

Remark 1. Running a linear program using Matlab for $P = 400$ we can get a lower bound of 4.591 on C.

5   Offline Problems

5.1   The Unbounded Model

The offline problem is clearly polynomially solvable for the unbounded model. The algorithm computes the maximum load, and then opens a single color with capacity equal to the maximum load. Clearly all the intervals can be colored using this color, and we obtain a feasible solution whose cost equals the maximum load, which is a lower bound on the optimal cost. Hence, we can conclude the following.

Proposition 1. *The offline problem of the unbounded model is polynomially solvable.*

5.2   The Bounded Model

First, we can show that the resulting offline problem for the bounded model is NP-hard, using a reduction from the 3-Partition problem.

Theorem 6. *The offline problem of the bounded model is NP-hard in the strong sense.*

Because of the fact that the bounded model problem is NP-hard, we turn our focus to designing an approximation algorithm for this problem. We define a *small request* to be a request with bandwidth that is at most $\frac{1}{2}$, and a *large request* to be a request whose bandwidth is strictly larger than $\frac{1}{2}$. Our algorithm uses disjoint sets of colors to color the small requests and the large requests.

For small requests we sort the intervals in non-decreasing order of their left end-point. Then, we use colors with maximum capacity 1 and color the intervals according to the First-Fit algorithm. After we color all the small requests, we compute the maximum load of the last color that is opened and we change its capacity to be this value of the maximum load.

Lemma 7. *The cost of colors that our algorithm uses to color the small requests is at most $2 \cdot \text{OPT}$.*

It remains to consider the large requests. Before presenting our algorithm, we consider the following algorithm. We sort the large requests in non-decreasing order of their left end-point. Then, we use colors with capacity 1 and color the intervals according to the First-Fit algorithm. We note that using First-Fit minimizes the number of colors that are used to color the large requests (when the intervals are sorted) and since the capacity of each color is 1 whereas in the optimal solution the capacity of each color is at least $\frac{1}{2}$, we conclude that this algorithm uses colors with total cost of at most $2 \cdot \text{OPT}$.

Let $\varepsilon > 0$ be a given constant such that $k = \frac{1}{2\varepsilon} - 1$ is an integer to be selected afterwards. Our algorithm for the large requests computes $k + 1$ solutions and picks the cheapest solution among these. The first solution is to pack all the large requests with a minimum number of unit capacity colors (using First-Fit on the sorted list of large requests). For each $j = 1, 2, \ldots, k$ we define $a_j = \frac{1}{2} + j\varepsilon$ and our $(j+1)$-th solution is constructed as follows. We partition the large requests into two classes: the first class consists of all large requests with bandwidth at most a_j, and the second class consists of all the remaining large requests. Each class is packed separately using its own set of colors. The capacity of the colors that are used for the first class is a_j, whereas the capacity of the colors that are used for the second class is 1. Each class is packed optimally using the minimum number of colors (using First-Fit on the sorted list of intervals from this class). We next show that the cheapest solution among the $k + 1$ solutions has a cost of at most $\left(\frac{8}{5} + O(\varepsilon)\right) \cdot \text{OPT}$.

Lemma 8. *The cheapest solution among the $k + 1$ solutions has a cost of at most $\left(\frac{8}{5-2\varepsilon}\right) \cdot \text{OPT}$.*

Proof. We prove that the algorithm colors the large requests with total cost of at most $\frac{8}{5-2\varepsilon} \cdot \text{OPT}$ and the approximation ratio of the algorithm is at most $\frac{8}{5-2\varepsilon}$. Let $a_0 = \frac{1}{2}$ and $a_{k+1} = 1$. Let ρ be the competitive ratio of the algorithm, we prove that $\rho \leq \frac{8}{5-2\varepsilon}$. Denote by X_j the number of colors that OPT opens with capacity in the interval $(a_j, a_{j+1}]$, for $j = 0, 1, 2, \ldots, k$. Then, OPT $\geq \sum_{j=0}^{k} a_j \cdot X_j$. We assume that the cheapest solution among the $k + 1$ solutions costs at least $\rho \cdot \text{OPT}$.

Since two intersecting large requests cannot be colored by the same color in any solution, we can compute upper bounds on the number of colors of each capacity used by the algorithm in each one of the cases. Note that our first solution can pack all the large requests using at most $\sum_{j=0}^{k} X_j$ colors, and therefore the cost of this solution is at most $\sum_{j=0}^{k} X_j$. Since we assume that the cheapest solution among the $k + 1$ solutions costs at least $\rho \cdot \text{OPT}$, we conclude that $\sum_{j=0}^{k} X_j \geq \rho \cdot \text{OPT}$. Next, consider the $(j+1)$-th solution for $j \geq 1$. The intervals of the first class can be colored using at most $\sum_{i=0}^{k} X_i$ colors each with capacity a_j (since this amount of colors suffices to color all the large requests). The intervals of the second class can be colored using at most $\sum_{i=j}^{k} X_i$ unit capacity colors. Therefore, the cost of the $(j+1)$-th solution is at most $a_j \cdot \left(\sum_{i=0}^{k} X_i\right) + \sum_{i=j}^{k} X_i$. Since we assume that the cheapest solution among the $k + 1$ solutions costs at least $\rho \cdot \text{OPT}$, we conclude that $a_j \cdot \left(\sum_{i=0}^{k} X_i\right) + \sum_{i=j}^{k} X_i \geq \rho \cdot \text{OPT}$.

We next consider the following set of inequalities (these inequalities hold by our assumption):

$$\text{OPT} \geq \sum_{j=0}^{k} a_j \cdot X_j \tag{3}$$

$$\sum_{j=0}^{k} X_j \geq \rho \cdot \text{OPT} \tag{4}$$

$$a_j \cdot \left(\sum_{i=0}^{k} X_i \right) + \sum_{i=j}^{k} X_i \geq \rho \cdot \text{OPT} \qquad \forall j = 1, 2, \ldots, k. \tag{5}$$

We construct the following inequality: we multiply (4) by $y_0 = a_0 - \sum_{i=1}^{k}(a_i - a_{i-1}) \cdot a_i$, and for each $j = 1, 2, \ldots, k$ we multiply the j-th constraint of (5) by $y_j = a_j - a_{j-1} = \varepsilon$, and we add up all the resulting inequalities. The left hand side of the resulting inequality is exactly $\sum_{j=0}^{k} a_j \cdot X_j$. This is so because the coefficient of X_j in the resulting inequality is $y_0 + \sum_{i=1}^{j} y_i(a_i + 1) + \sum_{i=j+1}^{k} y_i a_i = a_0 - \sum_{i=1}^{k}(a_i - a_{i-1}) \cdot a_i + \sum_{i=1}^{k}(a_i - a_{i-1})a_i + \sum_{i=1}^{j}(a_i - a_{i-1}) = a_j$. By (3), we conclude that the left hand side of the resulting inequality is at most OPT. The right hand side of the resulting inequality is $\rho \cdot \text{OPT} \cdot \sum_{i=0}^{k} y_i$. We note also that the coefficients y_j are non-negative. To see this last claim note that for $j \geq 1$, $y_j = \varepsilon > 0$ and for $j = 0$, $y_0 = a_0 - \sum_{i=1}^{k}(a_i - a_{i-1}) \cdot a_i = \frac{1}{2} - \sum_{i=1}^{k} \varepsilon \cdot \left(\frac{1}{2} + i\varepsilon \right) = \frac{1 - k\varepsilon}{2} - \varepsilon^2 \sum_{i=1}^{k} i = \frac{1 - k\varepsilon}{2} - \varepsilon^2 \cdot \frac{k(k+1)}{2} \geq \frac{1 - \frac{1}{2\varepsilon}\varepsilon}{2} - \varepsilon^2 \cdot \frac{1}{8\varepsilon^2} = \frac{1}{8} > 0$. Therefore, the inequality $\text{OPT} \geq \sum_{j=0}^{k} a_j \cdot X_j \geq \rho \cdot \text{OPT} \cdot \sum_{i=0}^{k} y_i$ holds and we get

$$\rho \leq \frac{1}{\sum_{i=0}^{k} y_i} = \frac{1}{\frac{1 - k\varepsilon}{2} - \varepsilon^2 \cdot \frac{k(k+1)}{2} + k\varepsilon} = \frac{8}{5 - 2\varepsilon}.$$

This completes the proof. □

By Lemma 8, we obtain a solution for the large requests with colors of total cost at most $\left(\frac{8}{5} + O(\varepsilon) \right) \cdot \text{OPT}$. We would like to argue that by picking ε as an infinitesimally small positive number we obtain an $\frac{8}{5}$ approximation algorithm. However, picking such a value of ε will increase dramatically the time complexity of our algorithm. To avoid these bad consequences we note the following lemma.

Lemma 9. *There is a polynomial time algorithm that emulates the solution returned by our previous algorithm for infinitesimally small value of ε.*

The following corollary is a direct consequence of Lemmas 8 and 9.

Corollary 1. *There is a polynomial time algorithm that colors the large requests with colors of total cost that is at most $\frac{8}{5} \cdot \text{OPT}$.*

Finally we combine the results for small requests and large requests.

Theorem 7. *There is an approximation algorithm with ratio $\frac{18}{5} = 3.6$ for the variable sized interval coloring problem in the bounded model.*

References

1. U. Adamy and T. Erlebach. Online coloring of intervals with bandwidth. In *Proceedings of the First International Workshop on Approximation and Online Algorithms (WAOA'03)*, LNCS 2909, pages 1–12, 2003.
2. J. Aspnes, Y. Azar, A. Fiat, S. A. Plotkin, and O. Waarts. On-line routing of virtual circuits with applications to load balancing and machine scheduling. *Journal of the ACM*, 44(3):486–504, 1997.
3. S. K. Baruah, G. Koren, D. Mao, B. Mishra, A. Raghunathan, L. E. Rosier, D. Shasha, and F. Wang. On the competitiveness of on-line real-time task scheduling. *Real-Time Systems*, 4(2):125–144, 1992.
4. M. Chrobak and M. Ślusarek. On some packing problems relating to dynamical storage allocation. *RAIRO Journal on Information Theory and Applications*, 22:487–499, 1988.
5. E. G. Coffman, M. R. Garey, and D. S. Johnson. Approximation algorithms for bin packing: A survey. In D. Hochbaum, editor, *Approximation algorithms*. PWS Publishing Company, 1997.
6. J. Csirik. An online algorithm for variable-sized bin packing. *Acta Informatica*, 26:697–709, 1989.
7. J. Csirik and G. J. Woeginger. On-line packing and covering problems. In *A. Fiat and G. J. Woeginger, editors*, Online Algorithms: The State of the Art, LNCS 1442, pages 147–177, 1998.
8. L. Epstein and M. Levy. Online interval coloring and variants. In *Proceedings of the 32nd International Colloquium on Automata, Languages and Programming (ICALP'05)*, LNCS 3580, pages 602–613, 2005.
9. L. Epstein and M. Levy. Online interval coloring with packing constraints. In *Proceedings of the 30th International Symposium on Mathematical Foundations of Computer Science (MFCS'05)*, LNCS 3618, pages 295–307, 2005.
10. D. K. Friesen and M. A. Langston. Variable sized bin packing. *SIAM J. Comput.*, 15:222–230, 1986.
11. T. R. Jensen and B. Toft. *Graph coloring problems*. Wiley, 1995.
12. H. A. Kierstead. The linearity of first-fit coloring of interval graphs. *SIAM Journal on Discrete Mathematics*, 1(4):526–530, 1988.
13. H. A. Kierstead and J. Qin. Coloring interval graphs with First-Fit. *SIAM Journal on Discrete Mathematics*, 8:47–57, 1995.
14. H. A. Kierstead and W. T. Trotter. An extremal problem in recursive combinatorics. *Congressus Numerantium*, 33:143–153, 1981.
15. F. D. Murgolo. An efficient approximation scheme for variable-sized bin packing. *SIAM J. Comput.*, 16(1):149–161, 1987.
16. N. S. Narayanaswamy. Dynamic storage allocation and online colouring interval graphs. In *Proceedings of the 10th Annual International Conference on Computing and Combinatorics (COCOON'04)*, LNCS 3106, pages 329–338, 2004.
17. S. V. Pemmaraju, R. Raman, and K. R. Varadarajan. Buffer minimization using max-coloring. In *Proceedings of the 15th Annual ACM-SIAM Symposium on Discrete Algorithms (SODA'04)*, pages 562–571, 2004.
18. S. S. Seiden. An optimal online algorithm for bounded space variable-sized bin packing. *SIAM Journal on Discrete Mathematics*, 14(4):458–470, 2001.
19. S. S. Seiden. On the online bin packing problem. *Journal of the ACM*, 49(5):640–671, 2002.
20. S. S. Seiden, R. van Stee, and L. Epstein. New bounds for variable-sized online bin packing. *SIAM Journal on Computing*, 32(2):455–469, 2003.

A Simpler Linear-Time Recognition of Circular-Arc Graphs

Haim Kaplan and Yahav Nussbaum

School of Computer Science, Tel Aviv University,
Tel Aviv 69978, Israel
{haimk, nuss}@post.tau.ac.il

Abstract. We give a linear time recognition algorithm for circular-arc graphs. Our algorithm is much simpler than the linear time recognition algorithm of McConnell [10] (which is the only linear time recognition algorithm previously known). Our algorithm is a new and careful implementation of the algorithm of Eschen and Spinrad [4, 5]. We also tighten the analysis of Eschen and Spinrad.

1 Introduction

A *Circular-arc graph* is an intersection graph of arcs on the circle. That is, every vertex is represented by an arc, such that two vertices are adjacent if and only if the corresponding arcs intersect. Many subclasses of circular-arc graphs have also been studied such as *proper circular-arc graphs* [3], and *unit circular-arc graphs* [8]. An extensive overview of circular-arc graphs can be found at the book by Spinrad [13]. Recent applications of circular-arc graphs are in modeling ring networks [15] and item graphs of combinatorial auctions [2].

The first polynomial time algorithm for circular-arc recognition was given by Tucker [16]. This algorithm splits into one of two cases according to whether $\bar{G}$ is bipartite (G is co-bipartite). In case $\bar{G}$ is not bipartite the algorithm finds an odd length induced cycle in $\bar{G}$, and further splits into one of two subcases according to whether the cycle it found is of length 3 or of length at least 5. Using Tucker's terminology we refer to the first case where G is co-bipartite as Case I. We refer to the subcase where we found in $\bar{G}$ a cycle of length 3, and therefore we found in G an independent set of size 3, as Subcase IIa. We refer to the case where we found in $\bar{G}$ an induced cycle of length at least 5 as Subcase IIb.

Tucker showed how to implement his algorithm in $O(n^3)$ time. One of the bottlenecks in Tucker's implementation is a preprocessing phase where we identify containment relations between the neighborhoods of the vertices. Specifically, for every pair of vertices v and w we determine whether the neighborhood of v is contained in the neighborhood of w or vice versa. Furthermore, Tucker runs his algorithm recursively on particular graphs and this recursive structure also leads to cubic running time.

Spinrad [12] simplified Case I in Tucker's algorithm – the case where G is co-bipartite. Spinrad reduced this case to the problem of recognizing two dimensional posets [14]. We construct the poset using particular relations between the

L. Arge and R. Freivalds (Eds.): SWAT 2006, LNCS 4059, pp. 41–52, 2006.

vertices of G. Two vertices are related in the poset if their corresponding arcs are either disjoint, one is contained in the other, or together they cover the circle. The relations between the arcs are determined from the relations between the neighborhoods of the vertices. In case G is a circular-arc graph then from any two total orders that represent the poset we can construct a representation for G. Spinrad showed that this algorithm runs in $O(n^3)$ time.

Eschen and Spinrad [4, 5] gave an $O(n^2)$ algorithm for recognizing circular-arc graphs by addressing the two bottlenecks in Tucker's implementation. Eschen and Spinrad show how to compute neighborhood containment relations in $O(n^2)$ time. Specifically, they construct four graphs such that if G is indeed circular-arc graph then each of the four graphs is either an interval graph or a chordal bipartite graph. These graphs are constructed such that the neighborhood of v contains the neighborhood of w in G, if and only if the neighborhood of v contains the neighborhood of w in each of these graphs. The quadratic time bound follows since one can compute neighborhood containment relations in interval graphs and chordal bipartite graphs in quadratic time [4, 5, 9].

Eschen and Spinrad also showed that in Case I of the algorithm, when G is co-bipartite, we can use the same reduction to determine all pair of arcs that can cover circle in a model of G in $O(n^2)$ time. Since this was the only bottleneck in Spinrad's algorithm for this case, we obtain an $O(n^2)$ implementation of Case I. To implement Subcases IIa and IIb in $O(n^2)$ time, they changed the recursive structure of Tucker's algorithm. They show how to implement the algorithm such that each recursive call is on a co-bipartite graph (Case I) and therefore does not trigger further recursion. Since the sum of the sizes of the graphs in all recursive calls is proportional to the size of G, the quadratic bound follows.

Recently, McConnell [10] presented the first recognition algorithm for circular-arc graphs that runs in linear time. The algorithm reduced the problem to an interval graph recognition problem where specific intersection types between the intervals are specified. McConnell's algorithm uses the same preprocessing stage of Eschen and Spinrad where it computes neighborhood containment relations. To establish the linear time bound, McConnell tightens the analysis of Eschen and Spinrad's preprocessing stage. He shows that this preprocessing stage can be implemented in linear time since we are interested only in neighborhood containment relations between adjacent vertices, and the associated chordal bipartite graphs cannot be too large.

McConnell's algorithm is quite involved. Its most complicated computation is to find a partition of a graph into a particular kind of modules called Δ modules. Those Δ modules are used to turn the input circular-arc graph into an interval graph with specific types of intersections between the intervals, and to find a representation for this interval graph. McConnell first presents an implementation that runs in $O(m + n \log n)$ time. To get the linear time bound a more complicated partitioning procedure has to be adapted from the linear time transitive orientation algorithm [11] which is by itself quite involved. This algorithm also uses probe interval graphs to find pairs of arcs that can cover the circle in linear time.

Hsu [6] presented a different recognition algorithm for circular-arc graphs that runs in $O(mn)$ time and reduces the problem to recognition of circle graphs.

1.1 Our Contribution

We give a careful implementation of the recognition algorithm of Eschen and Spinrad that in fact runs in linear time. Our implementation first either finds an independent set of size 3, and then we can apply Subcase IIa of the algorithm, or it concludes that the graph has $\Theta(n^2)$ edges. In the latter case the implementation of Eschen and Spinrad in fact runs in linear time.

Eschen and Spinrad find in Subcase IIa a particular maximal independent set and place the corresponding arcs one the circle. We show how to find the independent set and place its arcs on the circle in linear time. Our implementation then continues as the implementation of Eschen and Spinrad, but we tighten their analysis to show that the running time is linear. Our main new insight is that each subgraph considered by the algorithm while placing and ordering the arcs on the circle is dense. That is, the number of edges that each subgraph contains is quadratic in the size of its vertex set. Furthermore, the total size of these subgraphs is linear in the size of the input graph.

Our algorithm also performs a preprocessing phase where neighborhood containment relations are computed. As proved by McConnell [10] this can be done in linear time. As all previous algorithms, we also require a postprocessing verification step where we check that the representation we obtain is indeed a representation of G. McConnell [10] gave a straightforward linear time implementation of this postprocessing step, which traverse the circular-arc model and extract all the pairs of intersecting arcs from it. The model correspondes to the graph G, only if the intersections of arcs fit the adjacencies of vertices.

We describe a linear time implementation of Subcase IIa. Subcase IIb can also be implemented in linear time in a similar way. We do not describe it here since we apply Subcase IIb only when we are sure that G has $\Theta(n^2)$ edges. Our implementation is much simpler than McConnell's algorithm.

2 Preliminaries

We consider a finite simple graph $G = (V, E)$, Where $|V| = n$ and $|E| = m$. We represent graphs using adjacency-lists. For a vertex v in a graph, the *(closed) neighborhood* of v, denoted by $N[v]$ is the set of all vertices adjacent to v together with v itself. For a set of vertices U we define $N_U[v]$ to be $N[v] \cap U$.

A *circular-arc model* of a graph G is a mapping from the vertices of G to arcs on the circle, such that two vertices are adjacent if and only if the corresponding arcs intersect. A graph G is a *circular-arc graph* if it has a circular-arc model. Note that a circular-arc graph may have more than one model.

We represent a single arc in a circular-arc model by its clockwise and counterclockwise endpoints. We assume that no arc covers the entire circle. We represent a circular-arc model by an ordered cyclic list of the endpoints of its arcs. To simplify we refer to the clockwise direction as *right* and to the counterclockwise direction as *left*, as we view them if we stand at the center of the circle.

There are three possible types of intersections between arcs x and y [6, 16]. Arcs x and y *cross* if each contains a single endpoint of the other. Arcs x and y *cover the circle* if each contains both endpoints of the other. One of the arcs x, y may *contain* the other. If x and y either cross or cover the circle, we say that x and y *overlap*.

For convenience, we refer to the vertices of G as arcs even before we decide if G is a circular-arc graph and find a model for it. We would say that two adjacent vertices intersect even before we have a model, because the arcs of adjacent vertices must intersect in every model. Hsu [6] showed that if G is a circular-arc graph then it has a model M such that for every pair of vertices v and u, the arc representing v in M contains the arc representing u in M if and only if $N[u] \subseteq N[v]$. So we would say that v contains u when $N[u] \subseteq N[v]$, even before we have found a model. This relation between u and v is the *neighborhood containment relation*. Additionally we would say that two vertices overlap when they intersect but do not contain each other.

A graph that can be partitioned into two independent sets is called *bipartite*. If G is not bipartite then it must have an odd-length induced cycle. If $\bar{G}$, the complement of G, is bipartite then G is *co-bipartite*, and is covered by two cliques.

A (0,1)-matrix is said to have the *circular-ones* property if its columns can be ordered such that the 1's in each row are circularly consecutive. Circular-ones arrangement can be found in $O(m + n + r)$ time [1, 7] where m is the number of columns, n is the number of rows, and r is the number of 1's.

3   Preprocessing

An arc representing a universal vertex can be placed on the circle in $O(1)$ time by placing its right endpoint anywhere on the circle and its left endpoint immediately to the right side of it. It is easy to find all universal vertices of G in linear time. Thus, we may assume that G does not have any universal vertices.

Let x be an arc that have the same neighborhood as another arc y that was already placed on the circle. The arc x can be placed on the circle by placing its endpoints next to the endpoints of y, in $O(1)$ time. McConnell [10] showed how to find vertices with the same neighborhood in linear time using a simple process called *radix partitioning*, which is similar to radix sort. Thus, we may assume that there are no two vertices in G that have the same neighborhood.

Before running our algorithm we preprocess the graph and for every pair of adjacent vertices v and u we check whether v contains u or u contains v, that is whether $N[v] \subseteq N[u]$ or $N[u] \subseteq N[v]$. Recall that Eschen and Spinrad [4, 5] showed how to compute neighborhood containment relations in $O(n^2)$ time, and McConnell [10] tighten the analysis to show that this can be done in linear time. For more details of this part, which are not complicated, see [10].

4   Splitting into Cases

Recall that the algorithms of Tucker [16] and Eschen and Spinrad [4, 5] split into one of three cases. *Case I* where G is co-bipartite, and *Case II* where $\bar{G}$ has an odd-length induced cycle. Case II splits further into two non-exclusive subcases. *Subcase IIa* in which we find three independent vertices in G, and *Subcase IIb* in which we find in $\bar{G}$ an induced cycle of odd length 5 or more. Our algorithm splits into one of these three cases as well. But we decide on the case to apply more carefully.

Let a_1 be a vertex of minimum degree in G. If $|N[a_1]| > \frac{n}{2}$ then every vertex of G has an edge to at least $\frac{n}{2}$ other vertices, so $m = \Theta(n^2)$. Otherwise, let Y be the set of arcs nonadjacent to a_1. We look for a pair of nonadjacent vertices in Y. For every vertex $y \in Y$ we traverse its adjacency list, and construct its restriction to Y. The time to traverse all the adjacency lists is $O(n + m)$. If for every $y \in Y$ we found that $N_Y[y] = Y$ then $m = \Theta(n^2)$ since we know that $|Y| \geq \frac{n}{2}$. Otherwise, we find nonadjacent pair of arcs $a_2, a_3 \in Y$.

So either we concluded that $m = \Theta(n^2)$, and therefore the $O(n^2)$ time bound of Eschen and Spinrad [5] is linear, or we found three independent vertices a_1, a_2, a_3, and we can apply Subcase IIa. In the rest of the paper we describe a linear time implementation of Subcase IIa.

The algorithm for Subcase IIa consists of the three stages of Tucker's algorithm. In Stage 1, we find a set of arcs that can be ordered around the circle and divide it into sections, such that no arc has its both endpoints in the same section. In Stage 2, we place every endpoint of every other arc in its section. And in Stage 3, we order the endpoints within each section. We describe each of these stages in the following three sections.

5   Stage 1: Dividing the Circle into Sections

The algorithm begins by finding an independent set of arcs, I, that can be embedded around the circle, in an order consistent with some model of G. This set of arcs divides the circle into sections, such that no arc has its two endpoints in the same section.

5.1   Finding a Maximal Independent Set

The algorithm of Tucker uses maximal independent set of arcs I of size at least 3 that obeys two requirements. First, no arc of I contains any other arc of G. Second, there is no arc $x \in I$ that has two nonadjacent arcs $y, z \notin I$ such that y and z overlap x and do not overlap any other arc in I. We begin by constructing a maximal independent set I' greedily, which satisfies the first requirement, and then change it to an independent set I that satisfies the second requirement as well.

Before constructing I', we eliminate any arc that contains another arc from G, since those arcs cannot be in I. Let G' be the subgraph of G without these arcs. Every pair of intersecting arcs in G' overlaps, since no arc of G' contains another. In order to construct I' we maintain a set J consisting of every arc in G' that is nonadjacent to any arc already in I'. For every arc in G' we maintain a counter of the number of arcs in I' that intersect it.

Let $\{a_1, a_2, a_3\}$ be the independent set that we found in Sect. 4. We initialize I' to consist of arcs $\{a'_1, a'_2, a'_3\}$ where a'_i is an arc from G' and may be either a_i or a minimal arc contained in a_i. The set I' is an independent set in G', since $\{a_1, a_2, a_3\}$ is an independent set in G. For every $a'_i \in I'$, we remove $N[a'_i]$ from J and increase the counters of the members of $N[a'_i]$.

As long as J is not empty, we pick an arbitrary arc $x \in J$ and add x to I'. We increase the counter of every arc y that overlaps x, and set $J = J \setminus \{y\}$. When J is empty, I' is a maximal independent set.

Next we construct I from I'. For every arc $x \in I'$ such that there are two nonadjacent arcs y_1 and y_2 in G' which overlaps only x in I', we add y_1 and y_2 to I. If such y_1 and y_2 do not exist, we add x to I. To do so in linear time, we find all arcs in G' that overlaps only x by scanning $N[x]$, and identifying all neighbors of x whose counter equals to one. Let $Y \subset N[x]$ consist of these neighbors. For every $y \in Y$ we scan $N[y]$ and construct $N_Y[y]$, if $N_Y[y] \neq Y$ then we find $y' \in Y \setminus N_Y[y]$ which is nonadjacent to y.

The following lemma proves that I satisfies the requirements stated above.

Lemma 1. *If G is a circular-arc graph then I is a maximal independent set in G and we cannot get a larger independent set by replacing an arc $y_1 \in I$ with two nonadjacent arcs $z_1, z_2 \notin I$ that intersect y_1.*

Proof. First note that I' is a maximal independent set in G, since it is a maximal independent set in G', and every arc in G which is not in G' contains an arc in G'.

We now prove that I is a maximal independent set in G. Assume otherwise, then there is an arc $z \notin I$ which is nonadjacent to every arc of I. We may assume that z is in G', as otherwise we can replace z by an arc which z contains. The arc z cannot be in I' because otherwise z or an adjacent arc would be inserted to I. Then, since I' is maximal independent set, z must overlap some $x \in I'$, such that $x \notin I$ and x was replaced by y_1, y_2 in I. It follows that $\{y_1, y_2, z\}$ is an independent set of three arcs that overlap x, but this is impossible since each of them should cover an endpoint of x, and x has only two endpoints.

We next prove that we cannot get a larger independent set by replacing an arc $y_1 \in I$ with two nonadjacent arcs $z_1, z_2 \notin I$ that overlap y_1. Assume that $y_1 \in I$ can be replaced by two nonadjacent arcs $z_1, z_2 \notin I$ that overlap y_1 but do not overlap any other arc in I. The arc y_1 cannot be a member of I', since otherwise we would add z_1, z_2 to I instead of y_1. Therefore there are arcs $x \in I'$ and $y_2 \in I$ such that y_1 and y_2 are nonadjacent and overlap x. Arcs z_1 and z_2 do not overlap any arc $x' \neq x$, $x' \in I'$, because if they do, they must overlap some arc different from y_1 in I. Since I' is maximal, the two arcs z_1 and z_2 must be adjacent to x. Again, we got independent set of three arcs $\{y_2, z_1, z_2\}$, that should overlap x, but x has only two endpoints.   □

Note that if G is not a circular-arc graph, I might not satisfy the requirements stated in Lemma 1. In this case our algorithm continues and will detect that G is not a circular-arc graph later on.

In Sect. 5.2 we show how to place the arcs of I on the circle. We label the arcs of I by $a_1, \ldots, a_{|I|}$ according to their cyclic order around the circle, where a_1 is some arbitrary arc in I. The endpoints of the arcs split the circle into sections. Each section is either an arc of I or a gap between two consecutive arcs of I. Let S be a section. The two endpoints of the arcs of I that define S are called *the endpoints of S*. We assume that a section contains its left endpoint, but does not contain its right endpoint. We denote by S_{2i} the section of arc a_i. We denote by S_{2i+1} the section which is the gap between S_{2i} and $S_{2(i+1)}$. Subscripts of arcs are modulo $|I|$ and subscripts of sections are modulo $2|I|$.

Let I^c bet the set of arcs of G not in I. For every $x \in I^c$, the arc x cannot be contained in an arc of I and is not universal. Furthermore, since I is a maximal independent set, the following lemma holds.

Lemma 2. *[16] Let $x \in I^c$. In a model of G consistent with the placement of I, the endpoints of x are in different sections.*

5.2 Placing the Independent Set Around the Circle

We now place the arcs of I around the circle. Tucker showed how to order the arcs of I around the circle, using the adjacencies between arcs in I and arcs in I^c, such that there exist a circular-arc model of G consistent with this order.

Lemma 3. *[16] If G is a circular-arc graph then there exists a model of G consistent with every cyclic order of I that satisfies the following two requirements: (1) For each arc $x \in I^c$, the neighborhood of x in I, $N_I[x]$, is consecutive around the circle, with the arcs that are contained in x in the middle and the arcs that overlap x in the ends. (2) For each pair of adjacent arcs $x, y \in I^c$, the union of their neighborhoods in I, $N_I[x] \cup N_I[y]$ is consecutive around the circle.*

Let $x \in I^c$. We define $D(x)$ to be the set consisting of every arc $y \in N_{I^c}[x]$ such that $N_I[x] \cap N_I[y] = \emptyset$. Let $D^m(x)$ be the subset of $D(x)$ consisting of every arc $y \in D(x)$ such that there is no $y' \in D(x)$ for which $N_I[y'] \subset N_I[y]$ (see Fig. 1). Eschen and Spinrad [5] proved the following.

Lemma 4. *[5] Assume that G is a circular-arc graph. Let P be an order of I that satisfies the second requirement of Lemma 3 with respect to every pair of arcs $x, y \in I^c$ such that $y \in D^m(x)$ and $x \in D^m(y)$. Then, P satisfies the second requirement of Lemma 3 with respect to every pair of adjacent arcs in I^c.*

We construct a matrix M such that from a circular-ones arrangement of M we can define the order of I. Every arc of I corresponds to a column of M and every requirement of Lemma 3 corresponds to a row. We arrange the matrix such that the ones in every row are cyclically consecutive. The order of the columns will give us an order of I that is consistent with the requirements of Lemma 3. Any arc $x \in I^c$ with $N_I[x] = I$ cannot affect the order of I according to the requirements of Lemma 3, so we ignore those arcs.

For each arc $x \in I^c$ we create a row that have 1's in the columns of the arcs in $N_I[x]$. This row forces the consecutiveness of $N_I[x]$. If G is a circular-arc graph then there are at most two arcs in I that x overlaps. For each such arc, z, we create a row that have 1's only in the columns of $N_I[x]\setminus\{z\}$. These rows will force $N_I[x]$ to be ordered so that the arcs that x contains are in the middle and the arcs that x overlaps are at the ends. If for some x there are more than two arcs in I that it overlaps then we halt since G is not a circular-arc graph. We created at most three rows for each arc, and a total of at most $3n$ rows with $3m$ ones.

In order to find $D(x)$ and neighborhood containment relations with respect to I, we decide for each pair of arcs $x, y \in I^c$ whether $N_I[x] \cap N_I[y] = \emptyset$ or $N_I[x] \subseteq N_I[y]$. To do so, we find a circular-ones arrangement [1, 7] of M. This arrangement gives us a preliminary cyclic order of the arcs of I. If such an arrangement does not exist then G is not a circular-arc graph, and we halt. For each pair of adjacent arcs in I^c we can detect if their neighborhoods in I do not intersect or one contains the other by looking at the first and last neighbors of both arcs in the cyclic order of I. We find the last neighbor of all arcs of I^c in the cyclic order by scanning the arcs of I starting from an arbitrary arc in the cyclic order. An arc $z \in I$ is the last neighbor of $x \in I^c$ if it is a neighbor of x, but the arc z' following z in the cyclic order is nonadjacent to x. We find the first neighbor in I of each $x \in I^c$ symmetrically.

Let $x \in I^c$ and consider the neighborhood containment relation restricted to I of the arcs in $D(x)$. In any circular-arc model of G, every $y \in D(x)$ covers one endpoint of x and stretches away from x. So $N_I[y]$ consists of a member of I next to x in the model, followed by zero or more members of I consecutively after it, in the direction in which y stretches. Therefore, the arcs of $D(x)$ form at most two chains with respect to the neighborhood containment relation restricted to I, each chain consisting of arcs that cover the same endpoint of x. So, there are at most two distinct neighborhoods in I for arcs in $D^m(x)$. For example, in the illustration of Fig. 1, $N_I[b]$ and $N_I[c]$ are the two distinct neighborhoods in I for arcs in $D^m(x)$.

For each arc of x, we go through $D(x)$ to find $D^m(x)$. We find $D^m(x)$ partitioned into two sets, each consisting of arcs with the same neighborhood in I. We consider the elements in $D(x)$ one by one, in an arbitrary order. While scanning $D(x)$ we maintain at most two sets of minimal elements with respect to the neighborhood containment relation restricted to I. We denote these sets by $M_1(x)$ and $M_2(x)$. If for the next arc $y \in D(x)$, we have that $N_I[y] = N_I[m_i]$ for $m_i \in M_i(x)$ we add y to $M_i(x)$. If $N_I[y] \subset N_I[m_i]$ for $m_i \in M_i(x)$, we replace $M_i(x)$ by $\{y\}$. If $N_I[m_i] \subset N_I[y]$ for $m_i \in M_i(x)$, we skip y. Otherwise, the relation does not form two chains and thus G is not a circular-arc graph and we halt. When we finish scanning $D(x)$, we have identified $D^m(x)$ partitioned into two sets $M_1(x)$ and $M_2(x)$, each set consist of all elements with the same neighborhood in I.

According to Lemma 4, for every pair of arcs $x, y \in I^c$ such that $x \in D^m(y)$ and $y \in D^m(x)$, we should add a row to M with 1's in the columns of $N_I[x] \cup N_I[y]$. Although there could be $\Omega(n^2)$ pairs $x, y \in I^c$ such that $x \in D^m(y)$ and $y \in D^m(x)$, the number of distinct sets $N_I[x] \cup N_I[y]$ is at most n. This is because for every arc $x \in I^c$, the members of $D^m(x)$ have at most two distinct neighborhoods in I. We identify these distinct rows to add to M as follows.

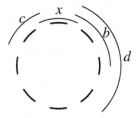

Fig. 1. $D(x)$ and $D^m(x)$. Arcs of I are drawn in boldface. $b, c, d \in D(x)$. Also, $b, c \in D^m(x)$ but $d \notin D^m(x)$, since $N_I[b] \subset N_I[d]$.

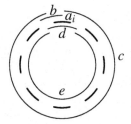

Fig. 2. U_i, W_i, A_i^e and A_i^c. Arcs of I are drawn in boldface. $b \in U_i, c \in W_i, d \in A_i^e$ and $e \in A_i^c$.

For every $x \in I^c$, we traverse every set $M_i(x)$ which is not empty. For each $y \in M_i(x)$ we check if $x \in D^m(y)$. If indeed $x \in D^m(y)$, we add a row to M with 1's in the columns of $N_I[x] \cup N_I[y]$. In this case we also set $M_i(x)$ to be empty and stop the traversal, since all other arcs in $M_i(x)$ have the same neighborhood in I as y. To check if $x \in D^m(y)$ in constant time, we pick an arbitrary arc z_i from each $M_i(y)$ that is not empty, and check if $N_I[x] = N_I[z_i]$.

Since we use the neighborhood of each arc to define at most two rows, we add to M at most n rows containing at most $2m$ ones. We can find circular-ones arrangement for M in $O(n + m)$ time. If such an order does not exist then G is not a circular-arc graph. Otherwise, we place the arcs of I in this order clockwise on the circle. We keep the section $S_1, \ldots, S_{2|I|}$ that are formed by the endpoints of arcs of I in an ordered cyclic list.

6 Stage 2: Placing the Endpoints of the Arcs in the Sections

Consider the order of I found in Sect. 5.2. For every arc $x \in I^c$, the members of $N_I[x]$ are consecutive on the circle. Since there are no universal arcs in G, and I is a maximal independent set, x cannot contain all arcs of I and $N_I[x] \neq \emptyset$. Also, x overlaps at most two arcs of I, since otherwise G is not a circular-arc graph and we should have detected it in Stage 1.

Let $x \in I^c$, the way we place the endpoints of x into their sections depends on the relation between x and the arcs of I. In most cases these relations suffice to determine the sections, and in the other cases we apply the algorithm recursively on an appropriate graph. The arc x satisfies one of the following cases (see Fig. 2).

- Arc x contains arc $a_i \in I$ and does not intersect any other arc in I. In this case the left endpoint of x is placed in S_{2i-1} and the right endpoint is placed in S_{2i+1}. For every $a_i \in I$ we accumulate all arcs that contain it and does not intersect any other arc of I in a set which we call A_i^e.
- Arc x overlaps $a_i \in I$ and does not intersect any other arc in I. For every $a_i \in I$ we accumulate these arcs in a set which we call U_i.

– Arc x intersects at least two arcs of I and does not intersect at least one. For all these arcs we identify in $N_I[x]$ the leftmost arc a_i, and the rightmost arc a_j. We do that as we identified the first and last neighbor of every arc in the preliminary order of I in Sect. 5.2. If x contains a_i then the left endpoint of x is in S_{2i-1}, if x overlaps a_i then this endpoint is in S_{2i}. Similarly, if x contains a_j then the right endpoint of x is in S_{2j+1}, if x overlaps a_j then this endpoint in S_{2j}. For every arc $a_i \in I$ we accumulate every arc that contains all arcs in I except a_i in a set which we call A_i^c.

– Arc x overlaps two consecutive arcs $a_i, a_{i+1} \in I$ and contains all other arcs of I. In this case, we place the left endpoint of x in $S_{2(i+1)}$ and the right endpoint of x in S_{2i}

– Arc x overlaps one arc $a_i \in I$ and contains all other arcs of I. For each $a_i \in I$ we accumulate these arcs in a set W_i.

At this point we placed the endpoints of all arcs in I^c into their sections except arcs in U_i and W_i for $i = 1, \ldots, |I|$. Consider any arc $a_i \in I$ and the associated sets U_i and W_i. Each arc in $U_i \cup W_i$ has one endpoint in S_{2i} and the other in S_{2i-1} or in S_{2i+1}. Furthermore, all arcs of U_i must form a clique, as otherwise we can get from I a larger independent set by replacing a_i by two nonadjacent arcs in U_i, contradicting Lemma 1.

We place the endpoints of the arcs of $U_i \cup W_i$ in the sections S_{2i-1}, S_{2i} and S_{2i+1} for each $a_i \in I$ separately, by solving a new problem recursively on a graph G_i. The graph G_i which we construct is identical to the graph that Eschen [4] constructs[1]. This graph is co-bipartite and therefore when we apply the algorithm to G_i, Case I applies and there would not be further recursion. We contribute the following observations. If G is a circular-arc graph then the recursive application of the algorithm on G_i takes time linear in the size of G_i. Furthermore, the sum of the sizes of all G_i's is proportional to the size of G.

Let C_i^a be the set of arcs $\{a_i\} \cup A_i^e \cup U_i$. The set C_i^a forms a clique in G, since U_i forms a clique and all A_i^e arcs intersect every arc that a_i intersects. The clique C_i^a consists of all arcs contained in the union of the sections S_{2i-1}, S_{2i} and S_{2i+1}. Let Q_i be the set of arcs adjacent to some but not all arcs in C_i^a.

To define G_i we first define a subgraph of G which we denote by G_i'. The graph G_i' will also be a subgraph of G_i. The graph G_i' is the subgraph induced by $C_i^a \cup Q_i \cup A_i^c \cup W_i$. Note that Q_i is not necessarily disjoint from W_i and A_i^c.

We find Q_i by scanning the adjacency list of each $x \in C_i^a$. We maintain a set Y of arcs encountered during the scan. For each arc $y \in Y$, we also keep a counter that counts the number of neighbors of y in C_i^a. When we finish scanning the adjacency list of every $x \in C_i^a$, the arcs of Q_i are exactly those arcs $y \in Y$ whose counters are smaller than $|C_i^a|$. We construct G_i' by scanning the adjacency list of each arc in it and restricting the list to contain only arcs inside G_i'.

The following lemma proves that all G_i''s are constructed in linear time.

Lemma 5. *[4] Every arc x participates in a constant number of graphs G_i'.*

---

[1] Note that this graph is different to the one from [5] which seems to have an error.

Proof. From the definition of the sets U_i, W_i, A_i^e, A_i^c, it follows that an arc x can belong to at most one such set. If $x \in Q_i$ for some i, then one of the arcs a_{i-1}, a_i, a_{i+1} must be the leftmost or the rightmost arc of $N_I[x]$ in the cyclic order of I. So, x can belong to at most constant number of sets Q_i. □

Let n_i' be the number of vertices in G_i', and let m_i' be the number of edges in G_i'. Every arc in G_i' covers at least one of the four endpoints of the three consecutive sections $S_{2i-1}, S_{2i}, S_{2i+1}$. Therefore, the arcs of G_i' are covered by four cliques, one for each endpoint. One of these cliques should have at least $\frac{n_i'}{4}$ vertices and therefore has at least $\frac{n_i'}{4}(\frac{n_i'}{4} - 1)$ edges. So we check if $m_i' \geq \frac{n_i'}{4}(\frac{n_i'}{4} - 1)$. If this inequality does not hold then G is not a circular-arc graph and we halt. Otherwise, we know that $m_i' = \Theta(n_i'^2)$.

We construct G_i from G_i' by adding a constant number of vertices and $O(n_i'^2) = O(m_i')$ edges. The vertices guarantee that any model of G_i can be embedded into a model of G, and the edges make all the vertices which are not in C_i^a a second clique. So if n_i and m_i denote the number of vertices and edges in G_i respectively, then we also have that $m_i = \Theta(m_i') = \Theta(n_i'^2) = \Theta(n_i^2)$. The details of the construction of G_i are as in Eschen [4].

Since $m_i = \Theta(n_i^2)$, the recursive application of our algorithm to G_i takes $O(m_i)$ time. Since each arc of G belong to at most a constant number of graphs G_i, then each edge of G must belong to at most constant number of graphs G_i. And therefore, $\sum m_i = O(m)$ and the linear time bound for Stage 2 follows.

7   Stage 3: Arranging the Endpoints in Each Section

We now know which sections contain the endpoints of every arc. Next we would arrange the endpoints inside each section. We follow Eschen and Spinrad's algorithm [4, 5], but provide a tighter analysis of it.

Our algorithm goes through the sections and tries to split each section S into ordered list of subsections. If S is split into subsections, then these subsections replace S in the cyclic order of sections. When we cannot split sections anymore then each section S has a corresponding section S' such that all arcs that have one endpoint in S have their other endpoint in S' and vice versa. We then use recursion to order the endpoints inside sections containing more than one endpoint.

Our initial list of sections, $S_1, \ldots, S_{2|I|}$, are the sections of Stage 2. Let n_i be the number of arcs that have an endpoint in S_i, and let m_i be the number of edges in G between these arcs. If G is a circular-arc graph then the arcs that have their right endpoint in S_i form a clique in G, since they all cover the left endpoint of S_i. Similarly, the arcs that have their left endpoint in S_i also form a clique in G. So for each of the initial sections m_i should be at least $\frac{n_i}{2}(\frac{n_i}{2} - 1)$. We check for all $i = 1, \ldots, 2|I|$ that indeed $m_i \geq \frac{n_i}{2}(\frac{n_i}{2} - 1)$, and if it does not hold for some i, then G is not a circular-arc graph. Note that since each arc has endpoints in two sections, $\sum m_i = O(m)$.

We split sections in the same way as Eschen and Spinrad [4, 5]. Intuitively, since the order of the endpoints inside a particular section S is not affected by any arc that does not have an endpoint in S, it suffices to determine the order between pairs of endpoints in the same section. Therefore the time it takes to split the sections is $O(\sum n_i^2)$. Since $O(\sum n_i^2) = O(\sum m_i) = O(m)$ this time is linear in the size of G. Details of this stage can be found in [4, 16].

References

1. K. S. Booth and G. S. Lueker. Testing for the consecutive ones property, interval graphs, and graph planarity using PQ-tree algorithms. *J. Comput. Syst. Sci.*, 13(3):335–379, 1976.
2. V. Conitzer, J. Derryberry, and T. Sandholm. Combinatorial auctions with structured item graphs. In *Proceedings of the Nineteenth National Conference on Artificial Intelligence*, pages 212–218, 2004.
3. X. Deng, P. Hell, and J. Huang. Linear-time representation algorithms for proper circular-arc graphs and proper interval graphs. *SIAM J. Comput.*, 25(2):390–403, 1996.
4. E. M. Eschen. *Circular-arc graph recognition and related problems*. PhD thesis, Department of Computer Science, Vanderbilt University, 1997.
5. E. M. Eschen and J. P. Spinrad. An $O(n^2)$ algorithm for circular-arc graph recognition. In *SODA '93: Proceedings of the fourth annual ACM-SIAM Symposium on Discrete algorithms*, pages 128–137, 1993.
6. W.-L. Hsu. $O(mn)$ algorithms for the recognition and isomorphism problems on circular-arc graphs. *SIAM J. Comput.*, 24(3):411–439, 1995.
7. W.-L. Hsu and R. M. McConnell. PC-trees and circular-ones arrangements. *Theor. Comput. Sci.*, 296(1):99–116, 2003.
8. M. C. Lin and J. L. Szwarcfiter. Efficient construction of unit circular-arc models. In *SODA '06: Proceedings of the seventeenth annual ACM-SIAM symposium on Discrete algorithm*, pages 309–315, 2006.
9. T.-H. Ma and J. P. Spinrad. Avoiding matrix multiplication. In *Graph-Theoretic Concepts in Computer Science: 16th International Workshop WG '90*, Lecture Notes in Computer Science 484, pages 61–71, 1991.
10. R. M. McConnell. Linear-time recognition of circular-arc graphs. *Algorithmica*, 37(2):93–147, 2003.
11. R. M. McConnell and J. P. Spinrad. Modular decomposition and transitive orientation. *Discrete Mathematics*, 201(1-3):189–241, 1999.
12. J. P. Spinrad. Circular-arc graphs with clique cover number two. *Journal of Combinatorial Theory Series B*, 44(3):300–306, 1988.
13. J. P. Spinrad. *Efficient Graph Representations*. Fields Institute Monographs 19. American Mathematical Society, 2003.
14. J. P. Spinrad and J. Valdes. Recognition and isomorphism of two dimensional partial orders. In *Automata, Languages and Programming, 10th Colloquium*, Lecture Notes in Computer Science 154, pages 676–686, 1983.
15. S. Stefanakos and T. Erlebach. Routing in all-optical ring networks revisited. In *Proceedings of the 9th IEEE Symposium on Computers and Communication*, pages 288–293, 2004.
16. A. C. Tucker. An efficient test for circular-arc graphs. *SIAM J. Comput.*, 9(1):1–24, 1980.

An $\mathcal{O}(\bullet^{\ 2 \cdot 75})$ Algorithm for Online Topological Ordering[*]

Deepak Ajwani, Tobias Friedrich, and Ulrich Meyer

Max-Planck-Institut für Informatik,
Saarbrücken, Germany

Abstract. We present a simple algorithm which maintains the topological order of a directed acyclic graph with n nodes under an on-line edge insertion sequence in $\mathcal{O}(n^{2.75})$ time, independent of the number of edges m inserted. For dense DAGs, this is an improvement over the previous best result of $\mathcal{O}(\min\{m^{\frac{3}{2}} \log n, m^{\frac{3}{2}} + n^2 \log n\})$ by Katriel and Bodlaender. We also provide an empirical comparison of our algorithm with other algorithms for online topological sorting.

1 Introduction

A topological order T of a given directed acyclic graph (DAG) $G = (V, E)$ (with $n := |V|$ and $m := |E|$) is a linear ordering of its nodes such that for all directed paths from $x \in V$ to $y \in V$ ($x \neq y$), it holds that $T(x) < T(y)$. There exist well known algorithms for computing the topological ordering of a DAG in $\mathcal{O}(m+n)$ in an offline setting (see e.g. [3]).

In the online variant of this problem, the edges of the DAG are not known in advance but are given one at a time. Each time an edge is added to the DAG, we are required to update the bijective mapping T.

The online topological ordering has been studied in the following contexts

- As an online cycle detection routine in pointer analysis [10].
- Incremental evaluation of computational circuits [2].
- Compilation [5,7] where dependencies between modules are maintained to reduce the amount of recompilation performed when an update occurs.

The naïve way of maintaining an online topological order, i.e., to compute it each time from scratch with the offline algorithm, takes $\mathcal{O}(m^2 + mn)$ time. Marchetti-Spaccamela et al. [6] (MNR) gave an algorithm that can insert m edges in $\mathcal{O}(mn)$ time. Alpern et al. proposed a different algorithm [2] (AHRSZ) which runs in $\mathcal{O}(\|\delta\| \log \|\delta\|)$ time per edge insertion with $\|\delta\|$ measuring the number of edges of the minimal node subgraph that needs to be updated. Note that not all edges of this subgraph need to be visited and hence even $\mathcal{O}(\|\delta\|)$ time per insertion is not optimal. Katriel and Bodlaender (KB) [4] analyzed a variant of the AHRSZ algorithm and obtained an upper bound of $\mathcal{O}(\min\{m^{\frac{3}{2}} \log n, m^{\frac{3}{2}} + n^2 \log n\})$ for a general DAG. In addition, they show

---

[*] Research partially supported by DFG grant ME 2088/1-3.

L. Arge and R. Freivalds (Eds.): SWAT 2006, LNCS 4059, pp. 53–64, 2006.

that their algorithm runs in time $\mathcal{O}(m \cdot k \cdot \log^2 n)$ for a DAG for which the underlying undirected graph has a treewidth k. Also, they give an $\mathcal{O}(n \log n)$ algorithm for DAGs whose underlying undirected graph is a tree. The algorithm by Pearce and Kelly [9] (PK) empirically outperforms the other algorithms for sparse random DAGs, although its worst-case runtime is inferior to KB.

We propose a simple algorithm that works in $\mathcal{O}(n^{2.75}\sqrt{\log n})$ time and $\mathcal{O}(n^2)$ space, thereby improving upon the results of Katriel and Bodlaender for dense DAGs. With some simple modifications in our data structure, we can get $\mathcal{O}(n^{2.75})$ time with $\mathcal{O}(n^{2.25})$ space or $\mathcal{O}(n^{2.75})$ *expected* time with $\mathcal{O}(n^2)$ space. We also demonstrate empirically that this algorithm clearly outperforms MNR, AHRSZ, and PK on a certain class of hard sequences of edge insertions, while being competitive on random edge sequences leading to complete DAGs.

Our algorithm is dynamic, as it also supports deletion. However, our analysis holds only for a sequence of insertions. Our algorithm can also be used for online cycle detection in graphs, as well. Moreover, it permits an arbitrary starting point, which makes a hybrid approach possible, i.e., using the PK or KB algorithm for sparse graphs and ours for dense graphs.

The rest of this paper is organized as follows. In Section 2, we describe the algorithm and the data structures involved. In Section 3, we give the correctness argument for our algorithm, followed by an analysis of its runtime in Sections 4 and 5. An empirical comparison with other algorithms follows in Section 6.

2   Algorithm

We keep the current topological order as a bijective function, $T : V \to [1..n]$. If we start with an empty graph, we can initialize T with an arbitrary permutation, otherwise T is the topological order of the starting graph, computed offline. In this and the subsequent sections, we will use the following notations: $d(u, v)$ denotes $|T(u) - T(v)|$, $u < v$ is a short form of $T(u) < T(v)$, $u \to v$ denotes an edge from u to v, and $u \rightsquigarrow v$ expresses that v is reachable from u. Note that $u \rightsquigarrow u$, but *not* $u \to u$.

Figure 1 gives the pseudo code of our algorithm. Throughout the process of inserting new edges, we maintain some data structures which are dependent on the current topological order. Inserting a new edge (u, v) is done by calling INSERT(u, v). If $v > u$, we do not change anything in the current topological order and simply insert the edge into the graph data structure. Otherwise, we call REORDER to update the topological order as well as the data structures dependent on it. As we will prove in Theorem 2, detecting $v = u$ indicates a cycle. If $v < u$, we first collect sorted sets A and B as defined in the code. If both A and B are empty, we swap the topological order of the two nodes and update the data structures. The query and the update operations are described in more detail along with our data structures in Section 2.1. Otherwise, we recursively call REORDER until everything inside is topologically ordered. To make these recursive calls efficient, we first merge the sorted sets $\{v\} \cup A$ and $B \cup \{u\}$ and using this merged list, compute the set $\{u' : (u' \in B \cup \{u\}) \wedge (u' > v')\}$ for each node $v' \in \{v\} \cup A$.

INSERT(u, v)

    ▷ Insert edge (u, v) and calculate new topological order
1   if $v \leq u$ then REORDER(u,v)
2   insert edge (u, v) in graph

REORDER(u, v)

    ▷ Reorder nodes between u and v such that $v \leq u$
1   if $u - v$ then report detected cycle and quit
2   $A := \{w : v \to w$ and $w \leq u\}$
3   $B := \{w : w \to u$ and $v \leq w\}$
4   if $A = \emptyset$ and $B = \emptyset$
      then ▷ Correct the topological order
5   swap u and v
6   update the data structure
      else ▷ Reorder node pairs between u and v
7   for $v' \in \{v\} \cup A$ in decreasing topological order
8   for $u' \in B \cup \{u\} \wedge u' \geq v'$ in increasing topological order
9   REORDER(u',v')

Fig. 1. Our algorithm

2.1 Data Structure

We store the current topological order, as a set of two arrays, storing the bijective mapping T and its inverse. This ensures that finding $T(i)$ and $T^{-1}(u)$ are constant time operations.

The graph itself is stored as an array of vertices. For each vertex we maintain two adjacency lists, which keep the incoming and outgoing edges separately. Each adjacency list is stored as an array of buckets of vertices. Each bucket contains at most t nodes for a fixed t. Depending on the concrete implementation of the buckets, the parameter t is later chosen to be approximately $n^{0.75}$ so as to balance the number of inserts and deletes from the buckets and the extra edges touched by the algorithm. The i-th bucket ($i \geq 0$) of a node u contains all adjacent nodes v with $i \cdot t < d(u, v) \leq (i + 1) \cdot t$. The nodes of a bucket are stored with node index (and not topological order) as their key. The bucket can be kept as a balanced binary tree or as an array of n-bits or as a hash-table of a universal hashing function. The bucket data structure should provide efficient support for the following three operations:

1. Insert: Inserting an element in a given bucket.
2. Delete: Given an element and a bucket, find out if that element exists in that bucket. If yes, delete the element from there and return 1. Else, return 0.
3. Collect-all: Copying all the elements from the bucket to some vector.

Depending on how we choose to implement the buckets, we get different run-times. This will be discussed in Section 5. We will now discuss how we do the insertion of an edge, computation of A and B, and updating the data-structure under swapping of nodes in terms of the above three basic operations.

Inserting an edge (u, v) means, inserting node v to the forward adjacency list of u and u to the backward adjacency list of v. This requires $\mathcal{O}(1)$ bucket inserts.

For given u and v, the set $A := \{w : v \to w \text{ and } w < u\}$ sorted according to the current topological order can be computed from the adjacency list of v by sorting all nodes of the first $\lceil d(u, v)/t \rceil$ outgoing buckets and choosing all w with $w < u$. This can be done by $\mathcal{O}(d(u, v)/t)$ collect-all operations on buckets collecting a total of $\mathcal{O}(|A| + t)$ elements. These elements are integers in the range $\{1 \dots n\}$ and can be sorted in $\mathcal{O}(|A| + t + \sqrt{n})$ time using a two-pass radix sort algorithm. The set B is computed likewise from the incoming edges.

When we swap two nodes u and v, we need to update the adjacency lists of u and v as well as that of all nodes w that are adjacent to u and/or v. First, we show how to update the adjacency lists of u and v. If $d(u, v) > t$, we have to build their adjacency lists from scratch. Otherwise, the new bucket boundaries will differ from the old boundaries by $d(u, v)$ and at most $d(u, v)$ nodes will need to be transferred between any pair of consecutive buckets. The total number of transfers are therefore bounded by $d(u, v)\lceil n/t \rceil$. Determining whether a node should be transferred can be done in $\mathcal{O}(1)$ using the inverse mapping T^{-1} and as noted above, a transfer can be done in $\mathcal{O}(1)$ bucket inserts and deletes. Hence, updating the adjacency lists of u and v needs $\min\{n, d(u, v)\lceil n/t \rceil\}$ bucket inserts and deletes.

Let w be a node which is adjacent to u or v. Its adjacency list needs to be updated only if u and v are in different buckets. This corresponds to w being in different buckets of the adjacency lists of u and v. Therefore, the number of nodes to be transferred between different buckets for maintaining the adjacency lists of all w's is the same as the number of nodes that need to be transferred for maintaining the adjacency lists of u and v, i.e., $\min\{n, d(u, v)\lceil n/t \rceil\}$.

Updating the mappings T and T^{-1} after such a swap is trivial and can be done in constant time. Thus, we conclude that swapping nodes u and v can be done by $\mathcal{O}(d(u, v)\lceil n/t \rceil)$ bucket inserts and deletes.

3   Correctness

Theorem 1. *The above algorithm returns a valid topological order after each edge insertion.*

Proof. For a graph with no edges, any ordering is a correct topological order, and therefore, the theorem is trivially correct. Assuming that we have a valid topological order of a graph G, we show that when inserting a new edge (u, v) using INSERT(u, v), our algorithm maintains the correct topological order of $G' := G \cup \{(u, v)\}$. If $u < v$, this is trivial.

We need to prove that $x < y$ for all nodes x, y of G' with $x \rightsquigarrow y$. If there was a path $x \rightsquigarrow y$ in G, Lemma 1 gives $x < y$. Otherwise (if there is no $x \rightsquigarrow y$ in G), the path $x \rightsquigarrow y$ must have been introduced to G' by the new edge (u, v). Hence $x < y$ in G' by Lemma 2 since there is $x \rightsquigarrow u \rightarrow v \rightsquigarrow y$ in G'. □

Lemma 1. *Given a DAG G and a valid topological order. If $u \rightsquigarrow v$ and $u < v$, then all subsequent calls to* REORDER *will maintain $u < v$.*

Proof. Let us assume the contrary. Consider the first call of REORDER which leads to $u > v$. Either this call led to swapping u and w with $v \leq w$ or it caused swapping w and v with $w \leq u$. Note that in our algorithm, a call of REORDER(u, v) leads to a swapping only if $A = \emptyset$ and $B = \emptyset$. Assuming that it was the first case (swapping u and w) caused by the call to REORDER(u, w), $A = \emptyset$. However, $x \in A$ for an x with $u \rightarrow x \rightsquigarrow v$, leading to a contradiction. The other case is proved similarly. □

Lemma 2. *Given a DAG G with $v \rightsquigarrow y$ and $x \rightsquigarrow u$, a call of* REORDER(u, v) *will ensure that $x < y$.*

Proof. The proof follows by induction on the recursion depth of REORDER(u, v). For leaf nodes of the recursion tree, $A = B = \emptyset$. If $x < y$ before this call happens, Lemma 1 ensures that $x < y$ will continue. Otherwise, $y := v$ and $x := u$. The swapping of u and v in line 5 gives $x < y$.

We assume this lemma to be true for calls of REORDER up to a certain tree level. If $A \neq \emptyset$, then there is a $\tilde{v}$ such that $v \rightarrow \tilde{v} \rightsquigarrow y$, otherwise $\tilde{v} := v = y$. If $B \neq \emptyset$, then there is a $\tilde{u}$ such that $x \rightsquigarrow \tilde{u} \rightarrow u$, otherwise $\tilde{u} := u = x$. Hence $\tilde{v} \rightsquigarrow y < x \rightsquigarrow \tilde{u}$. The **for**-loops of lines 7 and 8 will call REORDER$(\tilde{u}, \tilde{v})$. By the inductive hypothesis, this will ensure $x < y$. According to Lemma 1, further calls to REORDER will maintain $x < y$. □

Theorem 2. *The algorithm detects a cycle if and only if there is a cycle in the given edge sequence.*

Proof. "⇒": First, we show that within a call to INSERT(u, v), there are paths $v \rightsquigarrow v'$ and $u' \rightsquigarrow u$ for each recursive call to REORDER(u', v'). This is trivial for the first call to REORDER and follows immediately by the definition of A and B for all subsequent recursive calls to REORDER. This implies that if the algorithm indicates a cycle in line 1 of REORDER, there is indeed a cycle $u \rightarrow v \rightsquigarrow v' = u' \rightsquigarrow u$. In fact, the cycle itself can be computed using the recursion stack of the current call to REORDER.

"⇐": Consider the edge (u, v) of the cycle $v \rightsquigarrow u \rightarrow v$ inserted last. Since $v \rightsquigarrow u$ before the insertion of this edge, the topological order computed will have $v < u$ (Theorem 1) and therefore, REORDER(u, v) would be called. In fact, all edges in the path $v \rightsquigarrow u$ will obey the current topological ordering and by Lemma 1, it will remain so for all subsequent calls of REORDER. We prove by induction on the number of nodes in the path $v \rightsquigarrow u$ (including u and v) that whenever $v \rightsquigarrow u$ and REORDER(u, v) is called, it detects the cycle. A call of REORDER(u', v') with $u' = v'$ or REORDER(u', v') with $v' \rightarrow u'$ clearly reports a cycle. Consider a path $v \rightarrow x \rightsquigarrow y \rightarrow u$ of length $k > 2$ and the call of REORDER(u, v). As noted before, $v < x \leq y < u$ before the call to

REORDER(u, v). Hence $x \in A$ and $y \in B$ and a call to REORDER(y, x) will be made in the for loop of lines 7 and 8. As $y \rightsquigarrow x$ has $k - 2$ nodes in the path, the call to REORDER(y, x) (by our inductive hypothesis) will detect the cycle.   □

4   Runtime

Theorem 3. *Online topological ordering can be computed using* $\mathcal{O}(n^{3.5}/t)$ *bucket inserts and deletes,* $\mathcal{O}(n^3/t)$ *bucket collect-all operations collecting* $\mathcal{O}(n^2t)$ *elements, and* $\mathcal{O}(n^{2.5} + n^2t)$ *operations.*

Proof. Lemma 4 shows that REORDER is called $\mathcal{O}(n^2)$ times. Lemma 6 shows that the calculation of the sets A and B over all calls of REORDER can be done by $\mathcal{O}(n^3/t)$ bucket collect-all operations touching $\mathcal{O}(n^2t)$ edges, and $\mathcal{O}(n^{2.5} + n^2t)$ operations. In Lemma 9, we prove that all the updates can be done by $\mathcal{O}(n^{3.5}/t)$ bucket inserts and deletes.

As for lines 7 and 8, we first merge the two sorted sets A and B, which takes $\mathcal{O}(|A| + |B|)$ operations. For a particular node $v' \in \{v\} \cup A$, we can compute the set $V' = \{u' : (u' \in B \cup \{u\}) \wedge (u' > v')\}$ (as required by line 8) using this merged set in complexity $\mathcal{O}(1 + |V'|)$, which is also the number of calls of REORDER emanating for this particular node. Summing over the entire *for* loop of line 7, the total complexity of lines 7 and 8 is $\mathcal{O}(|A| + |B| + \#(\text{calls of REORDER emanating from here}))$. Since by Lemma 5, the summation of $|A| + |B|$ over all calls of REORDER is $\mathcal{O}(n^2)$ and by Lemma 4, the total number of calls to REORDER is also $\mathcal{O}(n^2)$, we get a total of $\mathcal{O}(n^2)$ operations for lines 7 and 8. Putting everything together, the theorem follows.   □

Lemma 3. REORDER *is local, i.e., a call to* REORDER(u, v) *does not affect the topological ordering of nodes w such that either $w < v$ or $w > u$ just before the call was made.*

Proof. This theorem can be proved by induction on the level of recursion tree of the call to REORDER(u, v). For the leaf node of the recursion tree, $|A| = |B| = 0$ and the topological order of u and v is swapped, not affecting the topological ordering of any other node.

We assume this lemma to be true up to a certain tree level. To see that it is valid even for a level higher, note that the arrays A and B contain elements w such that $v < w < u$. Since each call of REORDER in the **for**-loop of line 7 and 8 is from an element of A to an element of B and all of these calls are themselves local by our induction hypothesis, this call of REORDER is also local.   □

Lemma 4. REORDER *is called* $\mathcal{O}(n^2)$ *times.*

Proof. Let u and v be arbitrary nodes. Let us consider the first time, RE-ORDER(u, v) is called. If $A = B = \emptyset$, u and v will be swapped. Otherwise, REORDER(u', v') is called recursively for all $v' \in \{v\} \cup A$ and $u' \in B \cup \{u\}$ with $u' > v'$. The order in which we make these recursive calls and the fact that REORDER is local (Lemma 3) ensures that REORDER(u, v) is not called except as the last of these recursive calls. In this second call to REORDER(u, v), $A = B = \emptyset$. To see this consider all $v' \in A$ and $u' \in B$ (A and B from the first call of REORDER(u, v)). REORDER(u, v') and REORDER(u', v) must have been called within the **for**-loop of the first execution of REORDER(u, v) before this second call was made. Therefore it follows from Lemma 2 and Lemma 1 that before the second call, $u < v'$ and $u' < v$ for all $v' \in A$ and $u' \in B$. Hence u and v will be swapped at the latest in the second call of REORDER(u, v). Since REORDER(u, v) is only called if $v < u$, REORDER(u, v) will not be called again. Hence, REORDER(u, v) is called at most two times for each node pair (u, v). $\quad\square$

Lemma 5. *The summation of $|A| + |B|$ over all calls of* REORDER *is $\mathcal{O}(n^2)$.*

Proof. Consider arbitrary nodes u and v'. We prove that for all $v \in V$, $v' \in A$ happens only once over all calls of REORDER(u, v). This proves that $\sum |A| \leq n$, for all such calls of REORDER(u, v). Therefore summing up for all $u \in V$, $\sum |A| \leq n^2$ over all calls of REORDER.

In order to see that for all $v \in V$, $v' \in A$ happens only once over all calls of REORDER(u, v), observe that $v' \in A$ implies that $v' < u$ before REORDER(u, v) was called. In particular, $v' < u$ before the call of REORDER(u, v') in the **for**-loop of REORDER(u, v) (follows from the order of recursive calls) and by Lemma 2, $u < v'$ after this call. Therefore, $v' \notin A$ for a call of REORDER(u, w) for any node w after this call. The same is true for all calls of REORDER(u, w) before this call as otherwise $u < v'$ even before the beginning of the current call of REORDER(u, v) and $v' \notin A$ for the current call. Also, $v' \notin A$ for any of the recursive calls of this call to REORDER(u, v'). This follows from the order in which we make the recursive calls and the fact that REORDER is local (Lemma 3).

Analogously, it can be proved that for arbitrary nodes v and v' and for all $u \in V$, $v' \in B$ happens only once over all calls of REORDER(u, v). The proof for $\sum |B| \leq n^2$ follows similarly and it completes the proof for this lemma. $\quad\square$

Lemma 6. *Calculating the sorted sets A and B over all calls of* REORDER *can be done by $\mathcal{O}(n^3/t)$ bucket collect-all operations touching a total of $\mathcal{O}(n^2 t)$ elements and $\mathcal{O}(n^{2.5} + n^2 t)$ operations for sorting these elements.*

Proof. Consider the calculation of set A in a call of REORDER(u, v). As discussed before in Section 2.1, we look at the out adjacency list of u, stored in the form of buckets. In particular, we will need $\mathcal{O}(d(u, v)/t)$ bucket collect-all operations touching $\mathcal{O}(|A| + t)$ elements to calculate A. The additional worst-case factor of t stems from the last bucket visited. Summing up over all calls of REORDER, we get $\mathcal{O}\left(\sum d(u, v)/t\right)$ collect-alls touching $\sum (|A| + |B| + t)$ elements. Since $d(u, v) \leq n$ for every call of REORDER(u, v) and there are $\mathcal{O}(n^2)$ calls of RE-ORDER (Lemma 4), there are $\mathcal{O}(n^3/t)$ bucket collect-all operations. Also, since $\sum (|A| + |B|) = \mathcal{O}(n^2)$ by Lemma 5, the total number of elements touched is $\mathcal{O}(n^2 + \sum t) = \mathcal{O}(n^2 t)$. Since the keys are in the range $\{1 .. n\}$, we can use a two-pass radix sort to sort the elements collected from the buckets. The total sorting time over all calls of REORDER is $\sum (2(|A| + t) + \sqrt{n}) + \sum (2(|B| + t) + \sqrt{n}) = \mathcal{O}(n^{2.5} + n^2 t)$. $\quad\square$

Lemma 7. *Each node-pair is swapped at most once.*

Proof. REORDER(u, v) is called only when $v < u$. Once a swapping happens, $u < v$. By Lemma 1, it will remain so for all calls of REORDER thereafter. Therefore, REORDER(u, v) is never called again and u and v will not be swapped again. □

Lemma 8. $\sum d(u, v) = \mathcal{O}(n^{5/2})$ *where the summation is taken over all calls of* REORDER(u,v) *in which u and v are swapped.*

Proof. Let T^* denote the final topological ordering and

$$X(T^*(u), T^*(v)) := \begin{cases} d(u, v) & \text{if and when REORDER}(u, v) \text{ leads to a swapping} \\ 0 & \text{otherwise} \end{cases}$$

Since by Lemma 7 any node-pair is swapped at most once, the variable $X(i, j)$ is clearly defined. Next, we model a few linear constraints on $X(i, j)$, formulate it as the linear program and use this LP to prove that $\max\{\sum_{i,j} X(i, j)\} = \mathcal{O}(n^{5/2})$. By definition of $d(u, v)$ and $X(i, j)$,

$$0 \le X(i, j) \le n \text{ for all } i, j \in [1 .. n].$$

For $j \le i$, the corresponding edges $(T^{* \ -1}(i), T^{* \ -1}(j))$ go backwards and thus are never inserted at all. Consequently,

$$X(i, j) = 0 \text{ for all } j \le i.$$

Now consider an arbitrary node u, which is finally at position i, i.e., $T^*(u) = i$. Over the insertion of all edges, this node has been moved left and right via swapping with several other nodes. Strictly speaking, it has been swapped right with nodes at final positions $j > i$ and has been swapped left with nodes at final positions $j < i$. Hence, the overall movement to the right is $\sum_{j>i} X(i, j)$ and to left is $\sum_{j<i} X(j, i)$. Since the net movement (difference between the final and the initial position) must be less than n,

$$\sum_{j>i} X(i, j) - \sum_{j<i} X(j, i) \le n \text{ for all } 1 \le i \le n.$$

Putting all the constraints together, we aim to solve the following linear program.

$$\max \sum_{\substack{1 \le i \le n \\ 1 \le j \le n}} X(i, j) \text{ such that}$$

(i) $X(i, j) = 0$ for all $1 \le i \le n$ and $1 \le j \le i$
(ii) $0 \le X(i, j) \le n$ for all $1 \le i \le n$ and $i < j \le n$
(iii) $\sum_{j>i} X(i, j) - \sum_{j<i} X(j, i) \le n - 1$ for all $1 \le i \le n$

Note that these are necessary constraints, but not sufficient. But this is enough for our purpose as an upper bound to the solution of this LP will give an upper bound for the $\sum X(i,j)$ in our algorithm. In order to prove the upper bound on the solution to this LP, we consider the dual problem

$$\min \left[n \sum_{\substack{0 \le i < n \\ i < j < n}} Y_{i \cdot n + j} + n \sum_{0 \le i < n} Y_{n^2 + i} \right] \text{ such that}$$

(i) $Y_{i \cdot n + j} \ge 1$ for all $0 \le i < n$ and for all $j \le i$
(ii) $Y_{i \cdot n + j} + Y_{n^2 + i} - Y_{n^2 + j} \ge 1$ for all $0 \le i < n$ and for all $j > i$
(iii) $Y_i \ge 0$ for all $0 \le i < n^2 + n$

and the following feasible solution for the dual:

$$\begin{aligned}
Y_{i \cdot n + j} &= 1 & &\text{for all } 0 \le i < n \text{ and for all } 0 \le j \le i \\
Y_{i \cdot n + j} &= 1 & &\text{for all } 0 \le i < n \text{ and for all } i < j \le i + 1 + 2\sqrt{n} \\
Y_{i \cdot n + j} &= 0 & &\text{for all } 0 \le i < n \text{ and for all } j > i + 1 + 2\sqrt{n} \\
Y_{n^2 + i} &= \sqrt{n - i} & &\text{for all } 0 \le i < n.
\end{aligned}$$

This solution has a value of $n^2 + 2n^{\frac{5}{2}} + n \sum_{i=1}^{n} \sqrt{i} = \mathcal{O}(n^{\frac{5}{2}})$, which by the primal-dual theorem is a bound on the solution of the original LP.

In fact, it can be shown that there is a solution to primal LP whose value is $\mathcal{O}(n^{\frac{5}{2}})$, namely

$$\begin{aligned}
X(i,j) &= 0 & &\text{for all } 0 \le i < n \text{ and for all } 0 \le j \le i \\
X(i,j) &= n & &\text{for all } 0 \le i < n \text{ and for all } i < j \le i + \lceil \tfrac{\sqrt{1+8i}-1}{2} \rceil \\
X(i,j) &= 0 & &\text{for all } 0 \le i < n \text{ and for all } j > i + \lceil \tfrac{\sqrt{1+8i}-1}{2} \rceil.
\end{aligned}$$ $\square$

Lemma 9. *Updating the data structure over all calls of* REORDER *requires* $\mathcal{O}(n^{3.5}/t)$ *bucket inserts and deletes.*

Proof. Our data structure requires $\mathcal{O}(d(u,v) n/t)$ bucket inserts and deletes to swap two nodes u and v. Each node pair is swapped at most once (cf. Lemma 7). Hence, summing up over all calls of REORDER(u,v) where u and v are swapped, we need $\mathcal{O}(\sum d(u,v) n/t) = \mathcal{O}(n^{3.5}/t)$ bucket inserts and deletes using Lemma 8. $\square$

5   Bucket Data Structure

We get different runtimes and space requirements of our algorithm depending on the data structures of the buckets used:

(a) Balanced binary trees: Balanced binary trees give us $\mathcal{O}(1 + \log \tau)$ time insert and delete and $\mathcal{O}(1 + \tau)$ time collect-all operation, where τ is the number of elements in the bucket. Therefore, by Theorem 3, the total time required will be $\mathcal{O}(n^2 t + n^{3.5} \log n/t)$. Substituting $t = n^{0.75}\sqrt{\log n}$, we get a total time of $\mathcal{O}(n^{2.75}\sqrt{\log n})$. The total space requirement will be $\mathcal{O}(n^2)$ as a balanced binary tree needs $\mathcal{O}(t)$ nodes for storing at most t elements.

(b) n-bit array: A bucket that stores at most t elements can be kept as an n-bit array, where each bit is 0 or 1 depending on whether or not the element is present in the bucket. Also, we can keep a list of all elements in the bucket. To insert, we just flip the appropriate bit and insert at the end of the list. To delete, we just flip the appropriate bit. To collect all, we go through the list and for each element in the list, we check if the corresponding bit is 1 or 0. If it is 0, we also remove it from the list. This gives us constant-time insert and delete and the time for collect-all operation will be the total output size plus the total number of delete. Each delete is counted once in collect-all as we remove the corresponding element from the list after the first collect-all. By Theorem 3, the total time required will be $\mathcal{O}(n^2 t + n^{3.5}/t)$, giving us $\mathcal{O}(n^{2.75})$ for $t = n^{0.75}$. The total space requirement will be $\mathcal{O}(n)$ for each bucket, leading to a total of $\mathcal{O}(n^{2.25})$ for $\mathcal{O}(n^2/t)$ buckets.

(c) Uniform Hashing [8]: A data structure based on uniform hashing coupled with a list of elements in the bucket operated in the same way as the n-bit array will give an expected constant-time insert and delete and the same bound for collect-all as for the n-bit array. This gives an expected total time of $\mathcal{O}(n^2 t + n^{3.5}/t)$. With $t = n^{0.75}$ this yields an expected time of $\mathcal{O}(n^{2.75})$. Since the hashing based data structure as described in [8] takes only linear space, the total space requirement is $\mathcal{O}(n^2)$.

6   Empirical Comparison

In addition to the achieved worst-case bounds, we also implemented our algorithm (AFM) and compared it to David J. Pearce's [9] implementation of PK, MNR, and AHRSZ. The experiments were conducted on a 2.4 GHz Opteron machine with 8GB of main memory running Debian GNU/Linux.

On a random edge sequence, all the algorithms are quite fast and none of them encounters its worst-case behavior. Therefore, we consider a particular sequence of edges which we believe is a hard instance of the problem. This edge sequence is similar to the worst-case sequence given by Katriel and Bodlaender for their algorithm. On this sequence, PK, MNR and AHRSZ (the variant choosing the smallest permitted priority) face their worst-case of $\Omega(n^3)$ operations, while our algorithm (using n-bit array and quick sort) takes $\tilde{\mathcal{O}}(n^{2.5})$ time complexity. This sequence of edges is depicted in Fig. 2. For an example with n nodes, we divide the set of nodes into four blocks of different sizes: block 1 consist of nodes $[0 .. n/3)$, block 2 of nodes $[n/3 .. n/2)$, block 3 of nodes $[n/2 .. 2n/3)$, and block 4 of nodes $[2n/3 .. n)$. First, we insert $n - 4$ edges such that within each block, the vertices form a directed path from left to right. Then we insert the following edges,

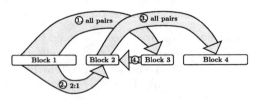

Fig. 2. Our hard-case graph

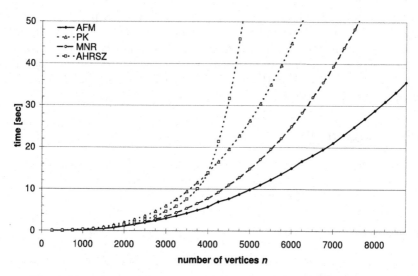

Fig. 3. Experimental data on a class of hard instances with varying n

(a) $\overrightarrow{\forall}\, j \in [0..n/3)\ \overleftarrow{\forall}\, k \in [0..n/6)$: add edge$(j, k + n/2)$,

(b) $\overrightarrow{\forall}\, j \in [0..n/6)$: add edge$(2j, j + n/3)$ and edge$(2j + 1, j + n/3)$,

(c) $\overrightarrow{\forall}\, j \in [0..n/6)\ \overleftarrow{\forall}\, k \in [0..n/3)$: add edge$(j + n/3, k + 2n/3)$,

(d) $\overrightarrow{\forall}\, j \in [0..n/6)\ \overleftarrow{\forall}\, k \in [0..n/6)$: add edge$(j + n/2, k + n/3)$,

where $\overrightarrow{\forall}$ denotes going from left to right in the **for**-loop and $\overleftarrow{\forall}$ the other way around. Similar sequences, which force AHRSZ to encounter its asymptotic worst-case complexity, can be chosen for all variants of AHRSZ.

Fig. 3 shows the runtimes of the four algorithms in consideration on the graphs described before. The discussed difference in the asymptotic behaviour is clearly visible. For $n = 8000$, AFM is 2 times faster than MNR, 3.6 times faster than PK, and 30 times faster than AHRSZ. Due to the more involved data structures, our implementation of AFM is a constant factor of 2-4 away from AHRSZ, MNR and PK on the random edge sequences that we tested [1]. For practical purposes, we believe that a hybrid approach would perform best. That is, one inserts the first $\mathcal{O}(n \log n)$ edges with either PK or KB and then inserts the remaining edges with our algorithm.

7   Discussion

We have presented the first $o(n^3)$ algorithm for online topological ordering. We also implemented this new algorithm and compared it with previous approaches, showing that for certain hard examples, it outperforms PK, MNR, and AHRSZ, while it is still competitive on random edge sequences leading to complete DAGs. The only non-trivial lower bound for this problem is by Ramalingam and Reps [11], who show that an adversary can force any algorithm maintaining explicit labels to need $\Omega(n \log n)$ time complexity for inserting $n-1$ edges. There is still a large gap between this, the trivial lower bound of $\Omega(m)$, and the upper bound of $\mathcal{O}(\min\{m^{1.5} + n^2 \log n, m^{1.5} \log n, n^{2.75}\})$. Bridging this gap remains an open problem.

Acknowledgements

The authors are grateful to David J. Pearce for providing us his code. Also, thanks are due to the anonymous referees for valuable comments and to Seth Pettie and Saurabh Ray for helpful discussions.

References

1. D. Ajwani, T. Friedrich, and U. Meyer. An $O(n^{2.75})$ algorithm for online topological ordering, 2006. arXiv:cs.DS/0602073.
2. B. Alpern, R. Hoover, B. K. Rosen, P. F. Sweeney, and F. Kenneth Zadeck. Incremental evaluation of computational circuits. In *Proceedings of the first annual ACM-SIAM Symposium on Discrete Algorithms*, pages 32–42, Philadelphia, PA, USA, 1990. Society for Industrial and Applied Mathematics.
3. T. Cormen, C. Leiserson, and R. Rivest. *Introduction to Algorithms*. The MIT Press, Cambridge, MA, 1989.
4. I. Katriel and H.L. Bodlaender. Online topological ordering. In *Proceeding of the 16th ACM-SIAM Symposium on Discrete Algorithms (SODA), Vancouver*, pages 443–450, 2005.
5. A. Marchetti-Spaccamela, U. Nanni, and H. Rohnert. On-line graph algorithms for incremental compilation. In *Proceedings of International Workshop on Graph-Theoretic Concepts in Computer Science (WG 93)*, volume 790 of *Lecture Notes in Computer Science*, pages 70–86, 1993.
6. A. Marchetti-Spaccamela, U. Nanni, and H. Rohnert. Maintaining a topological order under edge insertions. *Information Processing Letters*, 59(1):53–58, July 1996.
7. S. M. Omohundro, C. Lim, and J. Bilmes. The sather language compiler/debugger implementation. *Technical Report TR-92-017, International Computer Science Institute, Berkeley*, 1992.
8. A. Östlin and R. Pagh. Uniform hashing in constant time and linear space. In *Proceedings of the 35th Annual ACM symposium on theory of computing*, pages 622–628. ACM, 2003.
9. D. J. Pearce and P. H. J. Kelly. A dynamic algorithm for topologically sorting directed acyclic graphs. In *Proceedings of the Workshop on Efficient and experimental Algorithms*, volume 3059 of *Lecture Notes in Computer Science*, pages 383–398. Springer-Verlag, 2004.
10. D. J. Pearce, P. H. J. Kelly, and C. Hankin. Online cycle detection and difference propagation for pointer analysis. In *Proceedings of the third international IEEE Workshop on Source Code Analysis and Manipulation (SCAM03)*, 2003.
11. G. Ramalingam and T. W. Reps. On competitive on-line algorithms for the dynamic priority-ordering problem. *Information Processing Letters*, 51:155–161, 1994.

Dynamic Matching Markets and Voting Paths[*]

David J. Abraham[1] and Telikepalli Kavitha[2,**]

[1] Computer Science Department, Carnegie Mellon University, USA
dabraham@cs.cmu.edu
[2] Indian Institute of Science, Bangalore, India
kavitha@csa.iisc.ernet.in

Abstract. We consider a matching market, in which the aim is to maintain a *popular* matching between a set of applicants and a set of posts,
where each applicant has a preference list that ranks some subset of ac-
ceptable posts. The setting is dynamic: applicants and posts can enter
and leave the market, and applicants can also change their preferences
arbitrarily. After any change, the current matching may no longer be
popular, in which case, we are required to update it. However, our model
demands that we can switch from one matching to another only if there
is consensus among the applicants to agree to the switch. Hence, we need
to update via a *voting path*, which is a sequence of matchings, each more
popular than its predecessor, that ends in a popular matching. In this
paper, we show that, as long as some popular matching exists, there is
a 2-step voting path from any given matching to some popular matching. Furthermore, given any popular matching, we show how to find a
shortest-length such voting path in linear time.

1   Introduction

An instance of the *popular matching problem* consists of a bipartite graph $G = (\mathcal{A} \cup \mathcal{P}, E)$, together with a partition $E_1 \dot{\cup} E_2 \dots \dot{\cup} E_r$ of the edge set E. We call $\mathcal{A}$ the set of *applicants*, $\mathcal{P}$ the set of *posts*, and E_i the set of edges with rank i. If $(a, p) \in E_i$ and $(a, p') \in E_j$ with $i < j$, we say that a prefers p to p'. If $i = j$, then a is indifferent between p and p'. The ordering of posts adjacent to a is called a's preference list. We say that preference lists are strictly ordered if no applicant is indifferent between any two posts in its preference list.

A matching M of G is a subset of E, such that no two edges of M share a common endpoint. A node $u \in \mathcal{A} \cup \mathcal{P}$ is either unmatched in M, or matched to some node denoted by $M(u)$. We say an applicant a prefers matching M' to M if (i) a is matched in M' and unmatched in M, or (ii) a is matched in both M' and M, and a prefers $M'(a)$ to $M(a)$.

Matching M' is *more popular than* matching M, denoted by $M' \succ M$, if the number of applicants preferring M' to M is greater than the number of

---

[*] Part of this work was done while the authors were at MPII, Saarbrücken.
[**] This research was partially supported by a "Max Planck-India Fellowship" provided
by the Max Planck Society.

applicants preferring M to M'. A matching M is *popular* if there is no matching M' that is more popular than M.

It turns out that a popular matching may not always exist (see [3] for an example) - the reason being, of course, that the *more popular than* relation is not acyclic. The popular matching problem is to determine if a given instance admits a popular matching, and to find such a matching, if one exists. The first polynomial-time algorithms for this problem were given in [3]: when preference lists are strictly ordered, the problem can be solved in $O(n + m)$ time, where $n = |\mathcal{A} \cup \mathcal{P}|$ and $m = |E|$, and more generally, the problem can be solved in $O(m\sqrt{n})$ time. Note that when $E = E_1$, a matching is popular if and only if it has maximum cardinality. Hence, the popular matching problem is at least as hard as the problem of finding a maximum matching in a bipartite graph.

Problem Definition. In this paper, we consider a matching market where the aim is to *maintain* a popular matching. The setting is dynamic: applicants and posts can enter and leave the matching market, and applicants can change their preferences arbitrarily. More precisely, an instance of the *dynamic popular matching problem* consists of an instance G of the popular matching problem, together with an existing (possibly empty) matching M_0.

It turns out that we cannot ignore M_0 and simply compute a popular matching in G from scratch after each change, since any *particular* popular matching we find may not be more popular than M_0, and furthermore, it is possible that *no* popular matching is more popular than the existing matching M_0. Hence, in general, there may be no consensus amongst the applicants to move directly from M_0 to a popular matching. We show such an example below.

Consider Fig. 1. Let $M_0 = \{(a_1, p_5), (a_2, p_2), (a_3, p_3), (a_4, p_1)\}$. First note that M_0 is not popular, since it is less popular than $M_1 = \{(a_1, p_2), (a_2, p_3), (a_4, p_1)\}$ (even with a_3 unmatched). We can show using Lemma 2 from Section 2 that the only popular matchings are $M^* = \{(a_1, p_1), (a_2, p_2), (a_3, p_3), (a_4, p_4)\}$ and $N^* = \{(a_1, p_1), (a_2, p_3), (a_3, p_2), (a_4, p_4)\}$. However, it is clear that neither M^* nor N^* is more popular than M_0.

$$a_1 \; : \; p_1 \; p_2 \; p_5; \quad a_2 \; : \; p_3 \; p_2;$$
$$a_3 \; : \; p_3 \; p_2; \qquad a_4 \; : \; p_1 \; p_4.$$

Fig. 1. Instance that motivates the *voting-path* approach

In order to arrive at a popular matching by consensus, [3] introduced the following generalization of the more popular than relation: A matching M_k is *reachable* from M_0 if there is a sequence of matchings $\langle M_0, M_1, \ldots, M_k \rangle$, such that each matching is more popular than its predecessor. Such a sequence is called a length-k *voting path* from M_0 to M_k. (Note that the instance above has a length-2 voting path from M_0 to a popular matching, namely $\langle M_0, M_1, N^* \rangle$.)

There is no a priori reason to expect that such a voting path must exist: the *more popular than* relation is not acyclic, and so perhaps there are some matchings from which we cannot avoid cycling. Even if such a path does exist, it may have length exponential in the size of G, since there can be an exponential number of matchings. In this paper, we show the following result.

Theorem 1. *Let $\langle G, M_0 \rangle$ be an instance of the dynamic popular matching problem, where G admits a popular matching. Then G admits a voting path of length at most 2 from M_0 to some popular matching. Additionally, given any popular matching, we can find a shortest-length such voting path in only linear time.*

Hence, by using the popular matching algorithms in [3], we can solve the dynamic popular matching problem in $O(m + n)$ time when preference lists are strictly ordered, and more generally in $O(m\sqrt{n})$ time. This solves the problem of efficiently computing a shortest-length voting path to a popular matching, which was posed in [3]. We have also shown that such paths have length at most 2. A bound of 3 was claimed (without proof) in [3].

Interestingly, the improvement from 3 to 2 implies a connection to the famous result in graph theory that every tournament has a king [13]. The *more popular than* relation is a directed graph on an exponential number of vertices. This graph is not a tournament though, since for any pair of matchings, there is no guarantee that one is more popular than the other. However, even without these edges, the set of popular matchings collectively acts as a king, since every unpopular matching has a voting path of length at most 2 into this set.

Related Previous Work. The bipartite matching problem with a graded edge set is well-studied in both economics and computer science, see for example [1, 17, 21] and [5, 11, 2]. It models some important real-world problems, including the allocation of graduates to training positions [9], families to government-owned housing [20], and customers to rental DVDs [16].

Gardenfors [7] first introduced the notion of a popular matching in the context of the stable marriage problem. Of course, the *more popular than* concept can be traced back even further to the Condorcet voting protocol.

One drawback of the popularity criterion is that a popular matching may not always exist. However, in a recent work, Mahdian [14] showed that a popular matching exists with high probability, when (i) preference lists are randomly constructed, and (ii) the number of posts is a small multiplicative factor larger than of the number of applicants. Other recent work on popular matchings includes Mestre's [15] generalization of the efficient popular matching characterization in [3] to the case where each applicant vote carries a weight.

We remark that the result in our paper is analogous to a series of papers [12, 18, 19, 4] on decentralized mechanisms in the stable matching literature. The well-known mechanisms for stable matching, due to Gale/Shapley [6] and Irving [10], require a central body to collect preferences and dictate the final matching. Alternatively, in a decentralized setting, a blocking pair (i.e. a man and woman who prefer each other to their current partners) will *act* locally by divorcing their current partners and marrying each other. Knuth [12] showed that if the divorced partners also marry each other, this process may cycle. However, when divorced partners are not required to marry each other, and every blocking pair has some probability of acting next, Roth and Vande Vate [18] show by way of a potential argument that there is always a path to a stable matching.

In our setting, the analogue of a blocking pair is a coalition of applicants who prefer some matching M' to the current matching M, and have sufficient numbers to win a vote between M' and M. Although it is not too difficult to prove the existence of voting paths via a potential argument (at least for the restriction to strictly-ordered preference lists), in this paper we use more powerful techniques from matching theory, which also give the surprising length-2 bound. As with the result in [18], this means as long as every matching more popular than the current one has some probability of an up-or-down vote, then in the limit, a decentralized mechanism will lead to a popular matching.

2   Preliminaries: Length-0 Voting Paths

In this section, we review the algorithmic characterization of popular matchings given in [3]. This characterization can be used to determine if a given instance $\langle G, M_0 \rangle$ admits a length-0 voting path to a popular matching.

For exposition purposes, we create a unique strictly-least-preferred post $l(a)$ for each applicant a. In this way, we can assume that every applicant is matched, since any unmatched applicant a can be paired with $l(a)$. From now on then, matchings are $\mathcal{A}$-perfect. Also, without loss of generality, we assume that preference lists contain no gaps, i.e., if a is incident to an edge of rank i, then a is incident to an edge of rank $i - 1$, for all $i > 1$.

Let $G_1 = (\mathcal{A} \cup \mathcal{P}, E_1)$ be the graph containing only rank-one edges. Then [3, Lemma 3.1] shows that a matching M is popular in G only if $M \cap E_1$ is a maximum matching of G_1. Maximum matchings have the following important properties, which we use throughout the rest of the paper.

$M \cap E_1$ defines a partition of $\mathcal{A} \cup \mathcal{P}$ into three disjoint sets: a node $u \in \mathcal{A} \cup \mathcal{P}$ is *even* (resp. *odd*) if there is an even (resp. odd) length alternating path in G_1 (w.r.t. $M \cap E_1$) from an unmatched node to u. Similarly, a node u is *unreachable* if there is no alternating path from an unmatched node to u. Denote by $\mathcal{E}$, $\mathcal{O}$ and $\mathcal{U}$ the sets of even, odd, and unreachable nodes, respectively.

Lemma 1 (Gallai-Edmonds Decomposition). *Let $\mathcal{E}$, $\mathcal{O}$ and $\mathcal{U}$ be the sets of nodes defined by G_1 and $M \cap E_1$ above. Then*

(a) *$\mathcal{E}$, $\mathcal{O}$ and $\mathcal{U}$ are pairwise disjoint, and independent of the maximum matching $M \cap E_1$.*

(b) *In any maximum matching of G_1, every node in $\mathcal{O}$ is matched with a node in $\mathcal{E}$, and every node in $\mathcal{U}$ is matched with another node in $\mathcal{U}$. The size of a maximum matching is $|\mathcal{O}| + |\mathcal{U}|/2$.*

(c) *No maximum matching of G_1 contains an edge between a node in $\mathcal{O}$ and a node in $\mathcal{O} \cup \mathcal{U}$. Also, G_1 contains no edge between a node in $\mathcal{E}$ and a node in $\mathcal{E} \cup \mathcal{U}$.*

Using this node partition, we make the following definitions: for each applicant a, $f(a)$ is the set odd/unreachable posts amongst a's most-preferred posts[1]. Also, $s(a)$ is the set of a's most-preferred posts amongst all even posts. We refer to posts in $\cup_{a \in \mathcal{A}} f(a)$ as f-*posts* and posts in $\cup_{a \in \mathcal{A}} s(a)$ as s-*posts*. Note that f-posts and s-posts are disjoint, and that $s(a) \neq \emptyset$ for any a, since $l(a)$ is always even. Also note that there may be posts in $\mathcal{P}$ that are neither f-posts nor s-posts. The next lemma characterizes the set of all popular matchings.

Lemma 2 ([3]). *A matching M is popular in G iff (i) $M \cap E_1$ is a maximum matching of $G_1 = (\mathcal{A} \cup \mathcal{P}, E_1)$, and (ii) for each applicant a, $M(a) \in f(a) \cup s(a)$.*

Using this lemma, we check if M_0 is a popular matching in G or equivalently, if $\langle G, M_0 \rangle$ admits a length-0 voting path to a popular matching: $M_0 \cap E_1$ is a maximum matching of G_1 iff G_1 admits no augmenting path. Also, given that $M_0 \cap E_1$ is a maximum matching of G_1, it is trivial to compute the Gallai-Edmonds decomposition and then to check that each applicant a is matched to $M(a) \in f(a) \cup s(a)$. These checks can clearly be performed in linear time. Henceforth, we assume then that M_0 is not popular, for otherwise $\langle G, M_0 \rangle$ admits a length-0 voting path, and we are done. We also assume that G admits a popular matching, for otherwise no voting path can end in a popular matching.

We conclude this section with Fig. 2, which contains the algorithm from [3], based on Lemma 2, for solving the popular matching problem.

Popular-Matching($G = (\mathcal{A} \cup \mathcal{P}, E)$)

Construct the graph $G' = (\mathcal{A} \cup \mathcal{P}, E')$, where $E' = \{(a, p) : a \in \mathcal{A}$ and
$\qquad\qquad\qquad\qquad\qquad\qquad\qquad\quad p \in f(a) \cup s(a)\}$.
Construct a maximum matching M of $G_1 = (\mathcal{A} \cup \mathcal{P}, E_1)$.
  //Note that M is also a matching in G'.
Remove any edge in G' between a node in $\mathcal{O}$ and a node in $\mathcal{O} \cup \mathcal{U}$.
  //No maximum matching of G_1 contains such an edge.
Augment M in G' until it is a maximum matching of G'.
Return M if it is $\mathcal{A}$-perfect, otherwise return *"no popular matching"*.

Fig. 2. An $O(\sqrt{n}m)$-time algorithm for the popular matching problem (from [3])

3   Length-1 Voting Paths

In this section, we show that, given any popular matching of G, the problem of finding a length-1 voting path from M_0 to a popular matching, or proving that no such path exists, can be solved in linear time. First though, we work towards characterizing the set of all popular matchings that are more popular than M_0.

Let $r_a(p)$ be the rank of edge $(a, p) \in E$. Also, let $r_a(s(a))$ be the rank of any edge (a, p), where $p \in s(a)$. We define the *signature* of a matching M as the 4-tuple $(|A_f^M|, |A_m^M|, |A_s^M|, |A_l^M|)$ [2], where:

---

[1] In [3], $f(a)$ is defined as the set of rank-1 posts in a's preference list. We find the definition above more suitable.

[2] The subscripts $f, m, s,$ and l stand for *first, middle, second,* and *last*, respectively.

(i) $A_f^M = \{a \in \mathcal{A} | r_a(M(a)) = 1$, and a is even/unreachable, i.e. $a \in \mathcal{E} \cup \mathcal{U}\}$.
(ii) $A_m^M = \{a \in \mathcal{A} | 1 < r_a(M(a)) < r_a(s(a))\}$.
(iii) $A_s^M = \{a \in \mathcal{A} | r_a(M(a)) = r_a(s(a))\}$.
(iv) $A_l^M = \{a \in \mathcal{A} | r_a(M(a)) > r_a(s(a))\}$.

Note that an odd applicant $a \in \mathcal{O}$ can only belong to $A_s^M \cup A_l^M$, even if $r_a(M(a)) = 1$, since $a \notin A_f^M$ by definition, and $a \notin A_m^M$, since $r_a(s(a)) = 1$. Also note that for any even/unreachable applicant $a \in \mathcal{A} \setminus \mathcal{O}$, $r_a(s(a)) \neq 1$. Hence A_f^M, A_m^M, A_s^M, and A_l^M are pairwise disjoint and partition $\mathcal{A}$. So $|A_f^M| + |A_m^M| + |A_s^M| + |A_l^M| = |\mathcal{A}|$. Finally, note that $A_f^M = \{a \in \mathcal{A} : M(a) \in f(a)\}$. The partition of $\mathcal{A}$ by M into these four sets will guide us towards a popular matching that is more popular than M. Our first observation is that all popular matchings have the same signature, as shown by Lemma 3, where $\mathcal{F} = \cup_{a \in \mathcal{A}} f(a)$.

Lemma 3. *A matching M is popular iff its signature is $(|\mathcal{F}|, 0, |\mathcal{A}| - |\mathcal{F}|, 0)$.*

Proof. Suppose M is popular. Then by Lemma 2, $|A_m^M| = |A_l^M| = 0$, and so $|A_f^M| + |A_s^M| = |\mathcal{A}|$. Now, $|A_f^M| \leq |\mathcal{F}|$, since every applicant in A_f^M is matched with some post in $\mathcal{F}$. But since $M \cap E_1$ is a maximum matching of G_1, Lemma 1(b) requires that every post in $\mathcal{F}$ is matched with some applicant in A_f^M. Hence, $|A_f^M| = |\mathcal{F}|$, $|A_s^M| = |\mathcal{A}| - |\mathcal{F}|$, and M has signature $(|\mathcal{F}|, 0, |\mathcal{A}| - |\mathcal{F}|, 0)$.

Conversely, suppose M has signature $(|\mathcal{F}|, 0, |\mathcal{A}| - |\mathcal{F}|, 0)$. Then $|A_m^M| = |A_l^M| = 0$, and so for every $a \in \mathcal{A}$, $M(a) \in f(a) \cup s(a)$. It remains to prove that $M \cap E_1$ is a maximum matching of G_1. We have $|M \cap E_1| = |A_f^M| + |\{a \in A_s^M : a$ is odd$\}|$. Since $|A_f^M| = |\mathcal{F}|$, $|M \cap E_1| = |\mathcal{F}| + |\{a \in A : a$ is odd$\}|$. So, $|M \cap E_1| = |\{v \in \mathcal{A} \cup \mathcal{P} : v$ is odd$\}| + |\{p \in \mathcal{P} : p$ is unreachable$\}|$, and the result follows from Lemma 1(b). $\qquad \square$

Lemma 4. *For any matching M, $|A_f^M| + |A_m^M| \leq |\mathcal{F}|$.*

Proof. For each $a \in A_f^M$, $M(a)$ is odd/unreachable (i.e. belongs to $\mathcal{O} \cup \mathcal{U}$), for otherwise, G_1 contains an edge contradicting Lemma 1(c). Also, for each $a \in A_m^M$, $M(a)$ is odd/unreachable, since $s(a)$ contains a's most preferred *even* posts, and by definition of A_m^M, a prefers $M(a)$ to posts in $s(a)$ (i.e. $r_a(M(a)) > r_a(s(a))$). Hence, $|A_f^M| + |A_m^M| \leq |\mathcal{F}|$. $\qquad \square$

Finally, we come to the main technical lemma in this section, which characterizes popular matchings that are more popular than a given matching M.

Lemma 5. *Let M^* be a popular matching. Then M^* is more popular than M iff (i) $|A_f^M| + |A_m^M| < |\mathcal{F}|$, or (ii) $|A_m^M \cap A_f^{M^*}| > 0$, or (iii) $|A_l^M \cap A_s^{M^*}| > 0$.*

Proof. Let $\Delta(M^*, M)$ be the difference between the number of applicants who prefer M^* to M, and the number of applicants who prefer M to M^*. That is, $\Delta(M^*, M) = |(A_m^M \cup A_s^M \cup A_l^M) \cap A_f^{M^*}| + |A_l^M \cap A_s^{M^*}| - |(A_f^M \cup A_m^M) \cap A_s^{M^*}|$. Now, since M^* is popular, by Lemma 3 we have:

$$|A_f^{M^*}| = |(A_f^M \cup A_m^M \cup A_s^M \cup A_l^M) \cap A_f^{M^*}| = |\mathcal{F}|$$
$$= (|\mathcal{F}| - |A_f^M| - |A_m^M|) + |(A_f^M \cup A_m^M) \cap (A_f^{M^*} \cup A_s^{M^*})| \ .$$

Rearranging, we get $|(A_f^M \cup A_m^M) \cap A_s^{M^*}| = |(A_s^M \cup A_l^M) \cap A_f^{M^*}| - (|\mathcal{F}| - |A_f^M| - |A_m^M|)$. Hence $\Delta(M^*, M) = (|\mathcal{F}| - |A_f^M| - |A_m^M|) + |A_m^M \cap A_f^{M^*}| + |A_l^M \cap A_s^{M^*}|$. The theorem follows immediately, since $|A_m^M \cap A_f^{M^*}|$ and $|A_l^M \cap A_s^{M^*}|$ are both non-negative, while $|\mathcal{F}| - |A_f^M| - |A_m^M| \geq 0$ by Lemma 4.   □

Given $\langle G, M_0 \rangle$ and some popular matching M^*, we don't need Lemma 5 to determine if M^* is more popular than M_0 - instead, we can just count the number of applicants that prefer one matching to the other. Suppose, however, that M^* is not more popular than M_0 so that $|A_f^{M_0}| + |A_m^{M_0}| = |\mathcal{F}|$, $|A_l^{M_0} \cap A_s^{M^*}| = 0$, and $|A_m^{M_0} \cap A_f^{M^*}| = 0$. Our aim is to use Lemma 5 as a guide in finding a popular matching that is more popular than M_0, or proving that no such matching exists.

First we remark that $|A_f^{M_0}| + |A_m^{M_0}| = |\mathcal{F}|$, for otherwise, any popular matching, including M^*, is more popular than M_0 by Lemma 5. It follows that $A_m^{M_0} \neq \emptyset$ or $A_l^{M_0} \neq \emptyset$, for otherwise, M_0 has signature $(|\mathcal{F}|, 0, |\mathcal{A}| - |\mathcal{F}|, 0)$, contradicting the assumption from the previous section that M_0 is not popular.

Suppose $A_m^{M_0} \neq \emptyset$, so that there is an applicant $a \in A_m^{M_0} \cap A_s^{M^*}$. Since $a \in A_m^{M_0}$, a is even/unreachable. If a popular matching pairs a with a post in $f(a)$, then it must be more popular than M_0 by condition (ii) of Lemma 5. In order to test if there exists such a popular matching, we proceed as follows.

Let G' be the subgraph of G defined in Fig. 2 after step 3. So, G' contains all edges between applicants and their f-posts and s-posts, except those between nodes in $\mathcal{O}$ and nodes in $\mathcal{O} \cup \mathcal{U}$. Now, modify G' and M^* by removing all edges between this particular applicant a and posts in $s(a)$. Call the resulting structures G_a' and M_a^* respectively.

Lemma 6. *There exists a popular matching which pairs a with some post in $f(a)$ iff G_a' admits an augmenting path with respect to M_a^*.*

Proof. Suppose G_a' admits an augmenting path Q_a with respect to M_a^*. Since M^* is popular, the only unmatched applicant in M_a^* is a, and so $M_a^* \oplus Q_a$ matches a with some post in $f(a)$. We want to claim that $M_a^* \oplus Q_a$ is popular. First note that its signature is of the form $(k, 0, |\mathcal{A}| - k, 0)$ for some $k \geq 0$, since it is a matching in a subgraph of G', and G' only contains edges between applicants and their f-posts and s-posts. Now, $M_a^* \oplus Q_a$ matches all posts in $\mathcal{F}$, since every post matched in M_a^* is also matched in $M_a^* \oplus Q_a$. Recall that posts in $\mathcal{F}$ are incident in G' to rank-1 edges only, and furthermore, odd posts in $\mathcal{F}$ are only adjacent to even applicants, while unreachable posts in $\mathcal{F}$ are only adjacent to unreachable applicants. Hence, $k = |\mathcal{F}|$, and so by Lemma 3, $M_a^* \oplus Q_a$ is popular, since its signature is $(|\mathcal{F}|, 0, |\mathcal{A}| - |\mathcal{F}|, 0)$.

Conversely, suppose that G_a' admits no augmenting path with respect to M_a^*. Then, M_a^* is a maximum matching in G_a', which, since a is unmatched in M_a^*, means that there is no $\mathcal{A}$-perfect matching in G_a'. But by Lemma 2(b), every popular matching is an $\mathcal{A}$-perfect matching in G'. Hence, every popular matching must contain an edge in $G' \setminus G_a' = \{(a, p) : p \in s(a)\}$.   □

We now make use of Lemma 6. Begin by looking for an augmenting path in G_a' with respect to M_a^* by using depth-first search to construct the Hungarian tree

T_a (which is a tree consisting of alternating paths with respect to M_a^*) rooted at a. If we find such a path Q_a, then by Lemma 5, $\langle M_0, M_a^* \oplus Q_a \rangle$ is a length-1 voting path to a popular matching. Otherwise, we repeat this process with some other $a' \in A_m^M \cap A_s^{M^*}$, if any.

Suppose we find an augmenting path $Q_{a'}$ such that $M_{a'}^* \oplus Q_{a'}$ is popular. We claim that $Q_{a'}$ is edge disjoint from the set of edges in T_a, where $a \neq a'$.

For suppose otherwise. Then let e be the *first* edge in $Q_{a'}$ that is also in T_a. Now, G_a' admits an alternating path from a through e. However, since Q_a does not exist, this path cannot be extended in G_a' to end in some unmatched post. Hence, $Q_{a'}$ must contain an alternating path from e through the edge $(M^*(a), a)$ (this edge is missing in G_a'). But $M^*(a)$ is unmatched in G_a', and hence if we join the two alternating paths above, we get an augmenting path Q_a (from a to $M^*(a)$ through the edge e) in G_a'. This gives the required contradiction.

Since we only need to examine each edge a constant number of times, it is clear that we can determine in linear time if there is a popular matching that pairs some applicant in $A_m^{M_0} \cap A_s^{M^*}$ with one of its f-posts.

If there is no such applicant, we repeat this procedure with applicants $a \in A_l^{M_0} \cap A_f^{M^*}$, who must by definition be even/unreachable. Here, though our aim is to find a popular matching that satisfies Lemma 5(iii) by pairing a with a post in $s(a)$. For this to occur, a must be even, since by Lemmas 1 and 2(i), every unreachable applicant is matched by any popular matching to a post in $f(a)$.

Suppose we find such a matching $M_a^* \oplus Q_a$. Since a is even, $M^*(a)$ is odd, and so any popular matching must match $M^*(a)$ along a rank-1 edge to an even applicant. However, $M^*(a)$ may be unmatched in $M_a^* \oplus Q_a$, as we removed all edges between a and posts in $f(a)$ from G' (including $(a, M^*(a))$). Hence we may need to augment $M_a^* \oplus Q_a$ in G_1. But every odd node $M^*(a)$ is adjacent to at least one other even applicant along a rank-1 edge, namely its predecessor in the odd length alternating path from a vertex unmatched w.r.t. $M^* \cap E_1$ to $M^*(a)$ in G_1). Hence, such an augmentation always exists. And it is easy to see that here too we only examine each edge a constant number of times.

By Lemmas 5 and 6, the above algorithm finds a length-1 voting path from M_0 to some popular matching, or proves that no such path exists. Also, given a popular matching, we have just shown that the algorithm runs in linear time.

4   Length-2 Voting Paths

In this section, we show that, given any popular matching M^* of G, we can find a length-2 voting path from M_0 to some popular matching in linear time. We will assume that M_0 admits no shorter such voting path. In particular, this means M^* is not more popular than M_0, and $A_m^{M_0} \neq \emptyset$ or $A_l^{M_0} \neq \emptyset$.

Suppose $A_m^{M_0} \neq \emptyset$. Let $a \in A_m^{M_0}$ and let T_a be the Hungarian tree associated with a, as described in Section 3. In the next lemma, we give a sufficient condition for the existence of a length-2 voting path from M_0 to M^*.

Lemma 7. *Suppose there exists an applicant $a' \in T_a$ such that $M_0(a') \notin s(a')$ and $M^*(a') \in s(a')$. Then there exists a length-2 voting path from M_0 to M^*.*

Proof. Our aim is to find a matching M_1 such that (i) M_1 is more popular than M_0, and (ii) $a' \in A_l^{M_1}$. This last condition guarantees that M^* is more popular than M_1 by Lemma 5(iii), giving us the length-2 voting path $\langle M_0, M_1, M^* \rangle$.

Before constructing M_1, we first need to show that $a' \in A_f^{M_0} \cup A_m^{M_0}$. We have $a' \notin A_l^{M_0}$, for otherwise $a' \in A_l^{M_0} \cap A_s^{M^*}$ and M^* is more popular than M_0 by Lemma 5(iii) - a contradiction. Suppose then $a' \in A_s^{M_0}$. By definition we have $M_0(a') \notin s(a')$, and so $M_0(a')$ must be odd/unreachable and belongs to $\mathcal{F}$. But M_0 matches all posts in $\mathcal{F}$ to applicants in $A_f^{M_0} \cup A_m^{M_0}$, for otherwise M^* is more popular than M_0 by Lemma 5(i) - a contradiction. Hence, $a' \in A_f^{M_0} \cup A_m^{M_0}$.

Now we need to show that G_a' contains no edge between a' and $l(a')$. Suppose it does. Then since $l(a')$ is strictly the least preferred post of a', we have $s(a') = \{l(a')\}$. By definition, no other applicant is adjacent to $l(a')$, and so $l(a')$ is a leaf node in T_a with parent a'. It follows from the construction of T_a that $l(a')$ is unmatched in M_a^*, and hence M_a^* admits an augmenting path from a through a' to $l(a')$. This contradicts our assumption that M_0 admits no length-1 voting path to a popular matching. So, G_a' contains no edge between a' and $l(a')$.

Finally, we describe how to construct M_1. Add an edge between a' and $l(a')$ to G_a'. From above, we know G_a' admits an augmenting path Q_a from a through a' and ending in $l(a')$. Let $M_1 = M_a^* \oplus Q_a$. The signature of M_1 is $(|\mathcal{F}|, 0, |\mathcal{A}| - |\mathcal{F}| - 1, 1)$ by an argument similar to Lemma 6, except that here we have one applicant $a' \in A_l^{M_1}$. We now show M_1 is more popular than M_0.

$$\Delta(M_1, M_0) = |(A_m^{M_0} \cup A_s^{M_0} \cup A_l^{M_0}) \cap A_f^{M_1}| + |A_l^{M_0} \cap A_s^{M_1}| - |(A_f^{M_0} \cup A_m^{M_0})$$
$$\cap (A_s^{M_1} \cup A_l^{M_1})| \quad \{\text{NB: } |A_s^{M_0} \cap A_l^{M_1}| = 0, \text{ since } a' \notin A_s^{M_0}\}$$
$$\geq |(A_m^{M_0} \cup A_s^{M_0} \cup A_l^{M_0}) \cap A_f^{M_1}| - |(A_f^{M_0} \cup A_m^{M_0}) \cap (A_s^{M_1} \cup A_l^{M_1})|$$
$$= |(A_f^{M_0} \cup A_m^{M_0} \cup A_s^{M_0} \cup A_l^{M_0}) \cap A_f^{M_1}|$$
$$- |A_f^{M_0} \cap (A_f^{M_1} \cup A_s^{M_1} \cup A_l^{M_1})| - |A_m^{M_0} \cap (A_s^{M_1} \cup A_l^{M_1})|$$
$$= |\mathcal{F}| - |A_f^{M_0}| - |A_m^{M_0} \cap (A_s^{M_1} \cup A_l^{M_1})| \quad \{\text{since } |A_f^{M_1}| = |\mathcal{F}|\}$$
$$= |A_m^{M_0}| - |A_m^{M_0} \cap (A_s^{M_1} \cup A_l^{M_1})| \quad \{\text{by Lemmas 4 and 5(i)}\}$$
$$= |A_m^{M_0} \cap A_f^{M_1}| > 0 \quad \{\text{since } a \in A_m^{M_0} \cap A_f^{M_1}\} \qquad \square$$

In linear time, we can check if there exists an applicant $a' \in T_a$ such that $M_0(a') \notin s(a')$ and $M^*(a') \in s(a')$, and if so, we construct the matching M_1.

Suppose there is no such applicant in T_a. Our aim then is to find a different popular matching in which such an applicant exists. We will find such a matching by searching for a particular type of alternating cycle $\mathcal{C}$ in G_a' with respect to M^*. First though, we make some observations about T_a and G_a'.

By construction, posts in T_a are discovered along unmatched edges. Also, no post p is a leaf node in T_a, since then p would be unmatched in M_a^*, and M_a^* would admit an augmenting path, contradicting our assumption that M_0 admits no length-1 voting path to a popular matching. So, posts in T_a have degree 2.

By construction, an applicant $a'' \in T_a \setminus \{a\}$ is discovered along a matched edge. If a'' is even or unreachable, then a'' is incident to at least one unmatched child edge in G'_a, since $f(a'')$ and $s(a'')$ are non-empty and disjoint. If a'' is odd, then a'' must also be incident to at least one unmatched child edge in G'_a - since a'' is odd, G_1 admits an odd-length alternating path from an unmatched vertex to a'', and the last edge $e(a'') \in E_1$ on this path is unmatched by M^*.

Since every node in $T_a \setminus \{a\}$ is incident to at least one matching edge and one non-matching edge (w.r.t. M^*) in G'_a, we can build the following alternating path Q. Begin with an edge (a, p) where $p \in f(a)$. Let the successor of any post p in Q be its matched partner $M^*(p)$. The successor of any even/unreachable applicant a'' is any post in $f(a'')$ if $M^*(a'') \in s(a'')$, or any post in $s(a'')$ if $M^*(a'') \in f(a'')$. Finally, the successor of any odd applicant a'' is the post incident to $e(a'')$. Since $|T_a|$ is finite, this alternating path must form a cycle $\mathcal{C}$ by adding a post that is already in the path. This procedure takes linear time.

It is clear from Lemma 2 that $M^* \oplus \mathcal{C}$ is a popular matching. Now, if we can show that $M^* \oplus \mathcal{C}$ has some applicant $a' \in T_a \setminus \{a\}$ such that $M_0(a') \notin s(a')$ and $(M^* \oplus \mathcal{C})(a') \in s(a')$, we can use $M^* \oplus \mathcal{C}$ as the popular matching in Lemma 7.

First, we prove that $\mathcal{C}$ contains at least one applicant a' such that $M^*(a') \in f(a')$. Since $M^* \oplus \mathcal{C}$ matches a' with $s(a')$ by construction, the final step will be to show that $M_0(a') \notin s(a')$.

Lemma 8. $\mathcal{C}$ contains at least one applicant a' such that $M^*(a') \in f(a')$.

Proof. Note that the length of $\mathcal{C}$ is at least 4, since the predecessor and successor of each applicant are always distinct. Also note that if $\mathcal{C}$ contains an even/unreachable applicant, then either its predecessor or its successor is an f-post, whose partner in M^* is the required applicant. The only way that that $\mathcal{C}$ may not contain an f-post is if all the applicants in $\mathcal{C}$ are odd. We show this cannot happen.

Let a'' be the first odd applicant in Q that is in $\mathcal{C}$. By construction, $\mathcal{C}$ contains a subpath from a'' through $e(a'')$ to some post p that is unmatched in $M^* \cap E_1$. It follows that p is even, since odd/unreachable posts are matched in $M^* \cap E_1$ by Lemmas 1(b) and 2(i). Since p is matched in M^*, its partner $M^*(p)$ must be even (again by Lemmas 1(b) and 2(i)), and so $\mathcal{C}$ contains an even applicant. $\square$

Lemma 9. *Suppose there exists no applicant $a' \in T_a$ such that $M_0(a') \notin s(a')$ and $M^*(a') \in s(a')$. Then for each $a' \in T_a \setminus \{a\}$, $M_0(a') \notin s(a')$ iff $M^*(a') \in f(a')$.*

Proof. Let a' be any applicant in $T_a \setminus \{a\}$ such that $M_0(a') \notin s(a')$. Then by the assumption in the statement of the lemma, we have $M^*(a') \notin s(a')$, and so $M^*(a') \in f(a') \subseteq \mathcal{F}$, since M^* is popular. Hence, $|T_a \setminus \{a\} \cap (\mathcal{A} - A_s^{M_0})|$ is at most the number of f-posts in T_a, i.e., (1) $|T_a \setminus \{a\} \cap (\mathcal{A} - A_s^{M_0})| \leq |T_a \cap \mathcal{F}|$.

Now, since there is no augmenting path in T_a, we have that $|T_a \setminus \{a\} \cap \mathcal{A}| = |T_a \cap \mathcal{P}|$. Partitioning $\mathcal{A}$ into $\mathcal{A} \setminus A_s^{M_0}$ and $A_s^{M_0}$ and posts in T_a into $T_a \cap \mathcal{F}$ and $T_a \cap S$, where S is the set of all s-posts, we get $|T_a \setminus \{a\} \cap ((\mathcal{A} \setminus A_s^{M_0}) \cup A_s^{M_0})| = |T_a \cap \mathcal{F}| + |T_a \cap S|$. Note that no applicant in $A_s^{M_0}$ can be matched by M_0 to an odd/unreachable post, otherwise $|A_f^{M_0}| + |A_m^{M_0}| < |\mathcal{F}|$ and M^* would have been more popular than M_0 by Lemma 5(i). Hence each applicant in $A_s^{M_0}$ has to be

---

Voting-Path$(G = (\mathcal{A} \cup \mathcal{P}, E), M_0)$

> **if** M_0 has signature $(|\mathcal{F}|, 0, |\mathcal{A}| - |\mathcal{F}|, 0)$ **then**
> > **return** $\langle M_0 \rangle$
>
> Let M^* be any popular matching of G
> **if** M^* is more popular than M_0 **then**
> > **return** $\langle M_0, M^* \rangle$
>
> **for each** applicant $a \in A_m^{M_0} \cap A_s^{M^*}$ and **each** even applicant $a \in A_l^{M_0} \cap A_f^{M^*}$
> > Construct the Hungarian tree T_a with respect to M_a^*, including only un-
> > marked edges
> > Mark all edges in T_a
> > **if** T_a contains an augmenting path Q_a **then**
> > > Augment $M_a^* \oplus Q_a$ in G_1 from $M^*(a)$ // applies only when
> > > $\qquad\qquad\qquad\qquad\qquad\qquad\qquad\qquad\qquad$ $a \in A_l^{M_0} \cap A_f^{M^*}$
> > >
> > > **return** $\langle M_0, M_a^* \oplus Q_a \rangle$
>
> Let T_a be the first Hungarian tree constructed for any $a \in A_m^{M_0} \cap A_s^{M^*}$ or
> even $a \in A_l^{M_0} \cap A_f^{M^*}$
> **if** $\nexists \, a' \in T_a \setminus \{a\}$ such that $M_0(a') \notin s(a')$ and $M^*(a') \in s(a')$ **then**
> > Construct $\mathcal{C}$ in G_a' as described in Section 4
> > Let $M^* = M^* \oplus \mathcal{C}$
> > Let T_a be the Hungarian tree associated with a and the new M^*
> > Add $(a', l(a'))$ to T_a, and find the augmenting path Q_a in T_a
> > Let $M_1 = M_a^* \oplus Q_a$
> > Augment M_1 in G_1 from $M^*(a)$ // applies only when $a \in A_l^{M_0} \cap A_f^{M^*}$
> > **return** $\langle M_0, M_1, M^* \rangle$

Fig. 3. Our algorithm for finding a shortest-length voting path

matched by M_0 to one of its most preferred even posts, i.e., one of its s-posts, so $|T_a \setminus \{a\} \cap A_s^{M_0}| \leq |T_a \cap S|$. We thus get, (2) $|T_a \setminus \{a\} \cap (\mathcal{A} \setminus A_s^{M_0})| \geq |T_a \cap \mathcal{F}|$.

Combining (1) and (2), we have $|T_a \setminus \{a\} \cap (\mathcal{A} - A_s^{M_0})| = |T_a \cap \mathcal{F}| = |T_a \setminus \{a\} \cap A_f^{M^*}|$. That is, the number of applicants a' in $T_a \setminus \{a\}$ that satisfy $M^*(a') \in f(a')$ is equal to the number of applicants in $T_a \setminus \{a\}$ that satisfy $M_0(a') \notin s(a')$. But each applicant in $T_a \setminus \{a\}$ that satisfies $M_0(a') \notin s(a')$ has to satisfy $M^*(a') \in f(a')$ by the statement of the lemma. So the equivalence follows immediately. □

By Lemma 8, $\mathcal{C}$ contains at least one applicant a' such that $M^*(a') \in f(a')$. By Lemma 9, we have that $M_0(a') \notin s(a')$. Since $M^* \oplus \mathcal{C}$ matches a' with $s(a')$, $M^* \oplus \mathcal{C}$ satisfies the sufficient condition in Lemma 7. Hence there exists a length-2 voting path from M_0 to the popular matching $M^* \oplus \mathcal{C}$.

As in Section 3, if $A_m^{M_0} = \emptyset$, then $A_l^{M_0} \neq \emptyset$, and we perform an analogous procedure on the Hungarian tree associated with some $a \in A_l^{M_0}$. This finishes the proof of Theorem 1 (from Section 1). The overall algorithm is given in Fig. 3.

5   Conclusions

In this paper, we proved that whenever a popular matching exists, there is a voting path of length at most 2 from any matching to some popular matching.

We also gave a linear-time algorithm for finding a shortest-length such path, given any popular matching M^* in G. These results solve the dynamic matching market problem given in [3], and prove that a decentralized market can converge to a popular matching, when such a matching exists.

References

1. A. Abdulkadiroğlu and T. Sönmez. Random serial dictatorship and the core from random endowments in house allocation problems. *Econometrica*, 66(3):689–701, 1998.
2. D.J. Abraham, K. Cechlárová, D.F. Manlove, K. Mehlhorn. Pareto-optimality in house allocation problems. In *Proc. of 15th ISAAC*, pages 3-15, 2004.
3. D.J. Abraham, R.W. Irving, T. Kavitha, and K. Mehlhorn. Popular matchings. In *Proc. of 16th SODA*, pages 424-432, 2005.
4. E. Diamantoudi, E. Miyagawa, and L. Xue. Random paths to stability in the roommate problem. *Games and Economic Behavior*, 48(1):18-28, 2004.
5. S.P. Fekete, M. Skutella, and G.J. Woeginger. The complexity of economic equilibria for house allocation markets. *Information Processing Letters*, 88:219-223, 2003.
6. D. Gale and L.S. Shapley. College admissions and the stability of marriage. *American Mathematical Monthly*, 69:9–15, 1962.
7. P. Gardenfors. Match Making: assignments based on bilateral preferences. *Behavioural Sciences*, 20:166–173, 1975.
8. D. Gusfield and R.W. Irving The Stable Marriage Problem: Structure and Algorithms. *MIT Press*, 1989.
9. A. Hylland and R. Zeckhauser. The efficient allocation of individuals to positions. *Journal of Political Economy*, 87(2):293–314, 1979.
10. R.W. Irving. An efficient algorithm for the "stable roommates" problem. *Journal of Algorithms*, 6:577-596, 1985.
11. R.W. Irving, T. Kavitha, K. Mehlhorn, D. Michail, and K. Paluch. Rank-maximal matchings. In *In Proc. of 15th SODA*, pages 68–75. 2004.
12. D.E. Knuth. Stable marriage and its relation to other combinatorial problems. *CRM Proceedings and Lecture Notes*, volume 10, 1976.
13. H. Landau. On dominance relations and the structure of animal societies, III: the condition for secure structure. *Bulletin of Math. Biophysics*, 15(2):143-148, 1953.
14. M. Mahdian. Random popular matchings. To appear in *ACM-EC*, 2006.
15. J. Mestre. Weighted popular matchings. To appear in *ICALP*, 2006.
16. Netflix DVD Rental: see http://www.netflix.com.
17. A.E. Roth and A. Postlewaite. Weak versus strong domination in a market with indivisible goods. *Journal of Mathematical Economics*, 4:131–137, 1977.
18. A.E. Roth and J.H. Vande Vate. Random paths to stability in two-sided matching. *Econometrica*, 58:1475-1480, 1990.
19. A. Tamura. Transformation from arbitrary matchings to stable matchings. *Journal of Combinatorial Theory*, Series A, 62(2):310-323, 1993.
20. Y. Yuan. Residence exchange wanted: a stable residence exchange problem. *European Journal of Operational Research*, 90:536–546, 1996.
21. L. Zhou. On a conjecture by Gale about one-sided matching problems. *Journal of Economic Theory*, 52(1):123–135, 1990.

Sorting by Merging or Merging by Sorting?

Gianni Franceschini

Institute for Informatics and Telematics
Italian National Research Council
via Moruzzi 1, 56124 Pisa, Italy
gianni.franceschini@iit.cnr.it

Abstract. In the comparison model the only operations allowed on input elements are comparisons and moves to empty cells of memory. We prove the existence of an algorithm that, for any set of $s \leq n$ sorted sequences containing a total of n elements, computes the whole sorted sequence using $O(n \log s)$ comparisons, $O(n)$ data moves and $O(1)$ auxiliary cells of memory besides the ones necessary for the n input elements. The best known algorithms with these same bounds *are limited to the particular case $s = O(1)$.* From a more intuitive point of view, our result shows that it is possible to pass from merging to sorting in a seamless fashion, without losing the optimality with respect to any of the three main complexity measures of the comparison model. Our main statement has an implication in the field of adaptive sorting algorithms and improves [Franceschini and Geffert, Journal of the ACM, 52], showing that it is possible to exploit some form of pre-sortedness to lower the number of comparisons while still maintaining the optimality for space and data moves. More precisely, let us denote with $\mathrm{Opt}_M(X)$ the cost for sorting a sequence X with an algorithm that is optimal with respect to a pre-sortedness measure M. To the best of our knowledge, so far, for any pre-sortedness measure M, no full-optimal adaptive sorting algorithms were known (see [Estivill-Castro and Wood, ACM Comp. Surveys, 24], page 472). The best that could be obtained were algorithms sorting a sequence X using $O(1)$ space, $O(\mathrm{Opt}_M(X))$ comparisons and $O(\mathrm{Opt}_M(X))$ moves. Hence, the move complexity seemed bound to be a function of $M(X)$ (as for the comparison complexity). We prove that there exists a pre-sortedness measure for which that is false: the pre-sortedness measure *Runs*, defined as the number of ascending contiguous subsequences in a sequence. That follows directly from our main statement, since $\mathrm{Opt}_M(X) = O(|X| \log Runs(X))$.

1 Introduction

Sorting and merging have always been fundamental problems in computation. In this paper we prove a property that blurs the boundary between these two problems and seems to contradict the usual hierarchical relation that sees merging as a mere subproblem of sorting. Furthermore, as a consequence of our main statement we will improve the result in [1], showing that it is possible to exploit some form of pre-sortedness to lower the number of comparisons while still maintaining the optimality for space and data moves.

L. Arge and R. Freivalds (Eds.): SWAT 2006, LNCS 4059, pp. 77–89, 2006.
© Springer-Verlag Berlin Heidelberg 2006

The problem. We are given a total order $(\mathcal{U}, \leq)$, with a possibly infinite universe $\mathcal{U}$. The input element are drawn from $\mathcal{U}$ and they are considered atomic (no bit manipulation, hashing etc. . .). The only operations allowed on input elements are comparisons between two elements and moves to empty cells of memory. Hence, the complexity of an algorithm will be measured by the usual three metrics: *the number of comparisons, the number of moves* and *the number of cells of memory used,* besides the ones necessary to store the input elements. For any such metric, we will refer to the *optimality of an algorithm* with the usual meaning of *asymptotic optimality up to a constant factor.* We will call an algorithm *full-optimal* if it is *optimal with respect to the three metrics simultaneously.*

In the *sorting problem,* we are given a set of n elements from $\mathcal{U}$ and they have to be disposed in the ordered permutation induced by the relation $\leq$. We will deal with the following, related problems:

Problem 1 (Multiway Merging Problems). In the *balanced merging problem,* we are given $s \leq n$ sorted sequences of n/s elements each drawn from $\mathcal{U}$ and they have to be fused into a single sorted sequence. In the *unbalanced merging problem* the total number of input elements is still n but the s sorted subsequences to be fused can differ in their lengths.

An algorithm solving the two merging problems is supposed to exploit the pre-sortedness of the input elements in order to arrive to the final sorted sequence with less computational effort.

Seeing the parametric definition of the merging problems, natural questions arise: How far can we push the parameter s? How much marked is the boundary between merging and sorting problems? *Is there a full-optimal solution for the merging problems for any value of s?*

Let us enter more deeply into this matter trying to solve the *balanced merging problem* dropping the full-optimality constraint.

If we do not care about the number of moves performed, a space-optimal and comparison-optimal solution to the balanced merging problem for any value of s follows immediately by the existence of a full-optimal solution for the case $s = 2$. As we will see later, the research around the existence of an algorithm that could fuse two sequences of m elements each, using $O(m)$ comparisons, $O(m)$ moves and $O(1)$ auxiliary cells has been active and successful since the late sixties [2]. Given that fact, a space-optimal and comparison-optimal solution to the merging problem for any value of s is a simple variation of the Mergesort. It employs any full-optimal merging algorithm for $s = 2$ and starts the execution merging pairs of sorted sequences instead of pairs of elements. That approach performs $O(n \log s)$ comparisons and uses $O(1)$ auxiliary memory cells but we have to pay $O(n \log s)$ data moves.

If we give up on the number of comparisons, we can just use the recent full-optimal algorithm for the *sorting problem* in [1]. With that algorithm, we can just ignore the sorted sequences, and sort the whole sequence using $O(n)$ moves and $O(1)$ auxiliary cells plus the inevitable $O(n \log n)$ comparisons.

Finally, if we can exploit a linear auxiliary space, a solution for any value of s performing $O(n \log s)$ comparisons and $O(n)$ moves can be obtained using a dictionary that is searchable in $O(\log m)$ comparisons (when it contains $O(m)$ elements) and updatable in $O(1)$ moves, either amortized or in the worst case (e.g. $[3, 4, 5]$...). Any such dictionary can be simply used as a priority queue containing, at any time during the process, the smallest s elements among the ones still in the sorted sequences in input.

Previous work. As we said before, the research focusing on the existence of a full-optimal solution for the merging problems has been fervent since the sixties, bringing a lot of results for the special case where $s = O(1)$. To the best of our knowledge, *so far, even the existence of a full-optimal solution for the merging problems covering any range of values larger than $s = O(1)$ was unknown* (that is if we exclude the range $n^\epsilon \leq s \leq n$, for any real constant $\epsilon < 1$, because in that case a solution can be obtained with a plain application of the sorting algorithm in [1]). Perhaps the "nearest relative" of a full-optimal solution for the merging problems when $s = O(polylog(n))$ can be found in [6]. In this paper the authors present the first in-place sorting algorithm performing $O(n \log n)$ comparisons and $o(n \log n)$ data moves. Some of their techniques are suitable for the merging problem when $s = O(polylog(n))$ but unfortunately they use an internal buffer (see Section 2) of size $O(n)$ that disrupts irremediably the pre-sortedness of the input sequence (see Sections 2 and 3.2).

As a further witness of the intrinsic difficulty of finding full-optimal solutions for the merging problems, even for the particular case with $s = 2$, we will briefly review the main known results. The first solution was proposed in [2], in that seminal paper fundamental tools for space-optimality, like the *internal buffering* technique (see Section 2), were introduced. Unfortunately, the two-way merging algorithm of Kronrod contained an insidious error that ruins the correctness of the algorithm in the general case of input with repeated elements. After Kronrod, Horvath [7] devised a stable (i.e. the initial relative order of equal elements is maintained after the process) merging algorithm assuming that input elements could be modified. Subsequently, Trabb Pardo [8] removed this requirement. The error in Kronrod's work went undiscovered until [9], when a simpler way to stable merging was devised. In [10] an unstable modification of Kronrod algorithm is given. Later on, the same authors gave a stable algorithm in [11]. When $s = 2$, the lower bound for the number of comparisons in case of sequences of two different lengths m and n, with $m < n$, is $O(m \log(n/m))$. Symvonis achieved that lower bound in [12]. Subsequently, stable and unstable algorithms with the same asymptotic bound but better constant factors were proposed in [13] and [14].

As we will see, a direct consequence of our main statement pertains to the field of *adaptive sorting algorithms* (see [15] for a survey on the subject) and improve the result in [1]. In the adaptive sorting problem the complexity measures for the sorting algorithms are expressed as a function of a chosen pre-sortedness measure of the input sequence. With the development of this field, many pre-sortedness measures have been introduced together with a concept of optimality for any such measure. For example, the measure *Runs* is defined as the number of ascending contiguous subsequences of the input sequence. Using any in-place merging algorithm as the one in [9], it is possible to achieve adaptive sorting algorithms that are space-optimal, *Runs*-optimal *and with a number of moves of the same order of the number of comparisons*. Similar results can be obtained with other measures but, to the best of our knowledge, so far, for any pre-sortedness measure, *no adaptive sorting algorithm has been proven to be full-optimal*.

Our contribution. In this paper, we prove the following theorem:

Theorem 1. *There exists an algorithm $\mathcal{A}$ with the following property: For any $1 \leq s \leq n$ and for any set of s sorted sequences containing a total of n elements (drawn from $\mathcal{U}$), $\mathcal{A}$ computes the whole sorted sequence with $O(n \log s)$ comparisons, $O(n)$ moves and $O(1)$ auxiliary cells of memory.*

Obviously any solution for the *balanced merging problem* performs $\Omega(n)$ data moves. In the balanced merging problem the sorted subsequences are assumed to be of equal length. Hence, it is straightforward to prove that any solution for that problem performs $\Omega(n \log s)$ comparisons. Therefore, by Theorem 1, the following holds:

Corollary 1 (Full-optimal merging). *There exists a full-optimal solution for the* balanced merging problem *for any value of s.*

Concerning the *unbalanced merging problem*, if the lengths of the sorted subsequences are not considered in the complexity measures, that is if we aim to give complexity bounds that depend only on parameters n and s (as the definition of the problem implies), then the algorithm of Theorem 1 is a full-optimal solution for that problem too. Instead, if we are interested in evaluating the complexity of the unbalanced merging problem taking into account also the lengths of the sorted subsequences, let them be $n_1, n_2 \ldots n_s$, then simple approaches using less comparisons come immediately in mind. For example, a simple merging strategy performing at any time a binary merging operation between the two shortest runs at that time, would use $O\left(\Sigma_{i=1\ldots s} n_i \log(n/n_i)\right)$ comparisons. Unfortunately, an approach like this seems to require a number of data moves of the same order.

   Our main statement has a consequence involving the field of adaptive sorting algorithms. We will improve the result in [1], showing that it is possible to exploit some form of pre-sortedness to lower the number of comparisons while still maintaining the optimality for space and data moves. More precisely, let us denote with $\text{Opt}_M(X)$ the cost for sorting a sequence X with an algorithm that is optimal with respect to a pre-sortedness measure M. To the best of our knowledge, so far, for any pre-sortedness measure M, no full-optimal adaptive sorting algorithms were known (see [15], page 472). The best that could be obtained were algorithms sorting a sequence X using $O(1)$ space, $O(\text{Opt}_M(X))$ comparisons and $O(\text{Opt}_M(X))$ moves. Hence, the move complexity seemed bound to be a function of $M(X)$ (as for the comparison complexity). We prove that there exists a pre-sortedness measure for which that is false.

Corollary 2 (Full-optimal adaptive sorting). *There exists a pre-sortedness measure M and an algorithm $\mathcal{A}_M$ such that, for any sequence X of n elements, $\mathcal{A}_M$ sorts X with $O(Opt_M(X))$ comparisons, $O(n)$ moves and $O(1)$ auxiliary cells of memory.*

Consider the pre-sortedness measure *Runs*, defined as the number of ascending contiguous subsequences in a sequence. We know that $\text{Opt}_M(X) = O(|X| \log Runs(X))$, therefore Corollary 2 follows from Theorem 1.

   Finally, let us explain our contribution from a more intuitive point of view. In this paper we prove that it is possible to pass from merging to sorting in a seamless fashion, without losing the optimality with respect to any of the three main complexity measures. In light of this fact, we could say that merging is not a mere subroutine with limited power as in the well known *sorting by merging* approach. Instead, as we will see, *a sorting algorithm is a basic subroutine* for our full-optimal merging algorithm.

In Section 2, we will first make some assumptions for the sake of this preliminary presentation. Then, we will introduce some known tools that will be used in the algorithm. Finally, we will give a brief description of some of the obstacles posed by the problem and the basic, high level ideas to overcome them. In Section 3, we will describe our full-optimal *merging by sorting* algorithm.

2   Assumptions, Tools, Obstacles and Basic Ideas

Assumptions. In order to make this brief exposition more readable, we do not stress on formal details (we usually avoid the use of ceilings and floors). Moreover, let us assume that the n elements in input are distinct. Considering the case of repeated elements when the stability of the algorithm (that is the capability of maintaining in the output sequence the initial relative order of equal elements) is not a concern would fill the presentation with technicalities without adding anything valuable from a theoretical point of view. On the other hand, the problem of stability seems to be very insidious and some of the techniques used in our solution are particularly prone to destabilize the input during the merging process. A complete exposition will be given in the full version of this paper.

Tools. The first tool we use is the *internal buffering technique* [2]. Essentially, some of the elements are used as placeholders to simulate a working area in which the other input elements can be permuted efficiently. Usually, a set $\mathcal{A}$ of input elements is divided into two subsets $\mathcal{A}'$ and $\mathcal{B}$, where the latter is the buffer set and has cardinality $o(|\mathcal{A}|)$. Then, $\mathcal{A}'$ is processed efficiently with the aid of $\mathcal{B}$ that can be subsequently processed with a sub-optimal method. Finally, a last merging step brings us the final sorted sequence.

The second tool we use is the *bit stealing technique* [16]. This technique is very common and very simple: the value of a bit is encoded in the relative order of two distinct elements (e.g. the increasing order for 0). Stolen bits can be used pretty much as the normal ones. The important difference is that the costs of their use have to be carefully accounted in the comparison complexity (e.g. reading a word of stolen bits costs $O(\log n)$ comparisons in the worst case) and the move complexity (e.g. modifying a word of stolen bits could costs $O(\log n)$ moves in the worst case or $O(1)$ in amortized sense if the word is used as a binary counter [17]). The elements that back up the stolen bits and the other elements are divided and conquered with the same simple process used for the internal buffer.

The third tool we use is an *in-place linear time two-way merging*, any algorithm among the ones we briefly reviewed in Section 1 will be good for our purposes.

The fourth tool is the sorting algorithm in [1]. Following our new approach of *merging by sorting* we will use this algorithm to solve proper sub-problems of the main one.

The fifth tool is the well-known, basic technique for space-efficient block exchange. From a block $X = x_1 \ldots x_t$ of t consecutive elements we can obtain the reverse $X^R = x_t \ldots x_1$ in linear time and in-place simply exchanging x_1 with x_t, x_2 with x_{t-1} and so forth. Two consecutive blocks X and Y, possibly of different sizes, can be exchanged in-place and in linear time with three block reversals, since $YX = (X^R Y^R)^R$.

Some of the obstacles posed by the problem. At first sight, we could think that the main technique used in [1] could be used also with this problem. In that paper, the authors essentially used a dictionary that is searchable in $O(\log n)$ comparisons and updatable in $O(1)$ data moves in amortized sense. There were two major obstacles. First, encoding the auxiliary data used by the dictionary using stolen bits so that the decoding did not penalize the search. Second, embedding the dictionary into a large pool of buffer elements in order to achieve the space-optimality while maintaining the update in $O(1)$ data moves. Using those techniques to overcome the space inefficiency of the third sub-optimal approach we mentioned in Section 1 might seem to be the right way.

Unfortunately, that approach requires that any element inserted in the dictionary is coupled with the index of the sorted sequence it belonged to. That is unavoidable: when the minimum element x is removed from the dictionary, the new inserted element has to be the successor of x in the original sorted sequence. In the general case there is no way to predict which sorted subsequence the extracted element came from or in what position of the dictionary the new element will be inserted. Therefore, any element passing through the dictionary would have to be charged with $O(\log s)$ moves for the encoding of the index of its subsequence of origin.

There is another aspect of the solution in [1] that is not suitable for a direct use in the merging problem. Both the set of pairs of distinct elements for the bit stealing and the set of buffer elements are collected using *selection and partitioning algorithms*. Those algorithms disrupt the s sorted sequences in input, thereby nullifying any effort to exploit the pre-sortedness of the elements. As we will see, we are forced to use a smaller amount of buffer elements ($O(n/polylog(n))$ elements against the $O(n)$ elements used in [1]) in order not to compromise the pre-sortedness of the input. In the full version of this paper we will adapt the solution for the case of repeated elements. In that case we will have to use $O(n^\epsilon)$ stolen bits against the $O(n/\log n)$ used in [1].

Basic ideas. The starting intuition is that there has to be a different way to accomplish the task for anyone of three particular cases: (i) when s is $O(n^\epsilon)$ and $\Omega(polylog(n))$; (ii) when s is very small, that is $s = O(polylog(n))$; (iii) when s is very large, that is $s = \Omega(n^\epsilon)$.

The way we will solve the first case is just what the approach *merging by sorting* is all about: breaking the merging problem into sub-problems that can be easily solved with sorting algorithms. The merging problem will be divided into

(i) one sorting sub-problem of size $O(n/s)$ but with macro-elements of size s,
(ii) $\frac{n}{s^2}$ sorting sub-problems of size s^2 each and
(iii) $\frac{n}{s^2}$ binary merging sub-problems again of size s^2.

The details will be given in Section 3.1.

In the second case, it would be reasonable to think that the solution could be found extending the usual approach of two-way merging to the case of $polylog(n)$-way merging. The major obstacle to overcome consists in the fact that when $s = \omega(1)$ there does not seem to be a way to use less than $O(n)$ buffer elements. As we mentioned in Section 1, that is what prevents the sorting algorithm in [6] from being also a full-optimal solution for the merging problem when $s = O(polylog(n))$. As we will see, a way to solve the problem in this second range of values of s consists in *breaking the sorted subsequences* into pieces as if they were linked lists. That kind of technique could have also been used to make the algorithm in [6] stable.

The third case can be seen as the simple base case of the merging by sorting approach, The natural intuition is that s is so large that the solution for the problem has to be more similar to a sorting algorithm than a classical merging algorithm in which, at any step, the currently smallest element among the sorted subsequences is selected. As a matter of fact, since we are interested in asymptotic optimality up to a constant factor, *for any fixed constant ϵ* this base case can be solved simply applying the full-optimal sorting algorithm in [1].

3 Merging by Sorting

Let R be the input sequence of n elements and $R_1, R_2, R_3 \ldots R_{s-2}, R_{s-1}, R_s$ be the s sorted sequences composing R. We will distinguish among three main ranges of values for the parameter s and the solutions for these three cases will be given in the next sections.

For the sake of simplicity, in any of these sub-sections we will first describe the solution assuming s sorted sequences of the same length. After that, we will point out in the proof the necessary changes for sorted sequences of generic lengths. As we will see, these changes are very simple for the second range of values of s and almost null for the first and last ones.

3.1 What if $\log^2 \bullet \leq \bullet \leq \bullet^\epsilon$?

As we will see soon, we are going to need some "simulated resources". We need $O(\frac{n}{s} \log n) = O(\frac{n}{\log n})$ stolen bits. We can collect the $O(\frac{n}{s} \log n)$ pairs of distinct elements simply taking the first $O(\log n)$ sorted subsequences $R_1, R_2, \ldots$. We need also $O(\frac{n}{s})$ buffer elements that can be collected in the same way.

Let t be the index of the first remaining sorted subsequence. Since $s - t = \Theta(s)$, for the sake of simplicity we are going to pretend that, after collecting stolen bits and buffer elements, we are left with two new objects. These are the sequences B and E containing the wanted buffer elements and pairs of encoding elements, respectively. All the objects so far introduced are initially laid out in memory in the following way: $EBR_1R_2R_3 \ldots R_{s-2}R_{s-1}R_s$.

Let us consider each sorted subsequence as logically divided into n/s^2 *blocks of s contiguous elements each*. We have four main phases.

(i) We sort all the blocks according to their first elements (since any block is already sorted internally, its first element is also its smallest one). We can use the mergesort with the in-place linear time two-way merging algorithm we chose in Section 2. However, in order to stay within our resource bounds, we must sort only the set of the first elements of the blocks. Hence, we collect this set exchanging each one of its members with one buffer element in B. Finally, to maintain the connection between the set of the first elements of the blocks and the remaining parts of the blocks, we associate a pair of back and forward encoded pointers (of $O(\log n)$ stolen bits each) to each pair *[first element, remaining elements of the block]*. This encoded information can be easily maintained up to date during the execution of mergesort and subsequently used to bring the blocks in sorted order in only $O(\frac{n}{s})$ "block moves".

(ii) Let $B_1 B_2 \ldots B_{\frac{n}{s}-1} B_{\frac{n}{s}}$ be the sorted sequence of the blocks after the first step. Let us logically form $\frac{n}{s^2}$ groups $G_1, G_2, \ldots, G_{\frac{n}{s^2}}$ of s contiguous blocks each. In this phase we sort each group G_i using the algorithm in [1].

(iii) Let $G'_1 G'_2 \ldots G'_{\frac{n}{s^2}}$ be the sequence of sorted groups we obtained after the second phase. In this phase, we apply the in-place two-way merging algorithm we chose in Section 2 in a left-to-right "chained" fashion starting with the two first sorted groups G'_1, G'_2 then with G'_2, G'_3, $G'_3, G'_4 \ldots$ and so forth until the pair $G'_{\frac{n}{s^2}-1}, G'_{\frac{n}{s^2}}$ is merged.

(iv) We sort the buffer elements and the elements used to steal bits. Then we merge that sorted sequence with the one obtained after the execution of the third phase.

Lemma 1. *For any $\log^2 n \leq s \leq n^\epsilon$ and for any set of s sorted sequences containing a total of n elements (drawn from $\mathcal{U}$), we can compute the whole sorted sequence with $O(n \log s)$ comparisons, $O(n)$ moves and $O(1)$ auxiliary cells of memory.*

Proof sketch: In the first phase, we use the normal in-place binary mergesort over the set of the first elements of the blocks. That alone would cost $O(n/s) = o(n)$ comparisons and moves. However, at any basic step of the binary mergesort we have to decode and re-encode a constant number of pointers with $O(\log n)$ stolen bits each, for a total cost of $O((n/s) \log n)$ moves and comparisons. That is $O(n)$ by the hypothesis over the values of s. The final permutation of the blocks costs $O(n)$ moves and $O((n/s) \log n) = o(n)$ comparisons. In the second phase $O(\frac{n}{s^2})$ groups of s^2 elements each are internally sorted using the full-optimal sorting algorithm in [1]. For each group, $O(s^2 \log s)$ comparisons and $O(s^2)$ moves are spent, for a total of $O(n \log s)$ comparisons and $O(n)$ moves. The third phase exploits a combinatorial property that is a generalization of the one introduced in [2]. Basically, after the second phase, any element in the sequence $G'_1 G'_2 \ldots G'_{\frac{n}{s^2}}$ may be at most one group (excluding its own one) *above* its final position in the corresponding sorting sequence. Therefore, the "chained", left-to-right application of the binary in-place merging yields the sorted sequence in $O(n)$ moves and comparisons. Finally, the fourth phase easily conquers the sub-problem for the elements used to simulate the resources. If the s sorted subsequences have generic lengths, we have to add a simple pre-processing phase executed before the four phases we described. This additional phase is needed in order to ensure the initial assumption about the presence of n/s sorted blocks of s contiguous elements each. The additional phase is a scan of the sequence in input. Starting from the left end of the sequence we consider s contiguous elements at a time. If they are in sorted order we go to the next s, otherwise we sort them. Since the input sequence is composed by s sorted subsequences, it is straightforward to prove that the pre-processing phase performs $O(n + s^2 \log s)$ comparisons and moves. The constant ϵ can be chosen so that the cost of the pre-processing phase is $O(n)$. After the pre-processing phase the algorithm continues with the remaining four phases unchanged.   □

3.2 What if • • $\log^2$ • ?

We are going to show how to solve the problem just for the case $s \leq \frac{\log n}{\log \log n}$. When $\frac{\log n}{\log \log n} < s < \log^2 n$, we can solve the problem simply applying iteratively the solution for the case $s = \frac{\log n}{\log \log n}$ as if we were sorting the sequence by mergesort (a $\frac{\log n}{\log \log n}$-way mergesort algorithm). Let $g = \frac{\log n}{\log \log n}$. The g-way mergesort would scan the n elements $O(\log s/ \log g)$ times, performing a total of $O((\log s/ \log g)n \log g) = O(n \log s)$ comparisons and $O((\log s/ \log g)n) = O(n)$ moves.

Breaking the subsequences. Let us divide any sorted subsequence R_i into p *contiguous blocks* of size $\log^2 n$. The jth block of R_i will be denoted by R_i^j. Block R_i^1 contains the smallest $\log^2 n$ elements of R_i, block R_i^2 the second smallest $\log^2 n$ elements and so forth. Initially the blocks of R_i are laid out contiguously and in sorted order, (i.e. $R_i = R_i^1 R_i^2 \ldots R_i^{p-1} R_i^p$).

The blocked subsequences can be naturally seen as s doubly-linked lists of p macro-elements each (the blocks). In the following we will freely refer to a generic R_i as a list or as a sorted subsequence. As we will see, the introduction of those simple lists will be of great help. By allowing the possibility of merging s linked lists of sorted elements instead of s unbreakable sorted sequences, the need for buffer elements drops from $O(s \times n/s) = O(n)$ units to $O(s \times \text{block-size}) = O(s \log^2 n)$ only.

As in the previous case, some $(o(n))$ elements will be devoted to placeholding or encoding duties. However, this time the numbers are slightly different, especially for what concern the buffer elements. The quantities of stolen bits and buffer elements we have to collect depend on the lists:

- Since the blocks are part of doubly-linked lists that will be scattered in the n locations of memory, they are going to need *succ* and *pred* pointers of $O(\log n)$ bits each.
- We are going to iteratively extract the smallest elements of lists. We will need $\log^2 n$ buffer elements for any list so that the extraction can be reduced to an exchange with a buffer element.

Therefore, we need $O(\frac{n}{\log n})$ stolen bits and $s \log^2 n$ buffer elements. The corresponding elements can be collected in the same way we did in Section 3.1. As we already did in that section, we are going to pretend for the sake of presentation that the number of sorted subsequences to be processed is still s and that we are

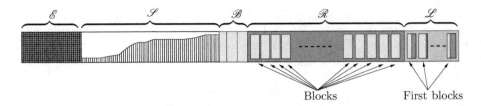

Fig. 1. The memory layout

left with two new objects: the sequences B and E containing the wanted buffer elements and the encoded bits. All the objects so far introduced are initially laid out in memory in the following way: $EBR_1R_2R_3\ldots R_{s-2}R_{s-1}R_s$.

Zones and their invariants. Now we will describe the layout of the memory right before the merging phase begins and the invariants defined over the layout that will be maintained during the computation. During the merging phase, the n cells of memory are partitioned into *five zones*. We will list them from the leftmost to the rightmost one (see Fig. 1).

The *encoding zone* $\mathcal{E}$. It contains the elements for stealing bits. This zone is static. During the merging phase there will be a lot of activity here, due to the continuous execution of encoding/decoding-related comparisons and moves. However, the boundaries of that zone will never change.

The *sorted zone* $\mathcal{S}$. At any time during the merging phase, it contains the $|\mathcal{S}|$ smallest elements (with the exception of buffer elements and the ones for stealing bits, of course) in sorted order.

The *buffer zone* $\mathcal{B}$. At any time of the merging phase, it contains a subset of the collected buffer elements. Initially $\mathcal{B}$ contains all the buffer elements but the extractions from the lists will make its size shrink or enlarge during the merging phase. Moreover, that zone will move toward the right end of the memory during the whole merging phase, because of the "pressure" by the ever-growing zone $\mathcal{S}$. At the end of the computation $\mathcal{B}$ will have regained all the buffer elements and will end up at the right end of the memory.

The *block zone* $\mathcal{R}$. At any time of the merging phase and for any list R_i, it will contain the currently remaining blocks of R_i with the exception of the first (the one with the smallest elements). Since the merging process will continuously take away batches of elements from the tops of the lists, the meaning of "currently remaining blocks" of a list should be clear. The left boundary of $\mathcal{R}$ will move toward right. At the end of the computation, this zone will be empty, given that the objective of the merging phase is to have all the elements of the lists transferred into $\mathcal{S}$ in sorted order.

The *leading zone* $\mathcal{L}$. This zone has fixed boundaries and comprises $s\log^2 n$ locations. At any time during the merging phase, the remaining elements of the first block of any list R_i are stored contiguously, in sorted order and padded with a sufficient number of buffer elements (i.e. if l is the number of the remaining elements of the first block of list R_i then the elements are laid out as $f_1f_2\ldots f_{\log^2 n - l}r_1r_2\ldots r_l$, where the r_is are the element of R_i). The leading zone will be at the center of the merging process since it is in that zone that the s-way choices will be made.

Merging phase. We have the problem of how to keep track of the position of the smallest element of any list. We are treating the case $s \leq \frac{\log n}{\log \log n}$ and therefore we cannot count on a lower bound for s. That excludes completely any solution involving encoding by bit stealing the value of any such pointer (the decoding would cost more than the wanted $O(\log s)$ comparisons). Since we want to maintain all these elements trapped into the leading zone $\mathscr{L}$, each pointer to one of them needs only $O(\log |\mathscr{L}|) = O(\log(s \log^2 n)) = O(\log \log n)$ bits. Since $s \leq \frac{\log n}{\log \log n}$, we can maintain a small balanced tree of such small pointers into a *constant number of auxiliary cells* of memory (which we are allowed to use). For any $1 \leq i \leq s$, together with the small pointer to the smallest element belonging to the list R_i, we are going to maintain also a small integer with the value of the number of buffer elements in the block of R_i currently contained in $\mathscr{L}$ ($O(\log \log n)$ bits are needed in that case too). A similar approach has been used also in [6] and is the combination of two classic basic techniques: integer packing and merging by selection tree.

The tree will be used to guide the iterative selection of the currently smallest element. In the merging phase the following steps will be executed until all the elements of the lists end up in sorted order in zone $\mathscr{S}$.

1. Find the smallest element among the ones contained into the leading zone $\mathscr{L}$.
2. Exchange this element with the first (leftmost) element of the buffer zone $\mathscr{B}$. (This implicitly enlarge and shrink by one position the sorted zone $\mathscr{S}$ and the buffer zone $\mathscr{B}$, respectively)
3. If the block in $\mathscr{L}$ corresponding to the just exchanged element, now contains only buffer elements, we load into $\mathscr{L}$ the next block in its list.

Lemma 2. *For any $s < \log^2 n$ and for any set of s sorted sequences containing a total of n elements (drawn from $\mathcal{U}$), we can compute the whole sorted sequence with $O(n \log s)$ comparisons, $O(n)$ moves and $O(1)$ auxiliary cells of memory.*

Proof sketch: The costs of bringing the zones in their initial state before the merging phase are within our target bounds. Basically, we have to do s block exchanges to move $R_1^1, R_2^1 \ldots, R_{s-1}^1, R_s^1$ into the last $s \log^2 n$ locations (that is into the leading zone $\mathscr{L}$). The total cost is $O(s \log^2 n)$ moves for the block exchanges plus $O(s \log n)$ moves for updating the *succ* and *pred* pointers of any block involved in the exchange.

Now we have to consider the costs of the merging phase. In step 1 we use the small tree to find the smallest element in $\mathscr{L}$. The small tree has s nodes, its pointers are completely contained into a constant number of auxiliary locations and is fully balanced. Therefore searching and updating tre tree costs $O(\log s)$ comparisons. Step 2 is a mere exchange of elements (we can maintain the starting location of the five zones into as many auxiliary locations). Finally, step 3 consists in an access to the small tree, two block exchanges (we first exchange the block b in $\mathscr{L}$ with the next one in its list, then we exchange again the block b, now in $\mathscr{R}$, with the first block in $\mathscr{R}$, thus enlarging and shrinking of $\log^2 n$ positions the sizes, respectively, of $\mathscr{B}$ and $\mathscr{R}$) and the update of $O(\log n)$ stolen bits (linked lists informations). Since this operation can be charged on $\Omega(\log^2 n)$ steps gone without block transfers, the amortized costs of step 3 is $O(1)$ comparisons and moves for any moved element.

The changes for the case of sorted subsequences of generic lengths are minimal and straightforward. That is because the algorithm is already capable to manage the fact that during the evolution of the merging process the remainders of the sorted subsequences are going to differ in length. Therefore, starting from an initial input sequence with sorted subsequences of generic length is just a special case of the common situations managed during the merging process.   □

3.3   What if • • • $^\epsilon$?

For any fixed real constant $\epsilon < 1$, if $s > n^\epsilon$ then the problem can be easily solved by applying the sorting algorithm in [1] to the whole input sequence R, completely ignoring its pre-sortedness. We would like to point out that this can be seen as a *base case* of the *merging by sorting* approach in which the sorting subroutine can solve the main problem of merging by itself. Similarly, for the dual companion of the *sorting by merging* approach, the particular case in which there are only a constant number of sorted sub-sequences in the input sequence can be solved directly by the merging subroutine.

Lemma 3. *For any $n^\epsilon < s \leq n$ and for any set of s sorted sequences containing a total of n elements (drawn from $\mathcal{U}$), we can compute the whole sorted sequence with $O(n \log s)$ comparisons, $O(n)$ moves and $O(1)$ auxiliary cells of memory.*

Proof. In [1], it has been proven that there exists a full-optimal solution for the sorting problem using $O(n \log n)$ comparisons, $O(n)$ moves and $O(1)$ auxiliary cells. If $s > n^\epsilon$, for a fixed real constant ϵ, that solution is also a full-optimal solution for the balanced merging problem. Since we care about the parameter s only, the case of sorted subsequences with generic lengths has the same solution.

Finally, we can conclude that Theorem 1 is proven by Lemmas 1, 2 and 3.

References

1. Franceschini, G., Geffert, V.: An In-Place Sorting with $O(n \log n)$ Comparisons and $O(n)$ Moves. Journal of the ACM **52** (2005) 515–537
2. Kronrod, M.A.: Optimal ordering algorithm without operational field. Soviet Math. Dokl. **10** (1969) 744–746
3. Levcopoulos, C., Overmars, M.H.: A balanced search tree with $O(1)$ worst-case update time. Acta Informatica **26**(3) (1988) 269–277
4. Andersson, A., Lai, T.W.: Comparison–efficient and write–optimal searching and sorting. In Hsu, W.L., Lee, R.C.T., eds.: Proceedings of International Symposium on Algorithms (ISA '91). Volume 557 of LNCS., Berlin, Germany, Springer (1991) 273–282
5. Fleischer, R.: A simple balanced search tree with $O(1)$ worst-case update time. Lecture Notes in Computer Science **762** (1993) 138–146
6. Katajainen, J., Pasanen, T.: In-place sorting with fewer moves. Inform. Process. Lett. **70** (1999) 31–37
7. Horvath, E.C.: Stable sorting in asymptotically optimal time and extra space. Journal of the ACM **25**(2) (1978) 177–199

8. Pardo, L.T.: Stable sorting and merging with optimal space and time bounds. SIAM Journal on Computing **6**(2) (1977) 351–372
9. Salowe, J., Steiger, W.: Simplified stable merging tasks. Journal of Algorithms **8**(4) (1987) 557–571
10. Huang, B.C., Langston, M.A.: Practical in-place merging. Communications of the ACM, CACM **31**(3) (1988) 348–352
11. Huang, B.C., Langston, M.A.: Stable set and multiset operations in optimal time and space. In ACM, ed.: PODS '88. Proceedings of the Seventh ACM SIGACT-SIGMOD-SIGART Symposium on Principles of Database Systems: March 21–23, 1988, Austin, Texas, New York, NY 10036, USA, ACM Press (1988) 288–293
12. Symvonis, A.: Optimal stable merging. Comput. J. **38** (1995) 681–90
13. Geffert, V., Katajainen, J., Pasanen, T.: Asymptotically efficient in-place merging. Theoret. Comput. Sci. **237** (2000) 159–81
14. Chen, J.: Optimizing stable in-place merging. Theor. Comput. Sci. **1-3**(302) (2003) 191–210
15. Estivill-Castro, V., Wood, D.: A survey of adaptive sorting algorithms. ACM Comp. Surveys **24**, **4** (1992) 441–476
16. Munro, J.: An implicit data structure supporting insertion, deletion, and search in $O(\log^2 n)$ time. J. Comput. System Sci. **33** (1986) 66–74
17. Cormen, T.H., Leiserson, C.E., Rivest, R.L., Stein, C.: Introduction to Algorithms. MIT Press (2001)

Finding the Position of the •-Mismatch and Approximate Tandem Repeats

Haim Kaplan[1], Ely Porat[2], and Nira Shafrir[1]

[1] School of Computer Science, Tel Aviv University, Tel Aviv 69978, Israel
{haimk, shafrirn}@post.tau.ac.il
[2] Department of Mathematics and Computer Science, Bar-Ilan University,
Ramat-Gan 52900, Israel
porately@cs.biu.ac.il

Abstract. Given a pattern P, a text T, and an integer k, we want to find for every position j of T, the index of the k-mismatch of P with the suffix of T starting at position j. We give an algorithm that finds the exact index for each j, and algorithms that approximate it. We use these algorithms to get an efficient solution for an approximate version of the tandem repeats problem with k-mismatches.

1 Introduction

Let P be a pattern of length m and let T be a text of length n. Let $T(i, \ell)$ denote the substring of T of length ℓ starting at position i.[1] In the k-*mismatch problem* we determine for every $1 \leq j \leq n - m + 1$, if $T(j, m)$ matches P with at most k mismatches. In case $T(j, m)$ does not match P with at most k mismatches we compute the position $k(j)$ in P of the k-mismatch. In case $T(j, m)$ matches P with at most k mismatches we compute the position of the last mismatch if there is at least one mismatch.

Several classical results are related to the k-mismatch problem. Abrahamson [1], gave an algorithm that finds for each $1 \leq j \leq n - m + 1$, the number of mismatches between $T(j, m)$ and P. The running time of Abrahamson's algorithm is $O(n\sqrt{m \log m})$. Amir et. al. [2], gave an algorithm that for each $1 \leq j \leq n - m + 1$, determines if the number of mismatches between $T(j, m)$ and P is at most k. running time of this algorithm is $O(n\sqrt{k \log k})$. Both of these algorithms do not give any information regarding the position of the last mismatch or the position of the k-mismatch. This information is useful for applications that want to know not only if the pattern matches with at most k-mismatches, but also want to know how long is the prefix of the pattern that matches with at most k-mismatches.

The major technique used by the algorithms of Abrahamson and of Amir et. al. is convolution. Lets fix a particular character $x \in \Sigma$. Suppose we want to compute for every $1 \leq j \leq n - m + 1$, the number of places in which an x in P does not coincide with an x in T when we align P with $T(j, m)$. We can

---

[1] We always assume that $i \leq n - m + 1$ when we use this notation.

L. Arge and R. Freivalds (Eds.): SWAT 2006, LNCS 4059, pp. 90–101, 2006.

perform this task by computing a convolution of a binary vector $P(x)$ of length m, and a binary vector $T(x)$ of length n as follows. The vector $P(x)$ contains 1 in every position where P contains the character x and 0 in all other positions. The vector $T(x)$ contains 1 in every position where T does not contain x and 0 in every position where T contains x. We can perform the convolution between $P(x)$ and $T(x)$ in $O(n \log m)$ time using the Fast Fourier Transform. So if P contains only $|\Sigma|$ different characters we can count for each $1 \leq j \leq n - m + 1$, the number of mismatches between $T(j, m)$ and P in $O(|\Sigma| n \log m)$. We do that by performing $|\Sigma|$ convolutions as described above, one for each character in P, and add up the mismatch counts.

There is a simple deterministic algorithm for the k-mismatch problem that runs in $O(nk)$ time and $O(n)$ space of Landau and Vishkin [8]. They construct a suffix tree for the text and the pattern, with a data structure for lowest common ancestor (LCA) queries, to allow constant-time jumps over equal substrings in the text and pattern. The algorithm of Landau and Vishkin finds for each j the position of the k-mismatch (or the last mismatch if there are less than k mismatches) between $T(j, m)$ and P in $O(k)$ time. It does that by performing at most k LCA queries on the appropriate substrings of the text and the pattern. We give an alternative algorithm that runs in $O(nk^{\frac{2}{3}} \log^{1/3} m \log k)$ time and linear space.

To see why the bound of $O(nk^{\frac{2}{3}} \log^{1/3} m)$, may be natural, consider a pattern of length $m = O(k)$. In this case, we can solve the problem using the method of Abrahamson [1]. We divide the pattern into $k^{\frac{1}{3}} / \log^{1/3} k$ blocks, each block of size $z = O(k^{\frac{2}{3}} \log^{1/3} k)$. By applying the algorithm of Abrahamson with the first block as the pattern, we determine in $O(n\sqrt{z \log z}) = O(nk^{\frac{1}{3}} \log^{2/3} k)$ time, the number of mismatches of each text location with the first block. Similarly, by applying the method of Abrahamson to each of the subsequent $k^{\frac{1}{3}} / \log^{1/3} k$ blocks of the pattern, and accumulating the number of mismatches for each text position, we know in $O(nk^{\frac{2}{3}} \log^{1/3} k)$ time for each text position, which block contains the k-mismatch. Moreover we also know for each text position the number of mismatches in the blocks preceding the one that contains the k-mismatch. With this information, we can find for each text position the k-mismatch in the relevant block in $O(k^{\frac{2}{3}} \log^{1/3} k)$ time by scanning the block character by character looking for the appropriate mismatch. It is not clear how to get a better bound even for this simple example.

We also define the *approximate k-mismatch problem*. This problem have an additional accuracy parameter ϵ. The task is to determine for every $1 \leq j \leq n - m + 1$ a position $k(j)$ in P such that the number of mismatches between $T(j, k(j))$ and $P(1, k(j))$ is at least $(1 - \epsilon)k$ and at most $(1 + \epsilon)k$, or report that there is no such position.

We give a deterministic and randomized algorithms for the *approximate k-mismatch problem*. We describe the deterministic algorithm in Section 3. The running time of this algorithm is $O((n/\epsilon^3)\sqrt{k} \log^3 m)$. In Sect. 4, we give a randomized algorithm with running time of $O(\frac{n}{\epsilon^2} \log n \log^3 m \log k)$. The randomized algorithm guarantees that for each j the number of mismatches between

$T(j, k(j))$ and $P(1, k(j))$ is at least $(1 - \epsilon)k$ and at most $(1 + \epsilon)k$ with high probability.[2]

A position $k(j)$ computed by our algorithms for the *approximate k-mismatch problem* may not contain an actual mismatch. That is, the character $k(j)$ of P may in fact be the same as character $j + k(j) - 1$ of T. We can change both algorithms such that $k(j)$ would always be a position of a mismatch in O(n) time as follows. For a string S we denote by S^R the string obtained by reversing S. We build a suffix tree for T^R and P^R, with a data structure for lowest common ancestor (LCA) queries in constant time. For each position j in T we perform an LCA query for the suffixes $(P(1, k(j)))^R$ of P^R and $(T(1, j + k(j) - 1))^R$ of T^R. Let h be the string depth of the resulting node. Clearly h is the length of the longest common prefix of $(P(1, k(j)))^R$ and $(T(1, j + k(j) - 1))^R$, and $k(j) - h$ is the position of the last mismatch between P and $T(j, m)$ prior to position $k(j)$. We change $k(j)$ to $k(j) - h$.

In Sect. 5, we use our algorithms for the k-mismatch problem to solve an approximate version of the k-mismatch tandem repeats problem. The *exact tandem repeats problem* is defined as follows. Given a string S of length n, find all substrings of S of the form uu. Main and Lorentz [9] gave an algorithm that solves this problem in $O(n \log n + z)$ time, where z is the number of tandem repeats in S. Repeats occur frequently in biological sequences, but they are usually not exact. Therefore algorithms for finding approximate tandem repeats were developed. The *k-mismatch tandem repeats problem* is defined as follows. Given a string S and a parameter k find all substrings uv of S such that $|u| = |v| > k$ and the number of mismatches between u and v is at most k. The best known algorithm for this problem is due to Landau, Schmidt and Sokol [7] and it runs in $O(nk \log(n/k) + z)$ time, where z is the number of k-mismatch tandem repeats.

We define the *approximate k-mismatch tandem repeats problem* which is a relaxation of the k-mismatch tandem repeats problem. In this relaxation we require that the algorithm will find all substrings uv of S such that $|u| = |v| > k$ and the number of mismatches between u and v is at most k, but we also allow the algorithm to report substrings uv such that the number of mismatches between u and v is at most $(1 + \epsilon)k$. Using our algorithm for the k-mismatch problem we get an algorithm for approximate k-mismatch tandem repeats that runs in $O((n/\epsilon)k^{\frac{2}{3}} \log^{1/3} n \log k \log(n/k) + z)$ time. Using our deterministic algorithm for the approximate k-mismatch problem we get an algorithm for approximate k-mismatch tandem repeats that runs in $O((n/\epsilon^4)\sqrt{k} \log^3 n \log(n/k) + z)$ time. We can also use the randomized algorithm of Sect. 4 and get an algorithm that reports all k-mismatch tandem repeats with high probability, and possibly tandem repeats with up to $(1+\epsilon)k$ mismatches in $O(\frac{n}{\epsilon^3} \log^3 n \log k \log(n/k) + z)$ time.

Preliminaries: A string s is *periodic* with period u, if $s = u^j w$, where $j \geq 2$ and w is a prefix of u. The *period* of s is the shortest substring u such that $s = u^j w$ and w is a prefix of u.

---

[2] By high probability we mean probability that is polynomially small in n.

A *break* of s is an aperiodic substring of s. An ℓ-*break* is a break of length ℓ. We choose a parameter $\ell < k$ (the value of ℓ will be decided later). We use the method of [3] to find a partition of the pattern into ℓ-breaks separated by substrings shorter than ℓ, or periodic substrings with period of length at most $\ell/2$. We call the substrings that separate the breaks *periodic stretches*.

In Sect. 2, we show how to solve the k-mismatch problem when the pattern P contains at most $2k$ ℓ-breaks in the time and space bounds mentioned above. In case the pattern P contains more than $2k$ ℓ-breaks, we reduce it to the case where P contains $2k$ ℓ-breaks as follows.

Assume P contains more than $2k$ ℓ-breaks and let P' be the prefix of P with exactly $2k$ ℓ-breaks. We run our algorithm using P' rather than P. Our algorithm also finds all positions in T that match P' with at most k mismatches. Amir et. al. [2] proved that at most n/ℓ positions of the text T match P' with at most k mismatches. After running our algorithm and finding these positions we use the algorithm of Landau and Vishkin [8] to check whether each of these positions matches the original pattern P with at most k mismatches, and to find the location of the k-mismatch in case it does not. The total time it takes to check all of these positions is $O(nk/\ell)$. Therefore we assume from now on that the pattern P contains at most $2k$ ℓ-breaks, and that the running time of our algorithm is $\Omega(nk/\ell)$.

2   Finding the Position of the •-Mismatch

We describe an algorithm that solves the problem in $O(nk^{\frac{3}{4}} \log^{1/4} m)$ time and $O(n)$ space. In the full version of this paper we show how to add another level of recursion to this algorithm and get an algorithm whose running time is $O(nk^{\frac{2}{3}} \log^{1/3} m \log k)$ and uses $O(n)$ space.

Recall that we assume that the pattern contains $O(k)$ breaks, which are substrings of length at most ℓ, and at most $2k$ periodic stretches. Let A be a periodic stretch let x be its period, $|x| \le \ell/2$. Let x' be the lexicographically first cyclic rotation of x. We call x' the *canonical period* of A. We can write $A = yx'^i z, i \ge 0$, where y is a prefix of x, (y may be empty), and z is a prefix of x' which may be empty. Let $A' = x'^i$. We add y and z to the set of breaks. We redefine the term *break* to include also the above substrings. The string A' is the new periodic stretch. We added to the set of breaks a total of $O(k)$ substrings each of length at most ℓ. After this preprocessing, the set of all different periods of the periodic stretches of the pattern contains only canonical periods, and thus it doesn't contain two periods that are cyclic rotations one of the other. In addition, all periodic stretches with period u are of the form $u^i, i > 0$.

Choosing a prefix of the pattern: We now show how to choose a prefix S of the pattern for which we can find the position of the k-mismatch with $T(j, |S|)$ or determine that S matches $T(j, |S|)$ with less than k-mismatches. We also prove that S cannot match $T(j, |S|)$ with at most k-mismatches in too many positions j. We assume that P contains $O(k)$ breaks, which are substrings of length at most ℓ, and at most $2k$ periodic stretches. All periodic stretches are of the form u^i, where u is a canonical period. We partition each periodic stretch into segments of length ℓ. We ignore the segments that are not fully contained in a periodic stretch.

Let S be the shortest prefix of P that satisfies at least one of the following criteria, or P itself if no prefix of P satisfies at least one of these criteria.

1. S contains a multiset A of $2k$ segments of periodic stretches, such that at most k/ℓ are of the same canonical period.
2. S contains a multiset of $2k$ characters in which each character appears at most k/ℓ times.

We use the following definitions. Let C be the set of canonical periods of the periodic stretches in S. We define a period $u \in C$ to be to *frequent* in S, if there are more than k/ℓ segments in the above partition with period u and *rare* otherwise. Similarly, we define a character to be *frequent* in S, if it appears more than k/ℓ times in S, and *rare* otherwise. The prefix S has the following properties.

1. C contains at most 2ℓ frequent periods. If C contains more than 2ℓ frequent periods, then we can obtain a shorter S satisfying (1) by taking the shortest prefix that contains k/ℓ segments of each of exactly 2ℓ frequent periods. By a similar argument, the total number of segments of periodic stretches that belong to rare periods in S is at most $2k$.
2. S contains at most 2ℓ frequent characters. Furthermore, the total number of occurrences of rare characters in S is at most $2k$.

We add to the set of breaks all rare periodic stretches. By property 1 we added $O(k)$ breaks of length at most ℓ. Following these changes, S contains $O(k)$ breaks. The set C of periods of the periodic stretches is of size $O(\ell)$.

Finding the position of the k-mismatch in S: Next we show how to find the position of the k-mismatch of each location of the text T with a prefix S of the pattern chosen as in Sect. 2. Recall that S contains $O(k)$ breaks and at most $2k$ periodic stretches, and satisfies Properties 1 and 2.

We partition the pattern into at most $O(k/y)$ *substrings* each contains at most y breaks, at most y rare characters and at most y periodic stretches. First we compute for each text position j the substring $W(j)$ of P that contains the k-mismatch of P with $T(j, m)$, or determine that P matches $T(j, m)$ with less than k-mismatches.

To do that we process the substrings sequentially from left to right, maintaining for each text position j the cumulative number of mismatches of the text starting at position j with the substrings processed so far. We denote this cumulative mismatch count of position j by $r(j)$. Let the next substring W of P that we process start at position i of the pattern. For each text position j, we compute the number of mismatches of $T(j, |W|)$ with W and denote it by $c(j)$. (We show below how to do that.) Then, for each text position j for which we haven't yet found the substring that contains the k-mismatch, we update the information as follows. If $r(j) + c(j + i) < k$, we set $r(j) = \dot{r}(j) + c(j + i)$. Otherwise, $r(j) + c(j + i) \geq k$, and we set $W(j)$ to be W.

We now show how to find the number of mismatches between a substring W of S and $T(j, |W|)$ for every $1 \leq j \leq n - |W| + 1$. We do that by separately counting the number of mismatches between occurrences of frequent characters in W and the corresponding characters of $T(j, |W|)$, and the number of mismatches between occurrences of rare characters in W and the corresponding characters of $T(j, |W|)$. Then we add these two counts.

By Property 2, W contains at most 2ℓ frequent characters. For each frequent character x we find the number of mismatches of the occurrences of x in W with the corresponding characters in $T(j, |W|)$ for all j, by performing a convolution as described in the introduction. We perform $O(\ell)$ convolutions for each of the $O(k/y)$ substrings, so the total time to perform all convolutions is $O((k/y)\ell n \log m)$.

It remains to find the number of mismatches of rare characters in W with the corresponding characters in $T(j, |W|)$. We do that using the algorithm of Amir et. al. [2]. This algorithm counts the number of mismatches of a pattern which may contain don't care symbols with each text position. The running time of this algorithm is $O(n\sqrt{g \log m})$, where g is the number of characters in the pattern that are not don't cares. We run this algorithm with a pattern which we obtain from W by replacing each occurrence of a frequent character by a don't care symbol, and the text T. We obtain for each j the number of mismatches between rare characters in W and the corresponding characters in $T(j, |W|)$. Since W contains at most y rare characters, the running time of this application of the algorithm of Abrahamson is $O(n\sqrt{y \log m})$. So for all $O(k/y)$ substrings this takes $O((k/y)n\sqrt{y \log m}) = O(n(k/y^{1/2})\sqrt{\log m})$ time.

We now show how to find the position of the k-mismatch within the substring $W(j)$ that contains it for each text position j. We assume that each substring contains y breaks and y periodic stretches. Each periodic stretch is of the form u^i, where $u \in C$, and $|C| \leq 2\ell$.

We begin by finding for each text position which periodic stretch or break contains the k-mismatch. We find it by performing a binary search on the periodic stretches and breaks in $W(j)$. We do the binary search simultaneously for all text positions j. After iteration h of the binary search, for each text position we focus on an interval of $y/2^h$ consecutive breaks and periodic stretches in $W(j)$ that contain the k-mismatch between $W(j)$ and the corresponding substring of $T(j, m)$. In particular after $\log y$ iterations, we know for each text position which periodic stretch or break contains the k-mismatch.

At the first iteration of the binary search we compute the number of mismatches in the first $y/2$ of the periodic stretches and breaks of $W(j)$. From this number we know if the k-mismatch is in the first $y/2$ breaks and periodic stretches or in the last $y/2$ breaks and periodic stretches of $W(j)$. In iteration h, let $I(j)$ be the interval of $y/2^h$ consecutive breaks and periodic stretches in $W(j)$ that contains the k-mismatch between $W(j)$ and the corresponding piece of $T(j, m)$. We compute the number of mismatches between the first $y/2^{h+1}$ breaks and periodic stretches in $I(j)$ and the corresponding part of $T(j, m)$. Using this count we know if to proceed with the first half of $I(j)$ or the second half of $I(j)$.

We describe the first iteration of the binary search. Subsequent iterations are similar. We count the number of mismatches in each of the first $y/2$ breaks in $W(j)$ and $T(j, m)$ by comparing them character by character in $y\ell/2$ time for a specific j, and $ny\ell/2$ total time. To count the number of mismatches in each of the first $y/2$ periodic stretches we process the different periods in C one by one. For each period $u \in C$ and each text position j we count the number of mismatches in periodic stretches of u among the first $y/2$ periodic stretches of $W(j)$. The sum of these mismatch counts over all periods $u \in C$ gives us the total number of mismatches in the first $y/2$ periodic stretches of $W(j)$ and $T(j, m)$ for every text position j.

Let $u \in C$. We compute the number of mismatches of u with each text location using the algorithm of Abrahamson [1] in $O(n\sqrt{\ell \log \ell})$ time. We build a data structure that consists of $|u|$ prefix sums arrays $A_i, i = 1, \cdots, |u|$, each of size $n/|u|$. We use these arrays to find the number of mismatches of periodic stretches of u among the first $y/2$ periodic stretches of $W(j)$ for all text positions j. The total size of the arrays is $O(n)$.

The entries of array A_i correspond to the text characters at positions β such that β modulo $|u| = i$ modulo $|u|$. The first entry of array A_i contains the number of mismatches between $T(i, |u|)$ to u that was computed by the algorithm of Abrahamson. Entry j in A_i contains the number of mismatches between $T(i, j|u|)$ and u^j. It is easy to see that based on entry $j-1$, entry j in A_i can be computed in $O(1)$ time. Suppose we need to find the number of mismatches of $T(i+j|u|, r|u|)$ with a periodic stretch u^r. The number of mismatches can be computed in $O(1)$ time given A_i. If $j = 0$, then the number of mismatches is $A_i[r]$. If $j > 0$, then the number of mismatches is $A_i[j + r] - A_i[j]$.

In each iteration of the binary search we repeat the procedure above for every $u \in C$. Since $|C| = O(\ell)$ we compute the number of mismatches of all periodic stretches in the first $y/2$ periodic stretches of $W(j)$ for all j, in $O(n\ell^{3/2}\sqrt{\log \ell})$ time. Summing up over all iterations the time of counting the number of mismatches within breaks and the time of counting the number of mismatches within periodic stretches, we obtain that the binary search takes $O(n\ell^{3/2}\sqrt{\log \ell} \log y) + O(ny\ell)$ time.

We now know for each text position which periodic stretch or break contains the position of the k-mismatch. If the k-mismatch is contained within a break we find it in $O(\ell)$ time by scanning the break character by character. If the k-mismatch is contained in a periodic stretch, then we find it as follows. For each $u \in C$ we build the $n/|u|$ prefix sum arrays A_i, as described above. We then compute the position of the k-mismatch, for all text position for which the k-mismatch occurs with a periodic stretch of period u. Given such text position, we perform a binary search on the appropriate prefix sum array to locate a segment of length $|u|$ within the periodic stretch that contains the k-mismatch. The binary search is performed on a sub-array of length at most $m/|u|$ in $O(\log m)$ time. At the end of the binary search, we found the segment of length $|u| < \ell$ that contains the k-mismatch, we search in this segment sequentially in $O(\ell)$ time to find the k-mismatch. We repeat this process for all the periods in C.

Summing over all stages we obtain that the total running time of the algorithm is $O((k/y)n\ell\log m) + O(n(k/y^{1/2})\sqrt{\log m}) + O(n\ell^{3/2}\sqrt{\log\ell}\log y) + O(ny\ell)$. The space used by the algorithm is $O(n)$.

To complete the analysis we prove in the full version of this paper that if S is not equal to P, then T contains at most n/ℓ positions that match S with at most k mismatches. In these cases we use the algorithm of Landau and Vishkin [8] to find the position of the k-mismatch (or the last mismatch if there are less than k-mismatches) of each of these positions with the pattern in $O(nk/\ell)$ time. We also recall that we have to take into account the overhead of $O(nk/\ell)$ time of the reduction in Sect. 1 to a pattern with at most $O(k)$ breaks and periodic stretches.

So if we add the extra $O(nk/\ell)$ overhead to the overall running time and choose ℓ and y to balance the expressions (and thereby minimize the running time) we get that $\ell = k^{1/4}/\log^{1/4} m, y = \sqrt{k\log m}$ and a running time of $O(nk^{3/4}\log^{1/4} m)$.

3   Approximate •-Mismatch

In this section we sketch how to obtain an algorithm for the approximate k-mismatch problem whose running time is $O(n(1/\epsilon^3)\sqrt{k}\log^3 m)$. The algorithm is similar to the algorithm of Sect. 2. The main difference is that instead of using convolutions or the algorithm of Abrahamson [1] (that uses convolutions), to count the number of mismatches of various parts of the pattern and the text, we use the algorithm of Karloff [6]. Given a pattern P and a text T, the algorithm of Karloff [6], finds for every text position $1 \le j \le n-m+1$, a number $g(j)$ such that $m(j) \le g(j) \le (1+\epsilon)m(j)$, where $m(j)$ is the exact number of mismatches between P and $T(j,m)$.

We choose a prefix S to satisfy the first of the two criteria of Sect. 2. We partition S into $O((1/\epsilon)k/y)$ substrings each containing at most ϵy breaks and at most ϵy periodic stretches. We use the algorithm of Karloff [6] to approximately count the number of mismatches of each text position and each substring of P in $O(n/\epsilon^2\log^3 m)$ time. Then we know for each j which substring of P contains the k-mismatch with $T(j,m)$. We then search within the substring by a binary search as in Section 2. Here we set $\ell = \sqrt{k}/\log k$, and $y = \sqrt{k}$, so $y\ell = k/\log k$, and therefore the total length of the breaks within each substring is at most $\epsilon y\ell = \epsilon k/\log k$. This allows us to ignore the breaks when looking for the position within a substring.

4   A Randomized Algorithm for Approximate •-Mismatch

We assume w.l.o.g. that the alphabet Σ consists of the integers $\{1, \cdots, |\Sigma|\}$. The algorithm computes signatures for substrings of the pattern and the text. These signatures are designed such that from the signatures of two strings we can quickly approximate the number of mismatches between the two strings. We construct a random string R of *sparsity* k by setting $R[i]$ to 0 with probability

$(1 - \frac{1}{k})$, and setting $R[i]$ to be a random integer with probability $\frac{1}{k}$, for every $i = 1, \cdots, |R|$. We choose a random integer from a space Π of size polynomial in n. For a string W and a random string R with sparsity k, we define the signature of W with respect to R as $Sig_k(W, R) = \sum_{i=1}^{|W|} W[i]R[i]$.

Let W_1 and W_2 be two strings of the same length. If W_1 and W_2 agree in all positions where $R[i] \neq 0$, then $Sig_k(W_1, R) = Sig_k(W_2, R)$. On the other hand, if W_1 and W_2 disagree in at least one position i where $R[i] \neq 0$, then $Sig_k(W_1, R) = Sig_k(W_2, R)$ with probability at most $\frac{1}{|\Pi|}$. Let us call the latter event a *bad event*. Our algorithm compares sub-quadratic number of signatures so by choosing Π large enough, we can make the probability that a bad event ever happens polynomially small. Therefore, we assume in the rest of the section that such event does not happen.

For $k \geq 2$ we define an algorithm A_k as follows. The input to A_k consists of a substring S of T and a substring W of P such that S and W are of the same length. Let y be the true number of mismatches between S and W. The algorithm A_k either detects that $y > 2k$, or detects that $y < k$, or returns an estimate y' of y. The algorithm A_k works as follows. Let $q = \frac{c}{\epsilon^2} \log n$ for some large enough constant c that we determine later, and let $b = |W| = |S|$. Algorithm A_k takes q random strings $R_1, \cdots, R_q$ of length b and sparsity k and compares $Sig_k(W, R_i)$ and $Sig_k(S, R_i)$ for $i = 1, \ldots, q$. Let z be the number of equal pairs of signatures. If $z \geq (1-\epsilon)q(1-\frac{1}{k})^{k/2}$ then A_k reports that the number of mismatches between S and W is smaller than k. If $z \leq (1 + \epsilon)q(1 - \frac{1}{k})^{3k}$ then A_k reports that the number of mismatches between S and W is greater than $2k$. Otherwise let y' be the largest integer such that $z \leq q(1-\frac{1}{k})^{y'}$. We then return y' as our estimate of y.

Using standard Chernoff bounds we establish that A_k satisfies the following properties with high probability. (Proof omitted from this abstract.)

1. If $y \leq k/2$ then A_k reports that the number of mismatches is smaller than k.
2. If $y \geq 3k$ then A_k reports that the number of mismatches is larger than $2k$.
3. If $k \leq y \leq 2k$ then A_k gives an estimate y' to y.
4. Whenever A_k gives an estimate y' of y then $(1-\epsilon)y \leq y' \leq (1+\epsilon)y$. (This can happen if $k/2 < y < 3k$ and happens with high probability if $k \leq y \leq 2k$.)

For $k < 2$ we build a generalized suffix tree for P and T. We use this suffix tree to check whether the number of mismatches between a substring of P and a substring of T is at most 2, and if so to find it exactly, by the method of Landau and Vishkin. We shall refer to this procedure as A_0.

We are now ready to describe the algorithm. To simplify the presentation, we assume that k is a power of 2. Our algorithm compares substrings of P and T, by comparing their signatures using the algorithm A_j, for some $j \leq k$ which is a power of two, and we always compare substrings of length which is a power of two. We prepare all signatures required by for these applications of A_j in a preprocessing phase using convolutions as follows.

For any 2^j, $0 \leq j \leq \lfloor \log m \rfloor$, and for any 2^i, $0 \leq i \leq \log k$, we generate independently at random $q = \frac{c}{\epsilon^2} \log n$ strings $R_1, \cdots, R_q$, of sparsity 2^i and length 2^j. For each random string R_l of length 2^j, we compute the signature of every substring of T of length 2^j with R_l by a convolution of T and R_l. We compute the signature of every substring of P of length 2^j with R_l by a convolution of P and R_l. We compute a total of $\frac{c}{\epsilon^2} \log n \log m \log k$ signatures in $O(\frac{n}{\epsilon^2} \log n \log^2 m \log k)$ time.

We find the approximated location of the k-mismatch of $T(j, m)$ with P by a binary search as follows. To simplify the presentation we assume that m is a power of 2 and we show in the full version of the paper how to handle patterns whose length is not a power of 2. We compute the approximate number of mismatches y', between $P(1, m/2)$ and $T(j, m/2)$. We find y' by performing a binary search on $A_j(P(1, m/2), T(j, m/2))$, for $j = 0, 2, 4, \cdots, k$. We first apply $A_{\sqrt{k}}(P(1, m/2), T(j, m/2))$, if $A_{\sqrt{k}}$ reports that the number of mismatches is smaller than $\sqrt{k}/2$ we repeat the process for $j = 0, 2, 4, \cdots \sqrt{k}/2$. If $A_{\sqrt{k}}$ reports that the number of mismatches is larger than $2\sqrt{k}$, we repeat the process for $j = 2\sqrt{k}, \cdots, k$. Otherwise the algorithm gave us a good estimation y' of the number of mismatches between $P(1, m/2)$, and $T(j, m/2)$. Once we find y' we proceed as follows. If $y' > (1 + \epsilon)k$ we search recursively for the position of the k-mismatch in $P(1, m/2)$. If $y' < (1 - \epsilon)k$ we search recursively for the $k - y'$-mismatch in $P(m/2 + 1, m/2)$. If $(1 - \epsilon)k \leq y' \leq (1 + \epsilon)k$, the approximated k-mismatch is at position $m/2$ of the pattern and we are done.

It is easy to see that the running time of the search is $O(\frac{n}{\epsilon^2} \log n \log m \log \log k)$. The total running time of the algorithm is $O(\frac{n}{\epsilon^2} \log n \log^2 m \log k)$.

5   Approximate Tandem Repeats

We first describe the algorithm for exact tandem repeats. Then we describe the algorithm for the k-mismatch tandem repeats that runs in $O(nk \log(n/k) + z)$. Finally we show how to change this algorithm to get our algorithm. Let S be the input string of length n. Let $S[i \cdots j]$ be the substring of S that starts at position i and ends at position j, and recall that $S[i \cdots j]^R$ is the string obtained by reversing $S[i \cdots j]$. Let $S[i]$ be the character at position i.

We now describe the exact algorithm of Main and Lorentz [9]. Let $h = \lfloor n/2 \rfloor$. Let $u = S[1 \cdots h]$ be the first half of S, and let $v = S[h + 1 \cdots n]$ be the second half of S. The algorithm finds all tandem repeats that contain $S[h]$ and $S[h+1]$. That is repeats that are not fully contained in u and are not fully contained in v, and then calls itself recursively on u to find all tandem repeats contained in the first half of S, and calls itself recursively on v to find all tandem repeats contained in the second half of S.

The repeats that contain $S[h]$ and $S[h + 1]$ are classified into *left repeats* and *right repeats*. *Left repeats* are all tandem repeats zz where the first copy of z contains h. *Right repeats* are all tandem repeats zz where the second copy of z contains h. We describe how to find all left repeats. Right repeats are found similarly. We build a suffix tree that supports LCA queries in $O(1)$ time for S and S^R. The algorithm for finding left repeats in S has $n/2$ iterations. In the i-th iteration, we find all left repeats of length $2i$ as follows.

1. Let $j = h + i$.
2. Find the longest common prefix of $S[h \cdots n]$ and of $S[j \cdots n]$. Let ℓ_1 be the length of this prefix.
3. Find the longest common prefix of $S[1 \cdots h - 1]^R$ and of $S[1 \cdots j - 1]^R$. Let ℓ_2 be the length of this prefix.
4. If $\ell_1 + \ell_2 \geq i$ there is at least one tandem repeat of length $2i$. All left repeats of length $2i$, begin at positions $\max(h - \ell_2, h - i + 1), \cdots, \min(h + \ell_1 - i, h)$.

Using the suffix tree we can find each longest common prefix in $O(1)$ time. Therefore, we can find an implicit representation of all left repeats of length $2i$ in $O(1)$ time. The total time it takes to find all left and right repeats for $h = \lfloor n/2 \rfloor$ is $O(n)$, and the total running time of the algorithm is $O(n \log n + z)$.

The algorithm of [7] for finding k-mismatch tandem repeats is an extension of the algorithm of Main and Lorentz [9]. Here we stop the recursion when the length of the string is at most $2k$, and in each iteration we compute only repeats of length greater than $2k$. Given $h = \lfloor n/2 \rfloor$ and $i > k$ the algorithm for finding all k-mismatch left repeats of size $2i$ is as follows.

1. Let $j = h + i$.
2. We find the positions of the first $k + 1$ mismatches of $S[h \cdots n]$ and $S[j \cdots n]$ by performing $k + 1$ successive LCA queries on the suffix tree of S. Let ℓ_1 be the position of the $(k + 1)$-mismatch of the two strings.
3. Similarly, we find the positions of the first $k + 1$ mismatches of $S[1 \cdots h - 1]^R$ and $S[1 \cdots j - 1]^R$ by performing $k + 1$ successive LCA queries on a suffix tree of S^R. Let ℓ_2 be the position of the $(k + 1)$-mismatch of the two strings.
4. If $\ell_1 + \ell_2 \geq i$, the k-mismatch tandem repeats will be those at positions $\max(h - \ell_2, h - i + 1) \cdots \min(h + \ell_1 - i, h)$ that have at most k mismatches. We can find all these positions in $O(k)$ time by merging the sorted list of item 2 containing the positions of the mismatches that are in $[h \cdots h + i]$ with the sorted list of item 3 containing the positions of the mismatches that are in $[h \cdots h + i]$. All positions in a segment between two successive elements in the merged list either all correspond to tandem repeats or none does. (See [7, 5] for more details).

The time it takes to find all left and right k-mismatch tandem repeats for $h = \lfloor n/2 \rfloor$ is $O(nk)$, and the total running time of the algorithm is $O(nk \log(n/k) + z)$.

We are now ready to describe our approximate tandem repeats algorithm for ϵ and k. We use the algorithm of Sect. 2 (with minor modifications and with different scaling of ϵ we can also use the algorithms of Sect. 3, and Sect. 4 instead). The algorithm has the same steps as the algorithm of [7]. The only difference is in the way left (and right) tandem repeats are computed. Let $h = \lfloor n/2 \rfloor$. Let the string $P_h = S[h \cdots n]$ and let $T_h = S[h \cdots n]\$^{n/2}$ be the string which is the catenation of P_h and the string $\$^{n/2}$, where $\$$ is a new character that doesn't appear in S. The string $\$^{n/2}$ is used to make sure that the text is always longer than the pattern, we ignore mismatches that are caused by it. Let $P_{h-1}^R = S[1 \cdots h - 1]^R$ and let $T^R = S[1 \cdots n]^R$. We compute the left repeats as follows.

1. Compute the position of the i-mismatch between the text T_h and the pattern P_h, for $i = \epsilon k, 2\epsilon k, \cdots, k - \epsilon k, k$. We do that by running the algorithm of Sect. 2 once for every $i = \epsilon k, 2\epsilon k, \cdots, k - \epsilon k, k$. Let B_i be the vector that contains these positions. That is $B_i[r], r \geq h$ contains the position of the i-mismatch between $S[r \cdots n]$ and $S[h \cdots n]$.

2. Compute the position of the i-mismatch between the text T^R and the pattern P_{h-1}^R, for $i = \epsilon k, 2\epsilon k, \cdots, k - \epsilon k, k$ with the algorithm of Sect. 2. Let $B_i^R, i \in \{\epsilon k, 2\epsilon k, \cdots, k\}$ be the vector that contains these positions. That is $B_i^R[r], r \geq h$ contains the position of the i-mismatch between $S[1 \cdots r]^R$ and $S[1 \cdots h - 1]^R$.

3. For each $r > k$ we find all approximate tandem repeats of length $2r$ whose first half contains h as follows. The qth element in the sequence $B_{\epsilon k}[h + r], \cdots, B_k[h+r]$ contains the position of the $q\epsilon k$-mismatch between $S[h \cdots n]$ and $S[h+r \cdots n]$. The qth element in the sequence $B_{\epsilon k}^R[h+r-1], \cdots, B_k^R[h+r-1]$ contains the position of the $q\epsilon k$-mismatch between $S[1 \cdots h - 1]^R$ and $S[1 \cdots h + r - 1]^R$. We activate the procedure of [7] that we described in item 4 of the previous algorithm, on these sequences of $O(1/\epsilon)$ positions of mismatches in $O(1/\epsilon)$ time. It is easy to see that this algorithm produces all tandem repeats with at most k mismatches. The algorithm may also report tandem reports with at most $(1 + 2\epsilon)k$-mismatches.

Items 1 and 2 that take $O((1/\epsilon)nk^{2/3} \log^{1/3} n \log k)$ time dominated the running time of each recursive call.

Therefore the total time is $O((1/\epsilon)nk^{2/3} \log^{1/3} n \log k \log(n/k) + z)$.

Acknowledgements. We thank Uri Zwick for suggesting to use prefix sum arrays in Sect. 2.

References

1. Karl Abrahamson. Generalized string matching. *SIAM J. Comput.*, 16(6):1039–1051, 1987.
2. Amihood Amir, Moshe Lewenstein, and Ely Porat. Faster algorithms for string matching with k mismatches. *J. Algorithms*, 50(2):257–275, 2004.
3. Richard Cole and Ramesh Hariharan. Approximate string matching: A simpler faster algorithm. *SIAM J. Comput.*, 31(6):1761–1782, 2002.
4. M. Crochemore and W. Rytter. *Text Algorithms*. Oxford Univ. Press, New-York, 1994. pp. 27-31.
5. Dan Gusfield. *Algorithms on strings, trees and sequences: computer science and computational biology*. Cambridge Univ. Press, 1997.
6. Howard J. Karloff. Fast algorithms for approximately counting mismatches. *Inf. Process. Lett.*, 48(2):53–60, 1993.
7. Gad M. Landau, Jeanette P. Schmidt, and Dina Sokol. An algorithm for approximate tandem repeats. *Journal of Computational Biology*, 8(1):1–18, 2001.
8. G.M. Landau and U. Vishkin. Efficient string matching in the presence of errors. In *Proc. 26th IEEE Symposium on Foundations of Computer Science*, pages 126–136, Los Alamitos CA, USA, 1985. IEEE Computer Society.
9. Michael G. Main and Richard J. Lorentz. An o(n log n) algorithm for finding all repetitions in a string. *J. Algorithms*, 5(3):422–432, 1984.

Unbiased Matrix Rounding

Benjamin Doerr, Tobias Friedrich, Christian Klein, and Ralf Osbild

Max-Planck-Institut für Informatik, Saarbrücken, Germany

Abstract. We show several ways to round a real matrix to an integer one such that the rounding errors in all rows and columns as well as the whole matrix are less than one. This is a classical problem with applications in many fields, in particular, statistics.

We improve earlier solutions of different authors in two ways. For rounding matrices of size $m \times n$, we reduce the runtime from $O((mn)^2)$ to $O(mn \log(mn))$. Second, our roundings also have a rounding error of less than one in all initial intervals of rows and columns. Consequently, arbitrary intervals have an error of at most two. This is particularly useful in the statistics application of controlled rounding.

The same result can be obtained via (dependent) randomized rounding. This has the additional advantage that the rounding is unbiased, that is, for all entries y_{ij} of our rounding, we have $E(y_{ij}) = x_{ij}$, where x_{ij} is the corresponding entry of the input matrix.

1 Introduction

In this paper, we analyze a rounding problem with strong connections to statistics, but also to different areas in discrete mathematics, computer science, and operations research. We show how to round a matrix to an integer one such that rounding errors in intervals of rows and columns are small.

Let m, n be positive integers. For some set S, we write $S^{m \times n}$ to denote the set of $m \times n$ matrices with entries in S. For real numbers a, b let $[a..b] := \{z \in \mathbb{Z} \mid a \leq z \leq b\}$. We show the following.

Theorem 1. *For all $X \in [0,1)^{m \times n}$ a rounding $Y \in \{0,1\}^{m \times n}$ such that*

$$\forall b \in [1..n], \ i \in [1..m] : \left| \sum_{j=1}^{b} (x_{ij} - y_{ij}) \right| < 1,$$

$$\forall b \in [1..m], \ j \in [1..n] : \left| \sum_{i=1}^{b} (x_{ij} - y_{ij}) \right| < 1,$$

$$\left| \sum_{i=1}^{m} \sum_{j=1}^{n} (x_{ij} - y_{ij}) \right| < 1$$

can be computed in time $O(mn \log(mn))$.

This result extends the famous rounding lemma of Baranyai [3] and several results on controlled rounding in statistics by Bacharach [2] and Causey, Cox and Ernst [7].

L. Arge and R. Freivalds (Eds.): SWAT 2006, LNCS 4059, pp. 102–112, 2006.

1.1   Baranyai's Rounding Lemma and Applications in Statistics

Baranyai [3] used a weaker version of Theorem 1 to obtain his well-known results on coloring and partitioning complete uniform hypergraphs. He showed that any matrix can be rounded such that the errors in all rows, all columns and the whole matrix are less than one. He used a formulation as flow problem to prove this statement. This yields an inferior runtime than the bound in Theorem 1. However, algorithmic issues were not his focus.

In statistics, Baranyai's result was independently obtained by Bacharach [2] (in a slightly weaker form) and again independently by Causey, Cox and Ernst [7]. There are two statistical applications for such rounding results. Note first that instead of rounding to integers, our result also applies to rounding to multiples of any other base (e.g., multiples of 10). Such a rounding can be used to improve the readability of data tables.

The main reason, however, to apply such a rounding procedure is confidentiality protection. Frequency counts that directly or indirectly disclose small counts may permit the identification of individual respondents. There are various methods to prevent this [25], one of which is *controlled rounding* [9]. Here, one tries to round an $(m + 1) \times (n + 1)$-table $\tilde{X}$ given by

$$
\begin{array}{c|c}
(x_{ij})_{\substack{i=1\ldots m \\ j=1\ldots n}} & \left(\sum_{j=1}^{n} x_{ij}\right)_{i=1\ldots m} \\
\hline
\left(\sum_{i=1}^{m} x_{ij}\right)_{j=1\ldots n} & \sum_{i=1}^{m}\sum_{j=1}^{n} x_{ij}
\end{array}
$$

to an $(m + 1) \times (n + 1)$-table $\tilde{Y}$ such that additivity is preserved, i.e., the last row and column of $\tilde{Y}$ contain the associated totals of $\tilde{Y}$. In our setting we round the $m \times n$-matrix X defined by the mn inner cells of the table $\tilde{X}$ to obtain a controlled rounding.

The additivity in the rounded table allows to derive information on the row and column totals of the original table. In contrast to other rounding algorithms, our result also permits to retrieve further reliable information from the rounded matrix, namely on the sums of consecutive elements in rows or columns. Such queries may occur if there is a linear ordering on statistical attributes. Here an example. Let x_{ij} be the number of people in country i that are j years old. Say Y is such that $\frac{1}{1000}Y$ is a rounding of $\frac{1}{1000}X$ as in Theorem 1. Now $\sum_{j=20}^{40} y_{ij}$ is the number of people in country i that are between 20 and 40 years old, apart from an error of less than 2000. Note that such guarantees are not provided by the results of Baranyai [3], Bacharach [2], and Causey, Cox and Ernst [7].

1.2   Unbiased Rounding

Section 4, we present a randomized algorithm computing roundings as in Theorem 1. It has the additional property that each matrix entry is rounded up with probability equal to its fractional value. This is known as randomized rounding [20] in computer science and as unbiased controlled rounding [8, 15] in statistics. Here, a controlled rounding is computed such that the expected values of each table entry (including the totals) equals its fractional value in the original table.

To state our result more precisely, we introduce the following notation. For $x \in \mathbb{R}$ write $\lfloor x \rfloor := \max\{z \in \mathbb{Z} \mid z \leq r\}, \lceil x \rceil := \min\{z \in \mathbb{Z} \mid z \geq r\}$ and $\{x\} := x - \lfloor x \rfloor$.

Definition 1. *Let $x \in \mathbb{R}$. A random variable y is called* randomized rounding *of x, denoted $y \approx x$, if $\Pr(y = \lfloor x \rfloor + 1) = \{x\}$ and $\Pr(y = \lfloor x \rfloor) = 1 - \{x\}$. For a matrix $X \in \mathbb{R}^{m \times n}$, we call an $m \times n$ matrix-valued random variable Y randomized rounding of X if $y_{ij} \approx x_{ij}$ for all $i \in [1..m], j \in [1..n]$.*

We then get the following randomized version of Theorem 1.

Theorem 2. *Let $X \in [0,1)^{m \times n}$ be a matrix having entries of binary length at most ℓ. Then a randomized rounding Y fulfilling the additional constraints that*

$$\forall b \in [1..n], \ i \in [1..m] : \sum_{j=1}^{b} x_{ij} \approx \sum_{j=1}^{b} y_{ij},$$

$$\forall b \in [1..m], \ j \in [1..n] : \sum_{i=1}^{b} x_{ij} \approx \sum_{i=1}^{b} y_{ij},$$

$$\sum_{i=1}^{m} \sum_{j=1}^{n} x_{ij} \approx \sum_{i=1}^{m} \sum_{j=1}^{n} y_{ij}$$

can be computed in time $O(mn\ell)$.

For a matrix with arbitrary entries $x_{ij} := \sum_{d=1}^{\ell} x_{ij}^{(d)} 2^{-d} + x'_{ij}$ where $x'_{ij} < 2^{-\ell}$ and $x_{ij}^{(d)} \in \{0,1\}$ for $i \in [1..m], j \in [1..n], d \in [1..\ell]$, we may use the ℓ highest bits to get an approximate randomized rounding. If (before doing so) we round the remaining part x'_{ij} of each entry to $2^{-\ell}$ with probability $2^{\ell} x'_{ij}$ and to 0 otherwise, we still have that $Y \approx X$, but we introduce an additional error of at most $2^{-\ell} mn$ in the constraints of Theorem 2.

1.3   Other Applications

One of the most basic rounding results states that any sequence $x_1, \ldots, x_n$ of numbers can be rounded to an integer one $y_1, \ldots, y_n$ such that the rounding errors $|\sum_{j=a}^{b} (x_j - y_j)|$ are less than one for all $a, b \in [1..n]$. Such roundings can be computed efficiently in linear time by a one-pass algorithm resembling Kadane's scanning algorithm (described in Bentley's Programming Pearls [5]). Extensions in different directions have been obtained in [11, 12, 17, 21, 23]. This rounding problem has found a number of applications, among others in image processing [1, 22].

Theorem 1 extends this result to two-dimensional sequences. Here the rounding error in arbitrary intervals of a row or column is less than two. In [14] a lower bound of 1.5 is shown for this problem. Thus an error of less than one as in the one-dimensional case cannot be achieved.

Rounding a matrix while considering the errors in column sums and partial row sums also arises in scheduling [6, 18, 19, 24]. For this, however, one does not need our result in full generality. It suffices to use the linear-time one-pass algorithm given in [14]. This algorithm rounds a matrix having unit column sums and can be extend to compute a quasi rounding for arbitrary matrices. While this algorithm keeps the error in all initial row intervals small, for columns only the error over the whole column is considered.

1.4   Knuth's Two-Way Rounding

In [17], Knuth showed how to round a sequence of n real numbers x_i to $y_i \in \{\lfloor x_i \rfloor, \lceil x_i \rceil\}$ such that for two given permutations $\sigma_1, \sigma_2 \in S_n$, we have both $|\sum_{i=1}^{k}(x_{\sigma_1(i)} - y_{\sigma_1(i)})| \leq n/(n+1)$ and $|\sum_{i=1}^{k}(x_{\sigma_2(i)} - y_{\sigma_2(i)})| \leq n/(n+1)$ for all k. Knuth's proof uses integer flows in a certain network [16]. On account of this his worst-case runtime is quadratic.

One application Knuth mentioned in [17] is that of matrix rounding. For this, simply choose a permutation σ_1 that enumerates the x_{ij} row by row, and a permutation σ_2 that enumerates the x_{ij} column by column. Applying Knuth's algorithm to these permutations gives a rounding with errors smaller than one in all initial row and column intervals.

2   Preliminaries

In this section, we provide two easy extensions of the result stated in the introduction. First, we immediately obtain rounding errors of less than two in arbitrary intervals in rows and columns. This is supplied by the following lemma.

Lemma 1. *Let Y be a rounding of a matrix X such that the errors $|\sum_{j=1}^{b}(x_{ij} - y_{ij})|$ in all initial intervals of rows are at most d. Then the errors in arbitrary intervals of rows are at most $2d$, that is, for all $i \in [1..m]$ and all $1 \leq a \leq b \leq n$,*

$$\left| \sum_{j=a}^{b}(x_{ij} - y_{ij}) \right| \leq 2d.$$

This also holds for column intervals, i.e., if the errors $|\sum_{i=1}^{b}(x_{ij} - y_{ij})|$ in all initial intervals of columns are at most d', then the errors $|\sum_{i=a}^{b}(x_{ij} - y_{ij})|$ in arbitrary intervals of columns are at most $2d'$.

Proof. Let $i \in [1..m]$ and $1 \leq a \leq b \leq n$. Then

$$\left| \sum_{j=a}^{b}(x_{ij} - y_{ij}) \right| = \left| \sum_{j=1}^{b}(x_{ij} - y_{ij}) - \sum_{j=1}^{a-1}(x_{ij} - y_{ij}) \right|$$

$$\leq \left| \sum_{j=1}^{b}(x_{ij} - y_{ij}) \right| + \left| \sum_{j=1}^{a-1}(x_{ij} - y_{ij}) \right| \leq 2d. \qquad \square$$

From now on, we will only consider matrices having integral row and column sums. This is justified by the following lemma.

Lemma 2. *Assume that for any $X \in \mathbb{R}^{m \times n}$ with integral column and row sums a rounding $Y \in \mathbb{Z}^{m \times n}$ such that*

$$\forall b \in [1..n], \ i \in [1..m] : \left| \sum_{j=1}^{b} (x_{ij} - y_{ij}) \right| < 1, \tag{1}$$

$$\forall b \in [1..m], \ j \in [1..n] : \left| \sum_{i=1}^{b} (x_{ij} - y_{ij}) \right| < 1 \tag{2}$$

can be computed in time $T(m,n)$. Then for all $\tilde{X} \in \mathbb{R}^{m \times n}$ with arbitrary column and row sums a rounding $\tilde{Y} \in \mathbb{Z}^{m \times n}$ satisfying (1), (2) and

$$\left| \sum_{i=1}^{m} \sum_{j=1}^{n} (x_{ij} - y_{ij}) \right| < 1 \tag{3}$$

can be computed in time $T(m+1, n+1) + O(mn)$.

Proof. Given an arbitrary matrix $\tilde{X} \in \mathbb{R}^{m \times n}$, we add an extra row taking what is missing towards integral column sums and add an extra column taking what is missing towards integral row sums. Hence, let $X \in \mathbb{R}^{(m+1) \times (n+1)}$ be such that

$$x_{ij} = \tilde{x}_{ij} \qquad \text{for all } i \in [1..m], \ j \in [1..n],$$

$$x_{m+1,j} = \left\lceil \sum_{i=1}^{m} \tilde{x}_{ij} \right\rceil - \sum_{i=1}^{m} \tilde{x}_{ij} \qquad \text{for all } j \in [1..n],$$

$$x_{i,n+1} = \left\lceil \sum_{j=1}^{n} \tilde{x}_{ij} \right\rceil - \sum_{j=1}^{n} \tilde{x}_{ij} \qquad \text{for all } i \in [1..m],$$

$$x_{m+1,n+1} = \left\lceil \sum_{i=1}^{m} \tilde{x}_{i,n+1} \right\rceil - \sum_{i=1}^{m} \tilde{x}_{i,n+1} = \left\lceil \sum_{j=1}^{n} \tilde{x}_{m+1,j} \right\rceil - \sum_{j=1}^{n} \tilde{x}_{m+1,j}.$$

Clearly, X has integral row and column sums. Therefore it can be rounded to $Y \in \mathbb{Z}^{(m+1) \times (n+1)}$ satisfying (1) and (2) in time $T(m+1, n+1)$.

For (3), observe that if a row (resp. column) sum is integral, the rounding error in the row (resp. column) is 0. Then the rounding error in the whole matrix is also 0, if all row and column sums are integral. Using this and the triangle inequality, we get inequality (3) as follows.

$$\left| \sum_{i=1}^{m} \sum_{j=1}^{n} (x_{ij} - y_{ij}) \right| = \left| \sum_{i=1}^{m+1} \sum_{j=1}^{n+1} (x_{ij} - y_{ij}) - \sum_{i=1}^{m+1} (x_{i,n+1} - y_{i,n+1}) \right.$$

$$\left. - \sum_{j=1}^{n+1} (x_{m+1,j} - y_{m+1,j}) + (x_{m+1,n+1} - y_{m+1,n+1}) \right|$$

$$\leq 0 + 0 + 0 + |x_{m+1,n+1} - y_{m+1,n+1}| < 1.$$

By setting $\tilde{y}_{ij} = y_{ij}$ for all $i \in [1..m]$ and $j \in [1..n]$, we obtain the desired rounding $\tilde{Y} \in \mathbb{Z}^{m \times n}$. $\qquad \square$

3   Bitwise Rounding

In this section, we present an alternative approach which will lead to a superior runtime. It uses a classical result on rounding problems, namely, that the problem of rounding arbitrary numbers can be reduced to the one of rounding half-integral numbers. For $X \in \{0, \frac{1}{2}\}^{m \times n}$, our rounding problem turns out to be much simpler. In fact, it can be solved in linear time.

3.1   The Binary Rounding Method

The following rounding method was introduced by Beck and Spencer [4] in 1984. They used it to prove the existence of two-colorings of $\mathbb{N}$ having small discrepancy in all arithmetic progressions of arbitrary length and bounded difference.

Given arbitrary numbers that have to be rounded, they use their binary expansion and (assuming all of them to be finite) round 'digit by digit'. To do the latter, they only need to understand the corresponding rounding problem for half-integral numbers. That is, an ℓ-bit number $x = x' + \frac{1}{2}x'', x' \in \{0, \frac{1}{2}\}$ can be recursively rounded by rounding the $(\ell - 1)$-bit number x'' to $y'' \in \{0, 1\}$ and then rounding $x' + \frac{1}{2}y'' \in \{0, \frac{1}{2}, 1\}$ to $y \in \{0, 1\}$. The resulting rounding errors are at most twice the ones incurred by the half-integral roundings.

If some numbers do not have a finite binary expansion, one can use a sufficiently large finite length approximation. To get rid of additional errors caused by this, we invoke a slight refinement of the binary rounding method. In [10] it was proven that the extra factor of two can be reduced to an extra factor of $2(1 - \frac{1}{2r})$, where r is the number of rounding errors we want to keep small.

In our setting, the number of rounding errors is the number of all initial row and column intervals, i.e., $r = 2mn$. In summary, we have the following.

Lemma 3. *Assume that for any* $X \in \{0, \frac{1}{2}\}^{m \times n}$ *a rounding* $Y \in \{0, 1\}^{m \times n}$ *can be computed in time* T *that satisfies*

$$\forall b \in [1..n], \ i \in [1..m] : \left| \sum_{j=1}^{b} (x_{ij} - y_{ij}) \right| \leq D,$$

$$\forall b \in [1..m], \ j \in [1..n] : \left| \sum_{i=1}^{b} (x_{ij} - y_{ij}) \right| \leq D.$$

Then for all $\ell \in \mathbb{N}$ *and* $X \in [0, 1)^{m \times n}$ *a rounding* $Y \in \{0, 1\}^{m \times n}$ *such that*

$$\forall b \in [1..n], \ i \in [1..m] : \left| \sum_{j=1}^{b} (x_{ij} - y_{ij}) \right| \leq 2(1 - \tfrac{1}{4mn})D + 2^{-\ell}b,$$

$$\forall b \in [1..m], \ j \in [1..n] : \left| \sum_{i=1}^{b} (x_{ij} - y_{ij}) \right| \leq 2(1 - \tfrac{1}{4mn})D + 2^{-\ell}b$$

can be computed in time $O(\ell T)$.

3.2   Rounding Half-Integral Matrices

It remains to show how to solve the rounding problem for half-integral matrices. Based on Lemma 2, we can assume integrality of row and column sums.

Here is an outline of our approach. For each row and column, we consider the sequence of its $\frac{1}{2}$-entries and partition them into disjoint pairs of neighbors. From the two $\frac{1}{2}$s forming such a pair, exactly one is rounded to 1 and the other to 0. Thus, if such a pair is contained in an initial interval, it does not contribute to the rounding error.

To make the idea precise, assume some row contains exactly $2K$ entries of value $\frac{1}{2}$. We call the $(2k-1)$-th and $(2k)$-th $\frac{1}{2}$-entry of this row a *row pair*, for all $1 \leq k \leq K$. The $\frac{1}{2}$s of a row pair are mutually referred to as *row neighbors*. Similarly, we define *column pairs* and *column neighbors*. Figure 1(a) shows a half-integral matrix together with row and column pairs marked by boxes. Since each $\frac{1}{2}$ belongs to a row pair *and* a column pair, the task of rounding is non-trivial.

Our solution makes use of an auxiliary graph $\mathcal{G}_X$ which contains the necessary information about row and column neighbors. Each $\frac{1}{2}$-entry is represented by a vertex that is labeled with the corresponding matrix indices. Each pair is represented by an edge connecting the vertices that correspond to the paired $\frac{1}{2}$s. Figure 1(b) shows the auxiliary graph that belongs to the matrix of Figure 1(a).

We collect some properties of this auxiliary graph.

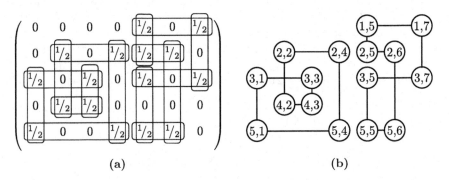

(a)   (b)

Fig. 1. Example for the construction of an auxiliary graph. (a) Input matrix X with its row and column pairs. (b) Auxiliary graph $\mathcal{G}_X$. Vertices are labeled with matrix indices and edges connect vertices of row and column pairs. $\mathcal{G}_X$ is a disjoint union of even cycles.

Lemma 4. *Let* $X \in \{0, \frac{1}{2}\}^{m \times n}$ *be a matrix with integral row and column sums.*

(a) Every vertex of $\mathcal{G}_X$ has degree 2.
(b) $\mathcal{G}_X$ is a disjoint union of even cycles.
(c) $\mathcal{G}_X$ is bipartite.

Proof. (a) Because of the integrality of the row and column sums, the number of $\frac{1}{2}$-entries in each row and column is even. Hence each $\frac{1}{2}$-entry has a row and a column neighbor. In consequence, each vertex is incident with exactly two edges. (b) The edge sequence of a path in $\mathcal{G}_X$ corresponds to an alternating sequence of row and column pairs. Therefore any cycle in $\mathcal{G}_X$ consists of an even number of edges. Since each vertex has degree two, $\mathcal{G}_X$ is a disjoint union of cycles. (c) Clearly, every even cycle is bipartite.   □

With this result, we are able to find the desired roundings.

Lemma 5. *Let* $X \in \{0, \frac{1}{2}\}^{m \times n}$ *and let* $V_0 \dot\cup V_1$ *be a bipartition of* $\mathcal{G}_X$. *Define* $Y = (y_{ij}) \in \{0, 1\}^{m \times n}$ *by*

$$y_{ij} = \begin{cases} 0, & \text{if } x_{ij} = 0 \\ 0, & \text{if } x_{ij} = \frac{1}{2} \text{ and } (i,j) \in V_0 \\ 1, & \text{if } x_{ij} = \frac{1}{2} \text{ and } (i,j) \in V_1. \end{cases}$$

Then Y *has the property that*

$$\forall b \in [1..n],\ i \in [1..m] : \left| \sum_{j=1}^{b} (x_{ij} - y_{ij}) \right| \leq \tfrac{1}{2}, \tag{4}$$

$$\forall b \in [1..m],\ j \in [1..n] : \left| \sum_{i=1}^{b} (x_{ij} - y_{ij}) \right| \leq \tfrac{1}{2}. \tag{5}$$

Proof. Because 0s of X are maintained in Y, it suffices to consider $\frac{1}{2}$-entries to determine the rounding error in initial intervals. Since the rounded values for the $(2k-1)$-th and $(2k)$-th $\frac{1}{2}$-entry sum up to 1 by construction, there is no error in initial intervals that contain an even number of $\frac{1}{2}$s, and an error of $\frac{1}{2}$ if they contain an odd number of $\frac{1}{2}$s.   □

After these considerations, we are able to present an algorithm that solves the problem in two steps: first we compute the auxiliary graph and afterwards the output matrix. To construct $\mathcal{G}_X$, we transform the input matrix X column by column from left to right. Of course, generating the labeled vertices is trivial. The column neighbors are detected just by numbering the $\frac{1}{2}$-entries within a column from top to bottom. When there are $2k$ such entries, we insert an edge between the vertices with number $2i-1$ and $2i$ with $1 \leq i \leq k$. The strategy to detect row neighbors is the same but we need more information. Therefore we store for each row the parity of its $\frac{1}{2}$-entries so far and, if the parity is odd, further a pointer to the last occurrence of $\frac{1}{2}$ in this row. Then, if the current $\frac{1}{2}$ is an even occurrence, we have a pointer to the preceding $\frac{1}{2}$, and are able to insert an edge between the corresponding vertices in $\mathcal{G}_X$.

The output matrix Y can be computed from X as follows. Every 0 in X is kept and every $\frac{1}{2}$-sequence that corresponds to a cycle in $\mathcal{G}_X$ is substituted by an alternating 0–1–sequence. By Lemma 4, this is always possible. It does not matter which of the two alternating 0–1 sequences we choose.

The graph $\mathcal{G}_X$ can be realized with adjacency lists (the vertex degree is always 2). The additional information per row can be realized by a simple pointer–array of length m (a special nil–value indicates even parity).

Since the runtime of each step is bounded by the size of the input matrix, the entire algorithm takes time $O(mn)$. In addition to the constant amount of space we need for each of the m rows, we store all k entries of value $\frac{1}{2}$ in the auxiliary graph. This leads to a total space consumption of $O(m + k)$. Summarizing the above, we obtain the following lemma.

Lemma 6. *Let* $X \in \{0, \frac{1}{2}\}^{m \times n}$. *Then a rounding* $Y \in \{0,1\}^{m \times n}$ *satisfying the inequalities (4) and (5) can be computed in time* $O(mn)$.

3.3   Final Result

By combining Lemma 3 and 6, we obtain the following result.

Theorem 3. *For all* $\ell \in \mathbb{N}$ *and* $X \in [0,1)^{m \times n}$ *a rounding* $Y \in \{0,1\}^{m \times n}$ *such that*

$$\forall b \in [1..n],\ i \in [1..m] : \left| \sum_{j=1}^{b} (x_{ij} - y_{ij}) \right| \leq 1 - \tfrac{1}{4mn} + 2^{-\ell}b,$$

$$\forall b \in [1..m],\ j \in [1..n] : \left| \sum_{i=1}^{b} (x_{ij} - y_{ij}) \right| \leq 1 - \tfrac{1}{4mn} + 2^{-\ell}b$$

can be computed in time $O(\ell mn)$.

For $\ell > \log_2(4mn \max\{m,n\})$ the above theorem together with Lemma 2 yields Theorem 1 in the introduction.

4   Unbiased Rounding

In this section we give a randomized algorithm that computes a randomized rounding satisfying Theorem 2. First observe, that the $\{0, \frac{1}{2}\}$ case has a very simple randomized solution. Whenever it has to round a cycle, it chooses one of the two alternating 0–1–sequences for each cycle uniformly at random. Then, each $x_{ij} = \frac{1}{2}$ is rounded up with probability $\frac{1}{2}$.

Now consider the output of the bitwise rounding algorithm using the randomized rounding algorithm for the half-integral case as subroutine. We adapt the proofs of [13] to show that this algorithm computes an unbiased controlled rounding.

Theorem 4. *Let* $X \in [0,1)^{m \times n}$ *be a matrix containing entries with binary representation of length at most* ℓ. *Let* Y *be a random variable modeling the output of the randomized algorithm. Then* $Y \approx X$ *and*

$$\forall b \in [1..n],\ i \in [1..m] : \sum_{j=1}^{b} y_{ij} \approx \sum_{j=1}^{b} x_{ij}, \tag{6}$$

$$\forall b \in [1..m],\ j \in [1..n] : \sum_{i=1}^{b} y_{ij} \approx \sum_{i=1}^{b} x_{ij}. \tag{7}$$

Proof. We prove $Y \approx X$ by induction. For $\ell = 1$ it is clear that $\Pr(y_{ij} = 1) = x_{ij}$. If $\ell > 1$, write $x_{ij} = x'_{ij} + \frac{1}{2}x''_{ij}$, where $x'_{ij} \in \{0, \frac{1}{2}\}$ and $x''_{ij} \in [0, 1)$ has bit-length $\ell - 1$. Let y''_{ij} be the rounding computed for x''_{ij}. Then $\Pr(y''_{ij} = 1) = x''_{ij}$ by induction. Now the algorithm will round $\tilde{x}_{ij} := x'_{ij} + \frac{1}{2}y''_{ij} \in \{0, \frac{1}{2}, 1\}$ to y_{ij}. If $y''_{ij} = 1$, then $\tilde{x}_{ij}$ will be rounded up with probability 1 if $x'_{ij} = \frac{1}{2}$ and with probability $\frac{1}{2}$ otherwise. If, on the other hand, $y''_{ij} = 0$, then $\tilde{x}_{ij}$ will be rounded up with probability x'_{ij}. Thus

$$\Pr(y_{ij} = 1) = x''_{ij}(\tfrac{1}{2} + x'_{ij}) + (1 - x''_{ij})x'_{ij} = x'_{ij} + \tfrac{1}{2}x''_{ij} = x_{ij}.$$

To prove equation (6), observe that $s_y := \sum_{j=1}^{b} y_{ij}$ is a rounding of $s_x := \sum_{j=1}^{b} x_{ij}$ by Lemma 3. We also have $E(s_y) = \sum_{j=1}^{b} E(y_{ij}) = s_x$ by linearity of expectation. But also $E(s_y) = \Pr(s_y = \lfloor s_x \rfloor)\lfloor s_x \rfloor + \Pr(s_y = \lfloor s_x \rfloor + 1)(\lfloor s_x \rfloor + 1)$, which is only possible if $s_y \approx s_x$. The proof of (7) is analogous. $\square$

References

1. T. Asano. Digital halftoning: Algorithm engineering challenges. *IEICE Trans. on Inf. and Syst.*, E86-D:159–178, 2003.
2. M. Bacharach. Matrix rounding problems. *Management Science (Series A)*, 12:732–742, 1966.
3. Zs. Baranyai. On the factorization of the complete uniform hypergraph. In *Infinite and finite sets (Colloq., Keszthely, 1973; dedicated to P. Erdős on his 60th birthday), Vol. I*, pages 91–108. Colloq. Math. Soc. János Bolyai, Vol. 10. North-Holland, Amsterdam, 1975.
4. J. Beck and J. Spencer. Well distributed 2-colorings of integers relative to long arithmetic progressions. *Acta Arith.*, 43:287–298, 1984.
5. J. L. Bentley. Algorithm design techniques. *Commun. ACM*, 27:865–871, 1984.
6. N. Brauner and Y. Crama. The maximum deviation just-in-time scheduling problem. *Discrete Appl. Math.*, 134:25–50, 2004.
7. B. D. Causey, L. H. Cox, and L. R. Ernst. Applications of transportation theory to statistical problems. *Journal of the American Statistical Association*, 80:903–909, 1985.
8. L. H. Cox. A constructive procedure for unbiased controlled rounding. *Journal of the American Statistical Association*, 82:520–524, 1987.
9. L. H. Cox and L. R. Ernst. Controlled rounding. *Informes*, 20:423–432, 1982.
10. B. Doerr. Linear and hereditary discrepancy. *Combinatorics, Probability and Computing*, 9:349–354, 2000.
11. B. Doerr. Lattice approximation and linear discrepancy of totally unimodular matrices. In *Proceedings of the 12th Annual ACM-SIAM Symposium on Discrete Algorithms (SODA)*, pages 119–125, 2001.
12. B. Doerr. Global roundings of sequences. *Information Processing Letters*, 92:113–116, 2004.
13. B. Doerr. Generating randomized roundings with cardinality constraints and derandomizations. In *23rd Annual Symposium on Theoretical Aspects of Computer Science*, 2006.

14. B. Doerr, T. Friedrich, C. Klein, and R. Osbild. Rounding of sequences and matrices, with applications. In *Third Workshop on Approximation and Online Algorithms*, volume 3879 of *Lecture Notes in Computer Science*, pages 96–109. Springer, 2006.

15. I. P. Fellegi. Controlled random rounding. *Survey Methodology*, 1:123–133, 1975.

16. L. R. Ford, Jr., and D. R. Fulkerson. *Flows in Networks*. Princeton University Press, 1962.

17. D. E. Knuth. Two-way rounding. *SIAM J. Discrete Math.*, 8:281–290, 1995.

18. Y. Monden. What makes the Toyota production system really tick? *Industrial Eng.*, 13:36–46, 1981.

19. Y. Monden. *Toyota Production System*. Industrial Engineering and Management Press, Norcross, GA, 1983.

20. P. Raghavan. Probabilistic construction of deterministic algorithms: Approximating packing integer programs. *J. Comput. Syst. Sci.*, 37:130–143, 1988.

21. K. Sadakane, N. Takki-Chebihi, and T. Tokuyama. Combinatorics and algorithms on low-discrepancy roundings of a real sequence. In *ICALP 2001*, volume 2076 of *Lecture Notes in Computer Science*, pages 166–177, Berlin Heidelberg, 2001. Springer-Verlag.

22. K. Sadakane, N. Takki-Chebihi, and T. Tokuyama. Discrepancy-based digital halftoning: Automatic evaluation and optimization. In *Geometry, Morphology, and Computational Imaging*, volume 2616 of *Lecture Notes in Computer Science*, pages 301–319, Berlin Heidelberg, 2003. Springer-Verlag.

23. J. Spencer. *Ten lectures on the probabilistic method*, volume 64 of *CBMS-NSF Regional Conference Series in Applied Mathematics*. Society for Industrial and Applied Mathematics (SIAM), Philadelphia, PA, 1994.

24. G. Steiner and S. Yeomans. Level schedules for mixed-model, just-in-time processes. *Management Science*, 39:728–735, 1993.

25. L. Willenborg and T. de Waal. *Elements of Statistical Disclosure Control*, volume 155 of *Lecture Notes in Statistics*. Springer, 2001.

Online, Non-preemptive Scheduling of Equal-Length Jobs on Two Identical Machines[*]

Michael H. Goldwasser and Mark Pedigo

Saint Louis University, Dept. of Mathematics and Computer Science
221 North Grand Blvd.; St. Louis, MO 63103-2007
{goldwamh, pedigom}@slu.edu

Abstract. We consider the non-preemptive scheduling of two identical machines for jobs with equal processing times yet arbitrary release dates and deadlines. Our objective is to maximize the number of jobs completed by their deadlines. Using standard nomenclature, this problem is denoted as $P2 \mid p_j = p, r_j \mid \sum \overline{U}_j$. The problem is known to be polynomially solvable in an offline setting.

In an online variant of the problem, a job's existence and parameters are revealed to the scheduler only upon that job's release date. We present an online, deterministic algorithm for the problem and prove that it is $\frac{3}{2}$-competitive. A simple lower bound shows that this is the optimal deterministic competitiveness.

Keywords: Algorithms, Online, Scheduling.

1 Introduction

We present an online, non-preemptive, deterministic algorithm for scheduling two machines in the following setting. Each job j is specified by three non-negative integer parameters, with r_j denoting its release time, p_j its processing time, and d_j its deadline. For this paper, we assume all processing times are equal, thus $p_j = p$ for a fixed constant p. In order to successfully complete a job j, the scheduler must devote a machine to it for p consecutive units of time during the interval $[r_j, d_j)$.

We examine an *online* model for the problem, in which the scheduler is oblivious to a job's existence and characteristics until that job's release time. We use competitive analysis to measure the performance of an algorithm $\mathcal{A}$ by comparing the quality of the schedule it produces to that of an optimal *offline* scheduler that has a priori knowledge of the entire instance [3]. Our main result is the presentation of a $\frac{3}{2}$-competitive, deterministic online algorithm for the two-machine problem. A simple lower bound shows that this is the best possible result.

Preliminaries and Notations. We let $x_j = d_j - p$ denote a job's *expiration time*, namely the last possible time it can be started, and we fix a canonical linear

---

[*] This material is based upon work supported by the National Science Foundation under Grant No. CCR-0417368.

L. Arge and R. Freivalds (Eds.): SWAT 2006, LNCS 4059, pp. 113–123, 2006.

ordering of jobs, $\prec$, such that $i \prec j$ implies $x_i \leq x_j$. As all job parameters are presumed to be integral, we consider time as a series of discrete steps. Our algorithm provides what has been termed *immediate notification* [8]. At the moment a job is released, the scheduler must either accept or reject that job. An accepted job need not be scheduled precisely at that moment, but the scheduler must guarantee that it will be successfully completed by its deadline. In the case where several jobs are released at a time t, we assume that they are considered in arbitrary order with the scheduler providing notification to each in turn.

We introduce notation $\text{FEASIBLE}(\mathcal{J}, t_1, t_2)$ to represent a combinatorial boolean property, namely whether a set $\mathcal{J}$ of released jobs can be achieved on two machines, given that one of the machines cannot be used prior to time t_1 nor the other prior to time t_2. We do not make any assumption about the relationship between t_1 and t_2, though conventionally we will order them with $t_1 \leq t_2$ when known. A further discussion of this feasibility property is provided in Section 3.

Related Work. In recent independent work, a different algorithm and analysis has been presented claiming the same $\frac{3}{2}$-competitive upper bound [5]. The single machine version of this problem is well studied. A deterministic, non-preemptive greedy algorithm is known to be 2-competitive with equal-length jobs, and this is the best possible deterministic result [2, 6]. If each job satisfies a *patience* requirement, namely that $d_j - r_j \geq (1 + \kappa) \cdot p_j$ for some constant $\kappa > 0$, then the deterministic greedy algorithm is $(1 + \frac{1}{\lfloor \kappa \rfloor + 1})$-competitive [7]. A similar algorithm achieves the same competitiveness while also providing immediate notification [8]. When considering randomized online algorithms, there exists a $\frac{5}{3}$-competitive algorithm [4] versus a $\frac{4}{3}$-competitive lower bound [6].

In the offline setting, checking the feasibility of a set of equal-length jobs with release dates and deadlines is polynomially solvable, even for an arbitrary number of identical machines [10, 11]. The optimization problem is polynomially solvable for any *fixed* number of machines, even with weighted utility $(Pm \mid p_j = p, r_j \mid \sum w_j U_j)$ [1], yet open with *arbitrary* number of machines, even when unweighted $(P \mid p_j = p, j \mid \sum U_j)$.

2   Algorithm Definition

The algorithm, $\mathcal{A}$, maintains a queue of jobs which have been accepted but not yet started. As the composition of the queue changes over time, we introduce the following notation. We let $Q_t^{\mathcal{A}}$ denote the queue as it exists at the onset of time step t. For each job j released at time t, we let $Q_{r_j}^{\mathcal{A}}$ denote the queue as it exists when j's release was considered (thus $Q_{r_j}^{\mathcal{A}} \supseteq Q_t^{\mathcal{A}}$ may contain newly accepted jobs which were considered prior to j). Job j is accepted into the system precisely if it is feasible to do so. Specifically, we check $\text{FEASIBLE}(Q_{r_j}^{\mathcal{A}} \cup \{j\}, c, \dot{c})$, where c (resp. $\dot{c}$) represents the time until which the first (resp. second) machine is committed to a currently running job. In the case where a machine is not running a job, we considered it trivially committed until time t.

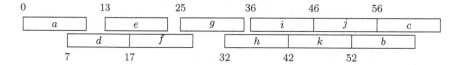

Fig. 1. The schedule produced by $\mathcal{A}$ on an instance with $p = 10$ and jobs: $a = \langle 0, 60 \rangle$, $b = \langle 0, 71 \rangle$, $c = \langle 0, 71 \rangle$, $d = \langle 3, 30 \rangle$, $e = \langle 3, 31 \rangle$, $f = \langle 3, 33 \rangle$, $g = \langle 3, 37 \rangle$, $h = \langle 3, 45 \rangle$, $i = \langle 3, 52 \rangle$, $j = \langle 3, 56 \rangle$, $k = \langle 38, 55 \rangle$

After considering all newly released jobs at time t, the scheduling policy is as follows. If neither machine is currently committed to a job and the queue is non-empty, the $\prec$-minimal job is started on an arbitrary machine. If one machine is committed to a job, yet the other is uncommitted (including the case when a job was just started by the preceding rule), a decision is made as to whether or not to start a job on the other machine. For the sake of analysis, we will refer to this as a *secondary decision*. Specifically, let $\dot{Q}_t^{\mathcal{A}}$ denote the queue at the point this decision is made and let $c > t$ denote the time until which the running machine is committed. We begin the $\prec$-minimal job of $\dot{Q}_t^{\mathcal{A}}$ on the available machine if the test $\mathrm{FEASIBLE}(\dot{Q}_t^{\mathcal{A}}, c, t + p + 1)$ fails. Intuitively, if there is enough flexibility, the algorithm prefers to idle for the moment, leaving open the possibility of starting a more urgent job should one soon arrive. Figure 1 shows the schedule produced by this algorithm on an example.

3   A Supplemental Feasibility Test

In this section, we discuss the feasibility test, $\mathrm{FEASIBLE}(\mathcal{J}, t_1, t_2)$. Such a feasibility condition is satisfied if and only if it can be achieved by scheduling jobs according to an earliest deadline first (EDF) rule.

A classic result of Jackson proves this for a single machine and a set of jobs with arbitrary processing times and deadlines [9]. With arbitrary job lengths and two or more machines, that argument no longer applies. However with equal-length jobs, the EDF schedule suffices. Since all jobs of $\mathcal{J}$ are presumed to have been released, any idleness in a feasible schedule beyond t_i on a machine M_i can be removed. Furthermore, if any job $j \in \mathcal{J}$ is started before some other job with earlier deadline, those two jobs of equal length can be transposed while still guaranteeing a feasible schedule. Based on this structure, we provide the following lemmas, specifically in the context of $\mathrm{FEASIBLE}(\mathcal{J}, t_1, t_2)$.

Lemma 1. *For arbitrary set $\mathcal{J}$, $t_1 \leq t_1'$ and $t_2 \leq t_2'$, $\mathrm{FEASIBLE}(\mathcal{J}, t_1', t_2')$ implies $\mathrm{FEASIBLE}(\mathcal{J}, t_1, t_2)$.*

Lemma 2. *For arbitrary set $\mathcal{J}$ and $t_1 \leq t_2$, let job f be $\prec$-minimal of $\mathcal{J}$. $\mathrm{FEASIBLE}(\mathcal{J}, t_1, t_2)$ if and only if $x_f \geq t$ and $\mathrm{FEASIBLE}(\mathcal{J} \setminus \{f\}, t_1 + p, t_2)$.*

Lemma 3. *Assume $\mathrm{FEASIBLE}(\mathcal{J}, t_1, t_2)$ for $t_1 \leq t_2$. Let $h \in \mathcal{J}$ be an element for which there are i other elements of $\mathcal{J}$ which precede it as per $\prec$-order. Then $x_h \geq t_1 + \lfloor \frac{i}{2} \rfloor \cdot p$.*

Lemma 4. *For arbitrary $\mathcal{J}$ and t, if* FEASIBLE$(\mathcal{J}, t, t+p+1)$ *yet not* FEASIBLE$(\mathcal{J}, t+1, t+p)$, *then there must exist a job* $f \in \mathcal{J}$ *with* $x_f = t$.

4   Competitive Analysis

We fix an arbitrary instance $\mathcal{I}$ and an optimal schedule, OPT, for that instance. For a job j achieved by OPT, we let s_j^{OPT} denote the time at which it is started; we similarly define $s_j^{\mathcal{A}}$ for jobs achieved by $\mathcal{A}$. Though OPT is not constructed in online fashion, we introduce a formal notation of Q_t^{OPT} to be symmetric to that of $Q_t^{\mathcal{A}}$. Namely Q_t^{OPT} consists of all jobs which are released strictly before time t yet started by OPT on or after time t. Our analysis is based upon a form of a potential argument. Specifically, we will soon define functions $\Phi_t^{\mathcal{A}}$ and Φ_t^{OPT} which measure the quality of the schedules being developed as of time t, by $\mathcal{A}$ and OPT respectively. We will view these two potential functions as payment to the respective schedules, with a handicap given to the online algorithm. Specifically, $\mathcal{A}$ will receive 3 units for each job it starts whereas OPT receives 2 units for each job. To prove a $\frac{3}{2}$-competitive ratio in the end, we must show that $\mathcal{A}$ collects at least as much payment as OPT. To properly compare the merits of the schedules at interim times, our potential functions contain full payment for jobs which have been started and a partial payment to account for the inherit potential of each queue.

Before defining the exact potential functions, we introduce some additional notations. We let $F_t^{\mathcal{A}}$ (resp. F_t^{OPT}) designate the set of jobs started strictly before time t by $\mathcal{A}$ (resp. OPT). We define $W_t^{\mathcal{A}} = Q_t^{\mathcal{A}} \setminus Q_t^{\mathrm{OPT}}$ and symmetrically, $W_t^{\mathrm{OPT}} = Q_t^{\mathrm{OPT}} \setminus Q_t^{\mathcal{A}}$, to denote jobs which are currently waiting in one queue but not the other (presumably because they were never accepted or were already started). Intuitively, our potential functions ignore jobs which are common to the two queues but account for the difference between those queues. However there is one more anomaly which may arise.

If we identify a job waiting in $W^{\mathcal{A}}$ which has an expiration time at least as large as some other job waiting in W^{OPT}, we choose not to let either job effect the potential functions. Intuitively, such a pairing can only be to the advantage of the algorithm. Therefore, for each time t we maintain a partial matching $M_t : W_t^{\mathcal{A}} \to W_t^{\mathrm{OPT}}$ (the precise rule for establishing a match is omitted from this version due to space limitations). We introduce notation $P_t^{\mathcal{A}}$ and P_t^{OPT} to identify respective subsets of the waiting jobs which do not participate in the matching.

Namely, we let $P_t^{\mathcal{A}} = \{j \in W_t^{\mathcal{A}} : M_t(j) \text{ is undefined}\}$. Similarly, we let $P_t^{\mathrm{OPT}} = \{j \in W_t^{\mathrm{OPT}} : M_t^{-1}(j) \text{ is undefined}\}$. A typical Venn diagram of the various sets we have defined is shown in Figure 2. We now define two potential functions as follows:

$$\Phi_t^{\mathcal{A}} = \begin{cases} 3 \cdot |F_t^{\mathcal{A}}| & \text{if } P_t^{\mathcal{A}} = \emptyset \\ 3 \cdot |F_t^{\mathcal{A}}| + 1 \cdot |P_t^{\mathcal{A}}| + 1 & \text{if } P_t^{\mathcal{A}} \neq \emptyset \end{cases}$$
$$\Phi_t^{\mathrm{OPT}} = 2 \cdot (|F_t^{\mathrm{OPT}} \cup P_t^{\mathrm{OPT}}|)$$

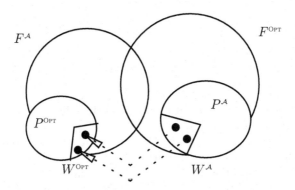

Fig. 2. Relationship among the sets used in analysis at a given time. The dashed lines represents matches between pairs of items from $W^{\mathcal{A}} \setminus P^{\mathcal{A}}$ to $W^{\text{OPT}} \setminus P^{\text{OPT}}$ respectively.

In effect, we take an advance credit of two for the first job in $P_t^{\mathcal{A}}$, and an advanced credit of one for each additional such job. This means that when these jobs are later scheduled, most provide a balance of two additional credits. The last job of P^{OPT} only provides a balance of one credit, yet we will see that an adversary cannot as readily exploit the existence of that last job. Looking at Φ_t^{OPT}, we make full payment for jobs which OPT has started as well as for unmatched jobs which OPT holds in P_t^{OPT}.

Our end goal is to show that $\Phi_\infty^{\text{OPT}} \leq \Phi_\infty^{\mathcal{A}}$. This inequality suffices to prove the $\frac{3}{2}$-competitiveness, since $P_\infty^{\mathcal{A}} = P_\infty^{\text{OPT}} = \emptyset$. Ideally, we would like to show that $\Phi_t^{\text{OPT}} \leq \Phi_t^{\text{OPT}}$ for all times t; unfortunately this cannot be assured.

Instead we break the overall time period into distinct regions, $[s, t)$, such that $\Phi_t^{\text{OPT}} \leq \Phi_t^{\mathcal{A}}$ at the end of each such region. We consider two types of regions: in an idle region both machines of $\mathcal{A}$ are idle, in a busy region at least one machine of $\mathcal{A}$ is at work throughout. If both machines of $\mathcal{A}$ are idle at a time s, we consider idle region $[s, t)$ where t is the next time at which either machine starts a job, if any.

Lemma 5. *For an idle region* $[s, t)$, $\Phi_t^{\mathcal{A}} = \Phi_s^{\mathcal{A}}$ *and* $\Phi_t^{\text{OPT}} = \Phi_s^{\text{OPT}}$.

Proof. As both machines idle at time s, $Q_s^{\mathcal{A}} = \emptyset$, and as they remain idle, no further jobs are released prior to time t. Thus $Q_t^{\mathcal{A}} = \emptyset$ as well. With the empty queue, sets $W_s^{\mathcal{A}}$, $P_s^{\mathcal{A}}$, $W_t^{\mathcal{A}}$ and $P_t^{\mathcal{A}}$ must be empty and the matchings $M_s(\cdot)$ and $M_t(\cdot)$ trivially empty. As no jobs are completed, $F_t^{\mathcal{A}} = F_s^{\mathcal{A}}$ and so we conclude that $\Phi_t^{\mathcal{A}} = \Phi_s^{\mathcal{A}}$. Because no new jobs are released throughout the region and no matches exist, the only jobs which OPT can schedule are those in $Q_s^{\text{OPT}} = P_s^{\text{OPT}}$. This implies that $F_t^{\text{OPT}} \cup P_t^{\text{OPT}} = F_s^{\text{OPT}} \cup P_s^{\text{OPT}}$. □

For a time s at which $\mathcal{A}$ starts executing a job, we define a busy region $[s, t)$ as follows. We define $t > s$ as the first subsequent time when either both machines are idle, or when one machine starts a job yet the other remains idle (that is, for at least one unit). A trivial consequence of this definition is that there is never

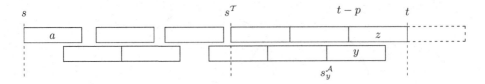

Fig. 3. Typical configuration of algorithm's two machines during a region $[s, t)$. In this example, $|\mathcal{R}| = 11$ and $|\mathcal{T}| = 5$.

a time when both machines are idle within such a region. For ease of exposition throughout the analysis of region $[s, t)$, we let $\mathcal{R}$ denote the set of jobs started by $\mathcal{A}$ during the region. We let a denote the first job of $\mathcal{R}$ to be started, namely at time s. We let z denote the last job of $\mathcal{R}$ to be started, namely at time $t - p$.

In the case where $|\mathcal{R}| = 1$, the region is composed trivially of job $a = z$. For $|\mathcal{R}| \geq 2$, we further let y denote the second-to-last job of $\mathcal{R}$ to be started. We note that y must run on the opposite machine as z and be executing at the time that z is started, as otherwise z would be starting at a time when the other machine is idle, contradicting our definition of t. Thus, $s_y^{\mathcal{A}} \leq s_z^{\mathcal{A}} = t - p < s_y^{\mathcal{A}} + p$.

For $|\mathcal{R}| \geq 2$, we define the *tail of the region*, denoted as $\mathcal{T}$, as follows. Let $s^{\mathcal{T}}$ be the latest time at which $\mathcal{A}$ starts a job of $\mathcal{R}$ following a time step $[s^{\mathcal{T}} - 1, s^{\mathcal{T}})$ at which a machine was idle. We note that $s^{\mathcal{T}}$ is well defined as at least one job of $\mathcal{R}$ must follow such an idle time step. In particular, if the second job of the region is started strictly after time s, then it suffices. Alternatively, the region begins with two jobs starting precisely at time s. Such a region could only follow an idle region, as a previous busy region could not have ended at such a time s. Having defined $s^{\mathcal{T}}$, tail $\mathcal{T} \subseteq \mathcal{R}$ is the set of jobs started by $\mathcal{A}$ during the interval $[s^{\mathcal{T}}, t)$. Figure 3 demonstrates a typical region.

Structural Properties of Busy Regions Produced by $\mathcal{A}$

Lemma 6. *If there exists a time t at which at least one machine is idle in $\mathcal{A}$ and a job j such that $r_j \leq t \leq x_j$, then j cannot be rejected by $\mathcal{A}$.*

Proof. By the algorithm definition, a machine is left idle at time t only if it is feasible to achieve its current queue even when that machine is left idle until time $t + p + 1$ or later. Since j could be feasibly achieved over the interval $[t, t + p]$ without disrupting the completion of any other jobs in the system at that time, it must be accepted. □

Lemma 7. *For busy region $[s, t)$, FEASIBLE$(Q_t^{\mathcal{A}}, t, t + p + 1)$.*

Proof. By the definition of a region, neither machine was committed to a job at the onset of time t, and at least one remains idle during the time $[t, t + 1)$. If both machines remain idle at time t, then $Q_t^{\mathcal{A}} = \emptyset$ and thus the lemma trivially true. Alternatively some job, f, starts on one machine at t, while the secondary decision is to remain idle. Based on the algorithm definition, it must be that the test FEASIBLE$(\dot{Q}_t^{\mathcal{A}}, t + p, t + p + 1)$ succeeded, and by Lemma 2,

$\text{FEASIBLE}(\dot{Q}_t^A \cup \{f\}, t, t+p+1)$. Since $Q_t^A \subseteq \{f\} \cup \dot{Q}_t^A$, this implies $\text{FEASIBLE}(Q_t^A, t, t+p+1)$.   □

Lemma 8. *For busy region $[s,t)$ and arbitrary time t', assume $j \neq a$ is the latest job of $\mathcal{R}$ to be started by $\mathcal{A}$ such that $x_j \geq t'$. In this case, j must start due to a secondary decision at some time s_j, and further it must be that $\dot{Q}_{s_j}^A \setminus \{j\} \subseteq Q_t^A$.*

Lemma 9. *For any $j \neq a$ started by $\mathcal{A}$ during busy region $[s,t)$, $x_j \leq t$.*

Proof. For contradiction, let j with $x_j \geq t+1$ be the latest such job to be started by $\mathcal{A}$. By Lemma 8, j is started through a secondary decision at a time s_j, and $\dot{Q}_{s_j}^A \setminus \{j\} \subseteq Q_t^A$. By Lemma 7, $\text{FEASIBLE}(Q_t^A, t, t+p+1)$ and thus $\text{FEASIBLE}(\dot{Q}_{s_j}^A \setminus \{j\}, t, t+p+1)$. Since $x_j \geq t+1$, we can schedule j over the interval $[t+1, t+p+1]$ while still achieving $\dot{Q}_{s_j}^A \setminus \{j\}$ starting the machines respectively at t and $t+p+1$. Therefore $\text{FEASIBLE}(\dot{Q}_{s_j}^A, t, t+1)$. Since $t \geq s_j + p$ and $t \geq c$ where c denotes the time until which the opposite machine was committed as j started, this demonstrates the feasibility of $\text{FEASIBLE}(\dot{Q}_{s_j}^A, c, s_j + p + 1)$. This contradicts the fact that $\mathcal{A}$ starts j with a secondary decision at s_j.   □

Lemma 10. *For busy region $[s,t)$, if $\exists j \in \mathcal{R}$ with $j \neq a$ and $x_j \geq s_y^A + p$, then $x_z \geq s_y^A + p$.*

Lemma 11. *For busy region $[s,t)$ with $|\mathcal{R}| \geq 2$, we consider the tail $\mathcal{T}$. If all jobs of $\mathcal{T}$ are released on or before time $s^{\mathcal{T}} - 1$, then $x_z \geq s_y^A + p$.*

Lemma 12. *For busy region $[s,t)$ with $|\mathcal{R}| \geq 2$, if $x_z \geq s_y^A + p$, then each of the following are true.*

(A) Any job k with $r_k \leq t - p \leq x_k$ must have been accepted by $\mathcal{A}$.
(B) There exist job $f \in Q_t^A$ such that $s_f^A = x_f = t$.

Comparing Progress of $\mathcal{A}$ Versus OPT over a Busy Region

For ease of exposition in the analysis of busy region $[s,t)$, we let set $\mathcal{G} = (F_t^{\text{OPT}} \cup P_t^{\text{OPT}}) \setminus (F_s^{\text{OPT}} \cup P_s^{\text{OPT}})$ denote those jobs which contribute to Φ^{OPT} during the region $[s,t)$. We let $\mathcal{G}_+$ (resp. $\mathcal{G}_-$) denote those jobs of $\mathcal{G}$ which were accepted (resp. rejected) by $\mathcal{A}$. We further let $\tilde{j}_t$ represent the element with which j is matched at time t, if such element exists, and j otherwise. An element's match often serves as a substitute in our analysis.

Lemma 13. *Each $j \in \mathcal{G}$ satisfies precisely one of the following conditions*

(A) $s \leq s_j^{\text{OPT}} < t$ and $j \notin P_s^{\text{OPT}}$;
(B) $s_j^{\text{OPT}} = t$, $\tilde{j}_s \in F_t^A \setminus F_s^A$ and $\tilde{j}_s \neq a$;
(C) $s_j^{\text{OPT}} \geq t$ and $\tilde{j}_s = a$.

Based upon the three conditions of Lemma 13, we can define another partition of the set $\mathcal{G}$ into sets $\mathcal{G}_A$, $\mathcal{G}_B$ and $\mathcal{G}_C$. We can superimpose this partition together with our partition of $\mathcal{G}_+$ and $\mathcal{G}_-$ to denote sets such as $\mathcal{G}_{A+} \equiv \mathcal{G}_A \cap \mathcal{G}_+$ and $\mathcal{G}_{A-} \equiv \mathcal{G}_A \cap \mathcal{G}_-$. Though we might use similar notations for $\mathcal{G}_B$ and $\mathcal{G}_C$, the next lemma shows that this is unnecessary.

Lemma 14. *All jobs of $\mathcal{G}_B$ and $\mathcal{G}_C$ are accepted by $\mathcal{A}$.*

Proof. For $j \in \mathcal{G}$, $r_j < t$, yet for $j \in \mathcal{G}_B \cup \mathcal{G}_C$, $x_j \geq s_j^{\text{OPT}} \geq t$. By the definition of the region, at least one machine of $\mathcal{A}$ is idle at time t. Therefore, we may apply Lemma 6 to assure the acceptance of j by $\mathcal{A}$. $\square$

Our next lemma specifically examines sets $\mathcal{G}_{A+}$ and $\mathcal{G}_{A-}$. We can further partition each of these sets based upon which machine OPT uses to achieve such a job. For machine m, we let $\mathcal{G}_{A+}^m$ denote the number of jobs of $\mathcal{G}_{A+}$ which are run by OPT, and $\mathcal{G}_{A-}^m$ the corresponding number from $\mathcal{G}_{A-}$.

Lemma 15. *For busy region $[s, t)$ and arbitrary machine m,*

(A) $|\mathcal{G}_B^m| + |\mathcal{G}_{A+}^m| + 2 \cdot |\mathcal{G}_{A-}^m| \leq |\mathcal{R}|$.
(B) Furthermore, if inequality (A) is tight and $|\mathcal{R}| \geq 2$, both of the following are true.
(1) Either m starts a job of $\mathcal{G}_{A-}^m$ at $t - p$ or is processing a job of $\mathcal{G}_A^m \cup \mathcal{G}_B^m$ during $[t, t+1)$.
(2) No job of $\mathcal{G}_{A+}^m$ can be started during the interval $[s^\tau, s_y^\mathcal{A} + p)$.

Proof (part A only). We use a basic counting argument to establish the lemma. We initially consider all elements of $\mathcal{R}$ to be unmarked. For each job $j \in \mathcal{G}_A$ started on m by OPT, we mark those jobs of $\mathcal{R}$ which are currently being processed at the time j starts. Because jobs have equal length, it is impossible for OPT to start two jobs on a single machine, both of which are started while a single job of $\mathcal{A}$ executes. Therefore, each job of $\mathcal{A}$ is marked at most once. By Lemma 6, if j were rejected by $\mathcal{A}$, two distinct jobs of $\mathcal{A}$ must be running at time s_j^{OPT}. Therefore two jobs of $\mathcal{R}$ are marked in association with j and $2 \cdot |\mathcal{G}_{A-}^m|$ jobs are marked overall in association with rejected jobs. Any job of $\mathcal{G}_{A+}^m$ must mark at least one further job of $\mathcal{R}$, since there is never a time during $[s, t)$ when both machines of $\mathcal{A}$ are idle. Therefore $|\mathcal{G}_{A+}^m| \leq |\mathcal{R}| - 2 \cdot |\mathcal{G}_{A-}^m|$ and thus $|\mathcal{G}_{A+}^m| + 2 \cdot |\mathcal{G}_{A-}^m| \leq |\mathcal{R}|$.

This establishes condition (A) in the case where $\mathcal{G}_B^m = \emptyset$. If $\mathcal{G}_B^m \neq \emptyset$, then by definition $\tilde{j}_s \in F_t^\mathcal{A} \setminus F_s^\mathcal{A}$ and $\tilde{j}_s \neq a$ for any $j \in \mathcal{G}_B^m$. Since $j \neq a$, this already requires that $|\mathcal{R}| \geq 2$. By Lemma 9, $x_{\tilde{j}_s} \leq t$ and thus $x_j \leq t$, as $j \prec \tilde{j}_s$ in the case when $j \in W_t^{\text{OPT}} \setminus P_t^{\text{OPT}}$. Such a job must then be started precisely at time t on m and so set $\mathcal{G}_B^m$ must consist of a single such element, if non-empty. Notice that if z was not marked by an element of $\mathcal{G}_A^m$, then $|\mathcal{G}_{A+}^m| + 2 \cdot |\mathcal{G}_{A-}^m| \leq |\mathcal{R}| - 1$ and so $|\mathcal{G}_B^m| + |\mathcal{G}_{A+}^m| + 2 \cdot |\mathcal{G}_{A-}^m| \leq |\mathcal{R}|$. If z is marked, the only way this can be accomplished is by a job which starts precisely at time $t - p$, given that $\tilde{j}_s$ must be started at time t. As $x_{\tilde{j}_s} \geq t \geq s_y^\mathcal{A} + p$, by combining Lemmas 10 and 12, we have that such a job starting at $t - p$ must have been a job accepted by $\mathcal{A}$. As this accepted job marks both y and z, again we find that $|\mathcal{G}_{A+}^m| + 2 \cdot |\mathcal{G}_{A-}^m| \le |\mathcal{R}| - 1$ and so $|\mathcal{G}_B^m| + |\mathcal{G}_{A+}^m| + 2 \cdot |\mathcal{G}_{A-}^m| \le |\mathcal{R}|$. This establishes part (A) in general.   □

Lemma 16. *For $u \in \mathcal{G}_+$, $\tilde{u}_s \in (F_t^{\mathcal{A}} \cup P_t^{\mathcal{A}}) \setminus (F_s^{\mathcal{A}} \cup P_s^{\mathcal{A}})$.*

Lemma 17. *For distinct $u, v \in \mathcal{G}$, $\tilde{u}_s$ and $\tilde{v}_s$ are distinct.*

Lemma 18. *If $\tilde{a}_s \in Q_t^{\text{OPT}}$ for region $[s, t)$, then $\tilde{a}_s \in W_t^{\text{OPT}}$ and $a \notin P_s^{\mathcal{A}}$.*

Lemma 19. *For busy region $[s, t)$, assume that $\tilde{a}_s \in Q_t^{\text{OPT}}$ and there exists some $b \in P_s^{\mathcal{A}} \cup W_t^{\mathcal{A}}$. We may legally establish a match, $M_t(b) = \tilde{a}_s$.*

Lemma 20. *If $\Phi_s^{\text{OPT}} \le \Phi_s^{\mathcal{A}}$ for busy region $[s, t)$, then $\Phi_t^{\text{OPT}} \le \Phi_t^{\mathcal{A}}$.*

Proof (sketch). For ease of exposition, we introduce notation $\Phi_{[s,t)}^{\mathcal{A}}$ to denote $(\Phi_t^{\mathcal{A}} - \Phi_s^{\mathcal{A}})$, and likewise $\Phi_{[s,t)}^{\text{OPT}} = (\Phi_t^{\text{OPT}} - \Phi_s^{\text{OPT}})$. Our goal is to show that $\Phi_{[s,t)}^{\text{OPT}} \le \Phi_{[s,t)}^{\mathcal{A}}$. In accordance with Lemmas 13 and 14, we rewrite $\Phi_{[s,t)}^{\text{OPT}}$ as,

$$
\begin{aligned}
\Phi_{[s,t)}^{\text{OPT}} &= 2 \cdot |\mathcal{G}| = 2 \cdot |\mathcal{G}_{A+}^{m_1} \cup \mathcal{G}_{A+}^{m_2} \cup \mathcal{G}_{A-}^{m_1} \cup \mathcal{G}_{A-}^{m_2} \cup \mathcal{G}_B^{m_1} \cup \mathcal{G}_B^{m_2} \cup \mathcal{G}_C| \\
&= (|\mathcal{G}_B^{m_1}| + |\mathcal{G}_{A+}^{m_1}| + 2 \cdot |\mathcal{G}_{A-}^{m_1}|) + (|\mathcal{G}_B^{m_2}| + |\mathcal{G}_{A+}^{m_2}| + 2 \cdot |\mathcal{G}_{A-}^{m_2}|) + |\mathcal{G}_C| \\
&\quad + |\mathcal{G}_{A+}^{m_1} \cup \mathcal{G}_{A+}^{m_2} \cup \mathcal{G}_B^{m_1} \cup \mathcal{G}_B^{m_2} \cup \mathcal{G}_C| \\
&= (|\mathcal{G}_B^{m_1}| + |\mathcal{G}_{A+}^{m_1}| + 2 \cdot |\mathcal{G}_{A-}^{m_1}|) + (|\mathcal{G}_B^{m_2}| + |\mathcal{G}_{A+}^{m_2}| + 2 \cdot |\mathcal{G}_{A-}^{m_2}|) + |\mathcal{G}_C| + |\mathcal{G}_+|
\end{aligned}
$$

By applying Lemma 15(A) to each of the two machines of OPT, we conclude that $\Phi_{[s,t)}^{\text{OPT}} \le 2 \cdot |\mathcal{R}| + |\mathcal{G}_C| + |\mathcal{G}_+|$. In contrast, we claim that $\Phi_{[s,t)}^{\mathcal{A}} \ge 2 \cdot |\mathcal{R}| + \delta + |\mathcal{G}_+|$, where δ is defined as

$$
\delta = \begin{cases} 1 & \text{if } P_s^{\mathcal{A}} = \emptyset \text{ and } P_t^{\mathcal{A}} \ne \emptyset \\ -1 & \text{if } P_s^{\mathcal{A}} \ne \emptyset \text{ and } P_t^{\mathcal{A}} = \emptyset \\ 0 & \text{otherwise} \end{cases}
$$

The δ term adjusts for the discontinuity inherent in our definition of $\Phi^{\mathcal{A}}$, based upon whether or not set $P^{\mathcal{A}}$ is empty. With that aside, each job of $\mathcal{R}$ results in a relative contribution of at least 2, as such job is added to $F^{\mathcal{A}}$ though perhaps removed from $P^{\mathcal{A}}$. The only other way in which a job can be removed from $P^{\mathcal{A}}$ is if it becomes matched. We consider the creation of a new match in one particular case. If $\tilde{a}_s \in Q_t^{\text{OPT}}$ and there exists some $b \in P_s^{\mathcal{A}} \cup W_t^{\mathcal{A}}$, we create a match, $M_t(b) = \tilde{a}_s$, with the validity of such a match established as per Lemma 19. In this special case there is indeed a relative loss of one unit due to job b. However as $\tilde{a}_s \in W_t^{\text{OPT}}$, we show that $a \notin P_s^{\mathcal{A}}$. If it were, then a would be unmatched at time s, thus $\tilde{a}_s = a$, yet since $a \in Q_t^{\text{OPT}}$ it is not possible that $a \in P_s^{\mathcal{A}} \subseteq W_s^{\mathcal{A}} = Q_s^{\mathcal{A}} \setminus Q_t^{\text{OPT}}$. Given that $a \notin P_s^{\mathcal{A}}$, its presence in $\mathcal{R}$ results in a profit of 3 rather than the minimally assumed profit of 2 and so this offsets the loss of 1 due to the matching of b.

Next, we consider the impact of set $\mathcal{G}_+$ on $\Phi_{[s,t)}^{\mathcal{A}}$. By Lemma 16, for any $u \in \mathcal{G}_+$ there exists $\tilde{u}_s \in (F_t^{\mathcal{A}} \cup P_t^{\mathcal{A}}) \setminus (F_s^{\mathcal{A}} \cup P_s^{\mathcal{A}})$ and by Lemma 17, those associated jobs are distinct. Each such job provides an additional point towards $\Phi_{[s,t)}^{\mathcal{A}}$ beyond the presumed $2 \cdot |\mathcal{R}|$, either providing 1 if in $P_t^{\mathcal{A}}$, or providing 3 rather than 2 if in $F_t^{\mathcal{A}}$. It is important to note that these $|\mathcal{G}_+|$ additional points are also distinct from the possible point associated with a in the preceding paragraph, as in that case neither a nor $\tilde{a}_s$ can lie in $\mathcal{G}_+$. We have thus far established that $\Phi_{[s,t)}^{\mathcal{A}} \geq 2 \cdot |\mathcal{R}| + \delta + |\mathcal{G}_+|$.

For the remainder of the proof, we focus on the expression $(\delta - |\mathcal{G}_C|)$, recalling that $|\mathcal{G}_C|$ is either 0 or 1. In the case that $(\delta - |\mathcal{G}_C|) \geq 0$ the theorem follows, as $\Phi_{[s,t)}^{\text{OPT}} \leq 2 \cdot |\mathcal{R}| + |\mathcal{G}_C| + |\mathcal{G}_+| \leq 2 \cdot |\mathcal{R}| + \delta + |\mathcal{G}_+| \leq \Phi_{[s,t)}^{\mathcal{A}}$.

If $(\delta - |\mathcal{G}_C|) < 0$ we undertake a detailed case analysis to counterbalance this deficit either by showing a gap in our upper bound on $\Phi_{[s,t)}^{\text{OPT}}$ due to a strict inequality when applying Lemma 15(A) to one or both of the two machines of OPT, or by showing a gap in our lower bound on $\Phi_{[s,t)}^{\mathcal{A}}$. Details are omitted here.   □

Theorem 1. *Algorithm $\mathcal{A}$ is $\frac{3}{2}$-competitive.*

Proof. We show that $\Phi_t^{\text{OPT}} \leq \Phi_t^{\mathcal{A}}$ by induction. Initially, $\Phi_0^{\text{OPT}} = \Phi_0^{\mathcal{A}} = 0$. Neither potential function changes during regions for which $\mathcal{A}$ remains completely idle, as per Lemma 5. The times during which $\mathcal{A}$ uses one or more machines can be partitioned into regions $[s,t)$, for which Lemma 20 applies, thereby extending the induction from time s to time t for each such region. We conclude that $\Phi_\infty^{\text{OPT}} \leq \Phi_\infty^{\mathcal{A}}$. Since the queues of both $\mathcal{A}$ and OPT are presumed to be empty at $t = \infty$, $P_\infty^{\mathcal{A}} = P_\infty^{\text{OPT}} = \emptyset$. Therefore, $\Phi_\infty^{\text{OPT}} \leq \Phi_\infty^{\mathcal{A}}$ is equivalent to $2 \cdot |F_\infty^{\text{OPT}}| \leq 3 \cdot |F_\infty^{\mathcal{A}}|$, thus $\frac{|\text{OPT}|}{|\mathcal{A}|} \leq \frac{3}{2}$.   □

Theorem 2. *For $m = 2$, no non-preemptive, deterministic algorithm can be better than $\frac{3}{2}$-competitive.*

Proof. Consider the release of a single job $j = \langle 0, 3p - 1 \rangle$. A deterministic algorithm must start j at some time $0 \leq t \leq 2p - 1$, or else have unbounded competitiveness. Yet if two identical jobs with parameters $\langle t + 1, t + 1 + p \rangle$ are subsequently released, one must be rejected. It is easy to verify that OPT can achieve all three jobs.   □

Acknowledgments

We gratefully acknowledge the correspondence of Jiří Sgall.

References

1. P. Baptiste, P. Brucker, S. Knust, and V. G. Timkovsky. Ten notes on equal-processing-time scheduling. *4OR: Quarterly J. Belgian, French and Italian Operations Research Societies*, 2(2):111–127, July 2004.
2. S. K. Baruah, J. R. Haritsa, and N. Sharma. On-line scheduling to maximize task completions. *J. Combin. Math. and Combin. Computing*, 39:65–78, 2001.

3. A. Borodin and R. El-Yaniv. *Online Computation and Competitive Analysis.* Cambridge University Press, New York, 1998.
4. M. Chrobak, W. Jawor, J. Sgall, and T. Tichý. Online scheduling of equal-length jobs: Randomization and restarts help. In J. Díaz, J. Karhumäki, A. Lepistö, and D. Sannella, editors, *Proc. 31st Int. Colloquium on Automata, Languages and Programming (ICALP)*, volume 3142 of *Lecture Notes in Computer Science*, pages 358–370, Turku, Finland, July 2004. Springer-Verlag.
5. J. Ding and G. Zhang. Online scheduling with hard deadlines on parallel machines. In *Proc. Second Int. Conference on Algorithmic Aspects in Information and Management*, Lecture Notes in Computer Science, Hong Kong, China, June 2006. Springer-Verlag. To appear.
6. S. Goldman, J. Parwatikar, and S. Suri. On-line scheduling with hard deadlines. *J. Algorithms*, 34(2):370–389, Feb. 2000.
7. M. H. Goldwasser. Patience is a virtue: The effect of slack on competitiveness for admission control. *J. Scheduling*, 6(2):183–211, Mar./Apr. 2003.
8. M. H. Goldwasser and B. Kerbikov. Admission control with immediate notification. *J. Scheduling*, 6(3):269–285, May/June 2003.
9. J. R. Jackson. Scheduling a production line to minimize maximum tardiness. Research Report 43, Management Science Research Project, University of California, Los Angeles, Jan. 1955.
10. B. B. Simons. Multiprocessor scheduling of unit length jobs with arbitrary release times and deadlines. *SIAM J. Comput.*, 12:294–299, 1983.
11. B. B. Simons and M. K. Warmuth. A fast algorithm for multiprocessor scheduling of unit-length jobs. *SIAM J. Comput.*, 18(4):690–710, 1989.

Paging with Request Sets

Leah Epstein[1], Rob van Stee[2,*], and Tami Tamir[3]

[1] Department of Mathematics, University of Haifa, 31905 Haifa, Israel
lea@math.haifa.ac.il
[2] Department of Computer Science, University of Karlsruhe,
D-76128 Karlsruhe, Germany
vanstee@ira.uka.de
[3] School of Computer Science, The Interdisciplinary Center, Herzliya, Israel
tami@idc.ac.il

Abstract. A generalized paging problem is considered. Each request is expressed as a set of u pages. In order to satisfy the request, at least one of these pages must be in the cache. Therefore, on a page fault, the algorithm must load into the cache at least one page out of the u pages given in the request. The problem arises in systems in which requests can be serviced by various utilities (e.g., a request for a data that lies in various web-pages) and a single utility can service many requests (e.g., a web-page containing various data). The server has the freedom to select the utility that will service the next request and hopefully additional requests in the future.

The case $u = 1$ is simply the classical paging problem, which is known to be polynomially solvable. We show that for any $u > 1$ the offline problem is NP-hard and hard to approximate if the cache size k is part of the input, but solvable in polynomial time for constant values of k. We consider mainly online algorithms, and design competitive algorithms for arbitrary values of k, u. We study in more detail the cases where u and k are small. We also give an algorithm which uses resource augmentation and which is asymptotically optimal for $u = 2$.

1 Introduction

Modern operating systems have multiple memory levels. In simple structures, a memory is organized in equally sized pages. The basic paging model is defined as follows. The system has a slow but large memory (e.g. disk) where all pages are stored. The second level is a small, fast memory (cache) where the system brings a page in order to use it. If a page which is not in the faster memory level is requested, a page fault occurs, and a page must be evicted in order to make room to bring in the new page. The processor must slow down until the page is brought into memory, and in practice, for many applications, the performance of the system depends almost entirely on the number of page caching (uploads to the cache). We define the cost of an algorithm as simply the total number of page uploads.

---

* Research supported by Alexander von Humboldt Foundation.

L. Arge and R. Freivalds (Eds.): SWAT 2006, LNCS 4059, pp. 124–135, 2006.

Traditional paging problems assume that at every step of a request sequence, there exists a unique page that can fulfill the needs of the system. This page must be loaded into the cache if it does not reside there already. Such a situation is plausible, however, often the need for a very specific page is not acute and the need is for a certain piece of *information* rather than a certain page.

For instance, on the *world wide web*, information is often mirrored across many websites (e.g. currency exchange rates). In such a situation, it makes sense to make a list of several places where the information can be found, and allow the system to conveniently choose from them. Another application is a *media broadcasting system*. In such a system, media files, which are the smallest media units that can be loaded into the system, are kept on disks. Media files are replicated and each is stored on several, not necessarily uniform, disks. The content of each disk is known. A broadcast is defined by a list of required media in some specific order (for example, list of songs to be transmitted). The goal is to broadcast all media while minimizing the number of disks loadings.

Let U_j denote the j-th request, that is, U_j is the set of pages containing the information required in the j-th request. In order to keep the running times of paging algorithms low, we fix the number of options given to the algorithm in every request to be a parameter u. Formally, for all j, we assume that $|U_j| = u$. The size of the cache is denoted by k. The request sequence is denoted by σ, and n denotes the total number of different pages that occur in σ. Given a set of requests $\{U_1, \ldots, U_j\}$, we say that a set S *covers* this set of requests if for all $i, 1 \le i \le j$, it holds that $S \cap U_i \neq \emptyset$. In online paging problems, each element in the request sequence arrives after the previous request was serviced (i.e. the decision on the eviction of another page was made). The competitive ratio is the asymptotic worst case ratio between the cost of an online algorithm and the cost of an optimal offline algorithm OPTwhich knows all requests in advance.

Related Work. For the classical offline problem, there exists a polynomial simple optimal algorithm LFD (Longest Forward Distance) designed by Belady [2]. LFD always evicts the page for which the time until the next request to it is maximal. Two common paging algorithms for the classical paging problem are LRU (Least Recently Used) and FIFO (First In First Out). LRU computes for each page which is present in the cache the last time it was used and evicts the page for which this time is minimum. FIFO acts like a queue, evicting the page that has spent the longest time in the cache. Variants of both are common in real systems. Although LRU outperforms FIFO in practice, LRU and FIFO are known to have the same competitive ratio of k. Further this ratio is known to be the best possible, see [14, 10]. Randomized algorithms were studied by [7, 1, 12]. See [9] for a survey on online paging problems.

One generalization of paging was studied by Fiat and Ricklin [8]. They studied the weighted paging problem (i.e., where each slot in the cache may have a distinct cost of replacing the page stored in it), and gave algorithms with a doubly exponential upper bound in k. They showed that the competitive ratio of any algorithm for this problem is at least $k^{\Omega(k)}$. For the special case where only two weights are allowed they have a $k^{O(k)}$-competitive algorithm.

As explained in Section 2, paging with request sets captures other well-studied problems such as set cover and hypergraph vertex cover [6]. The online problem captures dynamic versions of the above problems such as online vertex cover. However, in the studied version of online vertex cover (see e.g., [5]), the input graph is revealed vertex by vertex, while our problem induces a problem in which the graph is revealed edge after edge. Both our results and the results in [5] imply that the online problem is significantly harder than the offline one.

The online version of our problem is a special case of metrical task systems and specifically of their sub-class, *metrical service systems* (also called *forcing tasks systems*). For details, see [3, 11, 4].

Our Results. We show that unlike the classical paging problem, paging with request sets is NP-hard and in fact hard to approximate within a factor of $\Omega(u)$ unless P = NP. If k is fixed, then the problem can be solved via dynamic programming.

We further study the online problem. We show that natural extensions of LRU and FIFO are not competitive. We present competitive algorithms for all values of u and k. We consider the paging model described above as well as the same paging model with bypassing. Note that even though the competitive ratios of our algorithms are quite high for most variants (i.e. exponential in k), we show that in many of the cases this is unavoidable. This is similar to the generalization of paging considered in [8] that also results in high competitive ratios.

Finally, we present a simple online algorithm which uses resource augmentation. Here the offline algorithm that it is compared to is restricted to a cache size of $h < k$. This generalization was considered already by Sleator and Tarjan in [14]. We show that the competitive ratio of our algorithm tends to the optimal value of 2 for $u = 2$ and large values of k/h.

2   The Offline Problem

We begin by describing a dynamic program (DP) for the problem. For a set S of size exactly k, let $P_{j,S}$ denote the cost of servicing the first j requests of σ in a way that the cache content after the jth request is S. Since the jth request is the last to be serviced, $P_{j,S}$ is defined only for $S \cap U_j \neq \emptyset$ (or defined to be ∞ when this does not hold). An optimal solution can be obtained using the following dynamic program. Here we assume the optimal total cost is at least k.[1]

Initialization. For $j \leq k$, set $P_{j,S} = k$ if S covers $U_1, \ldots, U_j$. Else, set $P_{j,S} = \infty$.

In the initialization, we consider all possible ways to fill the cache. This requires k loads and therefore has the cost k. We assume that no page replacements are done as long as the cache has space to load pages, thus, we ignore these sets that do not cover the first j requests (by setting their price to ∞).

---

[1] This assumption can be removed by considering also sets of size smaller than k. We skip this technical extension.

Step. For $j > k$, if $S \cap U_j = \emptyset$, set $P_{j,S} = \infty$. Else set

$$P_{j,S} = \min \{P_{j-1,S}, \; 1 + \min\{P_{j-1,S'} | S = S' \cup \{x\} \setminus \{y\} \text{ for some } x \in U_j\}\}.$$

For $j > k$, when calculating $P_{j,S}$ for $S \cap U_j \neq \emptyset$ there are two cases. The first is when no upload is done for servicing the jth request, that is, the current content of the cache covers U_j. In this case the total cost of servicing the first j requests is equal to the cost of servicing the first $j - 1$ requests. The second case is when an upload is essential. If the content of the cache, S', does not cover U_j then some page $y \in S'$ is removed and a page $x \in U_j$ is inserted. The cost in this case is one for the current upload plus the cost of servicing the first $j - 1$ requests in a way that the cache content before the jth request is S'. Therefore, the minimal cost of servicing the first j requests is determined by the optimal S'. To calculate $P_{j,S}$, the minimum cost out of these two cases is considered.

The optimal cost for the whole sequence is the minimal value of $P_{|\sigma|,S}$ for some set S. The size of the DP table is polynomial in n: there are n^k possible sets S, for each such set the value $P_{j,S}$ is calculated for $|\sigma|$ different values of j. Each entry is calculated in time $O(uk)$.

Corollary 1. *Paging with request sets can be solved in time $O(|\sigma|ukn^k)$.*

Clearly, the above DP algorithm is polynomial only for constant k. While the offline traditional paging problem is known to be optimally solved for arbitrary k, this is not the case for the generalized problem. In particular, we show that our problem is NP-hard even for request sets of size 2.

Theorem 1. *Offline paging with request sets and an arbitrary cache size k is NP-hard even for $u = 2$.*

Proof. Reduction from *Vertex Cover* (VC). Given an instance for VC, $G = (V, E)$, and the question whether G has a vertex cover of size k, construct the following instance for paging with request sets. The sequence σ consists of $|E|$ requests; the jth request is $U_j = \{v_{j1}, v_{j2}\}$, where (v_{j1}, v_{j2}) is the jth edge (in arbitrary order) of E. It is easy to verify that it is possible to service all the requests at a total cost of k if and only if G has a vertex cover of size k. $\square$

This reduction can be generalized to show that for arbitrary sizes of sets the problem is as hard as set-cover. Thus, it cannot be approximated within factor $\Omega(\log n)$ [13]. The reduction from vertex cover can be extended for any instance with uniform size request sets, that is, when $|U_j| = u$ for all u.

Theorem 2. *Assuming $P \neq NP$, the optimal cost of paging with request sets for instances having request set of size u cannot be approximated in polynomial time within factor $(u - 1 - \varepsilon)$.*

Proof. We show an L-reduction from hypergraph vertex cover, for which this hardness result is known [6]. Let S be an instance of hypergraph vertex cover (HVC) with k nodes, and let *opt* be a minimal size vertex cover of S. Build an instance σ_s for our paging problem by listing the hyperedges of S in some order.

Consider a cache of size k. An optimal algorithm can service σ_s at cost $|opt|$ by placing the nodes of the vertex cover in the cache. Consider any algorithm that services σ_s. The set of vertices that are inserted into the cache along the whole sequence form an HVC. The cache size is selected to be k so no deletions are required. Therefore, the cost of servicing σ_s is the size of the HVC found by the algorithm.   □

3   The Online Problem

3.1   The Performance of Standard Algorithms

In this section we show that several reasonable versions of LRU and FIFO, adapted for paging with request sets, are not competitive. To generalize the algorithms, we need to define the behavior on a page fault. Specifically, we not only need to define the method of page eviction but also the method of choosing a page of the new request to be inserted to the cache.

The page eviction method of FIFO is identical to the original algorithm, that is, remove from the cache the page that was inserted first. For LRU, we say that a page is *used* if it appears in a request (but not necessarily downloaded), thus, LRU removes from the cache a page that appeared least recently in a request. Ties are broken arbitrarily. We mention that our non-competitiveness proof below is suitable also for other eviction methods of LRU, like removing the page that was least recently "essential", that is, loaded or serviced a request.

For analyzing the loading page method, we first consider a situation where the choice of a new page is arbitrary. In particular, it might be that the page inserted to the cache is the one that has been out of the cache the longest time (i.e., has never been in the cache, or has been evicted least recently). The same example is applicable to both LRU and FIFO.

Lemma 1. *The above versions of LRU and FIFO are not competitive for paging with request sets.*

Proof. Given $k \geq 2$, and $u \geq 2$, let $\{a_0, \ldots, a_{u+k-1}\}$ denote $u + k$ designated pages. For convenience of notation, define a_j for $j \geq u + k$ to be $a_{j \mod (u+k)}$.

The sequence of requests repeats the following subsequence $\sigma_0, \ldots, \sigma_{u+k-1}$. Request σ_i is defined to be $\{a_i, \ldots, a_{i+u-1}\}$. Both LRU and FIFO have the cache contents $\{a_{i+u}, \ldots, a_{i+u+k-1}\}$ before this request, where pages are listed in the order in which they will be evicted. The page that has been out of the cache for the longest time is $a_i = a_{i+u+k}$. Clearly each request is a fault.

However, OPT keeps in its cache the pages $a_{u \cdot \ell}$ for all $\ell \geq 0$ such that $u \cdot \ell < u + k$. The number of such pages is $\lceil \frac{k}{u} \rceil + 1$. Since $u \geq 2$, it is clear that this number never exceeds k. Thus, after loading these pages in its cache, OPT never makes another fault. This proves the lemma.   □

Next, consider versions of LRU and FIFO that prefer to insert into the cache a page that was evicted most recently. In this case we use a set of $k + 1$ pages

$\{a_0, \ldots, a_k\}$, and $u - 1$ additional pages $\{b_1, \ldots, b_{u-1}\}$. Similar to the previous example, we let a_j for $j \geq k + 1$ to be $a_{j \bmod(k+1)}$. The sequence repeats the sub-sequence of requests $\tau_0, \ldots, \tau_k$ where $\tau_i = \{a_i, b_1, \ldots, b_{u-1}\}$. The requests b_j are never inserted into the cache. Before τ_i arrives, the cache contains pages $\{a_{i+1}, a_{i+2}, \ldots, a_{i-1}\}$ (listed in the order in which they are to be evicted). LRU and FIFO fault on every request, whereas OPT keeps the pages $\{b_1, a_1, \ldots, a_{k-1}\}$ in its cache. We may conclude that LRU and FIFO are not competitive.

Since the standard algorithms fail, in the next subsections we design very different algorithms for our problem. These algorithms try to track the configuration of OPT in order to remain competitive.

3.2   A Competitive Algorithm

Our algorithm works in phases. A phase ends when it must be the case that OPT had a fault. Consider a single phase. Let C be a collection of sets $S_1, S_2, \ldots$ each of size at most k (cache size) such that each set S_i covers the requests presented so far in the phase. If C is empty then OPT must miss and a new phase begins. Otherwise try to make the cache of the online algorithm ONL as similar to the sets of C as possible. Specifically, ONL knows the set C. In every step (miss) ONL tries to exclude a set from C.

We use the following assumptions. Algorithms silently ignore requests which do not cause faults. In the analysis we may assume that each request is indeed a fault: requests which are not faults can only increase the optimal cost. Thus we simply remove non-fault requests from the request sequence before starting our analysis. Note that our algorithms may replace more than one page on a fault.

Our construction will lead to the following general theorem.

Theorem 3. *For paging with request sets of size u and a cache of size k, there exists an algorithm* $\text{ALG}_u(k)$ *which has a competitive ratio of* $\frac{u^{k+1} - u}{u - 1}$ *. Moreover, for any constant k, the competitive ratio of any online algorithm is* $\Omega(u^k)$.

We first describe a competitive online algorithm for the case $u = k = 2$. Below, we will show how to use this algorithm as a subroutine in more general cases.

The algorithm ALG works in phases. A phase is a subsequence of requests, where it can be proved that OPT has made a fault either on one of the requests of this phase (excluding the very first one) or on the first page of the next phase. In the sequel, we analyze the contents of the cache of OPT in the case that it did not make a fault in the current phase (except, possibly, on the first request). We denote the (pages of the) first request of a phase by $\{a, b\}$. Our algorithm will insert b in the cache. Throughout the phase, ALG always keeps one page out of $\{a, b\}$ in its cache. The situation of OPT is similar if it did not have a fault yet.

The phase continues until we know that OPT has a fault, or will have a fault on the current request. This request will start a new phase. We assume that OPT does not make a fault, until this leads to a contradiction. The easiest and most important case is the case where three independent (non-overlapping) sets are requested. This clearly implies a fault of OPT, and the request for the third independent set starts a new phase. There are two cases for the second request.

Request 2 is for an independent set. Denote request 2 by $\{c, d\}$. The algorithm loads c into the cache and has $\{b, c\}$ in the cache. From now on, ALG also keeps one page from $\{c, d\}$ in its cache during this phase. Thus, the remaining requests are serviced by loading a matching page (if possible). Since we assume that OPT does not make a fault, OPT now has one page from $\{a, b\}$ and one page from $\{c, d\}$ in its cache.

If a new independent set is requested ($\{e, f\}$), then OPT must have a fault and this request starts a new phase. As long as this does not happen, requests always have exactly one page in common with at least one of the first two requested pairs. The algorithm always loads such a page. Thus in this phase, it never loads a page outside of the set $\{a, b, c, d\}$. We analyze the options for the next requests.

Case A. If request 3 consists of one page from $\{a, b, c, d\}$ and one other page, then OPT must have that page from $\{a, b, c, d\}$ in its cache or make a fault. When this happens, ALG fixes this page in its cache and does not evict it anymore. Thus it mimics OPT in the case that OPT has not made a fault yet.

Suppose a is fixed (as a result of request $\{a, e\}$). Then a is not requested anymore in this phase (because ALG keeps it in its cache). We have two cases for request 4. If b is requested with an outside page, we have a third independent set unless this page is e. However, in this particular case OPT must have two pages out of $\{a, b, e\}$ to cover them and one page from $\{c, d\}$ which is not possible. So a new phase starts with this request (phase length is 3).

The other case is when request 4 contains c or d, together with b or an outside page. In this case, this page (c or d) is fixed in the cache. The entire cache is now fixed and therefore request 5 can start a new phase. Consequently, if Case A occurs, we defined a phase which consisted of at most four requests.

Case B. If request 3 consists of one page from $\{a, b\}$ and one page from $\{c, d\}$, then ALG has a choice which page to evict. In this case it initially evicts an arbitrary page, but keeps track of these type of requests, which are called *bridges*. If two bridges overlap in a page, say a, then OPT must either have a or make a fault – since it cannot have all of b, c and d in its cache. In such a case ALG fixes that page in its cache. It can be seen that if two bridges are requested in sequence, they must overlap in a page. From the first bridge, ALG only loads one page and so its state does not match the bridge. This means that the only non-overlapping bridge cannot be requested immediately afterwards.

Consider request 4. If it contains one request to an outside page, this brings us back to case 1, and we get a phase which consists of at most five requests. Otherwise, there are two bridges in succession, and therefore again ALG has one page fixed. This means that after at most four requests in a phase, at least one page in the cache is fixed, and the next request fixes the other page. Thus the maximum length of a phase is 5. ALG makes at most five uploads per at least one upload of OPT and therefore it is a most 5-competitive.

Request 2 is not independent. The proof goes along the lines of the previous case. Due to space constraints, it is omitted. In both cases, ALG makes at most five uploads per at least one uploads of OPT, so it is at most 5-competitive.

Generalization for Arbitrary Values of k, u. We first build an algorithm for $u = 2$ and $k > 2$, using induction on k. We denote the algorithm which works on a cache of size k by $\text{ALG}(k)$.

For $k = 3$, denote the first request in a phase by $\{a, b\}$ as before. Assume first that OPT has a in its cache. Then it has two 'free' places in its cache. For these places we can run the algorithm for $k = 2$. $\text{ALG}(3)$ loads a in its cache and calls a modified version of $\text{ALG}(2)$. This modified version runs for only one phase. When it returns, we know that OPT has made a fault (or is going to make a fault in the next step), *or* OPT did not load a after all. Now, $\text{ALG}(3)$ loads b and again calls $\text{ALG}(2)$. This time when it returns, we know that OPT has made a fault at some point, and we can start a new phase.

To improve the competitive ratio slightly, we can modify $\text{ALG}(2)$ further so that it indicates whether the next phase should start with the last request that it processed (in case that this request was for a third independent set) or with the next request after that (in case that this last request fixes the contents of the cache of ALG in the last possible way such that OPT does not have a fault yet). We then get a phase cost of at most $12 = 1 + 5 + 1 + 5$. These costs are the cost for loading a, the first call to $\text{ALG}(2)$, the cost for loading b and the second call to $\text{ALG}(2)$, respectively.

Generally, we find that $R(\text{ALG}(k)) = 2 + 2 \cdot R(\text{ALG}(k-1))$, where we can take as base case $R(\text{ALG}(2)) = 5.^2$ We find $R(\text{ALG}(k)) = 7 \cdot 2^{k-2} - 2$ for $k \geq 2$.

We can use a similar construction for $u > 2$. For the base case, we now do consider $k = 1$. The algorithm for this case loads the first page from the first request, say a_1 from request $\{a_1, a_2, \ldots, a_u\}$. On each fault after this, it loads the lowest-numbered page which is also in the new request, if possible. Note that a request which does not contain a certain page a_i immediately implies that OPT cannot have loaded a_i on the first request (unless it has a fault after this). Thus, as soon as we run out of pages from the first request in this way, we know that *OPT* must make (or has made) a fault and a new phase starts. This gives a u-competitive algorithm $\text{ALG}_u(1)$.

For the induction (on k), we call modified versions of simpler algorithms as before. In each induction step, we need to handle one request immediately (u times) before calling the simpler algorithm. Thus we find $R(\text{ALG}_u(k)) = u + u \cdot R(\text{ALG}_u(k-1)) = \frac{u^{k+1} - u}{u-1}$ since $R(\text{ALG}_u(1)) = u$. This proves the first half of Theorem 3.

3.3 Lower Bounds

In this section, we describe three different lower bounds for online algorithms. All lower bounds that we show use $u + k$ different pages. The request sequence is generated such that the online algorithm has a fault for every request, i.e.,

---

^{2} It is easy to give an algorithm of competitive ratio 2 for the case $k = 1$. This algorithm also works in phases, trying to guess the choice of OPT, that can be one of two pages in each phase. However, using $k = 1$ as a base case gives a ratio of 6 (instead of 5) for $k = 2$.

each request contains exactly the u pages that are absent from the cache of the algorithm.

Our lower bounds differ in the way that we define offline algorithms for this sequence. We begin by considering perhaps the simplest offline algorithm, which we will denote by OFF_1. OFF_1 checks which subset of u pages is least requested (in an arbitrarily long sequence of requests) and has the complement of this subset fixed in its cache. Each time the subset in question is requested anyway, it generates a cost of 2 for OFF_1: a cost of 1 to move to a different configuration (it is enough to replace one page) and again 1 to move back (on the next request). Since there are $(k + u)!/(k!u!)$ possible subsets, this gives us a lower bound of $(k + u)!/(2k!u!)$. This gives a lower bound of 3 for $u = k = 2$ (we improve this specific value later). Note also that if k is constant and u grows without bound, this lower bound becomes $\Omega(u^k)$. This proves the second statement of Theorem 3, and shows that our algorithm from the previous section is optimal (up to a constant factor) for constant k.

For small u however, this lower bound can be improved. We first consider the case $u = 2$ in more detail. In this case, we use $k+2$ pages to construct the request sequence. We use a different offline algorithm OFF_2, and define phases in such a way that OFF_2 has only one upload per phase. Given a configuration of OFF_2, that it has just before a phase starts, we now define a phase as a consecutive subsequence of requests that fixes the configuration to which OFF_2 moves in the beginning of this phase, paying 1 or 0 (in case it does not move). We make sure that OFF_2 does not need to change its configuration until the beginning of the next phase. It is possible to show that it takes at least $2k$ requests before the configuration of OFF_2 is fixed. At this point, immediately after the phase, OFF_2 can move to a new configuration, determined by the requests of the next phase.

This gives a lower bound of $2k$ on the competitiveness of any online algorithm (which fails on every request, by our design). In particular, for the case $u = k = 2$, we find a lower bound of 4, only 1 less than our upper bound.

For $u = 2$ and $k > 2$, it is possible to improve on this further. We use a third type of offline algorithm OFF_3. Instead of continuing a phase until only one option for OFF_3 is left, we maintain two options for OFF_3 throughout and only fix OFF_3 after all phases have been defined. We define the very first phase to have only $2k - 1$ requests, so that after this, OFF_3 still has (at least) two choices. Each later phase has $3k - 2$ requests. It is possible to show that we can maintain the invariant that OFF_3 has two choices.

The state to which OFF_3 should move at the beginning of each phase is determined scanning the sequence of phases starting from the end, choosing each time a configuration for OFF_3 out of the two possible ones, which is a function of the next configuration chosen for it. We find a lower bound of $3k - 2$.

We summarize some of the results obtained in the current section for $u = 2$ in the following table.

k	1	2	3	4	5
Lower bound	2	4	7	10	13
Upper bound	2	5	12	26	54

4   Paging with Bypassing

When bypassing is allowed, an algorithm is not forced to have in its cache a page from a request that it is going to service. Bypassing means that a page is used without loading it into the cache. The cost charged for this option is the same as for loading a page. Clearly, if a page is going to be used several times, it makes sense to load it into the cache. However, if a page is used only once, it is sometimes better to bypass it, so that the current contents of the cache can remain there. It is well known that for the standard problem, allowing bypassing increases the best competitive ratio by 1, i.e. it is $k + 1$ for a cache of size k. Note that the best algorithms for this problem are marking algorithms, such as LRU, and FIFO (which is not a marking algorithm), and that their competitive ratio increases by 1. Thus the best online algorithms do not actually make use of the bypassing option.

Consider first the first lower bound presented in section 3.3. We can construct the lower bound sequence in the same manner. This means that if an online algorithm bypasses a certain request, that request is repeated until it is no longer bypassed. Having the option of bypassing, OPT does not need to pay 2 each time the request is the exact complement set, but just 1, to bypass on it (or more specifically, to bypass one page of this request). This gives a lower bound of $(k + u)!/(k!u!)$.

Interestingly, despite the recursive construction that we use, here it is also possible to design an algorithm with a competitive ratio which is only 1 higher than for the case without bypassing for any k. Here, in order to complete a phase, we must make sure that OPT had either a fault or one bypass during this phase (the phases are not shifted as in the proof of the algorithm without bypassing).

For some fixed value of k, consider the outer phase of the recursion. There are three cases. If OPT bypasses the very first request, it has a fault in the current (outer) phase and we are done for this phase. If OPT loads page i from this request ($i = 1, \ldots, k - 1$), then by construction we know that it has a fault during the ith recursive call from the outer phase, or on the request which immediately follows it, which is also a part of the same outer phase. Finally, if OPT loads the kth page from this request, it can happen that it does not have a fault during the kth recursive call but instead on the very next request.

Therefore, we now construct the outer phase as before, but add one final request (without loading some page between the last recursive call and this request). This ensures that OPT has a fault in every outer phase. The inner phases can remain unchanged (we apply the old algorithm). This implies that the competitive ratio for any value of k increases by only 1.

This gives the following results for $u = 2$.

k	1	2	3	4	5
Lower bound	3	6	10	15	21
Upper bound	3	6	13	27	55

Thus for the case of bypassing, our algorithm is optimal for $u = k = 2$.

5 Resource Augmentation

Already in [14] the classical paging problem was studied in terms of resource augmentation. That is an extension of the usual competitive analysis, which allows an online algorithm to use greater resources than the optimal offline algorithm it is compared to. In this section, let h be the size of the cache used by an optimal offline algorithm and let $k > h$ be the size of cache used by an online algorithm. For standard online paging, the competitive ratio becomes constant if $h = \alpha k$ for fixed values of $\alpha < 1$ [14]. More precisely, it was shown in [14] that the best competitive ratio for this case is $\frac{k}{k-h+1}$.

In this section we focus on the case $u = 2$. We define a very simple algorithm which works in phases as before. The algorithm is defined for even values of k. The first $k/2$ requests are inserted completely into the cache. Note that this means that these first requests are independent. The next request starts a new phase (the cache contents are unmarked and all pages may be deleted). Let $\alpha = k/h$. We prove that for $\alpha > 2$, the algorithm has constant competitive ratio.

Theorem 4. *The competitive ratio of the above algorithm is at most $\frac{2\alpha}{\alpha-2}$ for even values of k. The competitive ratio tends to 2 as α grows.*

Proof. The sequence for a phase, including the first request of the next phase but not the first request of the current phase, contains k distinct pages. This holds since inside the phase at least $k - 2$ new pages are requested and kept in the cache, and two additional pages are the first request of the next phase. The two pages of the first request are not equal to any of the other pages, therefore we have a total of $k+2$ pages. Out of this amount, OPT can have at most h in its cache after the first request. One spot in the cache is taken by a page of the very first request. It needs to service $k/2$ additional requests upto and including the first request of the next phase. This means that it has at least $k/2 - h$ uploads. However, the algorithm has k uploads in each phase. □

For odd k, we need to define the algorithm slightly more carefully. In this case the first $\frac{k-1}{2}$ requests are inserted completely into the cache. On the request number $\frac{k+1}{2}$ of the phase, which is denoted $\{a, b\}$, only one page a is inserted into the cache. On the next request, if it contains page b, the algorithm evicts a and inserts b, and starts a new phase on the next request. If b does not belong to the next request, this request already starts a new phase. Using a proof very similar to the one above, it can be shown that the competitive ratio of this algorithm is $2\alpha'/(\alpha' - 2)$ where $\alpha' = (k + 1)/h$.

Proposition 1. *No online algorithm has a competitive ratio below 2 for any $k > 1$, even if $h = 1$.*

Proof. The sequence consists of requests as follows. The first request has two pages $\{a, b\}$. If the algorithm inserts both of them into the cache, the next request consists of two other pages. Otherwise, if only one page a is inserted into the cache, the next request is for the other page b together with a new page. This process is repeated. In the first option, OPT inserts one of the pages into its cache and in the second option it inserts b. In both cases the algorithm inserts two pages.   □

Next, we focus on the smallest not trivial case, $h = 2$, $k = 3$. We can show that this algorithm improves on our algorithm for $h = k = 2$. We also design a lower bound.

Proposition 2. *The competitive ratio of the above algorithm for the case $h = 2, k = 3$ is at most 4. Any algorithm for $h = 2, k = 3$ has competitive ratio at least $5/2$.*

References

1. Dimitris Achlioptas, Marek Chrobak, and John Noga. Competitive analysis of randomized paging algorithms. *Theoretical Comp. Science*, 234(1-2):203–218, 2000.
2. Laszlo A. Belady. A study of replacement algorithms for virtual storage computers. *IBM Systems Journal*, 5:78–101, 1966.
3. Allan Borodin, Nathan Linial, and Michael Saks, An optimal online algorithm for metrical task systems, *Proc. 19th ACM Symp. on Theory of Computing*, 1987.
4. Marek Chrobak and Lawrence L. Larmore, The server problem and on-line games, *DIMACS Series in Discrete Math. and Theoretical Comp. Science*, vol. 7, 1992.
5. Marc Demange and Vangelis Th. Paschos. On-line vertex-covering. *Theoretical Computer Science*, 332:83–108, 2005.
6. Irit Dinur, Venkatesan Guruswami, Subhash Khot, and Oded Regev. A new multilayered PCP and the hardness of hypergraph vertex cover. In *Proc. 35th ACM Symp. on Theory of Comp.*, pages 595–601, 2003.
7. Amos Fiat, Richard Karp, Michael Luby, Lyle A. McGeoch, Daniel Sleator, Neal E. Young. Competitive paging algorithms. *Journal of Algorithms*, 12:685–699, 1991.
8. Amos Fiat and Moty Ricklin. Competitive algorithms for the weighted server problem. *Theoretical Computer Science*, 130:85–99, 1994.
9. Sandy Irani. Competitive analysis of paging. In A. Fiat and G. Woeginger, editors, *Online Algorithms: The State of Art*, pages 52–73. Springer, 1998.
10. Anna Karlin, Mark Manasse, Lyle Rudolph, and Daniel Sleator. Competitive snoopy caching. *Algorithmica*, 3:79–119, 1988.
11. Mark Manasse, Lyle A. McGeoch, and Daniel Sleator, Competitive algorithms for server problems, *Journal of Algorithms*, vol. 11, pages 208-230, 1990.
12. Lyle McGeoch and Daniel Sleator. A strongly competitive randomized paging algorithm. *Algorithmica*, 6(6):816–825, 1991.
13. Ran Raz and Shmuel Safra. A sub-constant error-probability low-degree test, and sub-constant error-probability PCP characterization of NP. In *Proc. 29th ACM Symp. on Theory of Comp.*, pages 475–484, 1997.
14. Daniel Sleator and Robert E. Tarjan. Amortized efficiency of list update and paging rules. *Communications of the ACM*, 28:202–208, 1985.

Decentralization and Mechanism Design for Online Machine Scheduling

Birgit Heydenreich*, Rudolf Müller, and Marc Uetz

Maastricht University, Quantitative Economics, P.O.Box 616,
6200 MD Maastricht, The Netherlands
{b.heydenreich, r.muller, m.uetz}@ke.unimaas.nl

Abstract. We study the online version of the classical parallel machine scheduling problem to minimize the total weighted completion time from a new perspective: We assume that the data of each job, namely its release date r_j, its processing time p_j and its weight w_j is only known to the job itself, but not to the system. Furthermore, we assume a decentralized setting where jobs choose the machine on which they want to be processed themselves. We study this problem from the perspective of algorithmic mechanism design. We introduce the concept of a myopic best response equilibrium, a concept weaker than the dominant strategy equilibrium, but appropriate for online problems. We present a polynomial time, online scheduling mechanism that, assuming rational behavior of jobs, results in an equilibrium schedule that is 3.281-competitive. The mechanism deploys an online payment scheme that induces rational jobs to truthfully report their private data. We also show that the underlying local scheduling policy cannot be extended to a mechanism where truthful reports constitute a dominant strategy equilibrium.

1 Introduction

We study the online version of the classical parallel machine scheduling problem to minimize the total weighted completion time – $P \,|\, r_j \,|\, \sum w_j\, C_j$ in the notation of Graham et al. [1] – from a new perspective: We assume a strategic setting, where the data of each job, namely its release date r_j, its processing time p_j and its weight w_j is only known to the job itself, but not to the system. Any job j is interested in being finished as early as possible, and the weight w_j represents its indifference cost for spending one additional unit of time waiting. While jobs may strategically report false values $(\tilde{r}_j, \tilde{p}_j, \tilde{w}_j)$ in order to be scheduled earlier, the total social welfare is maximized whenever the weighted sum of completion times $\sum w_j\, C_j$ is minimized. Furthermore, we assume a restricted communication paradigm, referred to as *decentralization*: Jobs may communicate with machines, but neither do jobs communicate with each other, nor do machines communicate with each other. In particular, there is no central coordination authority hosting all the data of the problem. This leads to a setting

---

* Supported by NWO grant 2004/03545/MaGW 'Local Decisions in Decentralised Planning Environments'.

L. Arge and R. Freivalds (Eds.): SWAT 2006, LNCS 4059, pp. 136–147, 2006.

where the jobs themselves must select the machine to be processed on, and any machine sequences the jobs according to a (known) local sequencing policy.

The problem $P \mid r_j \mid \sum w_j C_j$ is well-understood in the non-strategic setting with centralized coordination. First, scheduling to minimize the weighted sum of completion times with release dates is NP-hard, even in the off-line case [2]. Second, no online algorithm for the single machine problem can be better than 2-competitive [3] regardless of the question whether or not P=NP, and lower bounds exist for parallel machines, too [4]. The best possible algorithm for the single machine case is 2-competitive [5]. For the parallel machine setting, the currently best known online algorithm is 2.61-competitive [6].

In the strategic setting, selfish agents trying to maximize their own benefit can do so by reporting strategically about their private information, thus manipulating the resulting schedule. In the model we propose, a job can report an arbitrary weight, an elongated processing time (e.g. by adding unnecessary work), and it can artificially delay its true release date r_j. We do not allow a job to report a processing time shorter than p_j, as this can easily be discovered and punished by the system, e.g. by preempting the job after the declared processing time $\tilde{p}_j$ before it is actually finished. Furthermore, as we assume that any job j comes into existence only at its release date r_j, it obviously does not make sense that a job reports a release date smaller than the true value r_j.

Our goal is to set up a mechanism that yields a reasonable overall performance with respect to the objective function $\sum w_j C_j$. To that end, the mechanism needs to motivate the jobs to reveal their private information truthfully. In addition, as we require decentralization, each machine must be equipped with a local sequencing policy that is publicly known, and jobs must be induced to select the machines in such a way that $\sum w_j C_j$ is not too large. Known algorithms with the best competitive ratio, e.g. [6, 7], crucially require central coordination to distribute jobs over machines. An approach by Megow et al. [8], developed for an online setting with release dates and stochastic job durations, however, turns out to be appropriate for being adopted to the decentralized, strategic setting.

Related Work and Contribution. Mechanism design in combination with the design of approximation algorithms for scheduling problems has been studied, e.g., by Nisan and Ronen [10], Archer and Tardos [11], and Kovacs [12]. In those papers, not the jobs but the machines are the selfishly behaving parts of the system, and their private information is the time they need to process the jobs. A scheduling model where the jobs are the selfish agents of the system has been studied by Porter [13]. He addresses a single machine scheduling problem, where the private data of each job consists of a release date, its processing time, its weight, and a deadline. In all mentioned papers, the only action of an agent (machine or job, respectively) is to reveal its private data; the resulting mechanisms are also called direct mechanisms. The model suggested in this paper does not give rise to a direct mechanism, since in addition to the revelation of private data, jobs must select the machine to be processed on.

In the algorithm of Megow et al. [8], jobs are locally sequenced according to an online variant of the well known WSPT rule [9], and arriving jobs are assigned to machines in order to minimize an expression that approximates the (expected) increase of the objective value. This algorithm achieves a competitive ratio of 3.281. The mechanism we propose develops their idea further. We present a polynomial time, decentralized online mechanism, called DECENTRALIZED LOCALGREEDY Mechanism. Thereby we provide also a new algorithm for the non-strategic, centralized setting, inspired by the MININCREASE Algorithm of [8], but improving upon the latter in terms of simplicity. We show that the DECENTRALIZED LOCALGREEDY Mechanism is 3.281-competitive. This coincides with the bound that is known for the non-strategic, centralized setting [7, 8]. The currently best known bound for the non-strategic setting, however, is 2.61 [6].

As usual in mechanism design, the DECENTRALIZED LOCALGREEDY Mechanism defines *payments* that have to be made by the jobs for being processed. Naturally, we require from an *online* mechanism that also the payments are computed online. Hence they can be completely settled by the time at which a job leaves the system. We also show that the payments result in a balanced budget. The payments induce the jobs to select 'the right' machines. Intuitively, the mechanism uses the payments to mimic a corresponding LOCALGREEDY online algorithm in the classical (non-strategic, centralized) parallel machine setting $P \mid r_j \mid \sum w_j C_j$. Moreover, the payments induce rational jobs to truthfully report about their private data. With respect to release dates and processing times, we can show that truthfulness is a dominant strategy equilibrium. With respect to the weights, however, we can only show that truthful reports are myopic best responses (in a sense to be made precise later). In addition, we show that there does not exist a payment scheme extending the allocation rule of the DECENTRALIZED LOCALGREEDY Mechanism to a mechanism where truthful reporting of all private information is a dominant strategy equilibrium.

This extended abstract is organized as follows. We formalize the model and introduce the required notation in Section 2. In Section 3 the LOCALGREEDY algorithm is defined. In Section 4, this algorithm is adapted to the strategic setting and extended by a payment scheme, yielding the DECENTRALIZED LOCALGREEDY Mechanism. Moreover, our main results are presented in that section. We analyze the performance of the mechanism in Section 5, mention a negative result in Section 6, and conclude with a short discussion in Section 7.

2   Model and Notation

The considered problem is online parallel machine scheduling with non-trivial release dates, with the objective to minimize the weighted sum of completion times, $P \mid r_j \mid \sum w_j C_j$. We are given a set of jobs $J = \{1, \ldots, n\}$, where each job needs to be processed on any of the parallel, identical machines from the set $M = \{1, \ldots, m\}$. The processing of each job must not be preempted, and each machine can process at most one job at a time. Each job j is viewed as a selfish agent and has the following private information: a release date $r_j \geq 0$, a processing time $p_j > 0$, and an indifference cost, or weight, denoted by $w_j \geq 0$. The release date denotes the time when the job comes into existence, whereas the weight represents the cost to a job for one additional unit of time spent waiting. Without loss of generality, we assume that the jobs are numbered in order of their release dates, i.e., $j < k \Rightarrow r_j \leq r_k$. The triple (r_j, p_j, w_j) is also denoted as the *type* of a job, and we use the shortcut notation $t_j = (r_j, p_j, w_j)$. By $T = \mathbb{R}_0^+ \times \mathbb{R}^+ \times \mathbb{R}_0^+$ we denote the space of possible types of each job.

Definition 1. *A* decentralized online scheduling mechanism *is a procedure that works as follows*

1. *Each job j has a release date r_j, but may pretend to come into existence at any time $\tilde{r}_j \geq r_j$. At that chosen release date, the job communicates to every machine reports $\tilde{w}_j$ and $\tilde{p}_j$ (which may differ from the true w_j and p_j)[1].*
2. *Machines communicate on the basis of that information a (tentative) completion time $\hat{C}_j$ and a (tentative) payment $\hat{\pi}_j$ to the job. This information is tentative due to the online situation. The values $\hat{C}_j$ and $\hat{\pi}_j$ can only change if later another job chooses the same machine.*
3. *Based on this response, the job chooses a machine. This choice is binding. The entire communication takes place at one point in time, namely $\tilde{r}_j$.*
4. *There is no communication between machines or between jobs.*
5. *Depending on later arrivals of jobs, machines may revise $\hat{C}_j$ and $\hat{\pi}_j$. Eventually, the mechanism leads to an (ex-post) completion time C_j and an (ex-post) payment π_j of each job.*

Hereby, we assume that jobs with equal reported release date arrive in some given order and communicate to machines in that order. Next, we define an online property of the payment scheme.

Definition 2. *If in a decentralized online scheduling mechanism for every job j payments to and from j are only made between time $\tilde{r}_j$ and time C_j, then we call the payment scheme of the mechanism an* online payment scheme

We assume that each job j prefers a lower completion time to a higher one and model this by the valuation $v_j(C_j \,|\, t_j) = -w_j C_j$. We assume *quasi-linear utilities*, that is, the utility of job j equals $u_j(C_j, \pi_j \,|\, t_j) = v_j(C_j \,|\, t_j) - \pi_j$, which is equal to $-w_j C_j - \pi_j$. In this model, the utility u_j is always negative. Therefore, we assume that a job has a constant and sufficiently large utility for 'being processed at all'. Note that the total social welfare is maximized whenever the weighted sum of completion times $\sum_{j \in J} w_j C_j$ is minimum, which is independent of whether we do or do not carry these constants with us.

The communication with machines, and the decision for a particular machine are called *actions* of the jobs; they constitute the strategic actions jobs can take in the non-cooperative game induced by the mechanism. A *strategy* s_j of a job j maps a type t_j to an action for every possible state of the system in which the job is required to take some action. A strategy profile is a vector $(s_1, \ldots, s_n)$ of strategies, one for each job. Given a mechanism, a strategy profile, and a realization of types t, we denote by $u_j(s, t)$ the utility that agent j receives.

---

[1] A job could even report different values to different machines. However, we prove existence of equilibria where the jobs do not make use of that option.

Definition 3. *A strategy profile* $s = (s_1, \ldots, s_n)$ *is called a* dominant strategy equilibrium *if for all jobs* $j \in J$, *all types* t *of the jobs, all strategies* $\tilde{s}_{-j}$ *of the other jobs, and all strategies* $\tilde{s}_j$ *that* j *could play instead of* s_j,

$$u_j((s_j, \tilde{s}_{-j}), t) \geq u_j((\tilde{s}_j, \tilde{s}_{-j}), t).$$

We could simplify notation if we restricted ourselves to *direct mechanisms*, that is mechanisms in which the only action of a job is to report its type. However, a decentralized online scheduling mechanism requires that jobs decide themselves on which machine they are scheduled. Since these decisions are likely to influence the utility of the jobs, they have to be modelled as actions in the game. Therefore, it is not sufficient to restrict oneself to direct mechanisms.

We will see that the mechanism proposed in this paper does not have a dominant strategy equilibrium, whatever modification we might apply to the payment scheme. However, a weaker equilibrium concept applies, which we define next. That definition uses the concept of the tentative utility, i.e., the utility a job would have if it was the last to be accepted on its machine.

Definition 4. *Given a decentralized, online scheduling mechanism as in Definition 1, a strategy profile* s, *and type profile* t. *Let* $\hat{C}_j$ *and* $\hat{\pi}_j$ *denote the tentative completion time and the tentative payment of job* j *at time* $\tilde{r}_j$. *Then* $\hat{u}_j(s, t) := \hat{C} w_j - \hat{\pi}_j$ *denotes* j's *tentative utility at time* $\tilde{r}_j$

If s and t are clear from the context, we will use $\hat{u}_j$ as short notation.

Definition 5. *A strategy profile* $(s_1, \ldots, s_n)$ *is called a* myopic best response equilibrium, *if for all jobs* $j \in J$, *all types* t *of the jobs, all strategies* $\tilde{s}_{-j}$ *of the other jobs and all strategies* $\tilde{s}_j$ *that* j *could play instead of* s_j,

$$\hat{u}_j((s_j, \tilde{s}_{-j}), t) \geq \hat{u}_j((\tilde{s}_j, \tilde{s}_{-j}), t).$$

2.1   Critical Jobs

For convenience of presentation, we make the following assumption for the main part of the paper. Fix some constant $0 < \alpha \leq 1$ that will be discussed later. Let us call jobs *critical* if $r_j < \alpha p_j$. Intuitively, a job is critical if it is long and appears comparably early in the system. The assumption we make is that such critical jobs do not exist, that is

$$r_j \geq \alpha p_j \quad \text{for all jobs } j \in J.$$

This assumption is a tribute to the desired performance guarantee, and in fact, it is well known that critical jobs must not be scheduled early to achieve constant competitive ratios [5, 7]. However, this assumption is only made due to cosmetic reasons. In Section 5.1, we show how to relax this assumption, and we discuss how critical jobs can be dealt with.

3   The LOCALGREEDY Algorithm

We next formulate an online scheduling algorithm that is inspired by the MIN-INCREASE Algorithm from Megow et al. [8]. For the time being, we assume that the job characteristics such as release date r_j, processing time p_j and indifference cost w_j are given. In the next section, we discuss how to turn this algorithm into a mechanism for the strategic, decentralized setting that we aim at.

The idea of the algorithm is that each machine uses (an online version of) the well known WSPT rule [9] locally. More precisely, each machine implements a priority queue containing the not yet scheduled jobs that have been assigned to the machine. The queue is organized according to WSPT, that is, jobs with higher ratio w_j/p_j have higher priority. In case of ties, jobs with lower index have higher priority. As soon as the machine falls idle, the currently first job from this priority queue is scheduled (if any). Given this local scheduling policy on each of the machines, any arriving job is assigned to that machine were the increase in the objective $\sum w_j\,C_j$ is minimal.

Algorithm 1. LOCALGREEDY algorithm

Local Sequencing Policy:
Whenever a machine becomes idle, it starts processing the job with highest (WSPT) priority among all jobs assigned to it.
Assignment:
(1) At time r_j job j arrives; the immediate increase of the objective $\sum w_j\,C_j$, given that j is assigned to machine i, is

$$z(j,i) := w_j\left[r_j + b_i(r_j) + \sum_{\substack{k\in H(j)\\ k\to i\\ k<j\\ S_k\geq r_j}} p_k + p_j\right] + p_j\sum_{\substack{k\in L(j)\\ k\to i\\ k<j\\ S_k>r_j}} w_k.$$

(2) Job j is assigned to machine $i_j \in \arg\min_{i\in M} z(j,i)$ with minimum index.

In the formulation of the algorithm, we utilize some shortcut notation. We let $j \to i$ denote the fact that job j is assigned to machine i. Let S_j be the time when job j eventually starts being processed. For any job j, $H(j)$ denotes the set of jobs that have higher priority than j, $H(j) = \{k \in J \,|\, w_k p_j > w_j p_k\} \cup \{k \leq j \,|\, w_k p_j = w_j p_k\}$. Note that $H(j)$ includes j, too. Similarly, $L(j) = J \setminus H(j)$ denotes the set of jobs with lower priority. At a given point t in time, machine i might be busy processing a job. We let $b_i(t)$ denote the remaining processing time of that job at time t, i.e., at time t machine i will be blocked during $b_i(t)$ units of time for new jobs. If machine i is idle at time t, we let $b_i(t) = 0$.

Clearly, the LOCALGREEDY algorithm still makes use of central coordination in Step (2). In the sequel we will introduce payments that allow to transform the algorithm into a decentralized online scheduling mechanism.

4   Payments for Myopic Rational Jobs

The payments we introduce can be motivated as follows: A job j pays at the moment of its placement on one of the machines an amount that compensates the decrease in utility of the other jobs. The final payment of each job j resulting from this mechanism will then consist of the immediate payment j has to make when selecting a machine and of the payments j receives when being displaced by other jobs. We will prove that utility maximizing jobs have an incentive to report truthfully and to choose the machine that the LOCAL-GREEDY Algorithm would have selected, too. Furthermore, the WSPT rule can be run locally on every machine and does not require communication between the machines. We will see in the next section that this yields a constant-factor approximation of the off-line optimum, given that the jobs behave rationally. The algorithm including the payments is displayed below as the DECENTRALIZED LOCALGREEDY Mechanism. Let the indices of the jobs be defined according to the reported release dates, i.e. $j < k \Rightarrow \tilde{r}_j \leq \tilde{r}_k$. Let $\tilde{H}(j)$ and $\tilde{L}(j)$ be defined analogously to $H(j)$ and $L(j)$ on the basis of the reported weights.

Algorithm 2. DECENTRALIZEDLOCALGREEDY Mechanism

Local Sequencing Policy:
Whenever a machine becomes idle, it starts processing the job with highest (WSPT) priority among all available jobs queuing at this machine.
Assignment:
(1) At time $\tilde{r}_j$ job j arrives and reports a weight $\tilde{w}_j$ and a processing time $\tilde{p}_j$ to all machines.
(2) Every machine i computes

$$\hat{C}_j(i) = \tilde{r}_j + b_i(\tilde{r}_j) + \sum_{\substack{k \in \tilde{H}(j) \\ k \to i \\ k < j \\ S_k \geq \tilde{r}_j}} \tilde{p}_k + \tilde{p}_j \quad \text{and} \quad \hat{\pi}_j(i) = \tilde{p}_j \sum_{\substack{k \in \tilde{L}(j) \\ k \to i \\ k < j \\ S_k > \tilde{r}_j}} \tilde{w}_k.$$

and informs j about both $\hat{C}_j(i)$ and $\hat{\pi}_j(i)$.
(3) Job j chooses a machine $i_j \in M$. Its tentative utility for being queued at machine i is $\hat{u}_j(i) := -w_j \hat{C}_j(i) - \hat{\pi}_j(i)$.
(4) The job is queued at i_j according to WSPT among all currently available jobs on i_j whose processing has not started yet. The payment $\hat{\pi}_j(i_j)$ has to be paid by j.
(5) The (tentative) completion time for every job k with $k \in \tilde{L}(j)$, $k \to i_j$, $k < j$, $S_k > \tilde{r}_j$ increases by $\tilde{p}_j$ due to j's presence. As compensation, k receives a payment of $\tilde{w}_k \tilde{p}_j$.

The DECENTRALIZEDLOCALGREEDY Mechanism together with the stated payments results in a balanced budget for the scheduler. That is, the payments paid and received by the jobs sum up to zero, since every arriving job immediately makes its payment to the jobs that are displaced by it. Notice that the payments are made online in the sense of Definition 2.

Theorem 1. *Regard any type vector t, any strategy profile s and any job j such that j reports $(\tilde{r}_j, \tilde{p}_j, \tilde{w}_j)$ and chooses machine $\tilde{m} \in M$. Then changing the report to $(\tilde{r}_j, \tilde{p}_j, w_j)$ and choosing a machine that maximizes its tentative utility at time $\tilde{r}_j$ does not decrease j's tentative utility under the DECENTRALIZED LOCALGREEDY Mechanism.*

Proof. We only give the idea here. For the single machine case, an arriving job j gains tentative utility $\tilde{p}_k w_j - \tilde{p}_j \tilde{w}_k$ from displacing an already present job k. WSPT assigns j in front of k if and only if $\tilde{p}_k \tilde{w}_j - \tilde{p}_j \tilde{w}_k > 0$. Thus, $\tilde{w}_j = w_j$ maximizes j's tentative utility. For $m > 1$, the theorem follows from the fact that j can select a machine itself. □

Lemma 1. *Consider any job $j \in J$. Then, under the* DECENTRALIZED LOCAL-GREEDY *Mechanism, for all reports of all other agents as well as all choices of machines of the other agents, the following is true:*
(a) If j reports $\tilde{w}_j = w_j$, then the tentative utility when queued at any of the machines will be preserved over time, i.e. it equals j's ex-post utility
(b) If j reports $\tilde{w}_j = w_j$, then selecting the machine that the LOCALGREEDY *Algorithm would have selected maximizes j's ex-post utility.*

Proof. See full version of the paper. □

Theorem 2. *Consider the restricted strategy space where all $j \in J$ report $\tilde{w}_j = w_j$. Then the strategy profile where all jobs j truthfully report $\tilde{r}_j = r_j$, $\tilde{p}_j = p_j$ and choose a machine that maximizes $\hat{u}_j$ is a dominant strategy equilibrium under the* DECENTRALIZED LOCALGREEDY *Mechanism.*

Proof. Let us start with $m = 1$. Suppose $\tilde{w}_j = w_j$, fix any pretended release date $\tilde{r}_j$ and regard any $\tilde{p}_j > p_j$. Let u_j denote j's (ex-post) utility when reporting p_j truthfully and let $\tilde{u}_j$ be its (ex-post) utility for reporting $\tilde{p}_j$. As $\tilde{w}_j = w_j$, the ex-post utility equals in both cases the tentative utility at decision point $\tilde{r}_j$ according to Lemma 1(a). Let us therefore regard the latter utilities. Clearly, according to the WSPT-priorities, j's position in the queue at the machine for report p_j will not be behind its position for report $\tilde{p}_j$. Let us divide the jobs already queuing at the machine upon j's arrival into three sets: Let $J_1 = \{k \in J \,|\, k < j, S_k > \tilde{r}_j, \tilde{w}_k/\tilde{p}_k \geq w_j/p_j\}$, $J_2 = \{k \in J \,|\, k < j, S_k > \tilde{r}_j, w_j/p_j > \tilde{w}_k/\tilde{p}_k \geq w_j/\tilde{p}_j\}$ and $J_3 = \{k \in J \,|\, k < j, S_k > \tilde{r}_j, w_j/\tilde{p}_j > \tilde{w}_k/\tilde{p}_k\}$. That is, J_1 comprises the jobs that are in front of j in the queue for both reports, J_2 consists of the jobs that are only in front of j when reporting $\tilde{p}_j$ and J_3 includes only jobs that queue behind j for both reports. Therefore, $\tilde{u}_j - u_j$ equals

$$- \sum_{k \in J_1 \cup J_2} w_j \tilde{p}_k - \sum_{k \in J_3} \tilde{p}_j \tilde{w}_k - w_j \tilde{p}_j - \left(- \sum_{k \in J_1} w_j \tilde{p}_k - \sum_{k \in J_2 \cup J_3} p_j \tilde{w}_k - w_j p_j \right)$$
$$= \sum_{k \in J_2} (p_j \tilde{w}_k - w_j \tilde{p}_k) - \sum_{k \in J_3} (\tilde{p}_j - p_j) \tilde{w}_k - w_j (\tilde{p}_j - p_j).$$

According to the definition of J_2, the first term is smaller than or equal to zero. As $\tilde{p}_j > p_j$, the whole right hand side becomes non-positive. Therefore $\tilde{u}_j \leq u_j$, i.e. truthfully reporting p_j maximizes j's ex-post utility on a single machine.

Let us now fix $\tilde{w}_j = w_j$ and any $\tilde{p}_j \geq p_j$ and regard any false release date $\tilde{r}_j > r_j$. There are two effects that can occur when arriving later than r_j. Firstly, jobs queued at the machine already at time r_j may have been processed or may have started receiving service by time $\tilde{r}_j$. But either j would have had to wait for those jobs anyway or it would have increased its immediate utility at decision point r_j by displacing a job and paying the compensation. So, j cannot gain from this effect by lying. The second effect is that new jobs have arrived at the machine between r_j and $\tilde{r}_j$. Those jobs either delay j's completion time and j looses the payment it could have received from those jobs by arriving earlier. Or the jobs do not delay j's completion time, but j has to pay the jobs for displacing them when arriving at $\tilde{r}_j$. If j arrived at time r_j, it would not have to pay for displacing such a job. Hence, j cannot gain from this effect either. Thus the immediate utility at time r_j will be at least as large as its immediate utility at time $\tilde{r}_j$. Therefore, j maximizes its immediate utility at time $\tilde{r}_j$ by choosing $\tilde{r}_j = r_j$. As $\tilde{w}_j = w_j$, it follows from Lemma 1(a) that choosing $\tilde{r}_j = r_j$ also maximizes the job's ex-post utility on a single machine.

For $m > 1$, note that on every machine, the immediate utility of job j at decision point $\tilde{r}_j$ is equal to its ex-post utility and that j can select a machine itself that maximizes its immediate utility and therefore its ex-post utility. Therefore, given that $\tilde{w}_j = w_j$, a job's ex-post utility is maximized by choosing $\tilde{r}_j = r_j$, $\tilde{p}_j = p_j$ and, according to Lemma 1(b), by choosing a machine that minimizes the immediate increase in the objective function.   □

Theorem 3. *Given the types of all jobs, the strategy profile where each job j reports $(\tilde{r}_j, \tilde{p}_j, \tilde{w}_j) = (r_j, p_j, w_j)$ and chooses a machine maximizing its tentative utility $\hat{u}_j$ is a myopic best response equilibrium under the* DECENTRALIZED LOCALGREEDY *Mechanism.*

Proof. Regard job j. According to the proof of Theorem 1, $\hat{u}_j$ on any machine is maximized by reporting $\tilde{w}_j = w_j$ for any $\tilde{r}_j$ and $\tilde{p}_j$. According to Theorem 2 and Lemma 1(b), $\tilde{p}_j = p_j$, $\tilde{r}_j = r_j$ and choosing a machine that maximizes j's tentative utility at time $\tilde{r}_j$ maximize j's ex-post utility if j truthfully reports $\tilde{w}_j = w_j$. According to Lemma 1(a) this ex-post utility is equal to $\hat{u}_j$ if j reports $\tilde{w}_j = w_j$. Therefore, any job j maximizes $\hat{u}_j$ by truthful reports and choosing the machine as claimed.   □

Given the restricted communication paradigm, jobs do not know at their arrival which jobs are already queuing at the machines and what reports the already present jobs have made. Therefore it is easy to see that for any non-truthful report of an arriving job about its weight, instances can be constructed in which this report yields a strictly lower utility for the job than a truthful report would have given. With arguments similar to those in the proof of Theorem 2, the same holds for false reports about the processing time and the release date.

5   Performance of the Mechanism

As shown in Section 4, jobs have a motivation to report truthfully about their data: According to Theorem 1, it is a myopic best response for a job j to report the true weight w_j, no matter what the other jobs do and no matter which

$\tilde{p}_j$ and $\tilde{r}_j$ are reported by j itself. Given a true report of w_j, it was proven in Theorem 2 that reporting the true processing time and release date as well as choosing a machine maximizing the tentative utility at arrival maximizes the job's ex-post utility. Therefore we will call a job *rational* if it truthfully reports w_j, p_j and r_j and chooses a machine maximizing its tentative utility $\hat{u}_j$. In this section, we will show that if all jobs are rational, then the DECENTRALIZED LOCALGREEDY Mechanism is 3.281-competitive.

5.1   Handling Critical Jobs

Recall that from Section 2.1 on, we assumed that no critical jobs exist, i.e. that $r_j \geq \alpha p_j$ for all jobs $j \in J$. We will now relax this assumption. Without the assumption, the DECENTRALIZEDLOCALGREEDY Mechanism as stated above does not yet yield a constant approximation factor; simple examples can be constructed in the same flavor as in [7]. In fact, it is well known that early arriving jobs with large processing times have to be delayed [5, 7, 8]. In order to achieve a constant competitive ratio, we also adopt this idea and use modified release dates as [7, 8]. To this end, we define the modified release date of every job $j \in J$ as $r'_j = \max\{r_j, \alpha p_j\}$, where $\alpha \in (0, 1]$ will later be chosen appropriately. For our decentralized setting, this means that a machine will not admit any job j to its priority queue before time $\max\{\tilde{r}_j, \alpha \tilde{p}_j\}$ if j arrives at time $\tilde{r}_j$ and reports processing time $\tilde{p}_j$. Moreover, machines refuse to provide information about the tentative completion time and payment to a job before its modified release date (with respect to the job's reported data). Note that this modification is part of the local scheduling policy of every machine and therefore does not restrict the required decentralization. Note further that any myopic rational job j still reports $\tilde{w}_j = w_j$ according to Theorem 1 and that a rational job reports $\tilde{p}_j = p_j$ as well as communicates to machines at the earliest opportunity, i.e. at time $\max\{r_j, \alpha p_j\}$, according to the arguments in the proof of Theorem 2. Moreover, the aforementioned properties concerning the balanced budget, the conservation of utility in the case of a truthfully reported weight, and the online property of the payments still apply to the algorithm with modified release dates.

5.2   Proof of the Competitive Ratio

It is not a goal in itself to have a truthful mechanism, but to use the truthfulness in order to achieve a reasonable overall performance in terms of the social welfare $\sum w_j C_j$. We derive a constant competitive ratio for the DECENTRALIZED LOCALGREEDY Mechanism by the following theorem:

Theorem 4. *Suppose every job is rational in the sense that it reports r_j, p_j, w_j and selects a machine that maximizes its tentative utility at arrival. Then the* DECENTRALIZED LOCALGREEDY *Mechanism is ϱ-competitive, with $\varrho = 3.281$.*

Proof. A rational job communicates to the machines at time $r'_j = \max\{r_j, \alpha p_j\}$ and chooses a machine i_j that maximizes its utility upon arrival $\hat{u}_j(i_j)$.

That is, it selects a machine i that minimizes

$$-\hat{u}_j(i) = w_j \hat{C}_j(i) + \hat{\pi}_j(i) = w_j \big[r'_j + b_i(r'_j) + \sum_{\substack{k \in H(j) \\ k \to i \\ k < j \\ S_k \geq r'_j}} p_k + p_j\big] + p_j \sum_{\substack{k \in L(j) \\ k \to i \\ k < j \\ S_k > r'_j}} w_k.$$

This, however, exactly equals the immediate increase of the objective value $\sum w_j C_j$ that is due to the addition of job j to the schedule. We now claim that we can express the objective value Z of the resulting schedule as $Z = \sum_{j \in J} -\hat{u}_j(i_j)$, where i_j is the machine selected by job j. Here, it is important to note that $-\hat{u}_j(i_j)$ does not express the total (ex-post) contribution of job j to $\sum w_j C_j$, but only the increase *upon arrival* of j on machine i_j. However, further contributions of job j to $\sum w_j C_j$ only appear when job j is displaced by some later arriving job with higher priority, say k. This contribution by job j to $\sum w_j C_j$, however, will be accounted for when adding $-\hat{u}_k(i_k)$.

Next, since we assume that any job maximizes its utility upon arrival, or equivalently minimizes $-\hat{u}_j(i)$ when selecting a machine i, we can apply an averaging argument over the number of machines, like in [8], to obtain:

$$Z \leq \sum_{i \in J} \frac{1}{m} \sum_{i=1}^{m} -\hat{u}_j(i).$$

The remainder of the proof utilizes the definitions of $\hat{u}_j(i)$ and particulary the fact that, upon arrival of job j on any of the machines i (at time r'_j), machine i is blocked for time $b_i(r'_j)$, which is upper bounded by r'_j/α. This upper bound is machine-independent, and follows from the definition of r'_j, since any job k in process at time r'_j fulfills $\alpha p_k \leq r'_k \leq r'_j$. Furthermore, the proof utilizes a lower bound on any (off-line) optimum schedule from Eastman et al. [14, Thm. 1]. For details, we refer to the full version of the paper. The resulting performance bound 3.281 is identical to the one of [8] (for deterministic processing times), when α is $(\sqrt{17m^2 - 2m + 1} - m + 1)/(4m)$.   □

6   Negative Result

Theorem 5. *There does not exist a payment scheme that extends the* LOCAL-GREEDY *algorithm to a truthful mechanism. Therefore, it is not possible to turn the* DECENTRALIZED LOCALGREEDY *Mechanism into a mechanism with a dominant strategy equilibrium in which all jobs report truthfully by only modifying the payment scheme.*

Proof. If the DECENTRALIZED LOCALGREEDY Mechanism can be turned into a truthful mechanism by only modifying the payment scheme, then the LO-CALGREEDY algorithm can be completed by a payment scheme to a truthful mechanism. Furthermore, we can show that a necessary condition for truthfulness, called weak monotonicity, is not satisfied by the LOCALGREEDY algorithm. Weak monotonicity has been introduced in [15].   □

7   Discussion

It would be interesting to find a constant competitive decentralized online scheduling mechanism such that there is a *dominant strategy equilibrium* in which the jobs report all data truthfully. As we have seen in Section 6, the LOCALGREEDY Algorithm cannot be extended by a payment scheme such that the resulting mechanism has the described properties. Furthermore, recall that the currently best known performance bound for the non-strategic, centralized setting is 2.61 [6]. This algorithm crucially requires a centralized distribution of jobs over machines, and therefore does not seem to be suited for decentralization. Nevertheless, it remains an interesting question to identify general rules for the transformation of centralized algorithms to decentralized mechanisms.

Acknowledgements. We thank the referees for some helpful remarks.

References

1. Graham, R.L., Lawler, E.L., Lenstra, J.K., Rinnooy Kan, A.H.G.: Optimization and approximation in deterministic sequencing and scheduling: A survey. Ann. Discr. Math. **5** (1979), 287–326
2. Lenstra, J.K., Rinnoy Kan, A.H.G., Brucker, P.: Complexity of machine scheduling problems. Ann. of Discr. Math. **1** (1977), 343–362
3. Hoogeveen, J.A., Vestjens, A.P.A.: Optimal on-line algorithms for single machine scheduling. In: Cunningham, W.H., McCormick, S.T., Queyranne, M., eds.: IPCO 1996, LNCS 1084 (1996), 404-414
4. Vestjens, A.P.A.: On-line Machine Scheduling. PhD thesis, Eindhoven University of Technology, Eindhoven, The Netherlands (1997)
5. Anderson, E.J., Potts, C.N.: Online scheduling of a single machine to minimize total weighted completion time. Math. Oper. Res. **29** (2004), 686-697
6. Correa, J.R., Wagner, M.R.: LP-based online scheduling: from single to parallel machines. In: Jünger, M., Kaibel, V., eds.: IPCO 2005. LNCS 3509 (2005), 196-209
7. Megow, N., Schulz, A.S.: On-line scheduling to minimize average completion time revisited. Oper. Res. Letters **32** (2004), 485-490
8. Megow, N., Uetz, M., Vredeveld, T.: Models and algorithms for stochastic online scheduling. Math. Oper. Res., to appear.
9. Smith, W.: Various optimizers for single stage production. Nav. Res. Log. Quarterly **3** (1956), 59-66
10. Nisan, N., Ronen, A.: Algorithmic mechanism design. Games and Economic Behavior **35** (2001), 166-196
11. Archer, A., Tardos, E.: Truthful mechanisms for one-parameter agents. In: Proc. 42nd FOCS. IEEE Computer Society (2001), 482-491
12. Kovacs, A.: Fast monotone 3-approximation algorithm for scheduling related machines. In: Brodal, G.S., Leonardi, S., eds.: ESA 2005. LNCS 3669 (2005), 616-627
13. Porter, R.: Mechanism design for online real-time scheduling. Proc. 5th ACM Conf. Electronic Commerce, ACM Press (2004), 61-70
14. Eastman, W.L., Even, S., Isaacs, I.M.: Bounds for the optimal scheduling of n jobs on m processors. Management Science **11** (1964), 268–279
15. Lavi, R., Mu'alem, A., Nisan, N.: Towards a characterization of truthful combinatorial auctions. In: Proc. 44th FOCS. IEEE Computer Society (2003), 574–583

Exponential Time Algorithms for the Minimum Dominating Set Problem on Some Graph Classes

Serge Gaspers[1], Dieter Kratsch[2], and Mathieu Liedloff[2]

[1] Department of Informatics, University of Bergen, N-5020 Bergen, Norway
serge.gaspers@ii.uib.no
[2] LITA, Université de Paul Verlaine - Metz, 57045 Metz Cedex 01, France
{kratsch, liedloff}@univ-metz.fr

Abstract. The Minimum Dominating Set problem remains NP-hard when restricted to chordal graphs, circle graphs and c-dense graphs (i.e. $|E| \geq cn^2$ for a constant c, $0 < c < 1/2$). For each of these three graph classes we present an exponential time algorithm solving the Minimum Dominating Set problem. The running times of those algorithms are $O(1.4173^n)$ for chordal graphs, $O(1.4956^n)$ for circle graphs, and $O(1.2303^{(1+\sqrt{1-2c})n})$ for c-dense graphs.

1   Introduction

During the last years there has been a growing interest in the design of exact exponential time algorithms. Woeginger has written a nice survey on the subject [19] emphasizing the major techniques used to design exact exponential time algorithms. We also refer the reader to the recent survey of Fomin et al. [9] discussing some new techniques in the design of exponential time algorithms. In particular they discuss treewidth based techniques, Measure & Conquer and memorization.

Known Results. A set $D \subseteq V$ of a graph $G = (V, E)$ is dominating if every vertex of $V \setminus D$ has a neighbor in D. The Minimum Dominating Set problem (MDS) asks to compute a dominating set of the input graph of minimum cardinality.

Exact exponential time algorithms for the Minimum Dominating Set problem have not been studied until recently. By now there is a large interest in this particular problem. In 2004 three papers with exact algorithms for MDS have been published. In [10] Fomin et al. presented an $O(1.9379^n)$ time algorithm for general graphs and algorithms for split graphs, bipartite graphs and graphs of maximum degree three with running time $O(1.4143^n)$, $O(1.7321^n)$, $O(1.5144^n)$, respectively. Exact algorithms for MDS on general graphs have also been given by Randerath and Schiermeyer [16] and by Grandoni [12]. Their running times are $O(1.8899^n)$ and $O(1.8026^n)$, respectively.

These algorithms have been significantly improved by Fomin et al. in [8] where the authors obtain the currently fastest exact algorithm for MDS. Their search tree algorithm is based on the so-called Measure & Conquer approach, and the upper bounds on the worst case running times are established by the use of non

L. Arge and R. Freivalds (Eds.): SWAT 2006, LNCS 4059, pp. 148–159, 2006.

standard measures. The MDS algorithm has running time $O(1.5263^n)$ and needs polynomial space. Using memorization one can speed up the running time to $O(1.5137^n)$ needing exponential space then. Both variants are based on algorithms for the minimum set cover problem where the input consists of a universe $\mathcal{U}$ and a collection $\mathcal{S}$ of subsets of $\mathcal{U}$. These algorithms need running time $O(1.2354^{|\mathcal{U}|+|\mathcal{S}|})$ and polynomial space, or running time $O(1.2303^{|\mathcal{U}|+|\mathcal{S}|})$ and exponential space [8].

Finally, Fomin and Høie used a treewidth based approach to establish an algorithm to compute a minimum dominating set for graphs of maximum degree three [7] with running time $O(1.2010^n)$.

It is known that the problem MDS is NP-hard when restricted to chordal graphs [5], and circle graphs [13]. Furthermore it is not hard to show that MDS is NP-hard for c-dense graphs.

Our Results. In this paper we study the Minimum Dominating Set problem on three graph classes and we obtain algorithms with a running time $O(\alpha^n)$ better than the best known running time for an algorithm solving MDS on general graphs, i.e. $O(1.5137^n)$.

In Section 3 we present an exact algorithm solving the MDS problem on chordal graphs in time $O(1.4173^n)$. In Section 4 an $O(1.4956^n)$ time algorithm to compute a minimum dominating set for circle graphs is established. In Section 5 we give an $O(1.2303^{n(1+\sqrt{1-2c})})$ time algorithm for c-dense graphs, i.e. for all graphs with at least cn^2 edges, where c is a constant with $0 < c < 1/2$.

Our algorithms rely heavily on the minimum set cover algorithms of Fomin et al. [8]. Furthermore the algorithms for chordal graphs and for circle graphs are treewidth based. Both of them use different algorithms for graphs of small treewidth, i.e. at most tn, and for graphs of large treewidth, i.e. larger than tn, where t is chosen to balance the running times of those two algorithms.

The algorithm for circle graphs relies on an upper bound of the treewidth of circle graphs in terms of the maximum degree which is interesting in its own. A related result for graphs of small chordality is provided in [4]. We are not aware of any previous result of this type for circle graphs.

2   Preliminaries

Let $G = (V, E)$ be an undirected and simple graph. For a vertex $v \in V$ we denote by $N(v)$ the neighborhood of v and by $N[v] = N(v) \cup \{v\}$ the closed neighborhood of v. For a given subset of vertices $S \subseteq V$, $G[S]$ denotes the subgraph of G induced by S. The maximum degree of a graph G is denoted by $\Delta(G)$ or by Δ if it is clear from the context which graph is meant.

A clique is a set $C \subseteq V$ of pairwise adjacent vertices. The maximum cardinality of a clique in a graph G is denoted by $\omega(G)$. A dominating set D of a graph $G = (V, E)$ is a subset of vertices such that every vertex of $V - D$ has at least one neighbor in D. The minimum cardinality of a dominating set of G is the domination number of G, and it is denoted by $\gamma(G)$.

Major tools of our paper are tree decompositions and treewidth of graphs. The notions have been introduced by Robertson and Seymour in [17].

Definition 1 (Tree decomposition). *Let $G = (V, E)$ be a graph. A tree decomposition of G is a pair $(\{X_i : i \in I\}, T)$ where each X_i, $i \in I$, is a subset of V and T is a tree with elements of I as nodes such that we have the following properties :*

1. $\cup_{i \in I} X_i = V$;
2. $\forall \{u, v\} \in E$, $\exists i \in I$ s.t. $\{u, v\} \subseteq X_i$;
3. $\forall i, j, k \in I$, if j is on the path from i to k in T then $X_i \cap X_k \subseteq X_j$.

The width of a tree decomposition is equal to $\max_{i \in I} |X_i| - 1$.

Definition 2 (Treewidth). *The* treewidth *of a graph G is the minimum width over all its tree decompositions and it is denoted by $tw(G)$.*

A tree decomposition is called *optimal* if its width is $tw(G)$.

Definition 3 (Nice tree decomposition). *A nice tree decomposition $(\{X_i : i \in I\}, T)$ is a tree decomposition satisfying the following properties:*

1. *every node of T has at most two children;*
2. *If a node i has two children j and k, then $X_i = X_j = X_k$ (i is a Join Node);*
3. *If a node i has one child j, then either*
   (a) *$|X_i| = |X_j| + 1$ and $X_j \subset X_i$ (i is a Insert Node);*
   (b) *$|X_i| = |X_j| - 1$ and $X_i \subset X_j$ (i is a Forget Node).*

Lemma 1 ([14]). *For a constant k, given a tree decomposition of a graph G of width k and $O(n)$ nodes, where n is the number of vertices of G, one can find a nice tree decomposition of G of width k and with at most $4n = O(n)$ nodes in $O(n)$ time.*

Structural and algorithmic properties of graph classes will be mentioned in the corresponding sections. For definitions and properties of graph classes not given in this paper we refer to [6, 11].

3   Domination on Chordal Graphs

In this section we present an exponential time algorithm for the minimum dominating set problem on chordal graphs.

A graph is *chordal* if it has no chordless cycle of length greater than 3. Chordal graphs are a well-known graph class with its own chapter in Golumbic's monograph [11]. Split graphs, strongly chordal graphs and undirected path graphs are well-studied subclasses of chordal graphs.

We shall use the clique tree representation of chordal graphs that we view as a tree decomposition of the graph. A tree T is as *clique tree* of a chordal graph $G = (V, E)$ if there is a bijection between the maximal cliques of G and the nodes of T such that for each $v \in V$ the cliques containing v induce a subtree of T. It is well-known that $tw(G) \geq \omega(G) - 1$ for all graphs. Furthermore the clique tree of a chordal graph G is an optimal tree decomposition of G, i.e. its width is $\omega(G) - 1$.

Lemma 2. *There is an $O^*(3^{tw(G)})$ time algorithm to compute a minimum dominating set on chordal graphs.*[1]

Proof. Alber et al. have shown in [1] that a minimum dominating set of a graph can be computed in time $O(4^l n)$ if a tree decomposition of width l of the input graph is known. Their algorithm uses a nice tree decomposition of the input graph and a standard bottom up dynamic programming on the tree decomposition. The crucial idea is to assign three different "colors" to the vertices of a bag:

- "black", meaning that the vertex belongs to the dominating set,
- "white", meaning that the vertex is already dominated,
- "gray", meaning that the vertex is not yet dominated.

Now let us assume that the input graph is chordal. A clique tree T of G can be computed in linear time [3]. By Lemma 1, a nice optimal tree decomposition of G can be computed from the optimal tree decomposition T in time $O(n)$ and it has at most $4n$ nodes. Since G is chordal every bag in the nice tree decomposition is a clique. Therefore no bag can have both a black vertex and a gray vertex. Due to this restriction there are at most $2^{|X|}$ possible so-called vector colorings of a bag X (instead of $3^{|X|}$ for general graphs).

Consequently the running time of a modification of the algorithm of Alber et al. to chordal graphs is $O^*(3^{tw(G)})$, where the only modification is to restrict allowed vector colorings of a bag such that black and gray vertices simultaneously are forbidden. □

The following theorem shows that graphs with sufficiently many vertices of high degree allow to speed up the MDS algorithm for general graphs.

Theorem 1. *Let $t > 0$ be a fixed integer. Then there is a $O(1.2303^{2n-t})$ time algorithm to solve the MDS problem if the input graph fulfills the condition $|\{v \in V : d(v) \geq t - 2\}| \geq t$.*

Proof. Let $t > 0$ be an integer and $G = (V, E)$ a graph fulfilling the conditions of the theorem. Let $T = \{v \in V : d(v) \geq t - 2\}$; thus $|T| \geq t$. Notice that for each minimum dominating set D of G either at least one vertex of T belongs to D, or $T \cap D = \emptyset$.

This allows to find a minimum dominating set of G by the following branching in two types of subproblems: "$v \in D$" for all $v \in T$, and "$T \cap D = \emptyset$". In both cases we shall apply the minimum set cover algorithm of [8] to solve the subproblem. Recall that the minimum set cover instance corresponding to the

---

[1] Modified big-Oh notation suppresses polynomially bounded factors.

MDS problem for G has universe $\mathcal{U} = V$ and $\mathcal{S} = \{N[u] : u \in V\}$, and thus $|\mathcal{U}| + |\mathcal{S}| = 2n$ [8]. Consequently the running time for a subproblem will be $O(1.2303^{2n-x})$, where x is the number of vertices plus the number of subsets eliminated from the original minimum set cover problem for the graph G.

Now let us consider the two types of subproblems. For every vertex $v \in T$, we choose v in the minimum dominating set and we execute the Minimum Set Cover algorithm presented in [8] on an instance of size at most $2n-(d(v)+1)-1 \le 2n-t$. Indeed, we remove from the universe $\mathcal{U}$ the elements of $N[v]$ and we remove from $\mathcal{S}$ the set corresponding to v. And we branch in the case "discard T": In this case we have an instance of set cover of size at most $2n - |T| = 2n - t$ since for every $v \in T$ we remove from $\mathcal{S}$ the set corresponding to each v.   □

Corollary 1. *There is an algorithm taking as input a graph G and a clique C of G and solving the* MDS *problem in time $O(1.2303^{2n-|C|})$.*

Proof. Note that every vertex in C has degree at least $|C| - 1$.   □

Our algorithm on chordal graphs works as follow: If the graph has a large treewidth then it necessarily has a large clique and we apply Corollary 1. Otherwise the graph has a small treewidth and we use Lemma 2.

Theorem 2. *There is an $O(1.4173^n)$ time algorithm to solve the* MDS *problem on chordal graphs.*

Proof. If $tw(G) \le 0.3174n$, by Lemma 2, MDS is solvable in time $O(3^{0.3174n}) = O(1.4173^n)$. Otherwise, $tw(G) > 0.3174n$ and using Corollary 1 we obtain an $O(1.2303^{2n-0.3174n}) = O(1.4173^n)$ time algorithm.   □

4   Domination on Circle Graphs

In this section, we present an exponential time algorithm for MDS on circle graphs in a treewidth based approach. For a survey on treewidth based exponential time algorithms we refer to [9].

Definition 4. *A circle graph is an intersection graph of chords in a circle. More precisely, G is a circle graph, if there is a circle with a collection of chords, such that one can associate in a one-to-one manner to each vertex a chord such that two vertices are adjacent if and only if the corresponding chords have a nonempty intersection. The circle and all its chords are called a circle model of the graph.*

Our algorithm heavily relies on a linear upper bound on the treewidth of circle graphs in terms of the maximum degree: $tw(G) \le 4\Delta(G) - 1$. This bound is interesting in its own and it is likely that such bounds for circle graphs or other graph classes can be used to construct exponential time algorithms for NP-hard problems on special graph classes in a way similar to our approach for domination on circle graphs.

The algorithm uses the treewidth to branch into two different approaches: one for "small treewidth" and one for "high treewidth". If there are many vertices of high degree in the input graph, Theorem 1 is used to continue, and if not, the treewidth is "small" and we use an $O^*(4^{tw(G)})$ algorithm to compute a minimum dominating set.

Theorem 3 ([1]). *Suppose the graph $G = (V, E)$ and a tree decomposition of width ℓ of G are given. Then there is an $O(4^\ell N)$ time algorithm to compute a minimum dominating set of G, where N is the number of nodes of the tree decomposition.*

We start with a brief summary of Kloks' algorithm to compute the treewidth of a circle graph [15]. Consider the circle model of a circle graph G. Go around the circle and place a new point (so-called *scanpoints*) between every two consecutive end points of chords. The treewidth of a circle graph can be computed by considering all possible triangulations of the polygon $\mathcal{P}$ formed by the convex hull of these scanpoints. The weight of a triangle in this triangulation is the number of chords in the circle model that cross this triangle. The weight of the triangulation $\mathcal{T}$ is the maximum weight of the triangles in $\mathcal{T}$. The treewidth of the graph is the minimum weight minus one over all triangulations of $\mathcal{P}$.

Theorem 4 ([15]). *There exists an $O(n^3)$ algorithm to compute the treewidth of circle graphs, that also computes an optimal tree-decomposition.*

We rely on the following technical definitions in our construction of a tree-decomposition of width at most $4\Delta(G) - 1$ for each circle graph G. The construction will be given in the proof of Theorem 5.

Definition 5. *A scanline $\tilde{s} = \langle \tilde{a}, \tilde{b} \rangle$ is a line segment connecting two scanpoints $\tilde{a}$ and $\tilde{b}$.*

To emphasize the difference between scanlines and chords we use different notations: A chord v connecting two end points c and d in the circle model of the graph is denoted $v = [c, d]$. We also use the following convention: two scanlines with empty intersection or intersecting in exactly one scanpoint are said to be *non-crossing*.

Definition 6. *Let $\tilde{s}_1$ and $\tilde{s}_2$ be two non-crossing scanlines. A scanline $\tilde{s}$ is between $\tilde{s}_1$ and $\tilde{s}_2$ if every path from a scanpoint of $\tilde{s}_1$ to a scanpoint of $\tilde{s}_2$ along the circle passes through a scanpoint of $\tilde{s}$.*

Definition 7. *A set S of* parallel *scanlines is a set of scanlines respecting*

(i) $|S| \leq 2$ and the scanlines of S are non-crossing, or
(ii) $|S| > 2$ and for every subset of three scanlines in S, one of these scanlines is between the other two.

The following theorem is one of the main results of this paper. It shows that the treewidth $tw(G)$ of circle graphs can be upper bounded by a linear function of the maximum degree $\Delta(G)$ of the graph G. Surprisingly, no linear bound seems to have been known prior to our work.

Theorem 5. *For every circle graph G holds $tw(G) \leq 4\Delta(G) - 1$.*

Proof. We construct a triangulation of the polygon $\mathcal{P}$ such that every triangle has weight at most 4Δ, i.e. it intersects at most 4Δ chords, and therefore the corresponding tree-decomposition has width at most $4\Delta - 1$.

Notice that by the definition of a circle graph, every chord intersects at most Δ other chords. The triangulation of the polygon $\mathcal{P}$ is obtained by constructing the corresponding set of scanlines S which is explained by the following procedures. Having described our algorithm, we will analyze the number of chords that cross each triangle and show that it is less than or equal to 4Δ.

1. Description of the algorithm

- **FirstCut().** Start with $S = \emptyset$. Choose a chord v in the circle model of the graph G. Call **ScanChord**(S, v). Call **ParaCuts**(S).

- **ScanChord**$(S, v = [a, b])$. Let $\tilde{c}$ and $\tilde{c}'$ (resp. $\tilde{d}$ and $\tilde{d}'$) be the two scanpoints closest to a (resp. b) on the circle such that the order of the points on the circle is $\tilde{c}, a, \tilde{c}', \tilde{d}', b$ and $\tilde{d}$. Now the algorithm adds the following three scanlines to S: $\tilde{s}_1 = \langle \tilde{c}, \tilde{d} \rangle, \tilde{s}_2 = \langle \tilde{c}', \tilde{d}' \rangle$ and $\tilde{s}_3 = \langle \tilde{c}, \tilde{d}' \rangle$. If $\tilde{c} = \tilde{d}$ (or $\tilde{c}' = \tilde{d}'$) then we add only the scanline $\tilde{s}_2$ (or $\tilde{s}_1$).

- **ParaCuts**(S). While S is not a maximal (by inclusion) parallel set of scanlines in $\mathcal{P}$, choose a chord v such that S remains parallel when calling **Scan-Chord**(S, v). Call **ScanChord**(S, v). If S is maximal parallel, every polygon inside $\mathcal{P}$ is delimited by one or two scanlines. We call the polygons that are delimited by one scanline *outer polygons*, and those that are delimited by two scanlines *inner polygons* (see Fig. 1). There are exactly two outer polygons now, one delimited by $\tilde{s}_1$ and the other one by $\tilde{s}_2$. Call **TriangOuter**$(S, \tilde{s}_1)$ and **Triang-Outer**$(S, \tilde{s}_2)$. For every inner polygon, call **TriangInner**$(S, \tilde{t}_1, \tilde{t}_2)$ where $\tilde{t}_1$ and $\tilde{t}_2$ are the two scanlines delimiting this polygon.

- **TriangOuter**$(S, \tilde{s} = \langle \tilde{a}, \tilde{b} \rangle)$. The scanline $\tilde{s}$ divides the polygon $\mathcal{P}$ into two parts. Call $\mathcal{P}_{\tilde{s}}$ the polygon delimited by $\tilde{s}$ and the part of $\mathcal{P}$ that does not contain any scanlines. Add a scanline between $\tilde{a}$ and every scanpoint of $\mathcal{P}_{\tilde{s}}$ except $\tilde{a}$ and $\tilde{b}$ to S.

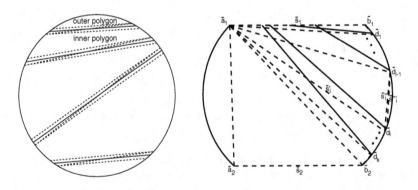

Fig. 1. ParaCuts   Fig. 2. TriangInner

- **TriangInner**$(S, \tilde{s}_1 = \langle \tilde{a}_1, \tilde{b}_1 \rangle, \tilde{s}_2 = \langle \tilde{a}_2, \tilde{b}_2 \rangle)$. Let the end points of $\tilde{s}_1$ and $\tilde{s}_2$ be ordered $\tilde{a}_1, \tilde{b}_1, \tilde{b}_2, \tilde{a}_2$ around the circle. W.l.o.g. assume that fewer chords cross the line $\tilde{a}_1, \tilde{a}_2$ than the line $\tilde{b}_1, \tilde{b}_2$. Now add a new scanline $\tilde{t} = \langle \tilde{a}_1, \tilde{a}_2 \rangle$ to S. Call **OuterParaCuts**$(S, \tilde{t})$. Go around the circle from $\tilde{b}_1$ to $\tilde{b}_2$ (without passing through $\tilde{a}_1$ and $\tilde{a}_2$). Every time one passes through an end point e_i, $i = 1, ..., k$, (where k is the number of chords that cross $\tilde{s}_1$ and $\tilde{b}_1, \tilde{b}_2$) of a chord v_i that crosses $\tilde{s}_1$, add the following scanlines to S:

  - $\tilde{s}_i' = \langle \tilde{a}_1, \tilde{d}_i \rangle$ with $\tilde{d}_i$ being the scanpoint immediately following e_i
  - $\tilde{s}_i'' = \langle \tilde{d}_i, \tilde{d}_{i-1} \rangle$ with $\tilde{d}_0 = \tilde{b}_1$
  - $\tilde{s}_i''' = \langle \tilde{d}_{i-1}, \tilde{d}_i' \rangle$ with $\tilde{d}_i'$ being the scanpoint just before $\tilde{d}_i$.

To triangulate the part of the polygon $\mathcal{P}$ delimited by $\tilde{s}_i'''$ that does not intersect any scanlines, execute **OuterParaCuts**$(S, \tilde{s}_i''')$. Finally, add the scanlines $\tilde{s}_3 = \langle \tilde{d}_k, \tilde{b}_2 \rangle$ and $\tilde{s}_4 = \langle \tilde{b}_2, \tilde{a}_1 \rangle$ to S (see Fig. 2). Execute **OuterParaCuts**$(S, \tilde{s}_3)$.

- **OuterParaCuts**$(S, \tilde{s} = \langle \tilde{a}, \tilde{b} \rangle)$. This procedure is similar to **ParaCuts** on the outer polygon delimited by $\tilde{s}$. Call $\mathcal{P}_{\tilde{s}}$ the polygon delimited by $\tilde{s}$ that does not contain any scanlines. Create a new set of scanlines $S' = \{\tilde{s}\}$. While S' is not a maximal (by inclusion) parallel set of scanlines for $\mathcal{P}_{\tilde{s}}$, choose a chord v in $\mathcal{P}_{\tilde{s}}$ such that S' remains parallel when calling **ScanChord**(S', v). Call **ScanChord**(S', v). After that there is exactly one outer polygon in $\mathcal{P}_{\tilde{s}}$, delimited by a scanline $\tilde{t}$. Call **TriangOuter**$(S', \tilde{t})$. For every inner polygon in $\mathcal{P}_{\tilde{s}}$, call **TriangInner**$(S', \tilde{t}_1, \tilde{t}_2)$ where $\tilde{t}_1$ and $\tilde{t}_2$ are the two scanlines delimiting this polygon. Add the set of new scanlines S' to S.

2. Analysis of the algorithm

In the main procedure, **FirstCut**, no scanlines are directly added to S.

Every time **ScanChord** is executed, one or three scanlines are added to S. They form at most two triangles: $\tilde{c}, \tilde{d}, \tilde{d}'$ and $\tilde{c}, \tilde{d}', \tilde{c}'$. Each of them intersects at most $\Delta + 1$ chords: v and the chords crossing v. Furthermore, at most Δ chords cross $\tilde{s}'$ and $\tilde{s}''$, precisely the chords that cross v.

In the procedure **ParaCuts**, no scanlines are directly added to S. Moreover, when it calls the procedures **TriangOuter** and **TriangInner**, the set S is maximal parallel, which is a necessary condition for these procedures.

When **TriangOuter** is called, two conditions are always respected:

(i) S is maximal parallel by inclusion, and
(ii) at most 2Δ chords cross $\tilde{s}$.

The condition (i) implies that every chord that intersects $\mathcal{P}_{\tilde{s}}$ crosses $\tilde{s}$. Together with condition (ii) we obtain that at most 2Δ chords intersect $\mathcal{P}_{\tilde{s}}$. So any triangulation of $\mathcal{P}_{\tilde{s}}$ produces triangles with weight at most 2Δ.

When **TriangInner** is called, three conditions are always respected:

(i) S is a maximal parallel set of scanlines, and
(ii) at most Δ chords cross one of the scanlines; suppose this is $\tilde{s}_2$
(iii) at most 2Δ chords cross $\tilde{s}_1$.

There are at most 3Δ chords inside the quadrilateral $\tilde{a}_1, \tilde{b}_1, \tilde{b}_2, \tilde{a}_2$ since there is no chord crossing both the lines $\tilde{a}_1, \tilde{a}_2$ and $\tilde{b}_1, \tilde{b}_2$ (because S is maximal parallel). As fewer chords cross $\tilde{a}_1, \tilde{a}_2$ than $\tilde{b}_1, \tilde{b}_2$, at most $3/2\Delta$ chords cross the new scanline $\tilde{t} = \langle \tilde{a}_1, \tilde{a}_2 \rangle$. So, when we call **OuterParaCuts**$(S, \tilde{t})$ the condition that $\tilde{t}$ intersects at most 2Δ chords is respected. For every end point e_i of a chord v_i that crosses $\tilde{s}_1$, we create two triangles: $\tilde{a}_1, \tilde{d}_{i-1}, \tilde{d}_i$ and $\tilde{d}_i, \tilde{d}_{i-1}, \tilde{d}'_i$. The first triangle intersects at most 4Δ chords: at most 2Δ chords that cross $\tilde{s}_1$ (but neither v_i nor v_{i-1}), at most Δ chords that cross v_{i-1} and at most Δ chords that cross v_i. Moreover, there are at most $2\Delta + 1$ chords that intersect $\tilde{s}''_i$ and at most 2Δ chords intersect $\tilde{s}'''_i$. So, the weight of the triangle $\tilde{d}_i, \tilde{d}_{i-1}, \tilde{d}'_i$ is at most $2\Delta + 1$ and when we call **OuterParaCuts**$(S, \tilde{s}'''_i)$ we respect the condition that the second parameter of the procedure is a scanline that crosses at most 2Δ chords.

After adding the scanlines $\tilde{s}_3$ and $\tilde{s}_4$ we obtain two more triangles: $\tilde{a}_1, \tilde{d}_k, \tilde{b}_2$ and $\tilde{a}_1, \tilde{b}_2, \tilde{a}_2$. The first one intersects at most $7/2\Delta$ chords: at most 2Δ that cross $\tilde{s}_1$, at most Δ that cross v_k and at most Δ that cross $\tilde{s}_2$ of which we have already counted $1/2\Delta$ crossing $\tilde{s}_1$. At most $5/2\Delta$ chords intersect the triangle $\tilde{a}_1, \tilde{b}_2, \tilde{a}_2$: at most 2Δ that intersect $\tilde{s}_1$ and at most Δ that intersect $\tilde{s}_2$ of which we have already counted $1/2\Delta$ crossing $\tilde{s}_1$. Moreover at most 2Δ chords cross $\tilde{s}_3$, so **OuterParaCuts**$(S, \tilde{s}_3)$ has valid parameters.

In the procedure **OuterParaCuts**, no scanlines are directly added to S. The following condition is always respected:

(i) at most 2Δ chords cross $\tilde{s}$.

During this procedure, we consider only the polygon $\mathcal{P}_{\tilde{s}}$. A new set of scanlines $S' = \{\tilde{s}\}$ is created and is made maximal parallel by inclusion by calling **Scan-Chord**. If $\{\tilde{s}\}$ is already maximal parallel, then **TriangOuter**$(S', \tilde{s})$ is called and the two conditions of that procedure are respected. If other scanlines had to be added to S' to make it maximal parallel, the procedure **TriangOuter**$(S', \tilde{t})$ is called for the outer polygon where $\tilde{t}$ is a scanline intersecting at most Δ chords. Moreover, the procedure **TriangInner**$(S, \tilde{t}_1, \tilde{t}_2)$ is called for the inner polygons. Every scanline delimiting the inner polygons intersects at most Δ chords, except $\tilde{s}$ that can intersect up to 2Δ chords. So, we respect the condition for **TriangInner** that one scanline intersects at most Δ chords and the other one at most 2Δ chords. Finally, S' is added to S which does not create any new triangles.

We have provided a recursive algorithm to triangulate the polygon $\mathcal{P}$ and have shown that the obtained triangulation does not contain triangles intersecting more than 4Δ chords. Thus the corresponding tree-decomposition of G has width at most $4\Delta - 1$. $\square$

Linear upper bounds for the treewidth in terms of the maximum degree seem to have an immediate use in the design of treewidth based exact algorithms. Using Theorem 6 we obtain an algorithm to compute a minimum dominating set for circle graphs in time $O(1.4956^n)$. The algorithm **DS-circle** is simple and also based on the algorithms of Theorem 3 and Theorem 1.

Theorem 6. *Given a circle graph $G = (V, E)$, algorithm **DS-circle** computes a minimum dominating set of G in time $O(1.4956^n)$.*

Algorithm DS-circle(circle graph $G = (V, E)$; circle model of G
Input: A circle graph G and its circle model.
Output: The domination number $\gamma(G)$ of G.
$\quad \lambda \leftarrow 0.2322$
$\quad X \leftarrow \emptyset$
$\quad$ Compute the treewidth $tw(G)$ of G using theorem 4
$\quad$ **while** $tw(G - X) \geq \lambda n$ **do**
$\quad\quad \lfloor\ X \leftarrow X \cup \{u\}$ where u is a vertex of $G - X$ of highest degree
$\quad$ **if** $|X| \geq \lambda n/4$ **then**
$\quad\quad \lfloor$ use the algorithm of Theorem 1 and return the result
$\quad$ **else**
$\quad\quad \lfloor$ use the algorithm of Theorem 3 and return the result

---

Proof. The algorithm constructs a vertex set $X = \{x_1, x_2, ..., x_k\}$ starting from an empty set by adding maximum degree vertices of the remaining graph to the set X until $tw(G - X) < \lambda n$.

When the vertex x_i is added to $X = \{x_1, x_2, ..., x_{i-1}\}$, we have $tw(G - X) \geq \lambda n$. The vertex $x_i \in V - X$ is of highest degree in $G - X$, i.e. $d(x_i) = \Delta(G - X)$. We have $d(x_i) > tw(G - X)/4$ by Theorem 5. Now, $d(x_i) > \lambda n/4$ because $tw(G - X) \geq \lambda n$. So, $\forall x_i \in X, d(x_i) > \lambda n/4$.

In the case $|X| \geq \lambda n/4$, we have a subset $X \subseteq V$ such that $\forall v \in X, d(v) > \lambda n/4$. So, according to Theorem 1, a minimum dominating set can be found in time $O(1.2303^{2n - \lambda n/4}) = O(1.4956^n)$.

In the other case, $|X| < \lambda n/4$ and $tw(G - X) < \lambda n$. As adding one vertex to a graph increases its treewidth at most by one, $tw(G) < \lambda n + \lambda n/4$. Using the algorithm of Theorem 3, a minimum dominating set is determined in time $O^*(4^{tw(G)}) = O(4^{(5\lambda/4)n}) = O(1.4956^n)$. $\qquad\square$

5 Domination on Dense Graphs

It is known that problems like Independent Set, Hamiltonian Circuit and Hamiltonian Path remain NP-complete when restricted to graphs having a large number of edges [18]. An easy way to show that a graph problem remains NP-complete for c-dense graphs, for any c with $0 < c < 1/2$, is to construct the graph G' by adding a sufficiently large complete graph as new component to the original graph G such that G' is c-dense. It is not hard to show that the MDS problem on c-dense graph is also NP-complete. A proof will be given in the full version of this paper. In this section we present an exponential time algorithm for MDS problem on c-dense graphs.

Definition 8. *A graph $G = (V, E)$ is said to be c-dense (or simply dense if there is no ambiguity), if $|E| \geq cn^2$ where c is a constant with $0 < c < 1/2$.*

The main idea of our algorithm is to find a large subset of vertices of large degree. Despite the approach of the previous sections, neither clique trees nor tree decompositions will be used here.

Lemma 3. *For some fixed $1 \le t \le n$, $1 \le t' \le n - 1$, any graph $G = (V, E)$ with $|E| \ge 1 + \dfrac{(t - 1)(n - 1) + (n - t + 1)(t' - 1)}{2}$ has a subset $T \subseteq V$ such that*

(i) $|T| \ge t$,
(ii) $\forall v \in T$, $d(v) \ge t'$.

Proof. Let $1 \le t \le n$, $1 \le t' \le n - 1$, and a graph $G = (V, E)$ such that there is no subset T with the previous properties. Then for any subset $T \subseteq V$ of size at least t, $\exists v \in T$ such that $d(v) < t'$. Then a such graph can only have at most $k = k_1 + k_2$ edges where : $k_1 = (t-1)(n-1)/2$ which corresponds to $t-1$ vertices of degree $n - 1$ and $k_2 = (n - t + 1)(t' - 1)$ which corresponds to $n - (t - 1)$ vertices of degree $t' - 1$. Observe that if one of the $n - (t-1)$ vertices has a degree greater than $t' - 1$ then the graph has a subset T with the required properties, a contradiction. □

Lemma 4. *Every c-dense graph $G = (V, E)$ has a set $T \subseteq V$ fulfilling*

(i) $|T| \ge n - \dfrac{\sqrt{9 - 4n + 4n^2 - 8cn^2} - 3}{2}$,

(ii) $\forall v \in T$, $d(v) \ge n - \dfrac{\sqrt{9 - 4n + 4n^2 - 8cn^2} + 1}{2}$.

Proof. We apply Lemma 3 with $t' = t-2$. Since we have a dense graph, $|E| \ge cn^2$. Using inequality $1 + ((t - 1)(n - 1) + (n - t + 1)(t - 3))/2 \ge cn^2$ we obtain that in a dense graph the value of t in Lemma 3 is such that $n + \frac{3 - \sqrt{9 - 4n + 4n^2 - 8cn^2}}{2} \le t \le n + \frac{3 + \sqrt{9 - 4n + 4n^2 - 8cn^2}}{2}$. □

Theorem 7. *For any c with $0 < c < 1/2$, there is a $O(1.2303^{n(1+\sqrt{1-2c})})$ time algorithm to solve the MDS problem on c-dense graphs.*

Proof. Combining Theorem 1 and Lemma 4 we obtain an algorithm for solving the Minimum Dominating Set problem in time

$$1.2303^{2n - (n - \frac{\sqrt{9 - 4n + 4n^2 - 8cn^2} - 3}{2})} = O(1.2303^{n(1+\sqrt{1-2c})}).$$ □

6   Conclusions

In this paper we presented several exponential time algorithms to solve the Minimum Dominating Set problem on graph classes for which this problem remains NP-hard. All these algorithms are faster than the best known algorithm to solve MDS on general graphs. We would like to mention that any faster algorithm for the Minimum Set Cover problem, i.e. of running time $O(\alpha^{|\mathcal{U}|+|\mathcal{S}|})$ with $\alpha < 1.2303$, could immediately be used to speed up all our algorithms.

Besides classes of sparse graphs (see e.g. [7]) two more classes are of great interest in this respect: split and bipartite graphs. For split graphs, combining ideas of [10] and [8] one easily obtains an $O(1.2303^n)$ algorithm. Unfortunately, despite our efforts we could not construct an exponential time algorithm to solve MDS on bipartite graphs beating the best known algorithm for general graphs.

References

1. Alber, J., H. L. Bodlaender, H. Fernau, T. Kloks and R. Niedermeier, Fixed parameter algorithms for dominating set and related problems on planar graphs, *Algorithmica* **33**, (2002), pp. 461–493.
2. Bertossi, A. A., Dominating sets for split and bipartite graphs., *Inform. Process. Lett.* **19**, (1984), pp. 37–40.
3. Blair, J. R. S. and B. W. Peyton, An introduction to chordal graphs and clique trees, *Graph theory and sparse matrix computation*, IMA Vol. Math. Appl., vol. 56, Springer, 1993, pp. 1–29.
4. Bodlaender, H. L. and D. M. Thilikos, Graphs with branchwidth at most three, *J. Algorithms* **32**, (1999), pp. 167–194.
5. Booth, K. S. and J. H. Johnson, Dominating sets in chordal graphs, *SIAM J. Comput.* **11**, (1982), pp. 191–199.
6. Brandstädt, A., V. Le, and J. P. Spinrad, *Graph classes: A survey*, SIAM Monogr. Discrete Math. Appl., Philadelphia, 1999.
7. Fomin, F.V., and K. Høie, Pathwidth of cubic graphs and exact algorithms, Technical Report 298, Department of Informatics, University of Bergen, Norway, 2005.
8. Fomin, F.V., F. Grandoni, D. Kratsch, Measure and conquer: Domination - A case study, *Proceedings of ICALP 2005*, LNCS **3380**, (2005), pp. 192–203.
9. Fomin, F.V., F. Grandoni, D. Kratsch, Some new techniques in design and analysis of exact (exponential) algorithms, *Bull. EATCS*, **87**, (2005), pp. 47–77.
10. Fomin, F.V., D. Kratsch, and G. J. Woeginger, Exact (exponential) algorithms for the dominating set problem, *Proceedings of WG 2004*, LNCS **3353**, (2004), pp. 245–256.
11. Golumbic, M. C., *Algorithmic graph theory and perfect graphs*, Academic Press, New York, 1980.
12. Grandoni, F., A note on the complexity of minimum dominating set, *J. Discrete Algorithms*, to appear.
13. Keil, J. M., The complexity of domination problems in circle graphs, *Discrete Appl. Math.* **42**, (1993), pp. 51–63.
14. Kloks, T., *Treewidth. Computations and approximation*, LNCS **842**, Springer-Verlag, Berlin, 1994.
15. Kloks, T., Treewidth of Circle Graphs, *Internat. J. Found. Comput. Sci.* **7**, (1996) pp. 111–120.
16. Randerath, B., and I. Schiermeyer, Exact algorithms for Minimum Dominating Set, Technical Report zaik-469, Zentrum fur Angewandte Informatik, Köln, Germany, 2004.
17. Robertson, N. and P. D. Seymour, Graph Minors. II. Algorithmic Aspects of Tree-Width, *J. Algorithms* **7**, (1986), pp. 309–322.
18. Schiermeyer, I., Problems remaining NP-complete for sparse or dense graphs, *Discuss. Math. Graph Theory* **15**, (1995), pp. 33–41.
19. Woeginger, G.J., Exact algorithms for NP-hard problems: A survey, *Combinatorial Optimization - Eureka, You Shrink!*, LNCS **2570**, (2003), pp. 185–207.

Exact Computation of Maximum Induced Forest

Igor Razgon*

Computer Science Department, University College Cork, Ireland
`i.razgon@cs.ucc.ie`

Abstract. We propose a backtrack algorithm that solves a generalized version of the Maximum Induced Forest problem (MIF) in time $O^*(1.8899^n)$. The MIF problem is complementary to finding a minimum Feedback Vertex Set (FVS), a well-known intractable problem. Therefore the proposed algorithm can find a minimum FVS as well. To the best of our knowledge, this is the first algorithm that breaks the $O^*(2^n)$ barrier for the general case of FVS. Doing the analysis, we apply a more sophisticated measure of the problem size than the number of nodes of the underlying graph.

1 Introduction

Exact exponential algorithms are techniques for solving intractable problems with better complexity than trivial brute-force exploring of all the possible combinations. Examples of such algorithms include: [9] for maximum independent set, [1] for chromatic number, [3] for 3-COLORABILITY, [2] for 3-SAT, [4] for dominating set, and others. A recent overview of exact algorithms is provided in [10].

In this paper we propose an $O^*(1.8899^n)$ exact algorithm for solving the following problem. Given a graph G and a subset K of its vertices, find a largest superset S of K such that the subgraph of G induced by S is acyclic. If $K = \emptyset$ then S is a Maximum Induced Forest (MIF) of G. The complement of S, $V(G)\backslash S$, is a minimum Feedback Vertex Set (FVS) of G, i.e. a set of vertices that participate in all the cycles of G. Computing a minimum FVS is a "canonical" intractable optimization problem, whose NP-complete version is mentioned in [7]. To the best of our knowledge, the proposed algorithm is the first that breaks the $O^*(2^n)$ barrier for the general case of FVS. Previous studies [6,8] describe exact algorithms only for special cases of FVS.

The proposed algorithm computes MIF using the "branch-and-prune" strategy ([10], Section 4). Using this strategy for computing of MIF is not straightforward. The reason is that selection of a new vertex for MIF does not necessarily cause additional pruning: a vertex can be pruned only if it induces a cycle with the "already selected" vertices. For graphs with a large girth, many vertices must be selected before at least one can be discarded.

---

* I would like to thank Fedor Fomin, who inspired me to investigate the problem, and an anonymous reviewer who suggested me a way of improving the result reported in the first version of the paper.

L. Arge and R. Freivalds (Eds.): SWAT 2006, LNCS 4059, pp. 160–171, 2006.

To overcome this difficulty, we analyze complexity of the algorithm by applying a more sophisticated measure than the number of vertices of the residual graph, a strategy suggested in [5]. In particular, we observe that all the "remaining" vertices can be partitioned into the vertices that have neighbours with the already selected vertices and the vertices that do not have them. We associate the vertices of the former class with weight 1 and the vertices of the latter class with weight 1.565, a constant guessed by a computational procedure. The proposed measure is the sum of weights of all the vertices of the residual graph. Further analysis yields an upper bound $O^*(1.50189^m)$, where m is the value of the applied measure for the input graph G. We then demonstrate that $m \leq 1.565n$, where n is the number of vertices of G, which results in an upper bound of $O^*((1.50189^{1.565})^n)$. Taking into account that $1.8898 < 1.50189^{1.565} < 1.8899$, this bound is transformed to $O^*(1.8899^n)$ by rounding.

The rest of the paper is organized as follows. Section 2 introduces the necessary terminology. Section 3 presents the proposed algorithm. Section 4 proves correctness of the algorithm and provides complexity analysis [1].

2 Preliminaries

A simple undirected graph is referred in this paper as *a graph*. A set of vertices of a graph G is denoted by $V(G)$. Given $S \subseteq V(G)$, we denote by $G[S]$ the subgraph of G induced by S and by $G \setminus S$ the subgraph of G induced by $V(G) \setminus S$. If S consists of a single vertex v, we write $G \setminus v$ rather than $G \setminus \{v\}$. Two vertices v and w of G are *S-connected* if they are adjacent or if there is a path $v, p_1, \ldots, p_m, w$, where $\{p_1, \ldots p_m\} \subseteq S$.

The set S is a maximum induced forest (MIF) if $G[S]$ is acyclic and S is the largest set subject to this property [2]. In addition, we introduce the notion of a T-MIF.

Definition 1. *Let $T \subseteq V(G)$. A T-MIF of G is a largest superset S of T such that $G[S]$ is acyclic.*

Clearly, the definition makes sense only when $G[T]$ is acyclic. Observe that a $\emptyset$-MIF of G is an ordinary MIF.

To present complexity of algorithms, we use the O^* notation [10], which suppresses the polynomial factor. For example, $O(n^2 2^n)$ is written as $O^*(2^n)$.

3 The Algorithm

In this section we present an algorithm for computing a MIF of a given graph. We start with extending our notation.

Let G be a graph and let T be a subset of its vertices. We recognize the following classes of vertices of $G \setminus T$.

---

[1] Due to space constraints, proofs of some technical lemmas are omitted.

[2] It is more convenient for us to represent a MIF as a set of vertices rather than a subgraph of G.

- *Boundary vertices* denoted by $Bnd(G,T)$. The set $Bnd(G,T)$ contains all vertices $v \in V(G \setminus T)$ such that v is adjacent to exactly one vertex of T.
- *Conflicting vertices* denoted by $Cnf(G,T)$. The set $Cnf(G,T)$ contains all vertices $v \in V(G \setminus T)$ such that v is adjacent to at least two vertices of the same connected component of $G[T]$.
- *Free vertices* denoted by $Free(G,T)$. The set $Free(G,T)$ contains all vertices $v \in V(G \setminus T)$ such that v is not adjacent to any vertex of T.

Now we are ready to introduce the algorithm $Main\_MIF$ (Algorithm 1). It gets as input a graph G and a subset K of $V(G)$. The algorithm returns (as we will prove further) a K-MIF of G. Clearly, setting K to $\emptyset$ will make the algorithm to return a MIF of G.

The algorithm $Main\_MIF$ starts with checking whether $G[K]$ is acyclic. If not, $FAIL$ is returned immediately (line 1 of Algorithm 1) because no K-MIF of G exists in this case. Otherwise, the function $Find\_MIF$ runs (line 2 of Algorithm 1).

Function $Find\_MIF$ is the main "search engine" of $Main\_MIF$. It is described in lines 3-31 of Algorithm 1. The function gets as input a graph G_1, and subsets T_1 and K_1 of $V(G_1)$. The function is supposed to return a $T_1 \cup K_1$-MIF of G_1 (provided that $G_1[T_1 \cup K_1]$ is acyclic).

If $T_1 \cup K_1 = V(G_1)$, $Find\_MIF$ returns $T_1 \cup K_1$ (lines 4 and 5 of Algorithm 1). Otherwise, the execution can be divided into four stages: selecting a vertex v of $V(G_1) \setminus (T_1 \cup K_1)$, a recursive call processing the case when v is added to $T_1 \cup K_1$, a recursive call processing the case when v is eliminated from G_1, and returning the maximum-size set among the ones returned by the above two recursive calls.

Selection of a vertex v is described in lines 7-11 of Algorithm 1. The vertex is taken from $Bnd(G_1, T_1)$ unless the set is empty. In this case, the function selects an arbitrary vertex that does not belong to $T_1 \cup K_1$.

Having selected a vertex v, the function adds it to $T_1 \cup K_1$ (line 12 of Algorithm 1). The addition is performed by function T_Update (Algorithm 2). Applying of T_Update in line 12 returns a triplet (G_2, T_2, K_2), in which $T_2 \cup K_2 = T_1 \cup K_1 \cup \{v\}$, v itself and all vertices of K_1 that are K_1-connected to v are "moved" to T_2, G_2 is obtained from G_1 by removing all vertices of $Cnf(G_1, T_2)$ because every one of them induces cycles being added to T_2. Function $Find\_MIF$ is applied recursively to (G_2, T_2, K_2) in line 13 and returns a set S_2.

The way a vertex v is selected and then added to $T_2 \cup K_2$ ensures that the inputs (G', T', K') of all recursive applications of $FindIndep$ have a number of invariant properties which are crucial for our analysis (Section 4). Two most important properties are that any connected component of G' contains at most one connected component of $G'[T']$, and that there are no edges between vertices of T' and vertices of K'.

Processing the case, where v is eliminated from G_1 (lines 14-29 of Algorithm 1), depends on the number of vertices of $V(G_1) \setminus (T_1 \cup K_1 \cup \{v\})$ that are K_1-connected to v. If there is at most one such vertex, the function decides that S_2

is a $T_1 \cup K_1$-MIF of G_1 and returns it (lines 15-16 of Algorithm 1). The case when there are exactly 2 such vertices is processed in lines 17-25. The set W of these two vertices is added to $T_1 \cup K_1$ by function K_Update (Algorith 3). This function returns $FAIL$ if $G_1[T_1 \cup K_1 \cup W]$ contains cycles. In this case, $Find\_MIF$ returns S_2. Otherwise, $Find\_MIF$ is applied recursively to the triplet returned by K_Update, returns a set S_3, and the largest set among S_2 and S_3 is returned in line 22. If the number of vertices of $V(G_1) \backslash (T_1 \cup K_1 \cup \{v\})$ that are K_1-connected to v is at least 3, $Find\_MIF$ returns the largest set among S_2 and S_3, where S_3 is returned by the recursive application of $Find\_MIF$ to $(G_1 \backslash v, T_1, K_1)$ (lines 26-28 of Algorithm 1).

Consider the intuition behind the decisions made by the algorithm in lines 15-25. For this purpose, assume that $K_1 = \emptyset$. That is, we consider the cases where v have 1 or 2 neighbours that are not in T_1. In the former case, let $w \in V(G_1) \backslash T_1$ be the considered neighbour of v. Observe that it is safe to add v to T_2. Really, any T_1-MIF S of G_1 that does not contain v has to contain w (otherwise, we get contradiction to the maximality of S). In this case replacing w by v in S, we get another T_1-MIF of G_1.

Assume now that v is adjacent to vertices $w_1, w_2 \in V(G_1) \backslash T_1$ and that v does not belong to any T_1-MIF of G_1. Then, any T_1-MIF of G_1 contains both w_1 and w_2, otherwise, arguing as for the previous case, we get a contradiction with our assumption. A subtle question is where to add w_1 and w_2. The point is that the invariant property that every component of G_1 contains at most one component of $G_1[T_1]$ should not be violated for the inputs of the subsequent recursive calls of the $Find\_MIF$ function. To satisfy this requirement, for example, when none of w_1 and w_2 have neighbours in T_1, the function K_Update adds them to K_2, not to T_2. That is, even if $K = \emptyset$ in the original input, it can be transformed to a non-empty set in one of subsequent recursive calls of $Find\_MIF$. Thus the necessity to handle the case when v is adjacent to exactly two "remaining" vertices is what caused the author to consider a generalized version of the MIF-problem.

4   Analysis

The analysis of the $Main\_MIF$ algorithm is organized as follows. In Section 4.1 we introduce the notion of a *Fair Configuration* (FC) and prove a number of properties of FCs. In Section 4.2 we define a search tree generated by function $Find\_MIF$ with the nodes corresponding to the inputs of the recursive calls of $Find\_MIF$. We prove that all these inputs are FCs. Then, based on properties of FCs, we prove correctness of $Main\_MIF$ (section 4.3) and analyze its complexity (section 4.4). Due to space constraints, proofs of some technical lemmas are omitted.

4.1   Fair Configurations and Their Properties

Definition 2. *Let G be a graph and let T and K be subsets of $V(G)$. A triplet (G, T, K) is a Fair Configuration (FC) if the following conditions hold:*

Algorithm 1. Main_MIF($\mathbf{G}, \mathbf{K}$)

1: **if** $G[K]$ contains cycles **then** Return $FAIL$
2: Return $Find\_MIF(G \setminus Cnf(G, K), \emptyset, K)$
3: **function** $Find\_MIF(G_1, T_1, K_1)$
4: **if** $T_1 \cup K_1 = V(G_1)$ **then**
5:    Return $T_1 \cup K_1$
6: **else**
7:    **if** $Bnd(G_1, T_1)$ is not empty **then**
8:       Select a vertex $v \in Bnd(G_1, T_1)$
9:    **else**
10:       Select an arbitrary vertex $v \in G_1 \setminus (T_1 \cup K_1)$
11:    **end if**
12:    $(G_2, T_2, K_2) \leftarrow T\_Update(G_1, T_1, K_1, v)$
13:    $S_2 \leftarrow Find\_MIF(G_2, T_2, K_2)$
14:    **switch** The number of vertices of $V(G_1) \setminus (T_1 \cup K_1 \cup \{v\})$
              that are K_1-connected to v
15:       **case** ≤ 1
16:          Return S_2
17:       **case** 2
18:          Let W be the set of vertices of $V(G_1) \setminus (T_1 \cup K_1 \cup \{v\})$
              that are K_1-connected to v
19:          **if** $K\_Update(G_1 \setminus v, T_1, K_1, W)$ does not return $FAIL$ **then**
20:             $(G_3, T_3, K_3) \leftarrow K\_Update(G_1 \setminus v, T_1, K_1, W)$
21:             $S_3 \leftarrow Find\_MIF(G_3, T_3, K_3)$
22:             Return the largest set of S_2 and S_3
23:          **else**
24:             Return S_2
25:          **endif**
26:       **case** ≥ 3
27:          $S_3 \leftarrow Find\_MIF(G_1 \setminus v, T_1, K_1)$
28:          Return the largest set of S_2 and S_3
29:    **end switch**
30: **end if**
31: **end function**

Algorithm 2. T_Update($\mathbf{G}, \mathbf{T}, \mathbf{K}, \mathbf{v}$)

1: Let S be the subset of vertices of K that are K-connected to v
2: $T' \leftarrow T \cup \{v\} \cup S$
3: $K' \leftarrow K \setminus (\{v\} \cup S)$
4: $G' \leftarrow G \setminus Cnf(G, T')$
5: Return (G', K', T')

Algorithm 3. K_Update(G, T, K, W)

1: **if** $G[T \cup K \cup W]$ contains cycles **then** Return $FAIL$
2: **if** $W \cap Bnd(G, T) = \emptyset$ **then**
3:     Return $(G \setminus Cnf(G, K \cup W), T, K \cup W)$
4: **else**
5:     Let v_1 be a vertex of W that belongs to $Bnd(G, T)$
6:     $(G', T', K') \leftarrow T\_Update(G, T, K, v_1)$
7:     $\{v_2\} \leftarrow W \setminus \{v_1\}$
8:     **if** $v_2 \in Bnd(G', T')$ **then**
9:         Return $T\_Update(G', T', K', v_2)$
10:     **else**
11:         Return $(G' \setminus Cnf(G', K' \cup \{v_2\}), T', K' \cup \{v_2\})$
12:     **end if**
13: **end if**

 - $K \subseteq Free(G, T)$;
 - $G[T]$ and $G[K]$ are acyclic;
 - $Cnf(G, T) = Cnf(G, K) = \emptyset$;
 - every connected component of G contains at most one connected component of $G[T]$.

Due to importance of the notion for the proposed analysis, we demonstrate it on an example.

Let G be the graph shown in Figure 1, the black circles represent the vertices of T, the crossed circles represent the vertices of K, the other vertices are represented by the white circles. Observe that (G, T, K) is a FC. Indeed, there are no edges between the vertices of T and K, both T and K induce acyclic subgraphs of G, no single vertex of $V(G) \setminus (T \cup K)$ makes cycles with T and K. Finally, every connected component of G contains at most one connected component of $G[T]$. Note that by definition of a FC, the last requirement is not necessary for $G[K]$. In our example, the component induced by vertices v_9 to v_{12} contains two components of $G[K]$.

Lemma 1. *Let (G, T, K) be a FC. Then $T \cup Bnd(G, T) \cup Free(G, T) = V(G)$.*[3]

Lemma 2. *Let (G, T, K) be a FC and let $v \in V(G) \setminus (T \cup K)$. Assume that one of the following properties holds:*

 - $v \in Bnd(G, T)$;
 - $Bnd(G, T) = \emptyset$.

Then $(G', T', K') = T\_Update(G, T, K, v)$ is a FC.

As a result of application of T_Update, some vertices change their "roles". This statement is described precisely in the following lemma.

---

[3] In other words, any vertex of $V(G) \setminus T$ is adjacent to at most one vertex of T.

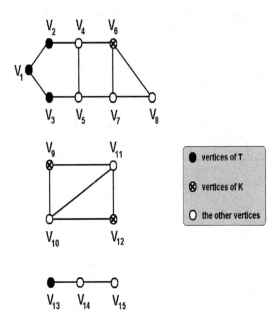

Fig. 1. A Fair Configuration (FC)

Lemma 3. *Let (G, T, K) be a FC and let $v \in V(G)$ be a vertex such that $(G', T', K') = T\_Update(G, T, K, v)$ is a FC.*

Let $w \in V(G) \setminus (T \cup K \cup \{v\})$ be a vertex, which is K-connected to v. Then

- $w \notin Free(G', T')$;
- *in addition, if $w \in Bnd(G, T)$ then $w \notin V(G')$.*

Lemma 4. *Let (G, T, K) be a FC. Let $S \subseteq V(G) \setminus (T \cup K)$ and $W \subseteq Free(G, T) \setminus K$ be two disjoint sets. Then $(G \setminus S \setminus Cnf(G \setminus S, K \cup W), T, K \cup W)$ is a FC.*

4.2   The Search Tree • •

In this section we define a *search tree ST* explored by $Main\_MIF$. The root of the tree is associated with a triplet $(G \backslash Cnf(G, K), \emptyset, K)$, where G and K constitute the input of $Main\_MIF$. Assume that a node x of ST is associated with a triplet (G_1, T_1, K_1). The structure of the subtree rooted by x depends on the execution of $Find\_MIF(G_1, T_1, K_1)$. If $T_1 \cup K_1 = V(G_1)$ then x is a leaf. Otherwise, x has a child associated with the triplet returned by $T\_Update(G_1, T_1, K_1, v)$, where v is the vertex selected by $Find\_MIF(G_1, T_1, K_1)$ in lines 7-11 of Algorithm 1. It is the only child if $Find\_MIF(G_1, T_1, K_1)$ executes line 16 or line 24. If line 22 is executed then x has the additional child associated with $K\_Update(G_1 \setminus v, T_1, K_1, \{v_1, v_2\})$; if line 28 is executed, the additional child is associated with the triplet $(G_1 \setminus v, T_1, K_1)$.

Lemma 5. *ST is of finite size.*

Lemma 6. *The triplet associated with every node of ST is a FC.*

4.3   Correctness Proof

In this section we will prove correctness of $Main\_MIF$ by demonstrating that $Main\_MIF(G, K)$ returns a K-MIF of G.

Lemma 7. *Let (G, T, K) be a FC and let $v \in V(G) \setminus (T \cup K)$. Let S be a $T \cup K$-MIF of G. Then either $v \in S$ or any cycle in $G[S \cup \{v\}]$ involves a vertex $w \in V(G) \setminus (T \cup K \cup \{v\})$, which is K-connected to v.*

Proof. Assume that $v \notin S$ and let $v, v_1, \ldots v_m$ be a cycle of $G[S \cup \{v\}]$ (clearly, v participates in any cycle of $G[S \cup \{v\}]$ because $G[S]$ is acyclic). If either v_1 or v_m belongs to $V(G) \setminus (T \cup K)$, we are done. Otherwise, note that $\{v_1, v_m\} \not\subseteq T$ because v cannot be connected to more than one vertex of T, by Lemma 1. It follows also that $\{v_1, \ldots, v_m\} \not\subseteq K$ because the opposite would mean that $v \in Cnf(G, K)$. Assume without loss of generality that $v_1 \in K$. Let i be the smallest index such that $v_i \notin K$, while $v_{i-1} \in K$. Note that by definition of a FC $v_i \notin T$ (because existence of an edge between K and T would follow otherwise). Thus $v_i \in V(G) \setminus (T \cup K)$ and the path $v, v_1, \ldots, v_i$, all intermediate vertices of which belong to K, certifies that v_i is K-connected to v. ■

Lemma 8. *Let G be a graph, $T \subset V(G)$ such that $G[T]$ is acyclic, and $U \subseteq V(G) \setminus T$. Assume that no T-MIF of G intersects with U. Then any T-MIF of $G \setminus U$ is a T-MIF of G.*

Lemma 9. *Let (G, T, K) be a FC and let $v \in V(G) \setminus (T \cup K)$. Let $(G', T', K') = T\_Update(G, T, K, v)$. Assume that at least one $T \cup K$-MIF of G contains v. Then any $T' \cup K'$-MIF of G' is a $T \cup K$-MIF of G.*

Lemma 10. *Let (G, T, K) be a FC. Let $v \in V(G) \setminus (T \cup K)$ be a vertex, which is K-connected to at most one vertex of $G \setminus (T \cup K \cup \{v\})$. Let $(G', T', K') = T\_Update(G, T, K, v)$. Then any $T' \cup K'$-MIF of G' is a $T \cup K$-MIF of G.*

Proof. If at least one $T \cup K$-MIF of G contains v, the statement follows from Lemma 9. Otherwise, let S be a $T \cup K$-MIF of G. By Lemma 7, every cycle of $G[S \cup \{v\}]$ contains a vertex of $V(G) \setminus (T \cup K \cup \{v\})$, which is K-connected to v. By the condition of the lemma, there is at most one such a vertex, say, w. Therefore w participates in all the cycles of $G[S \cup \{v\}]$, and removing of w breaks all the cycles. Clearly, $S \cup \{v\} \setminus \{w\}$ is a $T \cup K$-MIF of G containing v, in contradiction to our assumption. ■

Lemma 11. *Let (G, T, K) be a FC. Let $v \in V(G) \setminus (T \cup K)$ be a vertex, which is K-connected to exactly two vertices v_1 and v_2 of $G \setminus (T \cup K \cup \{v\})$. Then at least one of the following two statements is true.*

- *Let $(G_1, T_1, K_1) = T\_Update(G, T, K, v)$. Then any $T_1 \cup K_1$-MIF of G_1 is a $T \cup K$-MIF of G.*

- $K\_Update(G \setminus v, T, K, \{v_1, v_2\})$ *does not return $FAIL$. Moreover, let* $(G_2, T_2, K_2) = K\_Update(G \setminus \{v\}, T, K, \{v_1, v_2\})$. *Then any $T_2 \cup K_2$-MIF of G_2 is a $T \cup K$-MIF of G.*

Proof. Assume that the first statement does not hold. By Lemma 9, no $T \cup K$-MIF of G contains v. Let S be a $T \cup K$-MIF of G. According to Lemma 7, any cycle in $G[S \cup \{v\}]$ involves a vertex of $G \setminus (T \cup K \cup \{v\})$, which is K-connected to v. If S contains only one such vertex, say, v_1, then removing v_1 breaks all the cycles and $S \cup \{v\} \setminus \{v_1\}$ is a $T \cup K$-MIF of G in contradiction to our assumption. It follows that $\{v_1, v_2\} \subseteq S$. Clearly, S is a $T \cup K \cup \{v_1, v_2\}$-MIF of G because otherwise we get a contradiction with being S a $T \cup K$-MIF of G. It follows that any $T \cup K \cup \{v_1, v_2\}$-MIF of G is a $T \cup K$-MIF of G. Then, by Lemma 8, any $T \cup K \cup \{v_1, v_2\}$-MIF of $G \setminus v$ is a $T \cup K$-MIF of G. Consequently $(G \setminus v)[T \cup K \cup \{v_1, v_2\}]$ is acyclic and hence $K\_Update(G, T, K, \{v_1, v_2\})$ does not return $FAIL$. Furthermore, it follows from the description of K_Update that $T_2 \cup K_2 = T \cup K \cup \{v_1, v_2\}$ and G_2 is obtained from $G \setminus v$ by removing vertices that make cycles with $T \cup K \cup \{v_1, v_2\}$ in $G \setminus v$. Thus, any $T_2 \cup K_2$-MIF of G_2 is a $T \cup K$-MIF of G by Lemma 8 and the above reasoning. ∎

Theorem 1. *For any triplet (G_1, T_1, K_1) associated with a node of ST, $Find\_MIF(G_1, T_1, K_1)$ returns a $T_1 \cup K_1$-MIF of G_1.*

Proof. Let $x_1, x_2, \ldots$ be an order of nodes of ST such that children are ordered before their parents; existence of such an order follows from Lemma 5. The proof is by induction on the sequence. We also use the fact that the triplet associated with every node x of ST is a FC (Lemma 6).

Clearly, the statement holds for all leaves of ST and, in particular, for x_1. Consider a non-leaf node x_i, assuming validity of the theorem for all nodes placed before, and denote the FC associated with x_i by (G_1, T_1, K_1). The FCs associated with the children of x_i are exactly the inputs of the recursive calls performed by $Find\_MIF(G_1, T_1, K_1)$. By the induction assumption, these recursive calls work properly.

If the vertex v picked by $Find\_MIF(G_1, T_1, K_1)$ is K_1-connected to exactly one vertex of $V(G_1) \setminus (T_1 \cup K_1 \cup \{v\})$, the correctness follows from Lemma 10; in the case of two vertices, the correctness follows from Lemma 11; in the case of three or more vertices, the correctness follows from Lemmas 8 and 9. ∎

Corollary 1. *Let G be a graph and $K \subseteq V(G)$. If $G[K]$ is acyclic, $Main\_MIF(G, K)$ returns a K-MIF of G.*

4.4   Complexity Analysis

In this section we analyse the complexity of $Main\_MIF$ by deriving the upper bound on the number of nodes of ST. For the complexity analysis, we associate with every node x of ST the measure $Y(x) = c|Free(G_1, T_1) \setminus K_1| + |Bnd(G_1, T_1)|$, where (G_1, T_1, K_1) is the triplet associated with x, $c = 1.565$. In other words, the elements of $Free(G_1, T_1) \setminus K_1$ are assigned with weight c, the elements of $Bnd(G_1, T_1)$ are assigned with weight 1, $Y(x)$ is the sum of all the weights. The complexity analysis is structured as follows. For a given two nodes x and z such that x is the parent of z, we evaluate $Y(x) - Y(z)$. Based on the evaluation, we obtain an upper bound on the number of nodes of the subtree rooted at x. This upper bound is $O^*(\alpha(c)^{Y(x)})$, where $\alpha(c)$ is the constant depending on c. For $c = 1.565$, $\alpha(c) = 1.50189$. Then we notice that for the root node r, the value of $Y(r)$ is at most cn, where n is the number of vertices of the original graph. Thus we obtain the upper bound $O^*(1.50189^{1.565n})$. Taking into account that $1.8898 < 1.50189^{1.565} < 1.8899$, the upper bound obtained after rounding the base of the exponent is $O^*(1.8899^n)$.

The constant $c = 1.565$ was guessed by a binary search computational procedure that explored the range from 1.005 to 2 by steps of 0.005 and for every considered constant c computed $\alpha(c)^c$. The smallest value of this expression was obtained for $c = 1.565$.

Lemma 12. *Let x be a non-leaf node of ST associated with a triplet(G_1, T_1, K_1), let a node z associated with a triplet (G_2, T_2, K_2) be a child of x, and let $v \in V(G_1) \setminus (T_1 \cup K_1)$. Then the following statements hold.*

1. *If $(G_2, T_2, K_2) = T\_Update(G_1, T_1, K_1, v)$ then $Y(z) \leq Y(x) - ((c-1)|W| + 1)$, where W is the set of vertices of $V(G_1) \setminus (T_1 \cup K_1 \cup \{v\})$ that are K_1-connected to v.*
2. *If $(G_2, T_2, K_2) = K\_Update(G_1 \setminus v, T_1, K_1, \{v_1, v_2\})$, where$\{v_1, v_2\} \subseteq V(G_1) \setminus (T_1 \cup K_1 \cup \{v\})$, then $Y(z) \leq Y(x) - 3$.*
3. *If $(G_2, T_2, K_2) = (G_1 \setminus v, T_1, K_1)$ then $Y(z) \leq Y(x) - 1$.*

Proof. Let z be a child of x associated with (G_2, T_2, K_2) and assume that $(G_2, T_2, K_2) = T\_Update(G_1, T_1, K_1, v)$. Recall that all the triplets associated with the nodes of ST are FCs (Lemma 6). By definition of W and Lemma 1, the vertices of W can be partitioned into two subsets, $W_1 \subseteq Free(G_1, T_1) \setminus K_1$ and $W_2 \subseteq Bnd(G_1, T_1)$.

Observe that $|Free(G_2, T_2) \setminus K_2| \leq |Free(G_1, T_1) \setminus K_1| - |W_1|$. Really, $Free(G_2, T_2) \setminus K_2$ are the vertices of $V(G_2) \setminus (T_2 \cup K_2)$ that do not have neighbors in T_2. Taking into account that $T_1 \subset T_2$, $T_1 \cup K_1 \subset T_2 \cup K_2$, and $V(G_2) \subseteq V(G_1)$, it is clear that $Free(G_2, T_2) \setminus K_2 \subseteq Free(G_1, T_1) \setminus K_1$. Further, applying Lemma 3, we obtain that $Free(G_2, T_2) \setminus K_2 \subseteq (Free(G_1, T_1) \setminus K_1) \setminus W_1$. Considering that $W_1 \subseteq Free(G_1, T_1) \setminus K_1$, we get the desired inequality. A vertex of W_1 can be either removed from G_2 or added to $Bnd(G_2, T_2)$. In the former case the value of $Y(z)$ is decreased by c with respect to $Y(x)$, in the second case $Y(z)$ is decreased only by $c - 1$, because the weight of the vertex is changed from c to 1. We evaluate the maximal possible weight of $Y(z)$, hence we can assume that all vertices of W_1 are moved to $Bnd(G_2, T_2)$, decreasing $Y(z)$ by $(c-1)|W_1|$.

The vertices of W_2 are removed from G_2 by Lemma 3 and vertex v is moved from $Bnd(G_1, T_1)$ to T_2. These transformations decrease the value of $Y(z)$ with respect to $Y(x)$ by $|W_2| + 1$. Combining the above argumentation, and taking into account that $c = 1.565$, we see that $Y(z) \leq Y(x) - ((c-1)*|W_1|+|W_2|+1) \leq Y(x) - ((c-1)|W|+1)$, proving the first statement.

If the condition of the second statement holds then $|Free(G_2, T_2) \setminus K_2| + |Bnd(G_2, T_2)| \leq |Free(G_1, T_1) \setminus K_1| + |Bnd(G_1, T_1)| - 3$ because v is removed from G_2, v_1 and v_2 are moved to $T_2 \cup K_2$. Removing of anyone of these vertices decreases $Y(z)$ by at least 1. The second statement immediately follows.

The last statement is immediate when we observe that $|Free(G_2, T_2) \setminus K_2| + |Bnd(G_2, T_2)| \leq |Free(G_1, T_1) \setminus K_1| + |Bnd(G_1, T_1)| - 1$ if the condition of the last statement holds. ∎

Let m be an integer such that there is a node x of ST with $Y(x) = m$. We denote by $F(m)$ the maximum possible number of nodes of the subtree rooted at x.

Lemma 13. *For any node x of ST, $F(Y(x))$ is bounded by $O^*(1.50189^{Y(x)})$.*

Proof. Let (G_1, T_1, K_1) be the triplet associated with x. Recall that $Y(x) = c|Free(G_1, T_1) \setminus K_1| + |Bnd(G_1, T_1)|$. Clearly, $Y(x) \geq 0$.

Assume that $Y(x) = 0$. It is only possible when $V(G_1) = T_1 \cup K_1$. According to Algorithm 1, x is a leaf, hence the lemma holds for this case.

Assume now that $Y(x) > 0$. Clearly, x is a non-leaf. If x has only one child z then z is necessarily associated with $T\_Update(G_1, T_1, K_1, v)$ for some $v \in V(G_1) \setminus (T_1 \cup K_1)$. By Lemma 12, $Y(z) \leq Y(x) - 1$ (the equality holds when v is not K_1-connected to any vertex of $V(G_1) \setminus (T_1 \cup K_1 \cup \{v\})$). In this case, $F(Y(x)) = F(Y(z)) + 1 = F(Y(x) - l) + 1$, where $l \geq 1$.

If x has two children, z_1 and z_2, one of them, say z_1, is necessary associated with the triplet returned by $T\_Update(G_1, T_1, K_1, v)$. The node z_2 is associated either with $K\_Update(G_1 \setminus v, T_1, K_1, \{v_1, v_2\})$ or with $(G_1 \setminus v, T_1, K_1)$.

In the former case, $\{v_1, v_2\}$ is the set of vertices of $V(G_1) \setminus (T_1 \cup K_1 \cup \{v\})$ that are K_1-connected to v. By the first part of Lemma 12, $Y(z_1) \leq Y(x) - (2(c-1) + 1)$, by the second part of the same lemma, $Y(z_2) \leq Y(x) - 3$. Substituting $c = 1.565$, we obtain $F(Y(x)) = F(Y(z_1)) + F(Y(z_2)) + 1 = F(Y(x) - l_1) + F(Y(x) - l_2) + 1$, where $l_1 \geq 2.13$, $l_2 \geq 3$.

In the latter case, it follows from the description of Algorithm 1 that v is K_1-connected to at least 3 vertices of $V(G_1) \setminus (T_1 \cup K_1 \cup \{v\})$. Consequently, $Y(z_1) \leq Y(x) - (3(c-1) + 1)$ and $Y(z_2) \leq Y(x) - 1$ by the first and the last parts of Lemma 12. Arguing as for the previous two cases, we obtain $F(Y(x)) = F((Y(x) - l_1) + F(Y(x) - l_2) + 1$, where $l_1 \geq 2.695$, $l_2 \geq 1$.

The last recursive relation for $F(Y(x))$ yields the worst upper bound. Taking into account that the upper bound is exponential, we can ignore the additive constant because it contributes only a polynomial factor to the resulting bound. The upper bound following from the expression $F(Y(x)) = F(Y(x) - 1) + F(Y(x) - 2.695)$ is $O^*(\beta^{Y(x)})$, where β is the largest root of the equation $\beta^{2.695} = \beta^{1.695} + 1$. A simple computation shows that $1.50188 < \beta < 1.50189$. ∎

Theorem 2. *The Main_MIF algorithm, applied to a graph G with n vertices, takes $O^*(1.8899^n)$ time and a polynomial space.*

Proof. Let x be the root node of ST. It is associated with the triplet $(G, \emptyset, K)$. Then $Y(x) = c|Free(G, \emptyset) \setminus K| + |Bnd(G, \emptyset)| = c|Free(G, \emptyset) \setminus K| \leq cn$, where $c = 1.565$. It follows that ST has $O^*(1.50189^{1.565n}) < O^*(1.8899^n)$ nodes. The upper bound on the time-complexity of $Main\_MIF$ can be obtained by summing up the bounds on the processing time spent to every node of ST. Observe that processing of a node includes all the operations performed by $Find\_MIF$ except the recursive calls (whose processing time is related to other nodes of ST). The total time of these operations can be bounded by a polynomial multiplied to a number of nodes of ST. The resulting polynomial is suppressed by the O^* notation, hence $O^*(1.8899^n)$ is an upper bound on the time complexity of $Main\_MIF$.

Observe that $Find\_MIF$ has a polynomial space complexity because it is a backtrack-like procedure without explicit recording of results related to intermediate recursive calls. ∎

References

1. J. Byskov. Enumerating maximal independent sets with applications to graph colouring. *Operations Research Letters*, 32(6):547–556, November 2004.
2. E. Dantsin, A. Goerdt, E. Hirsch, R. Kannan, J. Kleinberg, C. Papadimitriou, P. Raghavan, and U. Schöning. A deterministic $(2-2/(k+1))^n$ algorithm for k-sat based on local search. *Theor. Comput. Sci.*, 289(1):69–83, 2002.
3. D. Eppstein. Improved algorithms for 3-coloring, 3-edge coloring and constraint satisfaction. In *SODA-2001*, pages 329–337, 2001.
4. F. Fomin, F. Grandoni, and D. Kratsch. Measure and conquer: Domination - a case study. In *ICALP*, pages 191–203, 2005.
5. F. Fomin, F. Grandoni, and D. Kratsch. Some new techniques in design and analysis of exact (exponential) algorithms. *Bulletin of the EATCS*, 87:47–77, 2005.
6. F. Fomin and A. Pyatkin. Finding minimum feedback vertex set in bipartite graphs. In *Report N 291, Department of Informatics, University of Bergen*, 2005.
7. R. Karp. Reducibility among combinatorial problems. In *Complexity of Computer Computations*, 1972.
8. V. Raman, S. Saurabh, and S. Sikdar. Improved exact exponential algorithms for vertex bipartization and other problems. In *ICTCS*, pages 375–389, 2005.
9. J. Robson. Algorithms for maximum independent sets. *Journal of Algorithms*, 7:425–440, 1986.
10. G. Woeginger. Exact algorithms for NP-hard problems: A survey. In *Combinatorial Optimization: "Eureka, you shrink", LNCS 2570*, pages 185–207, 2003.

Fast Subexponential Algorithm for Non-local Problems on Graphs of Bounded Genus

Frederic Dorn[1], Fedor V. Fomin[1,*], and Dimitrios M. Thilikos[2,**]

[1] Department of Informatics, University of Bergen, N-5020 Bergen, Norway
`frederic.dorn@ii.uib.no, fedor.fomin@ii.uib.no`
[2] Departament de Llenguatges i Sistemes Informàtics,
Universitat Politècnica de Catalunya, Barcelona, Spain
`sedthilk@lsi.upc.edu`

Abstract. We give a general technique for designing fast subexponential algorithms for several graph problems whose instances are restricted to graphs of bounded genus. We use it to obtain time $2^{O(\sqrt{n})}$ algorithms for a wide family of problems such as HAMILTONIAN CYCLE, Σ-EMBEDDED GRAPH TRAVELLING SALESMAN PROBLEM, LONGEST CYCLE, and MAX LEAF TREE. For our results, we combine planarizing techniques with dynamic programming on special type branch decompositions. Our techniques can also be used to solve parameterized problems. Thus, for example, we show how to find a cycle of length p (or to conclude that there is no such a cycle) on graphs of bounded genus in time $2^{O(\sqrt{p})} \cdot n^{O(1)}$.

1 Introduction

Many common computational problems are NP-hard and therefore do not seem to be solvable by efficient (polynomial time) algorithms. However, while NP-hardness is a good evidence for the intractability of a problem, in many cases, there is a real need for exact solutions. Consequently, an interesting and emerging question is to develop techniques for designing fast exponential or, when possible, sub-exponential algorithms for hard problems (see [15]).

The algorithmic study of graphs that can be embedded on a surface of small genus, and planar graphs in particular, has a long history. The first powerful tool for the design of sub-exponential algorithms on such graphs was the celebrated Lipton-Tarjan planar separator theorem [9, 10] and its generalization on graphs of bounded genus [7]. According to these theorems, an n-vertex graph of fixed genus can be "separated" into two roughly equal parts by a separator of size $O(\sqrt{n})$. This approach permits the use of a "divide and conquer" technique that provides subexponential algorithms of running time $2^{O(\sqrt{n})}$ for a wide range of combinatorial problems.

A similar approach is based on graph decompositions [6]. Here instead of separators one uses decompositions of small width, and instead of "divide and

---

[*] Additional support by the Research Council of Norway.

[**] Supported by the Spanish CICYT project TIN-2004-07925 (GRAMMARS).

conquer" techniques, dynamic programming (here we refer to tree or branch decompositions – see Section 2 for details). The main idea behind this approach is very simple: Suppose that for a problem $\mathcal{P}$ we are able to prove that for every n-vertex graph G of branchwidth at most ℓ, the problem $\mathcal{P}$ can be solved in time $2^{O(\ell(G))} \cdot n^{O(1)}$. Since the branchwidth of an n-vertex graph of a fixed genus is $O(\sqrt{n})$, we have that $\mathcal{P}$ is solvable on G in time $2^{O(\sqrt{n})} \cdot n^{O(1)}$.

For some problems like MINIMUM VERTEX COVER or MINIMUM DOMINATING SET, such an approach yields directly algorithms of running time $2^{O(\sqrt{n})} \cdot n^{O(1)}$ on graphs of bounded genus. However, for some problems, like HAMILTONIAN CYCLE, Σ-EMBEDDED GRAPH TSP, MAX LEAF TREE, and STEINER TREE, branchwidth arguments do not provide us with time $2^{O(\sqrt{n})} \cdot n^{O(1)}$ algorithms. The reason is that all these problems are "non-local" and despite many attempts, no time $2^{o(\ell(G) \log \ell)} \cdot n^{O(1)}$ algorithm solving these problems on graphs of branchwidth at most ℓ is known.

Recently, it was observed by several authors that if a graph G is not only of branchwidth at most ℓ but is also planar, then for a number of "non-local" problems the $\log \ell$ overhead can be removed [3, 5], resulting in time $2^{O(\sqrt{n})} \cdot n^{O(1)}$ algorithms on planar graphs. Similar result can be obtained by making use of separators [1].

It is a common belief that almost every technique working on planar graphs can be extended on graphs embedded on a surface of bounded genus. However, this is not always a straightforward task. The main difficulty in generalizing planar graph techniques [1, 3, 5] to graphs of bounded genus is that all these techniques are based on partitioning a graph embedded on a plane by a closed curve into smaller pieces. Deineko et al. use cyclic separators of triangulations [1], Demaine and Hajiaghayi use layers of k-outerplanar graphs [3], and Dorn et al. sphere cut decompositions [5]. But the essence of all these techniques is that, roughly speaking, the situation occurring in the "inner" part of the graph bounded by the closed curve can be represented in a compact way by Catalan structures. None of these tools works for graphs of bounded genus— separators are not cyclic anymore, nor are there sphere cut decompositions and k-outerplanarity in non-planar graphs.

In this paper we provide a method to design fast subexponential algorithms for graphs of bounded genus for a wide class of combinatorial problems. Our algorithms are "fast" in the sense that they avoid the $\log n$ overhead and also because the constants hidden in the big-Oh of the exponents are reasonable. The technique we use is based on reduction of the bounded genus instances of the problem to *planar* instances of a more general graph problem on planar graphs where Catalan structure arguments are still possible. Such a reduction employs several results from topological graph theory concerning graph structure and noncontractible cycles of non-planar embeddings.

Our techniques, combined with the excluded grid theorem for graphs of bounded genus and bidimensionality arguments [2] provide also faster *parameterized* algorithms. For example we introduce the first time $2^{O(\sqrt{p})} \cdot n^{O(1)}$ algorithm for parameterized p-CYCLE which asks, given a positive integer p and a

n-vertex graph G, whether G has a cycle of length at least p. Similar results can be obtained for other parameterized versions of non-local problems.

This paper is organized as follows. Towards simplifying the presentation of our results we decided to demonstrate how our approach works for the HAMIL-TONIAN CYCLE problem. Later, at the end of Section 4, we will explain how it can be applied to other combinatorial problems. We start with some basic definitions in Section 2 and some results from topological graph theory. Section 3 is devoted to the solution of HAMILTONIAN CYCLE problem (which asks if a given graph G has a cycle containing all its vertices) on torus-embedded graphs. These graphs already inherit all "nasty" properties of non-planar graphs and all difficulties arising on surfaces of higher genus appear for torus-embedded graphs. However, the case of torus-embedded graphs is still sufficiently simple to exemplify the minimization technique used to obtained reasonable constants in the exponent. In Section 4, we explain how the results on torus-embedded graphs can be extended for any graphs embedded in a surface of fixed genus. Also in this section we discuss briefly applications of our results to parameterized algorithms on graphs of bounded genus.

2   Definitions and Preliminary Results

In this section we will give a series of definitions and results that will be useful for the presentation of the algorithms in Sections 3 and 4.

Surface embeddible graphs. We use the notation $V(G)$ and $E(G)$, for the set of the vertices and edges of G. A *surface* Σ is a compact 2-manifold without boundary (we always consider connected surfaces). We denote by $\mathbb{S}_0$ the sphere $(x, y, z \mid x^2 + y^2 + z^2 = 1)$ and by $\mathbb{S}_1$ the torus $(x, y, z \mid z^2 = 1/4 - (\sqrt{x^2 + y^2} - 1)^2)$. A *line* in Σ is subset homeomorphic to $[0, 1]$. An *O-arc* is a subset of Σ homeomorphic to a circle. Whenever we refer to a Σ-*embedded graph* G we consider a 2-cell embedding of G in Σ. To simplify notations we do not distinguish between a vertex of G and the point of Σ used in the drawing to represent the vertex or between an edge and the line representing it. We also consider G as the union of the points corresponding to its vertices and edges. That way, a subgraph H of G can be seen as a graph H where $H \subseteq G$. We call by *region* of G any connected component of $(\Sigma \setminus E(G)) \setminus V(G)$. (Every region is an open set.) A subset of Σ meeting the drawing only in vertices of G is called G-*normal*. If an O-arc is G-normal then we call it *noose*. The length of a noose N is the number of its vertices and we denote it by $|N|$. Representativity [12] is the measure how dense a graph is embedded on a surface. The *representativity* (or *face-width*) **rep**(G) of a graph G embedded in surface $\Sigma \neq \mathbb{S}_0$ is the smallest length of a noncontractible noose in Σ. In other words, **rep**(G) is the smallest number k such that Σ contains a noncontractible (non null-homotopic in Σ) closed curve that intersects G in k points. Given a Σ-embedded graph G, its *radial graph* (also known as vertex-face graph) is defined as the the graph R_G that has as vertex set the vertices and the faces of G and where an edge exists iff it connects a face and a vertex incident to it in G (R_G is also a σ-embedded graph). If the intersection of a noose with any region results into a connected subset, then we call such a noose *tight*. Notice that each tight noose N in a Σ-embedded graph G, corresponds to some cycle C of its radial graph R_G (notice that the length of such a cycle is $2 \cdot |N|$). Also any cycle C of R_G is a tight noose in G. As it was shown by Thomassen in [14] (see also Theorem 4.3.2 of [11]) a shortest non-contractible cycle in a graph embedded on a surface can be found in polynomial time. By Proposition 5.5.4 of [11]) a noncontractible noose of minimum size is always a tight noose, i.e. corresponds to a cycle of the radial graph. Thus we have the following proposition.

Proposition 1. *There exists a polynomial time algorithm that for a given Σ-embedded graph G, where $\Sigma \neq \mathbb{S}_0$, finds a noncontractible tight noose of minimum size.*

The Euler genus of a surface Σ is $\mathbf{eg}(\Sigma) = \min\{2\mathbf{g}(\Sigma), \tilde{\mathbf{g}}(\Sigma)\}$ where $\mathbf{g}$ is the orientable genus and $\tilde{\mathbf{g}}$ the nonorientable genus. We need to define the graph obtained by *cutting along* a noncontractible tight noose N. We suppose that for any $v \in N \cap V(G)$, there exists an open disk Δ containing v and such that for every edge e adjacent to v, $e \cap \Delta$ is connected. We also assume that $\Delta \setminus N$ has two connected components Δ_1 and Δ_2. Thus we can define partition of $N(v) = N_1(v) \cup N_2(v)$, where $N_1(v) = \{u \in N(v): \{u, v\} \cap \Delta_1 \neq \emptyset\}$ and $N_2(v) = \{u \in N(v): \{u, v\} \cap \Delta_2 \neq \emptyset\}$. Now for each $v \in N \cap V(G)$ we *duplicate* v: (a) remove v and its incident edges (b) introduce two new vertices v^1, v^2 and (c) connect v^i with the vertices in $N_i, i = 1, 2$. v^1 and v^2 are vertices of the new G-normal O-arcs N_X and N_Y that border Δ_1 and Δ_2, respectively. We call N_X and N_Y *cut-nooses*. Note that cut-nooses are not necessarily tight (In other words, a cut-noose can enter and leave a region of G several times.) The following lemma is very useful in proofs by induction on the genus. The first part of the lemma follows from Proposition 4.2.1 (corresponding to surface separating cycle) and the second part follows from Lemma 4.2.4 (corresponding to non-separating cycle) in [11].

Proposition 2. *Let G be a Σ-embedded graph where $\Sigma \neq \mathbb{S}_0$ and let G' be a graph obtained from G by cutting along a noncontractible tight noose N on G. One of the following holds*

- G' can be embedded in a surface with Euler genus strictly smaller than $\mathbf{eg}(\Sigma)$.
- G' is the disjoint union of graphs G_1 and G_2 that can be embedded in surfaces Σ_1 and Σ_2 such that $\mathbf{eg}(\Sigma) = \mathbf{eg}(\Sigma_1) + \mathbf{eg}(\Sigma_2)$ and $\mathbf{eg}(\Sigma_i) > 0$, $i = 1, 2$.

Branchwidth. A *branch decomposition* of a graph G is a pair $\langle T, \mu \rangle$, where T is a tree with vertices of degree one or three and μ is a bijection from the set of leaves of T to $E(G)$. For a subset of edges $X \subseteq E(G)$ let $\delta_G(X)$ be the set of all vertices incident to edges in X and $E(G) \setminus X$. For each edge e of T, let $T_1(e)$ and $T_2(e)$ be the sets of leaves in two components of $T \setminus e$. For any edge $e \in E(T)$ we define the *middle set* as $\mathbf{mid}(e) = \bigcup_{v \in T_1(e)} \delta_G(\mu(v))$. The width of $\langle T, \mu \rangle$ is the maximum size of a middle set over all edges of T, and the *branch-width* of G, $\mathbf{bw}(G)$, is the minimum width over all branch decompositions of G. For a $\mathbb{S}_0$-embedded graph G, we define a *sphere cut decomposition* or *sc-decomposition* $\langle T, \mu, \pi \rangle$ as a branch decomposition such that for every edge e of T and the two subgraphs G_1 and G_2 induced by the edges in $\mu(T_1(e))$ and $\mu(T_2(e))$, there exists a tight noose O_e bounding two open discs Δ_1 and Δ_2 such that $G_i \subseteq \Delta_i \cup O_e$, $1 \leq i \leq 2$. Thus O_e meets G only in $\mathbf{mid}(e)$ and its length is $|\mathbf{mid}(e)|$. Clockwise traversing of O_e in the drawing G defines the cyclic ordering π of $\mathbf{mid}(e)$. We always assume that in an sc-decomposition the vertices of every middle set $\mathbf{mid}(e) = V(G_1) \cap V(G_2)$ are enumerated according to π. The following result follows from the celebrated ratcatcher algorithm due to Seymour and Thomas [13] (the running time of the algorithm was recently improved in [8]; see also [5]).

Proposition 3. *Let G be a connected $\mathbb{S}_0$-embedded graph without vertices of degree one. There exists an sc-decomposition of G of width $\mathbf{bw}(G)$. Moreover, such a branch decomposition can be constructed in time $O(n^3)$.*

3   Hamiltonicity on Torus-Embedded Graphs

The idea behind solving the HAMILTONIAN CYCLE PROBLEM on $\mathbb{S}_1$-embedded graphs is to suitably modify the graph G in such a way that the new graph G' is $\mathbb{S}_0$-embedded (i.e. planar) and restate the problem to an equivalent problem on G' that can be solved by dynamic programming on a sc-decomposition of G'. As we will see in Section 4, this procedure is extendable to graphs embedded on surfaces of higher genus.

Let G be an $\mathbb{S}_1$-embedded graph (i.e. a graph embedded in the torus). By Proposition 1, it is possible to find in polynomial time a shortest noncontractible (tight) noose N of G. Let G' be the graph obtained by cutting along N on G. By Proposition 2, G' is $\mathbb{S}_0$-embeddible.

Definition 1. *A cut of a Hamiltonian cycle C in G along a tight noose N is the set of disjoint paths in G' resulting by cutting G along N.*

Each cut-noose N_X and N_Y borders an open disk Δ_X and Δ_Y, respectively, with $\Delta_X \cup \Delta_Y = \emptyset$. Let $x_i \in N_X$ and $y_i \in N_Y$ be duplicated vertices of the same vertex in N.

Definition 2. *A set of disjoint paths $\mathbf{P}$ in G' is relaxed Hamiltonian if:*

(P1) Every path has its endpoints in N_X and N_Y.
(P2) Vertex x_i is an endpoint of some path P if and only if y_i is an endpoint of a path $P' \neq P$.
(P3) For x_i and y_i: one is an inner vertex of a path if and only if the other is not in any path.
(P4) Every vertex of $G' \setminus (N_X \cup N_Y)$ is in some path.

A *cut* of a Hamiltonian cycle in G is a relaxed Hamiltonian set in G', but not every relaxed Hamiltonian set in G' forms a Hamiltonian cycle in G. However, given a relaxed Hamiltonian set $\mathbf{P}$ one can check in linear time (by identifying the corresponding vertices of N_X and N_Y) if $\mathbf{P}$ is a cut of Hamiltonian path in G. Two sets of disjoint paths $\mathbf{P} = (P_1, P_2, \ldots, P_k)$ and $\mathbf{P}' = (P_1', P_2', \ldots, P_k')$ are *equivalent* if for every $i \in \{1, 2, \ldots, k\}$, the paths P_i and P_i' have the same endpoints and an inner vertex in one set is also an inner vertex in the other set.

Lemma 1. *Let G' be a $\mathbb{S}_0$-embedded graph obtained from a $\mathbb{S}_1$-embedded graph G by cutting along a tight noose N. The number of different equivalence classes of relaxed Hamiltonian sets in G' is $O(\frac{k^2}{2} 2^{3k-2} + 2^{3k})$, where k is the length of N.*

Proof. In [5] it is argued that the number of non-crossing paths with its endpoints in one noose corresponds to a number of algebraic terms, namely the Catalan numbers. Here we deal with two cut-nooses and our intention is to transform them into one cut-noose. For this, assume two vertices $x_i \in N_X$ and $y_j \in N_Y$ being two fixed endpoints of a path $P_{i,j}$ in a relaxed Hamiltonian set $\mathbf{P}$. We look at all possible residual paths in $\mathbf{P} \setminus P_{i,j}$ and we observe that no path crosses $P_{i,j}$ in the $\mathbb{S}_0$-embedded graph G'. So we are able to 'cut' the sphere $\mathbb{S}_0$ along $P_{i,j}$ and, that way, create a "tunnel" between Δ_X and Δ_Y unifying them to a single disk Δ_{XY}. Take the counter-clockwise order of the vertices of N_X beginning with x_i and concatenate N_Y in clockwise order with y_j the last vertex. We denote the new cyclic ordering by π_{XY} (see Figure 1 for an example). In π_{XY}, let a, b, c, d

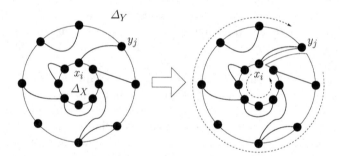

Fig. 1. Cut-nooses. In the left diagram, one equivalence class of relaxed Hamiltonian sets is illustrated. All paths have endpoints in N_X and N_Y. Fix one path with endpoints x_i and y_i. In the right diagram we create a tunnel along this path. The empty disks Δ_X and Δ_Y are united to a single empty disk. Thus, we can order the vertices bordering the disk to π_{XY}.

be four vertices where $x_i < a < b < c < d < y_j$. Notice that if there is a path $P_{a,c}$ between a and c, then there is no path between b and d since such a path either crosses $P_{a,c}$ or $P_{i,j}$. This means that we can encode the endpoints of each path with two symbols, one for the beginning and one for the ending of a path. The encoding corresponds to the brackets of an algebraic term. The number of algebraic terms is defined by the Catalan numbers. We say that $\mathbf{P}$ has a *Catalan structure*. With $k = |N_X| = |N_Y|$ and $x_i, y_j \in P_{i,j}$ fixed, there are $O(2^{2k-2})$ sets of paths having different endpoints and non-crossing $P_{i,j}$. An upper bound for the overall number of sets of paths satisfying (P1) is then $O(\frac{k^2}{2} 2^{2k-2} + 2^{2k})$ with the first summand counting all sets of paths for each fixed pair of endpoints x_i, y_j. The second summand counts the number of sets of paths when N_X and N_Y are not connected by any path. That is, each path has both endpoints in either only N_X or only N_Y. We now count the number of equivalent relaxed Hamiltonian sets **P** . Apparently, in a feasible solution, if a vertex $x_h \in N_X$ is an inner vertex of a path, then $y_h \in N_Y$ does not belong to any path and vice versa. With (P3), there are two more possibilities for the pair of vertices x_h, y_h to correlate with a path. With $|N_X| = |N_Y| = k$, the overall upper bound of equivalent sets of paths is $O(\frac{k^2}{2} 2^{3k-2} + 2^{3k})$.

We call a *candidate* **C** of an equivalence class of relaxed Hamiltonian sets to be a set of paths with vertices only in $N_X \cup N_Y$ satisfying conditions (P1)–(P3). Thus for each candidate we fix a path between N_X and N_Y and define the ordering π_{XY}. By making use of dynamic programming on sc-decompositions we check for each candidate **C** if there is a spanning subgraph of the planar graph G' isomorphic to a relaxed Hamiltonian set **P** such that **P** is equivalent to **C**.

Instead of looking at the HAMILTONIAN CYCLE problem on G we solve the RELAXED HAMILTONIAN SET problem on the $\mathbb{S}_0$-embedded graph G' obtained from G. Given a candidate **C**: a set of vertex tuples $\mathbf{T} = \{(s_1, t_1), (s_2, t_2), \dots, (s_k, t_k)\}$ with $s_i, t_i \in N_X \cup N_Y, i = 1, \dots, k$ and a vertex set $\mathbf{I} \subset N_X \cup N_Y$. Does there exist a relaxed Hamiltonian set **P** such that every (s_i, t_i) marks the endpoints of a path and the vertices of **I** are inner vertices of some paths? Our algorithm works as follows: first encode the vertices of $N_X \cup N_Y$ according to **C** by making use of the Catalan structure of **C** as it follows from the proof of Lemma 1. We may encode the vertices s_i as the 'beginning' and t_i as the 'ending' of a path of **C**. Using order π_{XY}, we ensure that the beginning is always connected to the next free ending. This allows us to design a dynamic programming algorithm using a small constant number of states. We call the encoding of the vertices of $N_X \cup N_Y$ *base encoding* to differ from the encoding of the sets of disjoint paths in the graph. We proceed with dynamic programming over middle sets of a rooted sc-decomposition $\langle T, \mu, \pi \rangle$ in order to check whether G' contains a relaxed Hamiltonian set **P** equivalent to candidate **C**. As T is a rooted tree, this defines an orientation of its edges towards its root. Let e be an edge of T and let O_e be the corresponding tight noose in $\mathbb{S}_0$. Recall that the tight noose O_e partitions $\mathbb{S}_0$ into two discs which, in turn, induces a partition of the edges of G into two sets. We define as G_e the graph induced by the edge set that corresponds to the "lower side" of e it its orientation towards the root. All paths of $\mathbf{P} \cap G_e$ start and end in O_e and $G_e \cap (N_X \cup N_Y)$. For each G_e, we encode the equivalence classes of sets of disjoint paths with endpoints in O_e. From the leaves to the root for a parent edge and its two children, we update the encodings of the parent middle set with those of the children (for an example of dynamic programming on sc-decompositions, see also [5]). We obtain the algorithm in Figure 2.

In the proof of the following lemma we show how to apply the dynamic programming step of **HamilTor**. The proof is technical, and can be found in the

---

Algorithm **HamilTor**

Input: $\mathbb{S}_1$-embedded graph G.

Output: Decision/Construction of the HAMILTONIAN CYCLE problem on G.

Preliminary Step: Cut G along a shortest noncontractible (tight) noose N and
      output the $\mathbb{S}_0$-embedded graph G' and the cut-nooses N_X, N_Y.

Main step: For all candidates **C** of relaxed Hamiltonian sets in G' {
    Identify the duplicated vertices in N_X, N_Y.
    If **C** is equivalent to a Hamiltonian cycle {
        Determine the endpoints (s_i, t_i) that build the first and last vertex in π_{XY}.
        Make a base encoding of the vertices of N_X and N_Y,
        marking the intersection of **C** and $N_X \cup N_Y$.
        Compute a rooted sc-decomposition $\langle T, \mu, \pi \rangle$ of G'.
        From the leaves to the root on each middle set O_e of T bordering G_e {
            Do dynamic programming — find all equivalence classes of sets of
            disjoint paths in G_e with endpoints in O_e and in $G_e \cap (N_X \cup N_Y)$
            with respect to the base encoding of N_X, N_Y.}
        If there exists a relaxed Hamiltonian set **P** in G' equivalent to **C**, then {
        Reconstruct **P** from the root to the leaves of T and
        output corresponding Hamiltonian cycle.} } }

Output "No Hamiltonian Cycle exists".

Fig. 2. Algorithm **HamilTor**

long version of this paper [4]. But we sketch the main idea here: For a dynamic programming step we need the information on how a tight noose O_e and $N_X \cup N_Y$ intersect and which parts of $N_X \cup N_Y$ are a subset of the subgraph G_e. Define the vertex set $\mathcal{X} = (G_e \setminus O_e) \cap (N_X \cup N_Y)$. G_e is bordered by O_e and $\mathcal{X}$. G_e is partitioned into several edge-disjoint components that we call *partial components*. Each partial component is bordered by a noose that is the union of subsets of O_e and $\mathcal{X}$. Let us remark that this noose is not necessarily tight. The partial components intersect pairwise only in vertices of $\mathcal{X}$ that we shall define as *connectors*. In each partial component we encode a collection of paths with endpoints in the bordering noose using Catalan structures. The union of these collections over all partial components must form a collection of paths in G_e with endpoints in O_e and in $\mathcal{X}$. We ensure that the encoding of the connectors of each two components fit. During the dynamic programming we need to keep track of the base encoding of $\mathcal{X}$. We do so by only encoding the vertices of O_e without explicitly memorizing with which vertices of $\mathcal{X}$ they form a path. With several technical tricks we can encode O_e such that two paths with an endpoint in O_e and the other in $\mathcal{X}$ can be connected to a path of **P** only if both endpoints in $\mathcal{X}$ are the endpoints of a common path in **C**.

Lemma 2. *For a given a sc-decomposition $\langle T, \mu, \pi \rangle$ of G' of width ℓ and a candidate* **C** $= (\mathbf{T}, \mathbf{I})$ *the running time of the main step of* **HamilTor** *on* **C** *is* $O(2^{5.433\ell} \cdot |V(G')|^{O(1)})$.

To finish the estimation of the running time we need some combinatorial results. The proof of the following two lemmata can be found in [4].

Lemma 3. *Let G be a $\mathbb{S}_1$-embedded graph on n vertices and G' the planar graph after cutting along a noncontractible tight noose. Then $\mathbf{bw}(G') \leq \sqrt{4.5} \cdot \sqrt{n} + 2$.*

Lemma 4. *Let G be a $\mathbb{S}_1$-embedded graph. Then $\mathbf{rep}(G) \leq \sqrt{4.5} \cdot \sqrt{n} + 2$.*

Putting all together we obtain the following theorem.

Theorem 1. *Let G be a graph on n vertices embedded on a torus $\mathbb{S}_1$. The* HAMILTONIAN CYCLE *problem on G can be solved in time $O(2^{17.893\sqrt{n}} \cdot n^{O(1)})$.*

Proof. We run the algorithm **HamilTor** on G. The algorithm terminates positively when the dynamic programming is successful for some candidate of an equivalence class of relaxed Hamiltonian sets and this candidate is a cut of a Hamiltonian cycle. By Propositions 1, Step 0 can be performed in polynomial time. Let k be the minimum length of a noncontractible noose N, and let G' be the graph obtained from G by cutting along N. By Lemma 1, the number of all candidates of relaxed Hamiltonian sets in G' is $O(2^{3k}) \cdot n^{O(1)}$. So the main step of the algorithm is called $O(2^{3k}) \cdot n^{O(1)}$ times. By Proposition 3, an optimal branch decomposition of G' of width ℓ can be constructed in polynomial time. By Lemma 2, dynamic programming takes time $O(2^{5.433\ell}) \cdot n^{O(1)}$. Thus the total running time of **HamilTor** is $O(2^{5.433\ell} \cdot 2^{3k}) \cdot n^{O(1)}$. By Lemma 4, $k \leq \sqrt{4.5} \cdot \sqrt{n} + 2$ and by Lemma 3, $\ell \leq \sqrt{4.5} \cdot \sqrt{n} + 2$, and the theorem follows.

4   Hamiltonicity on Graphs of Bounded Genus

Now we extend our algorithm to graphs of higher genus. For this, we use the following kind of planarization: We apply Proposition 2 and cut iteratively along shortest noncontractible nooses until we obtain a planar graph G'. If at some step G' is the disjoint union of two graphs G_1 and G_2, we apply Proposition 2 on G_1 and G_2 separately. The proof of the following lemma is in [4].

Lemma 5. *There exists a polynomial time algorithm that given a Σ-embedded graph G where $\Sigma \neq \mathbb{S}_0$, returns a minimum size noncontractible noose. Moreover, the length of such a noose, $\mathbf{rep}(G)$, is at most $\mathbf{bw}(G) \leq (\sqrt{4.5} + 2 \cdot \sqrt{2 \cdot \mathbf{eg}(\Sigma)})\sqrt{n}$.*

We examine how a shortest noncontractible noose affects the cut-nooses:

Definition 3. *Let $\mathcal{K}$ be a family of cycles in G. We say that $\mathcal{K}$ satisfies the 3-path-condition if it has the following property. If x, y are vertices of G and P_1, P_2, P_3 are internally disjoint paths joining x and y, and if two of the three cycles $C_{i,j} = P_i \cup P_j$, $(1 \leq i < j \leq 3)$ are not in $\mathcal{K}$, then also the third cycle is not in $\mathcal{K}$.*

Proposition 4. *(Mohar and Thomassen [11]) The family of Σ-noncontractible cycles of a Σ-embedded graph G satisfies the 3-path-condition.*

Proposition 4 is useful to restrict the number of ways not only on how a shortest noncontractible tight noose may intersect a face but as well on how it may intersect the vertices incident to a face. The proof of the following lemma can be found in [4].

Lemma 6. *Let G be Σ-embedded and F a face of G bordered by $V_1 \subseteq V(G)$. Let $\overline{F} := V_1 \cup F$. Let N_s be a shortest noncontractible (tight) noose of G. Then one of the following holds*

1) $N_s \cap \overline{F} = \emptyset$.
2.1) $N_s \cap F = \emptyset$ and $|N_s \cap V_1| = 1$.
2.2) $N_s \cap F = \emptyset$, $N_s \cap V_1 = \{x, y\}$, and x and y are both incident to one more face different than F which is intersected by N_s.
3) $N_s \cap F \neq \emptyset$ and $|N_s \cap V_1| = 2$.

We use Lemma 6 to extend the process of cutting along noncontractible tight nooses such that we obtain a planar graph with a small number of disjoint cut-nooses of small lengths. Let $g \leq \mathbf{eg}(\Sigma)$ be the number of iterations needed to cut along shortest noncontractible nooses such that they turn a Σ-embedded graph G into a planar graph G'. However, these cut-nooses may not be disjoint. In our dynamic programming approach we need pairwise disjoint cut-nooses. Thus, whenever we cut along a noose, we manipulate the cut-nooses found so far. After g iterations, we obtain the *set of cut-nooses* $\mathfrak{N}$ that is a set of disjoint cut-nooses bounding empty open disks in the embedding of G'. Let $L(\mathfrak{N})$ be the *length* of $\mathfrak{N}$ as the sum over the lengths of all cut-nooses in $\mathfrak{N}$. The proof of the following proposition can be found in [4].

Proposition 5. *It is possible to find, in polynomial time, a set of cut-nooses $\mathfrak{N}$ that contains at most $2g$ disjoint cut-nooses. $L(\mathfrak{N})$ is at most $2g\,\mathbf{rep}(G)$.*

We extend the definition of relaxed Hamiltonian sets from graphs embedded on a torus to graphs embedded on higher genus, i.e. from two cut-nooses N_X and N_Y to the set of cut-nooses $\mathfrak{N}$. For each vertex v in the vertex set $V(G)$ of graph G we define the vertex set D_v that contains all duplicated vertices $v_1, \ldots, v_f$ of v in $\mathfrak{N}$ along with v. Set $\mathfrak{D} = \bigcup_{v \in V(G)} D_v$.

Definition 4. *A set of disjoint paths $\mathbf{P}$ in G' is* relaxed Hamiltonian *if:*

(P1) Every path has its endpoints in $\mathfrak{N}$.
(P2) If a vertex $v_i \in D_v \in \mathfrak{D}$ is an endpoint of path P, then there is one $v_j \in D_v$ that is also an endpoint of a path $P' \neq P$. All $v_h \in D_v \setminus \{v_i, v_j\}$ do not belong to any path.
(P3) $v_i \in D_v$ is an inner path vertex if and only if all $v_h \in D_v \setminus \{v_i\}$ are not in any path.
(P4) Every vertex of the residual part of G' is in some path.

Similar to torus-embedded graphs, we order the vertices of $\mathfrak{N}$ for later encoding in a counterclockwise order $\pi_{\mathbf{L}}$ depending on the fixed paths between the cut-nooses of $\mathfrak{N}$. The proof of the following can be found in [4].

Lemma 7. *Let G' be the planar graph after cutting along $g \leq \mathbf{eg}(\Sigma)$ tight nooses of G along with its set of disjoint cut-nooses $\mathfrak{N}$. The number of different equivalence classes of relaxed Hamiltonian sets in G' is $2^{O(g \cdot (\log g + \mathbf{rep}(G)))}$.*

Given the order $\pi_{\mathbf{L}}$ of the vertices $\mathfrak{N}$ in the encoding of a candidate $\mathbf{C}$ of a relaxed Hamiltonian set. As in the previous section, we preprocess the graph G' and encode the vertices of $\mathfrak{N}$ with the base values. We extend the dynamic porgramming approach by analysing how the tight noose O_e can intersect several cut-nooses. The proofs of the next two statements can be found in [4].

Lemma 8. *Let G' be the planar graph after cutting along $g \leq \mathbf{eg}(\Sigma)$ shortest noncontractible nooses of G. For a given sc-decomposition $\langle T, \mu, \pi \rangle$ of G' of width ℓ and a candidate $\mathbf{C}$ the RELAXED HAMILTONIAN SET problem on G' can be solved in time $2^{O(g^2 \log \ell)} \cdot 2^{O(\ell)} \cdot n^{O(1)}$.*

Lemma 9. *Let G be a Σ-embedded graph with n vertices and G' the planar graph obtained after cutting along $g \leq \mathbf{eg}(\Sigma)$ tight nooses. Then, $\mathbf{bw}(G') \leq \sqrt{4.5} \cdot \sqrt{n} + 2g$.*

Lemmata 5, 7, 8 and 9 imply the following:

Theorem 2. *Given a Σ-embedded graph G on n vertices and $g \leq \mathbf{eg}(\Sigma)$. The HAMILTONIAN CYCLE problem on G can be solved in time $n^{O(g^2)} \cdot 2^{O(g\sqrt{g \cdot n})}$.*

Our dynamic programming technique can be used to design faster parameterized algorithms as well. For example, the parameterized p-CYCLE ON Σ-EMBEDDED GRAPHS problem asks for a given Σ-embedded graph G, to check for the existence of a cycle of length at least a parameter p. First, our technique can be used to find the longest cycle of G with $g \leq \mathbf{eg}(\Sigma)$ in time $n^{O(g^2)} \cdot 2^{O(g\sqrt{g \cdot n})}$. (On torus -embedded graphs this can be done in time $O(2^{17.957\sqrt{n}}n^3)$.) By combining this running time with bidimensionality arguments from [2] we arrive at a time $2^{O(g^2 \log p)} \cdot 2^{O(g\sqrt{g \cdot p})} \cdot n^{O(1)}$ algorithm solving the parameterized p-CYCLE ON Σ-EMBEDDED GRAPHS.

5   Conclusive Remarks

In this paper we have introduced a new approach for solving non-local problems on graphs of bounded genus. With some sophisticated modifications, this generic approach can be used to design time $2^{O(\sqrt{n})}$ algorithms for many other problems including Σ-EMBEDDED GRAPH TSP (TSP with the shortest path metric of a Σ-embedded graph as the distance metric for TSP), MAX LEAF TREE, and STEINER TREE, among others. Clearly, the ultimate step in this line of research is to prove the existence of time $2^{O(\sqrt{n})}$ algorithms for non-local problems on any graph class that is closed under taking of minors. Recently, we were able to complete a proof of such a general result, using results from the Graph Minor series. One of the steps of our proof is strongly based on the results and the ideas of this paper.

References

1. V. G. DEĬNEKO, B. KLINZ, AND G. J. WOEGINGER, *Exact algorithms for the Hamiltonian cycle problem in planar graphs*, Operations Research Letters, (2006), p. to appear.

2. E. D. DEMAINE, F. V. FOMIN, M. HAJIAGHAYI, AND D. M. THILIKOS, *Subexponential parameterized algorithms on graphs of bounded genus and H-minor-free graphs*, Journal of the ACM, 52 (2005), pp. 866–893.

3. E. D. DEMAINE AND M. HAJIAGHAYI, *Bidimensionality: new connections between FPT algorithms and PTASs*, in Proceedings of the 16th Annual ACM-SIAM Symposium on Discrete Algorithms (SODA 2005), New York, 2005, ACM-SIAM, pp. 590–601.

4. F. DORN, F. V. FOMIN, AND D. M. THILIKOS, *Fast subexponential algorithm for non-local problems on graphs of bounded genus*, manuscript, (2006). http://www.ii.uib.no/publikasjoner/texrap/pdf/2006-320.pdf.

5. F. DORN, E. PENNINKX, H. BODLAENDER, AND F. V. FOMIN, *Efficient exact algorithms on planar graphs: Exploiting sphere cut branch decompositions*, in Proceedings of the 13th Annual European Symposium on Algorithms (ESA 2005), vol. 3669 of LNCS, Springer, Berlin, 2005, pp. 95–106.

6. F. V. FOMIN AND D. M. THILIKOS, *New upper bounds on the decomposability of planar graphs*, Journal of Graph Theory, 51 (2006), pp. 53–81.

7. J. R. GILBERT, J. P. HUTCHINSON, AND R. E. TARJAN, *A separator theorem for graphs of bounded genus*, Journal of Algorithms, 5 (1984), pp. 391–407.

8. Q.-P. GU AND H. TAMAKI, *Optimal branch-decomposition of planar graphs in $O(n^3)$ time*, in Proceedings of the 32nd International Colloquium on Automata, Languages and Programming (ICALP 2005), vol. 3580 of LNCS, Springer, Berlin, 2005, pp. 373–384.

9. R. J. LIPTON AND R. E. TARJAN, *A separator theorem for planar graphs*, SIAM J. Appl. Math., 36 (1979), pp. 177–189.

10. ——, *Applications of a planar separator theorem*, SIAM J. Comput., 9 (1980), pp. 615–627.

11. B. MOHAR AND C. THOMASSEN, *Graphs on surfaces*, Johns Hopkins Studies in the Mathematical Sciences, Johns Hopkins University Press, Baltimore, MD, 2001.

12. N. ROBERTSON AND P. D. SEYMOUR, *Graph minors. VII. Disjoint paths on a surface*, J. Combin. Theory Ser. B, 45 (1988), pp. 212–254.

13. P. D. SEYMOUR AND R. THOMAS, *Call routing and the ratcatcher*, Combinatorica, 14 (1994), pp. 217–241.

14. C. THOMASSEN, *Embeddings of graphs with no short noncontractible cycles*, J. Combin. Theory Ser. B, 48 (1990), pp. 155–177.

15. G. WOEGINGER, *Exact algorithms for NP-hard problems: A survey*, in Combinatorial Optimization - Eureka, you shrink!, vol. 2570 of LNCS, Springer-Verlag, Berlin, 2003, pp. 185–207.

On the Approximation Hardness of Some Generalizations of TSP[*]

(Extended Abstract)

Hans-Joachim Böckenhauer[1], Juraj Hromkovič[1],
Joachim Kneis[2,**], and Joachim Kupke[1]

[1] Department of Computer Science, ETH Zurich, Switzerland
{hjb, juraj.hromkovic, joachim.kupke}@inf.ethz.ch
[2] Department of Computer Science, RWTH Aachen University, Germany
joachim.kneis@cs.rwth-aachen.de

Abstract. The aim of this paper is to investigate the approximability of some generalized versions of TSP which typically arise in practical applications. The most important generalization is TSP with time windows, where some vertices have to be visited after some specified opening time, but before some deadline. Our main results are as follows (assuming $P \neq NP$).

1. In contrast to the constant approximability of metric TSP, there is no polynomial-time $o(|V|)$-approximation algorithm for metric TSP with time windows.
2. Metric TSP with as few as two time windows is not approximable within ratio $2 - \varepsilon$.
3. There is no polynomial-time $o(|V|)$-approximation algorithm for TSP with a single time window and arbitrarily small violations of the triangle inequality.
4. Metric TSP with a prescribed linear order on some vertices can be solved in polynomial time with a constant approximation guarantee, even if the triangle inequality is violated by a constant factor.

Keywords: TSP with time windows, approximation, inapproximability.

1 Introduction

The traveling salesperson problem (TSP) is one of the most prominent optimization problems with numerous practical applications. Worst-case analyses show that it is indeed one of the hardest problems with respect to approximability because, provided that $P \neq NP$, there is no polynomial-time approximation algorithm for TSP with an approximation ratio bounded by a polynomial in the problem instance size.

---

[*] This work was partially supported by SNF grant 200021-109252/1. A full version of Section 3 will appear in *Theory of Computing Systems* [7].
[**] This work was done while this author was visiting ETH Zurich.

L. Arge and R. Freivalds (Eds.): SWAT 2006, LNCS 4059, pp. 184–195, 2006.

Our goal in this paper is to investigate the approximability of some generalizations of TSP. The most important of these generalizations is TSP with time windows, where for each vertex of the input graph, an opening and a closing point of time for a window is given, and the respective vertex must be visited while the window is open. This generalized problem naturally and frequently appears in a number of applications, the most prominent of which may be vehicle routing.

This importance has of course been recognized in operations research, and a multitude of both exact (yet exponential-time) algorithms and heuristics (yet without performance guarantee) have been proposed; for a survey, see [11]. It is due to the hardness of TSP with time windows that so far, it has not been possible to establish (reasonable) performance guarantees. Unaltered TSP being already one of the hardest known problems according to its worst-case approximability, there is no hope for (decent) approximability results regarding TSP with extensions. Beyond doubt, any extended TSP is at least as hard as unaltered TSP.

But it is a somewhat surprising fact that TSP is sometimes not quite as hard as it looks like, not only from a practical point of view, but also in worst-case analyses. Indeed, metric TSP (ΔTSP) can be solved efficiently with an approximation guarantee of 1.5 [10] and, using the concept of approximation stability [15, 6], quite a few papers have shown that even for a relaxation of the metricity constraint by relaxing the triangle inequality to the so-called β-triangle inequality for some $\beta > 1$, i.e.,

$$c(\{v_i, v_j\}) \leq \beta \cdot \left(c(\{v_i, v_z\}) + c(\{v_z, v_j\}) \right)$$

for any three vertices v_i, v_j, v_z, a constant approximation ratio can be achieved in polynomial time [2, 1, 4, 6]. More precisely, TSP on input instances satisfying a relaxed β-triangle inequality, Δ_βTSP for short, can be approximated in polynomial time within $\min\{\beta^2 + \beta, \frac{3}{2}\beta^2, 4\beta\}$.

Now, not only can we study TSP in the metric and the aforementioned "near-metric" case, but also the TSP generalizations which we will propose. We will give evidence that shifting step-by-step from ordinary TSP to TSP with different kinds of time windows results in an increase of hardness of the according metric and near-metric problems. The main generalizations we will consider are

1. k-OTSP, where up to k vertices are special in that they have to appear in a prescribed linear order in any feasible solution;
2. TSP with deadlines, where some vertices hold prescribed deadlines, i.e., points of time before which these vertices have to be visited (in other words, this is TSP with time windows, but all the windows open instantly when a tour begins and close independently).

A variation on TSP with deadlines was investigated in [3]. Here, the goal was to find a tour which obeys a maximum number of deadlines. In contrast, we will only look for such solutions that obey every given deadline.

We will also examine several generalizations of OTSP where the precedence constraints on the vertices are given by a set of paths or cycles rather than by one linear ordering. One of the most general cases is the one where precedence constraints are given by an arbitrary partial order on the vertices. Strong inapproximability results were given for this case in [9].

The main results of our paper are as follows:

1. There is no polynomial-time $o(|V|)$-approximation algorithm for metric TSP with deadlines.
2. Metric TSP with just two deadlines is not approximable within $2 - \varepsilon$.
3. There is no polynomial-time $o(|V|)$-approximation algorithm even for TSP with just one deadline in the case where $\beta > 1$, i.e., for arbitrarily small violations of the triangle inequality.
4. Metric TSP with a prescribed linear order on some vertices can be solved in polynomial time with a constant approximation guarantee, even for $\beta > 1$, i.e., if the triangle inequality is violated by a constant factor.

Please note that the inapproximability results directly carry over to the respective variations of TSP with time windows.

The paper is organized as follows: In Section 2, we will present our results for OTSP; Section 3 deals with TSP with deadlines and contains a parameterized approximation algorithm and the inapproximability results mentioned above, and Section 4 is devoted to TSP with generalized precedence constraints.

2 TSP with Prescribed Order on Some Vertices

In this section, we will present approximation algorithms for a generalization of Δ_βTSP, where a linear order on some of the vertices is prescribed and has to be obeyed by any feasible solution.

We start with a lemma describing the change in the cost of a path, in a graph satisfying the relaxed triangle inequality, when a subpath is replaced by the direct edge between its endpoints.

Lemma 1. *Let $G = (V, E)$ be a complete graph with edge weights $c\colon E \to \mathbb{Q}^+$ satisfying the Δ_β-inequality. Let $P := (v_1, \ldots, v_{k+1})$ be a (simple) path in G. Then $c(\{v_1, v_{k+1}\}) \leq \beta^{\log_2 k} \cdot \mathrm{cost}(P)$, where $\mathrm{cost}(P)$ denotes the sum of edge weights along the path.* □

A formal definition of the TSP variation to be investigated here follows.

Definition 1. *The input for the k-ordered TSP, or k-OTSP for short, consists of a complete graph $G = (V, E)$ with edge weights $c\colon E \to \mathbb{Q}^+$ and an ordered sequence of (distinct) vertices $v_1, \ldots, v_k \in V$, which we will call special vertices in what follows. The goal is now to find a Hamiltonian tour of minimum cost in G that contains the special vertices in their given order, i.e., if one edge were removed from this cycle, a path would result which contains the special vertices in their given order.*

If the given input graph obeys the triangle inequality, we denote the resulting problem by k-ΔOTSP; if it obeys the relaxed β-triangle inequality, we denote the problem by k-Δ_βOTSP.

The more general version of the problem, where also the number of special vertices is part of the input, will be denoted by ΔOTSP or Δ_βOTSP, resp.

Note that for $k \leq 3$, the order on the special vertices does not impose any constraint on the choice of the Hamiltonian tour, and such instances can be viewed as normal TSP instances. This implies that ΔOTSP in general is at least as hard as normal ΔTSP, and thus not approximable within $\frac{220}{219}$ unless $P = NP$ [16]. But in what follows, we will show that k-Δ_βOTSP admits a polynomial-time approximation algorithm with constant approximation ratio for any k and any $\beta \geq 1$.

Let us first show that k-ΔOTSP can easily be approximated within $\frac{5}{2}$.

Theorem 1. *For $k > 3$, Algorithm 1 is a polynomial-time $\frac{5}{2}$-approximation algorithm for k-ΔOTSP.*

Proof. Obviously, Algorithm 1 runs in polynomial time and outputs a feasible solution for k-ΔOTSP. Since any optimal solution H_{opt} respects the order given by the special vertices, the cost of C can be estimated as $cost(C) \leq cost(H_{opt})$, due to Lemma 1. Besides, $cost(D) \leq \frac{3}{2} \cdot cost(H_{opt})$ since Christofides' algorithm achieves an approximation ratio of $\frac{3}{2}$ on any metric input graph, and the cost of an optimal Hamiltonian tour in the graph induced by $\tilde{V}$ is, due to the triangle inequality, at most as much as $cost(H_{opt})$. □

Remark 1. Algorithm 1 relies on Christofides' algorithm to solve ΔTSP as it has the best known approximation guarantee for ΔTSP. Note that Theorem 1 can be generalized to arbitrary algorithms for ΔTSP. In fact, for any α-approximation algorithm for ΔTSP, we have a $(1 + \alpha)$-approximation algorithm for ΔOTSP.

It is possible to improve on this result for k-ΔOTSP if $k \leq 5$:

Theorem 2 ([5]). *There exist a polynomial-time 2-approximation algorithm for the 4-ΔOTSP and a polynomial-time 2.3-approximation algorithm for the 5-ΔOTSP.* □

Algorithm 1

Input: A complete graph $G = (V, E)$ with edge weights $c \colon E \to \mathbb{Q}^+$ and an ordered sequence of special vertices $v_1, \ldots, v_k \in V$.

1. Let $\tilde{V} := V \setminus \{v_2, \ldots, v_k\}$, and, using Christofides' algorithm, construct a Hamiltonian tour D on the subgraph of G induced by $\tilde{V}$. Let w be one of the neighbors of v_1 in D.
2. Combine the two cycles $C = (v_1, v_2, \ldots, v_k, v_1)$ and D to one Hamiltonian cycle H of G by replacing the edges $\{v_k, v_1\}$ in C and $\{v_1, w\}$ in D by the edge $\{v_k, w\}$.

Output: The Hamiltonian tour H.

We will now analyze the approximability of k-Δ_βOTSP. Since Δ_βTSP is a special case of Δ_βOTSP, the lower bounds from [8] carry over directly to Δ_βOTSP. However, applying the approach of simply concatenating the cycle C on the special vertices with an approximate TSP solution on the remaining vertices as used in Algorithm 1 to k-Δ_βOTSP does not lead to a constant approximation guarantee. These are the reasons why: First of all, direct edges from C might be too expensive as shown in Lemma 1. In order to overcome this difficulty, it would be possible to connect pairs of (adjacent) special vertices using the shortest path between them, but notice that these paths need not be vertex-disjoint. But even if we manage to "thin out" the paths (e.g., we might choose to always skip over a constant number of vertices in order not to have to use vertices more than once), the fact remains that shortest paths between special vertices will include some of the non-special vertices. Using paths between special vertices would thus make any approximate (non-ordered) TSP solution on the remaining non-special vertices more expensive than necessary.

Therefore, let us describe a slightly different approach. After computing an approximate TSP solution without respect to the order of special vertices, we will use $k + 1$ copies of this cycle, both to reach every vertex in the input graph and to arrange special vertices properly. Formally, we have Algorithm 2. Observe that it computes (disjoint) paths $H_0, \ldots, H_k$, all of which share the property that they result from a subpath of P by removing some vertices, but there is a bound on the number of contiguous vertices removed.

More precisely, H_{i-1} is a path from v_i to v_{i+1} such that two neighbors in it have a distance of at most $2k - 1$ edges in P because in the most adverse

---

Algorithm 2

Input: A complete graph $G = (V, E)$ with edge weights $c \colon E \to \mathbb{Q}^+$ that satisfy the Δ_β-inequality, and an ordered sequence of special vertices $v_1, \ldots, v_k \in V$.

1. Using some constant approximation guarantee algorithm, construct an approximate TSP solution C on (G, c), disregarding the order on $v_1, \ldots, v_k$.
2. Let P be one of the two paths which may be obtained by removing one of the edges incident with v_1 in C, and let $S = (w_1, \ldots, w_{n-k})$ be the sequence of non-special vertices in P, beginning with the non-special vertex closest to v_1 in P.
   Now, let $f \colon V \to \{-1, 0, \ldots, k\}$ be the function defined by $f(v_i) := -1$ for all $1 \le i \le k$ and $f(w_{i+1}) := i \bmod (k + 1)$ for all $0 \le i < n - k$, i.e., f is a cyclic $(k + 1)$-coloring of the non-special vertices.
3. For all $i \in \{1, \ldots, k - 1\}$, let H_{i-1} be the subpath in P from v_i to v_{i+1}, restricted to vertex v_i itself plus all vertices w (in this subpath) that have $f(w) = i - 1$.
4. Let H_{k-1} be the suffix of P that starts from v_k, restricted to v_k itself plus all vertices w (in this suffix) that have $f(w) = k - 1$, and let H_k be the reverse sequence of P, stripped of all vertices that appear in any of the H_i, $0 \le i < k$.
5. Let H be the concatenation of paths $H_0, \ldots, H_k$.

Output: The Hamiltonian tour H.

situation, H_{i-1} can only skip $k-2$ special vertices (i.e., all special vertices but v_i and v_{i+1}) plus k non-special vertices v with $f(v) \neq i-1$.

This observation, together with Lemma 1, yields the following estimation.

Lemma 2. *Let H, H_0, $\ldots$, H_k, and C denote the symbols from Algorithm 2. Then, $\mathrm{cost}(H_i) \leq \mathrm{cost}(C) \cdot \beta^{1+\log_2(k-1)}$ for all $i \in \{0,\ldots,k\}$ and $\mathrm{cost}(H) \leq (k+1) \cdot \mathrm{cost}(C) \cdot \beta^{1+\log_2(k-1)}$.* $\qquad\square$

Consequently, if in Step 1 of Algorithm 2, we use an algorithm for Δ_βTSP that guarantees an approximation ratio of at most α, Algorithm 2 is an $(\alpha(k+1)\beta^{1+\log_2(k-1)})$-approximation algorithm, i.e., a constant approximation algorithm whenever we fix both k and β. Substituting the approximation ratios from [2, 1, 4, 6] for α gives us our last result in this section.

Theorem 3. *The problem k-Δ_βOTSP admits a $\big((k+1) \cdot \min\{4\beta^{2+\log_2(k-1)}, \frac{3}{2}\beta^{3+\log_2(k-1)}, (\beta+1)\beta^{2+\log_2(k-1)}\}\big)$-approximation algorithm.* $\qquad\square$

3   Deadline TSP

In this section, we will analyze the approximability of TSP with deadlines. We start with the formal definition of this problem.

Definition 2. *Let $G = (V, E)$ be a complete graph with edge weights $c\colon E \to \mathbb{Q}^+$. We call (s, D, d) a deadline set for G if $s \in V, D \subseteq V\setminus\{s\}$ and $d\colon D \to \mathbb{Q}^+$. A vertex $v \in D$ is called deadline. A path $(v_0, v_1, \ldots, v_n)$ satisfies the deadlines iff $s = v_0$ and for all $v_i \in D$, we have $\sum_{j=1}^{i} c(\{v_{j-1}, v_j\}) \leq d(v_i)$. A cycle $(v_0, v_1, \ldots, v_n, v_0)$ satisfies the deadlines iff it contains a path $(v_0, v_1, \ldots, v_n)$ satisfying the deadlines.*

Definition 3. *The problem Δ_βDLTSP is defined as follows: For a given complete graph $G = (V, E)$ with edge weights $c\colon E \to \mathbb{Q}^+$ satisfying the Δ_β-inequality and deadlines (s, D, d) for G, find a minimum-weight Hamiltonian cycle satisfying all deadlines.*

If the number $|D|$ of deadline vertices is a constant k, we denote the resulting subproblem by k-Δ_βDLTSP. If $\beta = 1$, we omit β from the notation and obtain the problems ΔDLTSP and k-ΔDLTSP, respectively.

3.1   Algorithm

If only we knew the order in which the deadlines are visited in an optimal solution, we could try to start with this order and somehow insert the remaining vertices. As k is a constant, we can exhaustively try every permutation of deadlines to obtain this information. Unfortunately, inserting the remaining vertices into this sequence in an optimal way is still a hard problem. Therefore, Algorithm 3 just inserts them after the last deadline. This results in a 2.5-approximation.

Theorem 4. *Algorithm 3 solves k-ΔDLTSP with approximation ratio 2.5 in time $\mathcal{O}(k! \cdot p(|G|))$ for some polynomial p where G is the input graph.* $\qquad\square$

Algorithm 3

Input: A complete weighted graph $G = (V, E, c)$ and deadline set (s, D, d) for G.

> **for** every linear order π of all vertices in $D \cup \{s\}$ **do**
>   **if** π satisfies the deadlines **then**
>   Solve ΔOTSP with input G and π using Algorithm 1.

Output: The cheapest of all computed Hamiltonian cycles satisfying all deadlines, if one exists; an error message, otherwise.

Note that this algorithm only works if the Δ-inequality holds. If just the Δ_β-inequality holds, visiting the deadlines first does not guarantee that all of the deadlines are reached in time, since direct edges between the deadline vertices may be much more expensive than longer paths (cf. Lemma 1).

3.2   Lower Bounds for k-Δ_βDLTSP

At first sight, Algorithm 3 seems to have three major problems. First, it has running time exponential in the number of deadlines. Second, it only works if the Δ-inequality holds, i. e., for $\beta > 1$, it does not necessarily find a feasible solution, let alone a good one. And finally, its approximation ratio is only 2.5. We discuss whether polynomial time is possible in Section 3.3, the other problems are both handled here. First, we show that for all $\beta > 1$, no algorithm exists which would solve 1-Δ_βDLTSP with good approximation ratio.

Theorem 5. *Let $\beta > 1$. There is no polynomial-time algorithm for 1-Δ_βDLTSP with approximation ratio $\frac{1}{4}\beta^{\log_2(h|V|)}$ for any $0 < h < 1$.*

Proof. By means of a reduction, we will show that such an approximation algorithm could be used to solve the HAMILTONIAN PATH problem, i. e., the problem of deciding whether a given undirected graph contains a Hamiltonian path or not, which is NP-complete [13].

Let $\beta > 1$ and $0 < h < 1$, and let $G' = (V', E')$ be an input instance for the HAMILTONIAN PATH problem with $|V'| = n$. We construct a complete graph $G = (V, E, c)$ for the Δ_βDLTSP as follows:

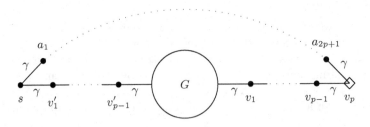

Fig. 1. Construction for 1-Δ_βDLTSP (Theorem 5)

Let $V := V' \cup \{v_1, \ldots, v_p\} \cup \{s, v'_1, \ldots, v'_{p-1}\} \cup \{a_1, \ldots, a_{2p+1}\}$ for a suitable $p \geq 1$ such that $\beta^{\log_2(p)} > \beta^{\log_2(h|V|)}$, and assign to every $e \in E$ the cost

$$c(e) := \begin{cases} 1 & \text{if } e \in E' \\ 2 & \text{if } e \in (E')^{\complement} \\ \gamma & \text{if } e \in \{\{v'_i, v'_{i+1}\} \mid i = 1, \ldots, p-2\} \cup \{\{v'_{p-1}, v\} \mid v \in V'\} \\ & \quad \cup \{\{v, v_1\} \mid v \in V'\} \cup \{\{v_i, v_{i+1}\} \mid i = 1, \ldots, p-1\} \\ & \quad \cup \{\{s, v'_1\}, \{s, a_1\}, \{v_p, a_{2p+1}\}\} \cup \{\{a_i, a_{i+1}\} \mid i = 1, \ldots, 2p\} \end{cases}$$

All other edges have maximal possible costs such that the Δ_β-inequality is satisfied. Here, γ may be chosen arbitrarily, provided

$$\gamma > \max\left\{ \left\lceil \frac{n+1}{2(\beta-1)} \right\rceil, \frac{n-1}{4p-1} \right\}.$$

Hence, we obtain $4p\gamma > \gamma + n - 1$, which will be useful later in this proof. This graph satisfies the Δ_β-inequality. Here, we set $d(v_p) := 2p\gamma + n - 1$.

Let W' be some path in V' and $W := (v'_1, \ldots, v'_{p-1}, W', v_1, \ldots, v_p)$. Obviously, W reaches v_p in time iff it spends at most time $n-1$ in V', thus $\text{cost}(W') \leq n-1$. The shortest path from s to v_p is $(s, v'_1, v'_2, \ldots, v'_{p-1}, v, v_1, v_2, \ldots, v_p)$ for some $v \in V'$ and costs exactly $2p\gamma$.

A path that visits some v'_i or v_j after v_p cannot reach this deadline because it causes an additional cost of at least $(\beta-1)2\gamma \geq (\beta-1)2\left\lceil \frac{n+1}{2(\beta-1)} \right\rceil > n-1$ as compared to the shortest path from s to v_p. If a path costs n or more in V before visiting any vertex in $\{v_1, \ldots, v_p\}$, the deadline will also be missed, regardless of the path to v_p. Finally, a path that visits some a_i before v_p will also violate this deadline.

Assume G' contains a Hamiltonian path P. Then, an optimal solution for k-Δ_βDLTSP is $(s, v'_1, \ldots, v'_{p-1}, P, v_1, \ldots, v_p, a_{2p+1}, \ldots, a_1, s)$. This cycle costs exactly $4p\gamma + \gamma + n - 1$.

Otherwise, an optimal solution cannot visit all vertices in V' before reaching v_p. Furthermore, it must visit all vertices v_i, v'_i for $i = 1, \ldots, p-1$ before v_p. Therefore, it must visit some vertex in V' after v_p. To do so, it must use some edge from a vertex in $\{v_p\} \cup \{a_i \mid i \in \{1, \ldots, 2p+1\}\}$ to some vertex $v \in V'$. It is not hard to see that such an edge costs at least $\beta^{\log_2(p)}p\gamma$. In order to leave V' again, another expensive edge must be used. Thus, an optimal solution costs at least $2\beta^{\log_2(p)}p\gamma + 4p\gamma + \gamma + n - 2$ if G does not contain a Hamiltonian path. This leads to the ratio

$$\frac{2\beta^{\log_2(p)}p\gamma + 4p\gamma + \gamma + n - 2}{4p\gamma + \gamma + n - 1} = \frac{2\beta^{\log_2(p)}p\gamma}{4p\gamma + \gamma + n - 1} + 1 - \frac{1}{4p\gamma + \gamma + n - 1}$$
$$> \frac{\beta^{\log_2(p)}}{4} > \frac{1}{4}\beta^{\log_2(h|V|)}.$$

Therefore, a polynomial-time algorithm for the 1-Δ_βDLTSP with approximation ratio $\frac{1}{4}\beta^{\log_2(h|V|)}$ could be used to solve the HAMILTONIAN PATH problem in polynomial-time. $\qquad \square$

Corollary 1. *Let $\beta > 1$. For k-Δ_βDLTSP, there is no polynomial-time algorithm with approximation ratio $\frac{1}{4}(\beta^{\log_2(h)} \cdot |V|^{\log_2(\beta)})$ for any $0 < h < 1$ and any $k \in \mathbb{N}$.*   □

We can use a similar construction to obtain a lower bound of $\frac{3}{2} - \varepsilon$ on the approximation ratio for 1-ΔDLTSP. As compared to the ratio $\frac{5}{2}$ of Algorithm 3, this leaves a rather large gap. The next theorem shows that for 2-ΔDLTSP we can raise the lower bound to $2 - \varepsilon$. Note that any bound larger than $2 + \frac{1}{219} + \varepsilon$ would directly imply a lower bound of $\frac{220}{219} + \varepsilon$ for ΔTSP, improving the currently best known lower bound from [16].

Theorem 6. *There is no polynomial-time algorithm for 2-ΔDLTSP with approximation ratio $2 - \varepsilon$ for any $\varepsilon > 0$.*   □

Corollary 2. *There is no polynomial-time algorithm for the k-ΔDLTSP with approximation ratio $2 - \varepsilon$ for any $\varepsilon > 0$ and any $k \geq 2$.*   □

3.3   Lower Bounds for • DLTSP

In the previous sections, we have seen that k-ΔDLTSP can be approximated within a factor of 2.5 and established a lower bound of $2 - \varepsilon$ on the approximation ratio. Now, we will show that the restriction of bounding the number of deadlines by a constant k is crucial to obtaining a constant approximation ratio.

In contrast to the situation where the number of deadlines is bounded, an unbounded number of them can make it hard to even find feasible solutions at all. In fact, if we assign to all the vertices of a (metric) graph the same deadline, namely the minimum length (which we now interpret as duration) of a Hamiltonian path, finding feasible solutions *means* solving the HAMILTONIAN PATH problem. Let us therefore say that ΔDLTSP$'$ is the same problem as ΔDLTSP, except that—as an additional part of the input—we are given one feasible solution. Obviously, ΔDLTSP$' \in NPO$. This is what makes the following theorem a strong inapproximability result.

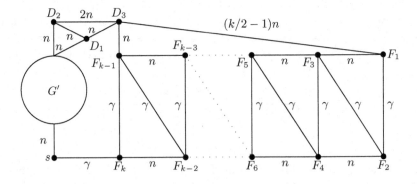

Fig. 2. Construction for ΔDLTSP$'$

Theorem 7. *There is no polynomial-time algorithm for ΔDLTSP$'$ with an approximation ratio $\frac{(1-\varepsilon)}{2}|V|$, for any $0 < \varepsilon < 1$.*

Proof idea. The HAMILTONIAN PATH problem can be reduced to ΔDLTSP$'$ by a construction as outlined in Figure 2. We set the deadlines $d(v) := 3n$ for all $v \in V'$, $d(D_1) := 4n$, $d(D_2) := 5n$, $d(D_3) := 7n$, $d(F_1) := 7n + (\frac{k}{2} - 1)n$, and $d(F_i) := d(F_{i-1}) + \gamma$ for all $i \in \{1, \ldots, k-1\}$.

Depending on whether a feasible solution contains a Hamiltonian path in G', it may visit F_{k-1}, F_{k-3}, $\ldots$, F_3, F_1 and thus avoid to run the zig-zag path F_1, F_2, $\ldots$, F_{k-1}, F_k. A full proof can be found in [7]. $\square$

4   Multiple Structure-Constrained TSP

In this section, we will consider generalizations of OTSP where instead of a single linear order constraint on a subset of vertices, multiple disjoint paths or cycles in the graph are given as precedence constraints. In the case that every vertex of the graph is contained in one of the constraints, i.e., if the constraints constitute a path or cycle cover of the graph, we will present an exact parameterized algorithm with running time polynomial in the size of the graph, but exponential in the number of paths or cycles in the cover. For the more general case where the path or cycle constraints do not cover the whole graph, we will propose a heuristic algorithm, which we conjecture to achieve a constant approximation ratio.

A useful application scenario for these problems may be, for example, routing of a vehicle which is supposed to transport commodities from several source vertices to several destination vertices: Here, it does not matter in which order transportation requests are serviced, but it does matter that sources need to be visited before destinations.

We will start with a formal definition of the cycle versions of these problems.

Definition 4. *Let $G = (V, E)$ be a complete graph with edge weights $c\colon E \to \mathbb{Q}^+$ and let (V, F) be a subgraph of G satisfying either of the following conditions.*

Cycle cover constraint. *Every vertex $v \in V$ is incident with some edge $f \in F$, and (V, F) is a collection of vertex-disjoint cycles, i.e., (V, F) is a cycle cover, that is, we have $d_F(v) = 2$ for all $v \in V$.*

Multiple cycle constraint. *(V, F) is a collection of vertex-disjoint cycles. (Isolated vertices may be treated as cycles of exactly one zero-cost loop, but we ordinarily do not count them in the number of cycles in F.)*

The respective multiple structure-constrained TSP is, given (G, c, F), to find a minimum-cost cycle T of all vertices in V such that T has the property that its restriction to each component of F (where shortcut edges are introduced in order to skip vertices from other components) is this component itself. We denote these problems by CYCCOVTSP and MCYCTSP, respectively.

Algorithm 4

Let there be k' cycles in F. For the sake of notation, assume that there are cycles of length one, all consisting in a single zero cost loop for every vertex uncovered by F. Counting these, let there be k cycles.

  while $k > 1$ **do**

    find the pair of cycles $\{C_1, C_2\}$ which minimizes

        $\text{cost}(\textit{cycle-merge}(C_1, C_2)) - \text{cost}(C_1) - \text{cost}(C_2)$

    merge C_1 and C_2

    recompute k

Note that in the case where F contains exactly one cycle, MCYCTSP coincides with OTSP. Indeed, MCYCTSP may offer a better intuition even for OTSP because it does not hide the fact that cyclically shifting the prescribed order on vertices does not actually alter a given problem instance.

Due to space constraints, we are not able to present the algorithms and proofs in detail in this extended abstract. Instead, we will just present the main results and give a short overview of the proof ideas.

Theorem 8. CYCCOVTSP *can be solved exactly in time* $\mathcal{O}((2n)^k \cdot k^2)$ *where* k *is the number of cycles in* F. □

Theorem 9. *For a constant number of constraining structures, MCYCTSP on metric graphs can be solved approximately in polynomial time with an approximation guarantee of* $\frac{5}{2}$. □

Note that in contrast to Theorem 9 and virtually all of our other results, Theorem 8 also applies to the general (i. e., non-metric) case. In proving Theorem 8, two more algorithms which are based on the idea of dynamic programming will be presented in the full version of this paper. In fact, they exploit the fact that CYCCOVTSP shares properties with problems like DNA alignment and longest common subsequence in words [14]. The basic idea is thus to cut the given cycles open, to give the resulting paths an orientation, and to align their vertices optimally with respect to the cost of the resulting path and cycle (which is yielded by closing the path).

Concludingly, this gives us an algorithm, *cycle-merge,* which motivates the heuristic approach of Algorithm 4 for (single or multiple) cycle-constrained metric TSP. (Bear in mind that the single cycle-constrained metric TSP *is* ΔOTSP.)

We have found no input instances where Algorithm 4 would yield an approximation ratio worse than 2. However, it seems to be difficult to prove an appropriate approximation guarantee.

Acknowledgments

The authors would like to thank Dirk Bongartz, Luca Forlizzi, and Sebastian Seibert for many fruitful discussions on the topic of this paper.

References

1. T. Andreae: On the Traveling Salesman Problem Restricted to Inputs Satisfying a Relaxed Triangle Inequality. *Networks* 38 (2001), pp. 59–67.
2. T. Andreae, H.-J. Bandelt: Performance guarantees for approximation algorithms depending on parameterized triangle inequalities. *SIAM Journal on Discrete Mathematics* 8 (1995), pp. 1–16.
3. N. Bansal, A. Blum, S. Chawla, A. Meyerson: Approximation algorithms for deadline-TSP and vehicle routing with time windows. *Proc. 36th ACM Symposium on the Theory of Computing (STOC'04)*, pp. 166–174.
4. M. Bender, C. Chekuri: Performance guarantees for TSP with a parametrized triangle inequality. *IPL* 73 (2000), pp. 17–21.
5. H.-J. Böckenhauer, D. Bongartz, L. Forlizzi, J. Hromkovič, J. Kneis, J. Kupke, G. Proietti, S. Seibert, W. Unger: *Approximation algorithms for the OTSP.* Unpublished manuscript, 2005.
6. H.-J. Böckenhauer, J. Hromkovič, R. Klasing, S. Seibert, W. Unger: Towards the notion of stability of approximation for hard optimization tasks and the traveling salesman problem. *TCS* 285 (2002), pp. 3–24.
7. H.-J. Böckenhauer, J. Hromkovič, J. Kneis, J. Kupke: On the parameterized approximability of TSP with deadlines. To appear in *Theory of Computing Systems*.
8. H.-J. Böckenhauer, S. Seibert: Improved lower bounds on the approximability of the traveling salesman problem. *RAIRO Theoretical Informatics and Applications* 34 (2000), pp. 213–255.
9. M. Charikar, R. Motwani, P. Raghavan, C. Silverstein: Constrained TSP and low-power computing. *Proc. 5th International Workshop on Algorithms and Data Structures (WADS'97)*, LNCS 2125 (1997), pp. 104–115.
10. N. Christofides: Worst-case analysis of a new heuristic for the travelling salesman problem. Technical Report 388, Graduate School of Industrial Administration, Carnegie-Mellon University, Pittsburgh, 1976.
11. J.-F. Cordeau, G. Desaulniers, J. Desrosiers, M. M. Solomon, F. Soumis: VRP with time windows. In: P. Toth, D. Vigo (eds.): *The Vehicle Routing Problem*, SIAM 2001, pp. 157–193.
12. L. Forlizzi, J. Hromkovič, G. Proietti, S. Seibert: On the stability of approximation for Hamiltonian path problems. *Proc. SOFSEM 2005: Theory and Practice of Computer Science*, LNCS 3381, Springer 2005, pp. 147–156.
13. M. R. Garey, D. S. Johnson: *Computers and Intractability – A Guide to the Theory of NP-Completeness*, Freeman, 1979.
14. D. Gusfield: *Algorithms on Strings, Trees, and Sequences*, Cambridge University Press 1997.
15. J. Hromkovič: Stability of approximation algorithms for hard optimization problems. *Proc. SOFSEM'99*, Springer LNCS 1725, 1999, pp. 29–47.
16. Ch. Papadimitriou, S. Vempala: On the approximability of the traveling salesman problem. *Proc. 32nd Ann. Symp. on Theory of Comp. (STOC '00)*, ACM 2000. Corrected full version available at http://www.cs.berkeley.edu/~christos/

Reoptimization of Minimum and Maximum Traveling Salesman's Tours

Giorgio Ausiello[1], Bruno Escoffier[2],
Jérôme Monnot[2], and Vangelis Th. Paschos[2]

[1] Dipartimento di Informatica e Sistemistica, Università di Roma
"La Sapienza", Roma, Italy
ausiello@dis.uniroma1.it.
[2] Lamsade, CNRS and Université Paris Dauphine, France
{escoffier, monnot, paschos}@lamsade.dauphine.fr.

Abstract. In this paper, reoptimization versions of the traveling salesman problem (TSP) are addressed. Assume that an optimum solution of
an instance is given and the goal is to determine if one can maintain a
good solution when the instance is subject to minor modifications. We
study the case where nodes are inserted in, or deleted from, the graph.
When inserting a node, we show that the reoptimization problem for
MinTSP is approximable within ratio 4/3 if the distance matrix is metric. We show that, dealing with metric MaxTSP, a simple heuristic is
asymptotically optimum when a constant number of nodes are inserted.
In the general case, we propose a 4/5-approximation algorithm for the
reoptimization version of MaxTSP.

1   Introduction

The traveling salesman problem (TSP) is one of the most interesting and paradig-
matic optimization problems. In both minimization and maximization versions,
TSP has been widely studied and a large bibliography is available (see, for example, the books [7, 11, 12]). As it is well known, both versions of TSP are NP-hard
but although in the case of MaxTSP the problem is approximable within constant ratio for all kinds of graphs [4, 9], in the case of MinTSP approximation
algorithms are known only for the metric case [5], i.e., when the graph distances
satisfy the triangle inequality.

In this paper, we address the *reoptimization* issue. We consider the case where
instances of a given optimization problem are subject to minor modifications.
The problem we are interested in consists, given an optimum solution on the initial instance, of trying to maintain efficiently a good solution when the instance
is slightly modified. This issue has already been studied for other optimization
problems such as scheduling problems (see [17, 2], or [3] for practical applications) and classical polynomial problems where the goal is to recompute the
optimum solution as fast as possible ([6, 10]). It has been recently considered for
MinTSP in [1]. The modifications for TSP consists in adding a new node to the

L. Arge and R. Freivalds (Eds.): SWAT 2006, LNCS 4059, pp. 196–207, 2006.

initial graph (we have a new city to visit), or removing one node from this graph (a city is dropped from the tour).

More precisely, we suppose that an n node graph G is given and an optimum solution of MinTSP for G has already been computed. In the problem-version we deal with, denoted MinTSP+ in the sequel, G is transformed into a graph G' by adding a new node v_{n+1} together with all edges connecting v_{n+1} to any node of G. How can we reuse the known optimum solution of MinTSP for G in order to compute a good approximate solution for G'? An analogous problem denoted MinTSP- consists of reoptimizing MinTSP when a node v in G is deleted together with all edges incident to it. In [1], Archetti, Bertazzi and Speranza show that both MinTSP+ and MinTSP- are NP-hard. Moreover they prove that if the simple *best insertion* rule is used for updating the previously known optimum tour, a (tight) $3/2$ approximate tour for MinTSP+ in metric case can be obtained whereas in the general case, they propose some instances leading to the claim that best insertion rule does not lead to a constant approximation; the same (tight) $3/2$ approximation ratio is obtained for MinTSP- in the metric case. In their paper, the authors of [1] were mainly motivated by the situation where a short amount of time is available for the reoptimization. However, another interesting question is to know if the knowledge of an optimum solution for a part of the input graph leads to strictly better approximation ratios for the whole of the graph than those achieved in the classical approximation framework.

In this paper we provide new insights for the reoptimization of MinTSP (for metric graphs), both in the case of a single update and in the case where k new nodes are inserted (denoted MinTSP+k). For MinTSP+ in metric case we show that by combining the best insertion heuristics with Christofides' algorithm the result of [1] can be outperformed, by achieving approximation ratio $4/3$. Moreover, it is possible to show that, for any k, MinTSP+k can be approximated asymptotically better than $3/2$, although, for large values of k, the approximation ratio converges to Christofides' bound. On the other hand, dealing with the general case, we prove that MinTSP+ is not constant approximable. We also study reoptimization of MaxTSP, by considering the problems MaxTSP+ and MaxTSP+k for the first time, both in the metric and in the general case (note that these problems are obviously **NP**-hard). In particular we show that, in the metric case, for any k, the best insertion rule is asymptotically optimum; in fact, for any k, MaxTSP+k can be approximated with ratio $\left(1 - \frac{O(k)}{\sqrt{n}}\right)$. In the general case we can exhibit a $4/5$-approximation algorithm, an improvement over the approximation ratio $61/81$, achieved in [4] (under the classical approximation paradigm).

The paper is organized as follows. In the next section, we provide basic definitions and notation. In Section 3, we address the reoptimization of MinTSP under single and multiple node insertions. Next, in Section 4, we consider the reoptimization of MaxTSP, first under single node insertion (both in the metric and in the general case) and subsequently under multiple insertions (in the metric case). Finally, in Section 5, some results concerning MinTSP- and MaxTSP- are provided. Concluding remarks are contained in Section 6.

2 Preliminaries

In this section we provide the formal definitions of the problems addressed in the paper, namely Min and MaxTSP+k, Min and MaxTSP-k. Then, we introduce three heuristics, Best Insertion, Longest Insertion, and Nearest Insertion, classically studied in the literature (see for instance [15, 7]) because they give rise to fast algorithms to solve TSP, and particularly suitable when dealing with reoptimization.

Definition 1 (MinTSP+k, MaxTSP+k). *We are given an instance* (I_{n+k}, T_n^*) *where* $I_{n+k} = (K_{n+k}, d)$, K_{n+k} *is a complete graph on* $n + k$ *nodes* $\{v_1, \cdots, v_{n+k}\}$, *with nonnegative weights* d *on the edges, and* T_n^* *is an optimum solution of MinTSP (resp. MaxTSP) on* $I_n = (K_n, d)$, *sub-instance of* I_{n+k} *induced by the nodes* $\{v_1, \cdots, v_n\}$.
Question : find a shortest (resp. longest) tour for the whole instance I_{n+k}.

Definition 2 (MinTSP-k, MaxTSP-k). *We are given an instance* (I_{n+k}, T_{n+k}^*) *where* $I_{n+k} = (K_{n+k}, d)$, K_{n+k} *is a complete graph on* $n + k$ *nodes* $\{v_1, \cdots, v_{n+k}\}$, *with nonnegative weights* d *on the edges, and* T_{n+k}^* *is an optimum solution of MinTSP (resp. MaxTSP) on* I_{n+k}.
Question : find a shortest (resp. longest) tour on $I_n = (K_n, d)$, *sub-instance of* I_{n+k} *induced by the nodes* $\{v_1, \cdots, v_n\}$.

For the case $k = 1$, we simply denote the problems MinTSP+, MaxTSP+, Min TSP- and MaxTSP-.

For TSP, a particular rapid way to get a tour is to iteratively insert nodes according to given rules, as the following classical ones.

Definition 3 (Nearest, Longest and Best Insertion rules). *Given a tour* T *on a graph* $G = [V, E]$, *and a node* $v \notin V$, *we insert* v *in the sequence of nodes of* T *as follows:*

- *Nearest Insertion: we find a node* v^* *minimizing* $d(u, v)$ *for* $u \in V$, *and insert* v *before or after* v^* *(choosing the best solution) in the tour;*
- *Longest Insertion: we find a node* v^* *maximizing* $d(u, v)$ *for* $u \in V$, *and insert* v *before or after* v^* *(choosing the best solution) in the tour;*
- *Best Insertion: we find an edge* $(u^*, v^*) \in T$ *optimizing* $(d(v, u) + d(v, w) - d(u, w))$ *for* $(u, v) \in T$, *and insert* v *between* u^* *and* v^*.

Concerning polynomial approximation of MinTSP in the metric case, it is shown in [15] that the behavior of Nearest and Best Insertions are quite different since the algorithms based on these two rules are a 2 and a $O(\log n)$ -approximation respectively.

Finally, when nodes are deleted, the most natural way to get a solution from a tour on the initial instance consists in taking the shortcut.

Definition 4 (Deletion). *Given a tour* T *on a graph* $G = [V, E]$, *and a node* $v \in V$, *Deletion consists in building a tour by deleting* v *in* T *(removing* (u, v) *and* (v, w) *from* T *and adding* (u, w)).

3   Reoptimizing Minimum TSP Under Node Insertions

In this section, we study the reoptimization problems where one node is inserted (MinTSP+) and several nodes are inserted (MinTSP+k). We show that we can improve the result of [1] proving that, in the metric case, MinTSP+ is approximable within ratio 4/3.

On the contrary, if the distance is not assumed to be metric, then the knowledge of an optimum solution in the initial instance is not useful at all in order to find an approximate solution of the final instance since MinTSP+ (and consequently MinTSP+k) is not constant approximable (unless $\mathbf{P \neq NP}$).

Finally, we generalize the result in the metric case by showing that when k nodes are inserted we get a $(3/2 - 1/(4k + 2))$-approximation algorithm.

3.1   One Node Insertion

When dealing with metric instances of MinTSP+, it is proved in [1] that Best Insertion gives a 3/2-approximate solution. Actually, we can show that Nearest Insertion also provides this bound. Of course, running Christofides' algorithm on the final instance gives directly also a 3/2-approximate solution. Here we show that a simple combination of Nearest (or Best) Insertion and Christofides' algorithm leads to a better approximation ratio.

Theorem 1. *In the metric case, MinTSP+ is approximable within ratio 4/3.*

Proof. Consider an optimum solution T_{n+1}^* on the whole instance I_{n+1}, and the solution T_n^* given to us on the sub-instance I_n.

Let v_i and v_j be the 2 neighbors of v_{n+1} in T_{n+1}^*, and let T_1 be the tour obtained from T_n^* with the Nearest Insertion rule.

Using the triangle inequality, we easily get $d(T_1) \leq d(T_{n+1}^*) + 2d(v_{n+1}^*, v_{n+1})$ where we recall that $d(v_{n+1}^*, v_{n+1}) = \min\{d(v_i, v_{n+1}) : i = 1, \cdots, n\}$. Thus

$$d(T_1) \leq d(T_{n+1}^*) + 2\max\{d(v_i, v_{n+1}), d(v_j, v_{n+1})\} \tag{1}$$

Now, consider the algorithm of Christofides ([5]) applied on I_{n+1}. This gives a tour T_2 of length at most $1/2 d(T_{n+1}^*) + MST(I_{n+1})$, where $MST(I_{n+1})$ is the value of a minimum spanning tree on I_{n+1}. Note that $MST(I_{n+1}) \leq d(T_{n+1}^*) - \max(d(v_i, v_{n+1}), d(v_j, v_{n+1}))$. Hence :

$$d(T_2) \leq \frac{3}{2}d(T_{n+1}^*) - \max(d(v_i, v_{n+1}), d(v_j, v_{n+1})) \tag{2}$$

We take the best solution between T_1 and T_2. A combination of equations (1) and (2) with coefficients 1 and 2 gives the expected result.

Obviously, if we apply Best Insertion instead of Nearest Insertion, the same result holds. Note that the running time of this algorithm is dominated by the one of Christofides' algorithm. □

In [1], it is shown that if the distance is not assumed to be metric, then Best Insertion is not constant approximate for MinTSP+. We strengthen this result by proving that this holds for any polynomial algorithm.

To do this, we need an intermediate result. Given a graph $G = [V, E]$ where $a, b, s, t \in V$, and an hamiltonian path of G from a to b, we consider the problem of determining if there exists an hamiltonian path from s to t. Using a slight modification of the result of [13], we can show that this problem, denoted by $SHP_{a,b,s,t}$ in the sequel, is **NP**-complete (proof omitted).

Lemma 1. $SHP_{a,b,s,t}$ is **NP**-complete.

This lemma leads to the following inapproximability result.

Theorem 2. In the general case, MinTSP+ is not $2^{p(n)}$-approximable, if $P \neq NP$, for any polynomial p.

Proof. We apply the general method described in [16]. Let $\rho > 1$. We start from an instance of $SHP_{a,b,s,t}$, i.e. a graph $G_n = [V, E]$ with n nodes, four nodes a, b, s, t, and an hamiltonian path P from a to b. We construct an instance (I_{n+1}, T_n^*) in the following way:

- If $(v_i, v_j) \in E$, then $d(v_i, v_j) = 1$.
- $d(a, b) = 1$ and $d(v_{n+1}, s) = d(v_{n+1}, t) = 1$.
- All the other edges have a weight $\rho(n + 1) + 1$.

It is clear that $T_n^* = P \cup \{(a, b)\}$ is an optimum solution of $I_n = (K_n, d)$ with cost $d(T_n^*) = n$. Thus, (I_{n+1}, T_n^*) is an instance of MinTSP+. Let T_{n+1}^* be an optimum solution of (K_{n+1}, d). Remark that any ρ-approximate solution allows us to decide if $d(T_{n+1}^*) = n + 1$. However $d(T_{n+1}^*) = n + 1$ iff there is a hamiltonian path from s to t in G_n. Setting $\rho = 2^{p(n)}$, we obtain the claimed result. □

3.2 • Node Insertions

When k nodes are inserted, we can generalize the result of Theorem 1 in the following way.

Theorem 3. In the metric case, MinTSP+k is approximable within ratio $3/2 - 1/(4k + 2)$.

Proof. Consider the given optimum solution T_n^*. We apply Nearest Insertion with a priority rule. In a first step, we sort the vertices to be inserted (and relabel them) in such a way that for all $p > n$, there exists v_j, $j < p$ such that $d(v_p, v_j) = \min\{d(v_i, v_l) : i \geq p, l < p\}$. Note that $d(v_p, v_j) \leq d_{max}(T_{n+k}^*)$, where $d_{max}(T_{n+k}^*)$ is a maximal weighted edge in T_{n+k}^*.

Then we insert the k vertices using Nearest Insertion.

For the analysis, note that when inserting node v_p, we increase the distance by $\Delta_p \leq 2d(v_p, v_j) \leq 2d_{max}(T^*_{n+k})$. We finally get an approximate solution T_1 such that

$$d(T_1) \leq d(T^*_n) + 2kd_{max}(T^*_{n+k}) \leq d(T^*_{n+k}) + 2kd_{max}(T^*_{n+k}) \qquad (3)$$

Christofides' algorithm gives a solution T_2 such that

$$d(T_2) \leq \frac{3}{2}d(T^*_{n+k}) - d_{max}(T^*_{n+k}) \qquad (4)$$

We take the best solution between T_1 and T_2. A combination of equations (3) and (4) with coefficients 1 and $2k$ gives $d(T) \leq \left(\frac{3}{2} - \frac{1}{4k+2}\right) d(T^*_{n+k})$.

Note that the computation time of T_1 is $O(k(n + k))$, hence the global complexity is dominated by running Christofides' algorithm.   □

4 Reoptimizing Maximum TSP Under Node Insertions

In this section, we consider the reoptimization of the maximization version of TSP. In the metric case, Best Insertion is a very good strategy since it is asymptotically optimum. Note that the usual MaxTSP problem in the metric case does not admit a PTAS (using [14]) and that the best algorithms for it are asymptotically 17/20 (deterministic, [4]) and 7/8 (randomized, [9]).

If the distance is not assumed to be metric, the situation is a bit more complicated. Longest and Best Insertion are only a 1/2-approximation. This situation is quite disappointing since we can easily prove that iterating Longest Insertion (from the empty graph) with a priority rule is already a 1/2-approximation for MaxTSP; however, we can get a polynomial algorithm achieving a ratio of 4/5. This shows that the knowledge of an optimum solution on the initial instance is useful since the best algorithm for the usual MaxTSP achieves an approximation ratio of 61/81 ([4]).

Finally, in section 4.2, we generalize the result in the metric case showing that if we insert a constant number of nodes, then iterating Best Insertion is also an asymptotically optimum strategy.

Note that the **NP**-hardness of all these problems is obvious since otherwise, starting from the empty graph, we could solve polynomially MaxTSP.

4.1 One Node Insertion

The central result of this section is the asymptotic optimality of Best Insertion. It is interesting to note that the behavior of Best and Longest Insertion are quite different for MaxTSP+ since Longest Insertion is only a 2/3-approximation, even asymptotically (proof omitted).

Proposition 1. *For MaxTSP+, in the metric case, Longest Insertion gives a 2/3-approximation, and this bound is tight (even if the graph has an arbitrary large number of nodes).*

Theorem 4. *In the metric case, Best Insertion is asymptotically optimum. More precisely, if the graph has n nodes, then Best Insertion is $(1 - O(1/\sqrt{n}))$-approximate.*

Proof (Sketch). Let T_n^* be an optimum solution on the initial instance I_n, T_{n+1}^* an optimum solution on the final instance I_{n+1}, and T the solution obtained by applying Best Insertion on T_n^*. Let $K = \sqrt{n}$ and $1 \le k \le K$.

Consider the following subsequence of nodes $(a_k, \cdots, a_1, v_{n+1}, b_1, \cdots, b_k)$ in T_{n+1}^*. Let J_k be the sub-instance of I_{n+1} induced by all the nodes but v_{n+1}, $a_1, a_2, \cdots, a_{k-1}$ and $b_1, b_2, \cdots, b_{k-1}$ (in particular J_1 is (K_n, d), the initial graph). We have :

$$d(T_{n+1}^*) \le d(v_{n+1}, a_1) + d(v_{n+1}, b_1) + \sum_{i=1}^{k-1} d(a_i, a_{i+1}) + \sum_{i=1}^{k-1} d(b_i, b_{i+1}) + opt(J_k)$$

where $opt(J_k)$ is the value of an optimum solution on J_k. Indeed, there is an hamiltonian path in T_{n+1}^* between a_k and b_k, the value of which is at most $opt(J_k)$.

Let $d_m^k(v)$ be the medium distance between a node v and the nodes in J_k, i.e., $d_m^k(v) = \frac{1}{|J_k|} \sum_{v_i \in J_k} d(v, v_i)$. Using the triangle inequality, we get that for any pair (u, v) of nodes (and for any k), $d(u, v) \le d_m^k(u) + d_m^k(v)$. Hence we get an upper bound on $d(T_{n+1}^*)$:

$$d(T_{n+1}^*) \le 2 \left(d_m^k(n+1) + \sum_{i=1}^{k-1} d_m^k(a_i) + \sum_{i=1}^{k-1} d_m^k(b_i) \right) + d_m^k(a_k) + d_m^k(b_k) + opt(J_k) \tag{5}$$

Now, our goal is to lower bound first $d(T_n^*)$ and then $d(T)$ in order to get the following inequality :

$$d(T) \ge \left(1 - \frac{O(k)}{n} \right) (d(T_{n+1}^*) - d_m^k(a_k) - d_m^k(b_k)) \tag{6}$$

To achieve this, first consider an optimum solution $T^*(J_k)$ (of value $opt(J_k)$) of J_k. Considering a particular subsequence $(v_1, \cdots, v_{2k-1})$ of $T^*(J_k)$, we insert the $2(k-1)$ nodes $a_1, a_2, \cdots, a_{k-1}$ and $b_1, b_2, \cdots, b_{k-1}$ in $T^*(J_k)$ in order to get the sequence $(v_1, a_1, v_2, \cdots, a_{k-1}, v_k, b_1, v_{k+1}, \cdots, b_{k-1}, v_{2k-1})$. Considering each node of J_k as v_1, we get with these insertions $n - 2(k-1)$ tours on I_n. After a careful counting of the edges appearing in these tours, one can show that:

$$d(T_n^*) \ge 2 \left(\sum_{i=1}^{k-1} d_m^k(a_i) + \sum_{i=1}^{k-1} d_m^k(b_i) \right) + \left(1 - \frac{O(k)}{n} \right) opt(J_k) \tag{7}$$

Now, we relate $d(T)$ and $d(T_n^*)$. Consider each of the n possible insertions of v_{n+1} in T_n^*. Since each edge of T_n^* is removed exactly once, we get that $nd(T) \ge (n-1)d(T_n^*) + 2\sum_{i=1}^n d(v_{n+1}, v_i)$. Using $\sum_{i=1}^n d(v_{n+1}, v_i) \ge \sum_{v \in J_k} d(v_{n+1}, v) = (n - 2(k-1))d_m^k(n+1)$, we get:

$$d(T) \ge \left(1 - \frac{1}{n} \right) d(T_n^*) + 2 \left(1 - \frac{2(k-1)}{n} \right) d_m^k(n+1) \tag{8}$$

From inequalities (5), (7) and (8), we can derive (6).

Inequality (6) is valid for any k. Let us write it for $k = 1, \cdots, K$, and consider the two following cases :

1. If, for some k, $d_m^k(a_k) + d_m^k(b_k) \leq \frac{1}{K} d(T_{n+1}^*)$, then we get

$$d(T) \geq \left(1 - \frac{O(k)}{n}\right)\left(1 - \frac{1}{K}\right) d(T_{n+1}^*)$$

   Since $k \leq K = \sqrt{n}$, we get $d(T) \geq \left(1 - O\left(\frac{1}{\sqrt{n}}\right)\right) d(T_{n+1}^*)$

2. In the other case, for any k, $d_m^k(a_k) + d_m^k(b_k) \geq \frac{1}{K} d(T_{n+1}^*)$. However, this is impossible. Indeed, by making the sum, we get $\sum_{k=1}^K d_m^k(a_k) + d_m^k(b_k) \geq d(T_{n+1}^*)$. But (details are omitted here), one can show that this would lead to $d(T_n^*) \geq 2\left(1 - \frac{O(K)}{n}\right) d(T_{n+1}^*)$, which is impossible for n large enough.

   $\square$

From Theorem 4, we get the following corollary.

Corollary 1. *MaxTSP+ admits a PTAS in the metric case.*

Proof. Let $\varepsilon > 0$. To get a $(1 - \varepsilon)$-approximation algorithm, we just have to apply Best Insertion on graphs with roughly $n \geq O(1/\varepsilon^2)$ nodes, and to solve optimally the other instances. $\square$

Unfortunately, if the triangle inequality is not assumed, Best Insertion has a much worse behavior (proof omitted).

Proposition 2. *For MaxTSP+, in the general case, Best Insertion and Longest Insertion give a 1/2-approximation, and this bound is tight (even if the graph has an arbitrary large number of nodes).*

However, we can use a more sophisticated algorithm to get a better approximation ratio.

Theorem 5. *MaxTSP+ is asymptotically approximable within ratio 4/5.*

Proof (Sketch). Assume n even; thus T_n^* is the sum of two perfect matchings M_1 and M_2 (if n is odd we can add the remaining edge to each matching. Details are omitted). Suppose $d(M_1) \geq d(M_2)$. We get:

$$d(M_1) \geq \frac{1}{2} d(T_n^*) \tag{9}$$

Let v_i and v_j be the neighbors of v_{n+1} in T_{n+1}^*. Consider $M^* = M_1 \cup \{(v_i, v_{n+1}), (v_{n+1}, v_j)\}$. Obviously, M^* can be found in polynomial time by guessing nodes v_i and v_j. Wlog., we can assume that M^* does not contain any cycle (otherwise, $(v_i, v_j) \in T_n^*$ and thus Best Insertion gives an optimum tour).

Now, consider $\mathcal{C} = \{C_1, \cdots, C_p\}$ a 2-matching (i.e., a partition of $\{v_1, \cdots, v_{n+1}\}$ into node disjoint cycles) of maximum weight among the 2-matchings satisfying (i) $\{(v_i, v_{n+1}), (v_{n+1}, v_j)\} \subset C_1$ and (ii) $|C_1| \geq 6$. Such a 2-matching can be found in polynomial time by testing all the possible subsequences of nodes $(v_{i''}, v_{i'}, v_i, v_{n+1}, v_j, v_{j'})$ (and thanks to the polynomiality of finding a maximum weight 2-matching, [8]). Obviously, we deduce:

$$d(\mathcal{C}) \geq d(T^*_{n+1}) \tag{10}$$

Applying the method of Serdyukov [18], we can iteratively for $i = 1, \cdots, p$, delete an edge $e_i \in C_i$, and add this edge to M^* in such a way that M^* does not contain any cycle. Note that in this method we can chose in C_1 a deleted edge not in M^* that does not create a cycle in P_1 (thanks to the length of C_1).

At the end, $P_1 = \cup_{i=1}^p (C_i \setminus \{e_i\})$ and $P_2 = M^* \cup_{i=1}^p \{e_i\}$ are two collection of node disjoint paths. Finally, we build two tours T_1 and T_2 by adding some edges to P_1 and P_2 respectively. Taking the best tour, and using inequalities (9) and (10), we get a tour T_3 with:

$$d(T_3) \geq \frac{3}{4} d(T^*_{n+1}) + \frac{1}{4}(d(v_i, v_{n+1}) + d(v_{n+1}, v_j)) \tag{11}$$

On the other hand, the Best Insertion gives a tour T_4 verifying:

$$d(T_4) \geq \frac{n-1}{n} d(T^*_n) \geq \frac{n-1}{n} d(T^*_{n+1}) - \frac{n-1}{n}(d(v_i, v_{n+1}) + d(v_{n+1}, v_j)) \tag{12}$$

Adding inequality (11) with coefficient $(n-1)/n$ and inequality (12) with coefficient $1/4$ we obtain a tour satisfying $d(T) \geq \frac{4n-4}{5n-4} d(T^*_{n+1})$. $\square$

4.2 • Node Insertions

When several nodes are inserted, we can iteratively use the Best Insertion rule to obtain an asymptotically optimum solution. This result is based on the following lemma.

Lemma 2. *If T_n is a ρ-approximation on the initial instance on n nodes G_n, then Best Insertion applied on T_n gives a $\rho \left(1 - \frac{O(1)}{\sqrt{n}}\right)$-approximate solution (in the metric case) on the instance G_{n+1} on $n+1$ nodes.*

Proof (Sketch). This is an easy generalization of the proof of theorem 4. Note that equation (5) and (7) still hold. Then, by taking into account that T_n is a ρ-approximation, we get, instead of equation (6):

$$d(T) \geq \rho \left(1 - \frac{O(k)}{n}\right)(d(T^*_{n+1}) - d^k_m(a_k) - d^k_m(b_k)) \tag{13}$$

The end of the proof is analogous, up to the factor ρ. $\square$

Theorem 6. *Iterated Best Insertion is a $\left(1 - \frac{O(k)}{\sqrt{n}}\right)$-approximation algorithm for MaxTSP+k in the metric case.*

Proof. Using proposition 2, we get, after k steps, a solution T_k such that:

$$d(T_k) \geq \left(1 - \frac{O(1)}{\sqrt{n}}\right)^k d(T^*_{n+k}) \geq \left(1 - \frac{O(k)}{\sqrt{n}}\right) d(T^*_{n+k})$$

$\square$

Using a similar proof as in corollary 1, we easily get the following result.

Corollary 2. *For any constant k (and even for any $k - o(\sqrt{n})$), MaxTSP+k admits a PTAS in the metric case.*

5   Node Deletions

Now, we give a few results concerning the reoptimization problems when nodes are deleted from the initial graph. Recall that in [1] it is shown that MinTSP- is **NP**-hard, even if distances are only 1 and 2, and that Deletion is a tight $3/2$-approximation in the metric case. Here, we show that MinTSP- is very hard to approximate if the triangle inequality doesn't not hold.

Dealing with MaxTSP-, we show that the problem is **NP**-hard, and that Deletion is a tight $1/2$-approximation algorithm (general and metric cases).

Proposition 3. *In the general case, MinTSP- is not $2^{p(n)}$-approximable, for any polynomial p, if $\boldsymbol{P \neq NP}$.*

Proof. The proof is a direct adaptation of the one of [1] showing that this problem is **NP**-hard. We consider the following problem, shown to be NP-complete in [13]: given a graph $G = [V, E]$ and an hamiltonian path P between two nodes a and b in G, determine if there's an hamiltonian cycle in G.

Given such an instance, we construct an instance on MinTSP-. The node set of the graph K_{n+1} is $V \cup \{v_{n+1}\}$, and the distances are:

- $d(v_i, v_j) = 1$ if $(v_i, v_j) \in E$;
- $d(v_{n+1}, a) = d(v_{n+1}, b) = 1$;
- Other distances are $\rho n + 1$.

The tour $T^*_{n+1} = P \cup \{(v_{n+1}, a), (v_{n+1}, b)\}$ is an optimum solution on $I_{n+1} = (K_{n+1}, d)$. Let T^*_n be an optimum solution on the instance I_n. Then $d(T^*_n) = n$ iff G has an hamiltonian cycle, and a ρ approximate solution allows to decide if $d(T^*_n) = n$. We get the lower bound setting $\rho = 2^{p(n)}$. $\square$

Proposition 4. *MaxTSP- is NP-hard, even if distances are only 1 and 2.*

Proof. In [1], it is shown that MinTSP- is NP-hard, even if distances are only 1 and 2. We have a trivial reduction from MinTSP- to MaxTSP- if distances are only 1 and 2: we just have to flip the distances between 1 and 2. Solving MinTSP- is equivalent to solve MaxTSP- with the new distances. $\square$

As a final result, let us remark that the deletion strategy has the same behavior in the metric case and in the general one.

Proposition 5. *For MaxTSP-, Deletion gives a 1/2-approximation, and this bound is tight (even if the graph has an arbitrary large number of nodes). These results hold in the general case as well as in the metric case.*

These results might be strengthened, but they seem to indicate that the knowledge of an optimum solution in the initial instance may not be really helpful to get good approximation ratios when nodes are deleted.

6   Conclusion

In this article we have proposed some complexity and approximability results for reoptimization versions of TSP. We have exhibited an interesting asymmetry between the maximization and the minimization versions: while we get an almost optimum tour by simply inserting the new node in the right position for MaxTSP+ (in the metric case), this is not true when dealing with the minimization version. One can even show that in order to get an almost optimum solution for MinTSP+, we need, on some instances, to change $n - o(n)$ edges from the initial optimum solution. This leads us to conjecture that MinTSP+ does not admit a PTAS.

Following our approach, an interesting generalization would be to consider TSP in a fully dynamic situation. Starting from a given solution (optimum or approximate) on an initial graph, the graph evolves (nodes are added and deleted), and the goal is to maintain efficiently, along this process, an approximate solution as good as possible. Some of our results can be easily generalized when starting from an approximate (instead of optimum) solution, and can be useful in such approach.

References

1. C. Archetti, L. Bertazzi, and M. G. Speranza. Reoptimizing the traveling salesman problem. *Networks*, 42(3):154–159, 2003.
2. M. Bartusch, R. H. Mohring, and F. J. Radermacher. Scheduling project networks with resource constraints and time windows. *Ann. Oper. Res.*, 16:201–240, 1988.
3. M. Bartusch, R. H. Mohring, and F. J. Radermacher. A conceptional outline of a dss for scheduling problems in the building industry. *Decision Support Systems*, 5:321–344, 1989.
4. Z.-Z. Chen, Y. Okamoto, and L. Wang. Improved deterministic approximation algorithms for Max TSP. *Information Processessing Letters*, 95:333–342, 2005.
5. N. Christofides. Worst-case analysis of a new heuristic for the traveling salesman problem. Technical Report 338, Grad School of Industrial Administration, Canergi-Mellon University, Pittsburgh, 1976.
6. D. Eppstein, Z. Galil, G. F. Italiano, and A. Nissenzweig. Sparsification - a technique for speeding up dynamic graph algorithms. *Journal of the ACM*, 44(5):669–696, 1997.

7. G. Gutin and A. P. Punnen. *The Traveling Salesman Problem and Its Variations.* Combinatorial Optimization. Kluwer Academic Publishers, 2002.
8. D. Hartvigsen. *Extensions of Matching Theory.* PhD thesis, Carnegie-Mellon University, 1984.
9. R. Hassin and S. Rubinstein. A 7/8-approximation algorithm for metric Max TSP. *Information Processessing Letters*, 81(5):247–251, 2002.
10. M. R. Henzinger and V. King. Maintaining minimum spanning trees in dynamic graphs. In P. Degano, R. Gorrieri, and A. Marchetti-Spaccamela, editors, *ICALP*, volume 1256 of *Lecture Notes in Computer Science*, pages 594–604. Springer, 1997.
11. D. S. Johnson and L. A. McGeoch. The traveling salesman problem : a case study. In E. Aarts and J. K. Lenstra, editors, *Local search in combinatorial optimization.* Wiley, 1997.
12. E. L. Lawler, J. K. Lenstra, A. H. G. Rinnooy Kan, and D. B. Shmoys. *The Traveling Salesman Problem: a guided tour of Combinatorial Optimization.* Discrete Mathematics and Optimization. Wiley, 1985.
13. C. H. Papadimitriou and K. Steiglitz. Some complexity results for the traveling salesman problem. In *STOC'76*, pages 1–9, 1976.
14. C. H. Papadimitriou and M. Yannakakis. The traveling salesman problem with distances one and two. *Mathematics of Operations Research*, 18:1–11, 1993.
15. D. J. Rosenkrantz, R. E. Stearns, and P. M. Lewis II. An analysis of several heuristics for the traveling salesman problem. *SIAM J. Comput.*, 6(3):563–581, 1977.
16. S. Sahni and T. F. Gonzalez. P-complete approximation problems. *J. ACM*, 23(3):555–565, 1976.
17. M. W. Schäffter. Scheduling with forbidden sets. *Discrete Applied Mathematics*, 72(1-2):155–166, 1997.
18. A. I. Serdyukov. An algorithm with an estimate for the traveling salesman problem of the maximum. *Upravlyaemye Sistemy (in Russian)*, 25:80–86, 1984.

The Node-Weighted Steiner Problem in Graphs of Restricted Node Weights

Spyros Angelopoulos

David R. Cheriton School of Computer Science, University of Waterloo
Waterloo ON N2L 3G1, Canada
sangelop@cs.uwaterloo.ca

Abstract. In this paper we study a variant of the Node-Weighted Steiner Tree problem in which the weights (costs) of vertices are restricted, in the sense that the ratio of the maximum node weight to the minimum node weight is bounded by a quantity α. This problem has applications in multicast routing where the cost of participating routers must be taken into consideration and the network is relatively homogenous in terms of the cost of the routers.

We consider both online and offline versions of the problem. For the offline version we show an upper bound of $O(\min\{\log \alpha, \log k\})$ on the approximation ratio of deterministic algorithms (where k is the number of terminals). We also prove that the bound is tight unless $P = NP$. For the online version we show a tight bound of $\Theta(\max\{\min\{\alpha, k\}, \log k\})$, which applies to both deterministic and randomized algorithms. We also show how to apply (and extend to node-weighted graphs) recent work of Alon *et al.* so as to obtain a randomized online algorithm with competitive ratio $O(\log m \log k)$, where m is the number of the edges in the graph, independently of the value of α. All our bounds also hold for the Generalized Node-Weighted Steiner Problem, in which only connectivity between pairs of vertices must be guaranteed.

1 Introduction

1.1 Problem Definition

The *Node-Weighted Steiner Tree* problem (NWS) is defined as follows. Given an undirected graph $G = (V, E)$ ($|V| = n, |E| = m$) with a cost function c on the edges and vertices, and a subset of vertices $K \subseteq V$ with $|K| = k$, (also called *terminals*), the goal is to find a minimum-cost tree which spans all vertices in K. The cost of the tree is defined as the sum of the costs of its edges and vertices. The *Generalized Node-Weighted Steiner* problem (GNWS) is defined along the same lines, with the exception that instead of a set of terminals, we are given a requirements function $r : V \times V \to \{0, 1\}$. The objective is to find a minimum-cost subgraph of G which provides connectivity for all pairs of vertices in the set $K = \{(v, u) \text{ with } r(v, u) = 1\}$. For uniformity, we call K the set of *terminal pairs*. Clearly, GNWS generalizes NWS.

L. Arge and R. Freivalds (Eds.): SWAT 2006, LNCS 4059, pp. 208–219, 2006.

Depending on whether K is known in advance, we distinguish between the offline and online versions of the problem. In the latter version, every time a new *request* appears (i.e., a terminal, or a pair of terminals) the algorithm must guarantee connectivity for the new request by buying irrevocably, if necessary, certain edges and vertices.

Both the offline and online version of this problem are variants of the well-known Steiner Tree Problem and Generalized Steiner Problem (GSP) which have been studied extensively in the literature (c.f. section 1.2 for some representative results on this problem).

In this paper we are interested in the variant of node-weighted Steiner problems in which some restriction is placed on the node weights. Specifically, let α denote the quantity $\max_{v,u \in V} c(v)/c(u)$ (we assume non-negative edge and vertex costs). We call α the *asymmetry* of the graph, since it provides an indication of the variance of the vertex costs. Our aim is to provide upper and lower bounds for both the approximation ratio and the competitive ratio as functions of α.

The (classic) Steiner problem has wide applications in multicast routing and network design (see e.g., [7]). On the other hand, NWS captures situations where the cost considerations include not only network links but also network nodes (e.g., routers), and the cost of the node simply reflects how much we must pay to have it included in the connectivity network. A network of relatively homogenous routers, in terms of their cost, is then modelled by a graph of small asymmetry. Our setting is largely motivated by the work of Faloutsos *et al.* [8]; in their work, the asymmetry is defined in the context of a directed graph, with only an edge-cost function, as the maximum ratio of the costs of the two directed edges between any two vertices in the graph, and reflects network homogeneity in terms of the edge costs of antiparallel links.

Summary of Our Results. We first show that unless P=NP, offline NWS cannot be approximated within a factor better than $O(\min\{\log \alpha, \log k\})$ (Theorem 1). The proof uses ideas from Berman's reduction (see section 1.2), however we reduce from Set Cover of Bounded Set Size; the result then follows from Trevisan's inapproximability result [19]. Theorem 2 shows that the hardness bound is asymptotically tight, by presenting an algorithm which is a combination of Ravi and Klein's algorithm for GNWS and the constant-factor approximation algorithm of Goemans and Williamson for GSP. In Theorem 3 we show an asymptotically tight bound of $\Theta(\max\{\min\{\alpha, k\}, \log k\})$ on the competitive ratio of (deterministic or randomized) online algorithms. Last, section 3.1 builds upon ideas found in the work of Alon *et al.* [3] which provides a general framework applicable to several online network optimization problems with edge cost functions. We show how a key lemma in their work can be extended to node-weighted problems which translates to a $O(\log m \log k)$-competitive randomized algorithm regardless of the asymmetry of the input graph. The bound is almost tight since the lower bound of $\Omega(\frac{\log m \log k}{\log \log m + \log \log k})$ shown for online Set Cover in [2] applies to online NWS as well.

1.2   Related Work

Berman (see ref. in [14]) showed an approximation-preserving reduction from Set Cover to NWS, which implies, using results of [9] and [15], that NWS cannot be approximated within a factor of $o(\log k)$ unless P=NP. We emphasize that the hardness result holds only when α is unbounded, in particular, it is required that α is as big as $\Omega(n)$. Klein and Ravi [14] presented an asymptotically optimal algorithm which guarantees an approximation ratio of $2 \ln k$. Guha and Khuller [11] improved the upper bound to $1.35 \ln k$ (approximately).

Concerning the classic Steiner Tree problem (where $w(v) = 0$, for all $v \in V$), Karp [13] showed very early that it is NP-hard, and is in fact APX-hard [5] [18]. Currently, the best upper bound for general graphs is 1.55 and is due to Robins and Zelikovsky [16]. For the Generalized Steiner Problem, the corresponding bound is 2 [10],[1]. In terms of online algorithms, Imase and Waxman [12] showed a tight bound of $\Theta(\log k)$ on the competitive ratio of online Steiner Tree, a result which Berman and Coulston [4] showed extends to the online GSP.

Faloutsos *et al.* [8] considered a somewhat related problem to ours, namely the online Steiner tree problem on directed graphs of bounded edge asymmetry. They showed that a simple greedy algorithm is $O(\min\{k, \beta \log k\})$ competitive, and they also proved a lower bound on the competitive ratio of every deterministic algorithm (which one can show extends to randomized algorithms) of $\Omega(\min\{k, \beta \log k / \log \beta\})$. Here β denotes the edge asymmetry, as defined earlier in this section. Their results are related to this work not only because they provide some motivation for our definition of node asymmetry, but also because NWS can be reduced to Steiner tree in directed graphs, as shown by Segev [17]. Note however, that applying this reduction by itself only cannot yield good bounds: in particular for offline NWS with asymmetry α, the reduction creates a directed graph of edge asymmetry β which can be as high as α (depending on the edge costs) and for such graphs it is not known how to approximate the Steiner Tree problem to a $o(\beta)$ factor (note that an obvious upper bound on the approximation ratio using this technique is $O(\beta)$).

1.3   Preliminaries

Given a (simple) path P between two vertices v, u in G, the cost of P is the sum of the costs of all vertices and edges in P, *excluding* the end vertices v and u. Note that a path of minimum cost can be computed in polynomial time, by constructing a directed graph $G' = (V, E')$ in which only edges have a cost, such that for every edge $e = (v, u) \in E$, $e' = (v, u) \in E'$, and $c(e') = c(e) + c(v)$. It is easy to see that a shortest path from v to u in G' translates to a minimum-cost path from v to u in G.

Given graph G, with edge and vertex weights, define the *shortest path completion of G* (or simply *path completion*), as the complete graph $G_p = (V, E')$ in which every vertex has the same cost as in V, and the cost of an edge $e = (v, u) \in E'$ is the cost of a minimum-cost path between v and u in G. Note that the path completion can be computed in polynomial time. Following a standard practice in the study of Steiner trees we observe that, when we need so, we can restrict our attention to the path completion of G, in the sense that a solution to NWS in G_p can be transformed in polynomial time to a solution to NWS in G without any increase to the solution's cost, and without affecting the approximation ratio (see, e.g., Theorem 3.2 in [20] which is cast in the context of the metric completion of the graph, but also can be applied in the case of the path completion, even though the latter does not necessarily give rise to metric distances). We can also use the following property, which is a folklore result for the classic Steiner Tree problem but applies to the node-weighted version as well: Let (G_p, K) be the input to NWS, then we can transform any solution (Steiner tree) T to a a tree T' which has total cost at most that of T, and for which the number of Steiner nodes (i.e., vertices which are not terminals) is at most $|K| - 2$. This follows from the observation that we can assume that all Steiner nodes have degree at least three, since Steiner nodes of degree two can be replaced by a single edge in the path completion of G.

Throughout this paper we assume that α is an integer, since we can always scale it to the nearest integer without affecting, asymptotically, the bounds. We assume that G is connected, since otherwise we can restrict the problem to the connected components of G. We will denote by c_{min}, c_{max} the minimum and maximum costs of vertices in V, respectively. Because G is part of the input in both the online and offline versions of the problem, we assume that α is known to the algorithm. Last, we note that all our lower bounds will be presented in terms of NWS, while all our algorithms will be described in the context of the more general GNWS problem.

2   Offline Algorithms

Theorem 1. *Any polynomial-time algorithm for NWS in graphs of asymmetry α has approximation ratio $\Omega(\min\{logk, \log\alpha\})$, unless $P = NP$.*

Proof. We present a polynomial-time reduction of a variation of the set cover problem in which all sets have bounded size, to NWS of bounded asymmetry. Consider an instance I of set cover which consists of a universe of elements, denoted by U and a collection $\mathcal{C}$ of sets, each containing *at most* B elements in U. The objective is to find a collection $C \subseteq \mathcal{C}$ of minimum cardinality such that for every $e \in U$ there exists $S \in C$ such that $e \in U$. The reduction is as follows: G is defined as a graph with a vertex v_S for each set $S \in \mathcal{C}$, and a vertex v_e for every element $e \in U$. Each v_e vertex has weight equal to 1 and each vertex v_s has weight B. All pairs of vertices of the form $v_S, v_{S'}$ are pairwise adjacent; furthermore, v_S and v_e are adjacent if and only if $e \in S$. All edge weights are zero. Last, we define the set of terminals K as the set of all v_e vertices in G (hence $k = |K| = |U|$). Note that this transformation gives rise to an instance I' of NWS in a graph of asymmetry B.

Denote by $OPT(I')$, $A'(I')$ the cost of the optimal algorithm and the cost of an approximation algorithm A' for NWS on the above instance I', respectively. Also denote by C'_{opt}, C' the set of vertices of the form v_s in (any fixed) optimal solution and the solution of A', on input I', respectively. It is easy to see that a Steiner tree for K in G corresponds to a set cover for the instance $(U, \mathcal{C})$, in that the set of vertices C' corresponds to a collection of sets (which we denote by C) which cover U. Define A as the set cover algorithm which, on instance I, selects all sets in C. Since all edges in G have zero weight,

$$A'(I') = k + B \cdot |C'| \quad \text{and} \quad A(I) = |C| = |C'|. \tag{1}$$

Likewise, we have

$$OPT(I') = k + B \cdot |C'_{opt}| \quad \text{and} \quad OPT(I) = |C_{opt}| = |C'_{opt}|. \tag{2}$$

where $OPT(I)$ is the optimal cost for the instance I of set cover. Since both C and C_{opt} are solutions for I, we have

$$B|C'_{opt}| = B|C_{opt}| \geq k \quad \text{and} \quad B|C'| = B|C| \geq k \tag{3}$$

Suppose that NWS is approximable within a factor of $o(\log \alpha) = o(\log B)$, then using (2) and (3) we get

$$A'(I') = o(\log B)OPT(I') = o(\log B)(k + B \cdot |C'_{opt}|)$$
$$= o(\log B)B \cdot |C'_{opt}|,$$

hence from (1) and (2) we get that $|C| = o(log B)|C_{opt}|$, which means that bounded-size set cover is approximable within a factor of $o(\log B)$, which implies that P=NP, by the inapproximability result of Trevisan [19], namely that set cover on instances with sets of size at most B is hard to approximate within a factor of $\ln B - O(\ln \ln B)$ unless $P = NP$. Since $B \leq k$ the result follows.   □

We now present an algorithm which has approximation ratio $O(\min(\log k, \log \alpha))$, thereby matching the lower bound of Theorem 1. We will assume that $k > \alpha$, since otherwise the lower bound of Theorem 1 is already known to be tight. The algorithm is a combination of the Klein-Ravi (KR) and Goemans-Williamson (GW) algorithms. The former works in iterations; in each iteration, a subset of currently *active* trees is merged into a single tree (i.e., a single connected component) by buying a vertex of smallest cost-efficiency, and minimum-cost paths from the vertex to the active trees in question. Here the term "active" reflects the fact that the tree contains terminals for which the requirement function is not satisfied by the current partial solution. More formally, a tree T is active iff there exist vertices $u, u \in V$, with $u \in T$, $v \notin T$ such that $r(v, u) = 1$. In addition, the cost-efficiency of a vertex v is defined as the minimum ratio, over all subsets of active trees, of the cost of v as well as the cost of the minimum-cost paths from v to each active tree in the subset, over the cardinality of the subset. In other words, the cost-efficiency is the minimum average cost paid so as to connect active subtrees by means of paths originating at v.

We emphasize that for the GSP instance (G'_p, r') we treat the graph as if the weights of the vertices do not exist, and that G'_p is the path completion of G'.

Algorithm Offline GNWS on input (G, r)

1 Execute the KR algorithm until at most k/α active trees remain
2 Create a new graph G' by contracting each tree in the partial solution to a single "supernode". Each supernode gets vertex weight zero.
3 Define a new requirements function r' on G' and solve GSP on instance (G'_p, r') using the GW algorithm

---

The requirement function r' is defined in the "natural" way. In particular, at the end of step 1, the set of edges and vertices which have been bought induces a forest Γ, with each tree T in the forest corresponding to a supernode v_T in the vertex set of G'. We define r' for supernodes $u, v \in G'$ to be $r'(u, v) = 1$ if and only if there exist vertices $\tilde{u}, \tilde{v} \in G$ which belong in trees T_1, T_2 in F, respectively, with $T_1 \neq T_2$, such that $r(\tilde{u}, \tilde{v}) = 1$. Informally, r' is determined by all vertices whose connectivity requirement has not been satisfied by the end of step 1. Let OPT denote the optimal solution cost for instance (G, r) of GNWS.

Theorem 2. *Algorithm Offline GNWS has approximation ratio $O(\min(\log k, \log \alpha))$.*

Proof. Consider the penultimate iteration of the KR algorithm in step 1. Since at least k/α active trees remain, the cost of the partial solution maintained up to that iteration, i,e edges and vertices bought by KR excluding terminals[1] is upper bounded by

$$2 \ln \frac{k}{k/a} OPT = O(\ln \alpha) OPT.$$

The above follows by the analysis of the KR algorithm (see Section 4 in [14], in particular the inequality following (4)).

The KR algorithm has the property that in every iteration it connects $q \geq 2$ active trees in a new component at an average cost of at most OPT/q, which implies that the cost of the last iteration in step 1 is bounded by OPT.

It remains to bound the cost due to step 3. Denote by $F = \{T_1, \ldots, T_l\}$ the forest of trees returned by Offline GNWS. First, note that the cost of edges in the optimal solution to the GSP problem on instance (G'_p, r') is bounded by OPT. Since GW is a 2-approximation algorithm for GSP, the cost of the edges in the forest F is at most 2OPT. Next, let k_i denote the number of terminal pairs in tree T_i (here, we stress that the term terminal refers to the instance (G'_p, r'), namely a supernode which corresponds to an active tree at the end of step 1 is considered a single terminal). By the argument in section 1.3 we can assume that the number of Steiner nodes in T_i is bounded by $2k_i - 2 \leq 2k_i$.

---

[1] The analysis of the KR algorithm in [14] makes wlog the assumption that all terminal costs are zero. For this purpose our analysis for step 1 of Offline GNWS does not take into account the cost of terminals; instead we simply add their contribution later in the proof.

Therefore the total weight of Steiner nodes in F is at most

$$\sum_{i=1}^{l} 2k_i \cdot c_{max} \leq \frac{k}{\alpha} c_{max} \leq k \cdot c_{min} \leq OPT.$$

Therefore the total node-weight of the forest F is at most 2OPT, and its total weight at most 4OPT.

Putting everything together, the cost of the solution returned by the algorithm is $O(\log \alpha)OPT$.   □

We note that the above algorithm is applicable, with the same performance guarantees, to a wider class of network design problems which can be formulated as cut-covering problems in which the family of cuts is defined by *proper* functions (see e.g., [10]). This follows from the following two facts: i) for such problems the KR algorithm upholds the properties we used in the proof of Theorem 2 ; and ii) the GW algorithm is still 2-approximation for the edge-weighted version of this problem.

3   Online Algorithms

Theorem 3. *The competitive ratio of deterministic online GNWS is* $\Theta(\max\{\min\{\alpha, k\}, \log k\})$.

Proof. For the lower bound, we consider the online NWS problem (the lower bound carries over to online GNWS). If $\alpha < \log k$, a lower bound of $\Omega(\log k)$ follows by the construction of Imase and Waxman [12]. Otherwise the adversary presents a graph G which is defined as follows: The vertex set of G consists of (disjoint) sets V_1 and V_2, with $|V_2| = k(k+1)/2$. Partition V_2 into $\left(\frac{k(k+1)}{2}\right)$ sets of size k each, and for each set define a vertex which is adjacent to those k vertices: this gives rise to the set V_1. No more vertices or edges exist in G. The weight of each vertex in V_1 is α, and each vertex in V_2 has unit weight, whereas the weight of all edges is zero. The adversary will present a nemesis sequence consisting of vertices in V_2, determined by a game against the algorithm. In the first round of the game, the adversary picks any two vertices in V_2 as the first two terminals in the sequence; the algorithm buys a vertex in V_1 to guarantee connectivity. In each subsequent round, the adversary presents a new terminal, namely a vertex in V_2, which is not adjacent to vertices bought in earlier rounds. The algorithm then must buy a new vertex in V_1 and the adversary repeats the game. Clearly, the game can go on for $k - 1$ rounds. At the end of the game, k terminals have been presented all of which are adjacent to a vertex in V_1, hence $OPT = \alpha + k$. On the other hand, the algorithm has bought $k - 1$ vertices in V_1, hence its cost is at least $\alpha(k - 1)$, and in this case the competitive ratio is $\Omega(\min\{\alpha, k\})$.

For the upper bound, we propose an algorithm, denoted by A, which works in two phases. Let $\bar{k}$ denote the number of pairs presented by the adversary thus far, and k the total number of pairs that will be presented eventually (A does not know k). The first phase is a greedy phase and lasts for as long as $\alpha > \overline{k}$: namely, the algorithm connects the two terminals in the pair by means of a minimum-cost path. The second phase begins at the point where $\overline{k}$ exceeds α, at which point the algorithm switches to the Berman and Coulston [4] (BC) algorithm for (edge-weighted) GSP. More precisely, the algorithm treats all edges bought during the greedy phase as having weight zero (meaning that it already paid for them). In addition, it ignores vertex weights (or alternatively, treats vertices as if they have zero weight). We remind the reader that once again, we work on the path completion graph.

The total cost due to the first phase is clearly $\min\{\alpha, k\}OPT$. For the second phase, since the BC algorithm is $O(\log k)$-competitive for GSP, it follows that the cost of the edges it buys is $O(\log k)OPT$. Observe also that since BC returns a forest of at most k trees, from the discussion in section 1.3 it follows that at most $2k$ vertices of the forest are Steiner vertices. Therefore the total node-cost of the forest is bounded by the quantity $C = 3k \cdot c_{max} \leq 3\alpha k \cdot c_{min}$. However, the latter contribution to the cost is in effect only when $k \geq \alpha$. Given that in such case $OPT \geq kc_{min} \geq \alpha c_{min}$, we have that $C \leq 3\alpha \cdot OPT = 3\min\{\alpha, k\} \cdot OPT$ Therefore, the total cost of A is

$$O(\log k \cdot OPT + \min\{\alpha, k\} \cdot OPT) = O(\max\{\min\{\alpha, k\}, \log k\} \cdot OPT). \qquad \square$$

Note: The lower bound of Theorem 3 can be extended to randomized algorithms, using a somewhat larger graph as part of the adversarial input. We sketch the proof (and omit certain details). Again, the adversarial input consists of a graph on vertices $V_1 \cup V_2$, $|V_1| = \binom{|V_2|}{k}$ and for every set S of k vertices in V_2 there is exactly one vertex in V_1 adjacent to S (and only S). We require that $|V_2| \geq 2k^2$. The edge and vertex weights remain the same. We define a probability distribution D on the sequence of terminals presented to any fixed deterministic online algorithm; from Yao's principle [21], the ratio of the average cost of the algorithm over the average optimal cost is a lower bound on the competitive ratio of every randomized algorithm against an oblivious adversary. In particular D chooses uniformly at random a subset of V_2 of cardinality k, in any order. It follows that every time the deterministic algorithm considers a terminal (except for the very first one), the probability it has already bought a vertex in V_1 which can guarantee connectivity is at most $\frac{k}{2k^2-k^2} = \frac{1}{k}$, and hence the probability that for each of the $k - 1$ terminals the algorithm must buy a new vertex in V_1 is bounded by $(1 - 1/k)^{k-1}$, thus the average cost of the algorithm is $\Omega(\alpha k)$, while the average optimal cost is still $\alpha + k$.

3.1 Randomized Online Algorithms for the General Case (Unbounded Asymmetry)

Theorem 3 suggests that in the case of large asymmetry, namely when $\alpha \in \Omega(k)$, the competitive ratio of any deterministic or randomized algorithm is disappointingly bad. However, the lower bound requires a construction of a graph with a large number of vertices. In fact, in this section we will show how to achieve a better upper bound when the number of vertices in the graph is subexponential on the total number of terminals. In particular, we present and analyze a randomized algorithm based on the framework of Alon *et al.* [3], which addresses broad classes of (edge-weighted) connectivity and cut problems[2]. We show how their approach can be extended to a variant of such problems in which nodes as well as edges are associated with a cost function. In this section we omit proofs due to space limitations, and only focus on how to adapt/modify the ideas of [3].

We will create an online algorithm for GNWS which consists of two distinct components. The first component maintains a fractional solution to the problem, i.e., a feasible weight assignment w for nodes and edges in the graph. Here, a feasible assignment is such that for any request, i.e., a pair of terminals (t_i, t_j) there is a flow from t_i to t_j of value at least 1, assuming that we treat the weights of both edges and vertices as capacities[3]. Note that nodes, and not only edges, are assigned capacities, in the sense that in any feasible flow the in-flow for any node cannot exceed the capacity (weight) of the node. When the algorithm receives a new request, the algorithm will update the weight assignments, by performing an appropriate *weight augmentation*. In particular, the algorithm may increase the weights of certain vertices and/or edges, but it will certainly not decrease any of the currently assigned weights. This guarantees that after each request is processed, a feasible fractional solution can be maintained; moreover this computation is accomplished in an online fashion. Naturally, one seeks a fractional solution of small cost (c.f., Lemma 2).

The second component of the algorithm is responsible for rounding the fractional solution to an integral one, and is performed during each step (i.e., after each request appears). This component is largely orthogonal to the first one, and as demonstrated in [3], randomized rounding can be applied successfully in a variety of connectivity/cut problems. We show how randomized rounding can yield an integral solution of cost within a factor of $O(\log k)$ from the cost of the fractional solution maintained by the first component of the algorithm (c.f., Lemma 3).

We start by describing the first component of the algorithm. Following [3], we can assume that all edges and vertices have costs in the interval $[1, m^2]$ (here we are also using the assumption that the graph is connected and hence $m \geq n - 1$). Initially the algorithm gives each edge and vertex a fractional weight of $1/m^3$. Suppose that a new request (t_1, t_2) arrives; the algorithm will then update the weight assignment (potentially increasing certain weights, but not decreasing any of them). More specifically, if the maximum flow from t_1 to t_2 is at least 1, we do nothing; otherwise we perform a weight augmentation step as follows. Find a minimum-weight $t_1 - t_2$ cut C, defined as a partition of V in sets S and $V \setminus S$ which separates t_1 and t_2. For clarity, we emphasize that the weight of C is

---

[2] [3] is focused on edge-weighted graphs, and the authors claim that the techniques can be extended to the vertex counterparts of the problems in which only vertices have costs. Here, we extend their major result to the case where both vertices and edges are associated with costs.

[3] The weights w should not be confused with the costs c of edges and vertices.

the sum of weights of edges in the cut. Denote by S_1 and S_2 the subsets of V and $V \setminus S$, respectively, for which each vertex in S_1 (resp. S_2) is incident with at least one edge in C. For every edge $e = (v, u) \in C$ let $\delta_e = w(e)/\max\{c(e), c(v), c(u)\}$. We increase $w(e)$ by $\delta(e)$. In addition, for every vertex $v \in S_1$ (resp. $u \in S_2$) we increase $w(v)$ (resp. $w(u)$) by $\sum_{e=(v,u):e \in C} \delta_e$ (resp. $\sum_{e=(v,u):e \in C} \delta_e$). We repeat the process, with a new weight augmentation step, until the maximum flow from t_1 to t_2 is at least 1. At an intuitive level, the augmentation is "balanced" in the sense that we increase the weights of certain vertices only as much as it is needed, and a vertex weight is increased proportionally to the weight increase of incident edges in the cut.

Note that at the end of the weight augmentation step associated with C the total edge-weight of C does not exceed the total vertex-weight of either S_1 or S_2. This guarantees that when the algorithm terminates, all connectivity demands are satisfied.

Lemma 1. *The number of weight augmentations steps performed by the algorithm is $O(\log m) \cdot OPT_f$, where OPT_f denotes the cost of the optimal fractional solution.*

Lemma 2. *The algorithm is $O(\log m)$ competitive for fractional node-weighted connectivity problems.*

The second part of the algorithm involves rounding the feasible fractional solution (which is maintained at each step) in an online fashion. Recall that when a request (t_1, t_2) appears we first compute a fractional, $O(\log m)$-competitive solution using the algorithm described earlier. The rounding method is simple and is based on the rounding employed in the context of the multicast problem in [3]. More specifically, the rounding method involves keeping $2\lceil \log(k'+1) \rceil$ independent random variables X_i distributed uniformly in the interval $[0, 1]$. Here, k' denotes the number of terminal pairs served by the online algorithm up to the current point, which implies that the number of the random variables increases as the algorithm serves more and more requests. Define the threshold θ as $\min_{j=1}^{2\lceil \log(k'+1) \rceil} \{X_j\}$. The algorithm will then update its current integral solution I by adding in I all edges e and vertices v of the graph with the property that $w(e) \geq \theta$ and $w(v) \geq \theta$, i.e., all vertices and edges whose weight exceeds the current value of the threshold (initially $I = \emptyset$).

The following Lemma follows easily using the ideas of Lemma 4.1 in [3]. The only change in the proof is that we have to account for the cost of the vertices added in the solution (and not only the edges), but essentially the proof remains the same.

Lemma 3. *Throughout the execution of the algorithm, the expected cost of the solution which the algorithm maintains is $O(\log k' \log m OPT)$, where OPT is the cost of the optimal integral solution so far. Furthermore, for any terminal pair $T = (t_1, t_2)$ the probability that the algorithm does not allocate a path that connects t_1 and t_2 is small, namely $1/k'^2$.*

Lemma 2 and Lemma 3 imply that the randomized algorithm is $O(\log k \log m)$-competitive (for integral solutions), with the condition that if the algorithm fails to allocate a path, a path has to be bought explicitly (since this happens infrequently, the overhead to the total cost is negligible).

4   Concluding Remarks

In this paper we studied the Steiner Tree Problem and Generalized Steiner Problem at the presence of node and edge weights, and presented upper and lower bounds on the performance of online and offline algorithms as a function of the node asymmetry α. Of course, as one may suggest, there are several possible ways to capture the variation of node weights in the graph; for instance, one may instead define α as $\max_{v,u \in V}\{|c(v) - c(u)|\}$. Or rather we could insists that the cost of a vertex v is within a factor of at most α from the cost of vertices adjacent to it, as well as edges incident with it. The latter definition takes into account, at least in a certain limited way, the cost of edges and is not biased towards the cost of nodes. It is easy to show that the upper and lower bounds shown in sections 2 and 3 carry over to the above definitions of asymmetry[4].

As argued in the introduction, GNWS reduces to a related variant of the directed Steiner problem, in which the asymmetry β of edge costs, as defined in [8], does not exceed α. Thus it would be very interesting to derive better upper bounds for the edge-asymmetric Steiner tree problem. In particular, can we show that the sublinear (in terms of k) upper bound of [6] for offline Steiner trees and forests in directed graphs can be improved assuming a known bound on the edge asymmetry? Can we get $o(\beta)$ approximation algorithms? The online version of this problem has been the topic of [8], but there still exists a gap between the known upper and lower bounds on the competitive ratio.

Last, since the asymmetric node-weighted and directed Steiner problems are related, we would like to study the combination of the two, namely the problem in which bounds on both vertex and edge asymmetry are known. Such a model would probably capture in a more realistic way practical network design problems.

References

1. A. Agrawal, P. N. Klein, and R. Ravi. When trees collide: An approximation algorithm for the generalized steiner tree problem on networks. *SIAM Journal on Computing*, 24:440–456, 1995.
2. N. Alon, B. Awerbuch, Y. Azar, N. Buchbinder, and J. Naor. The online set cover problem. In *Proceedings of the 35th annual ACM Symposium on the Theory of Computation*, pages 100–105, 2003.
3. N. Alon, B. Awerbuch, Y. Azar, N. Buchbinder, and J. Naor. A general approach to online network optimization problems. In *Proceedings of the 15th Annual ACM-SIAM Symposium on Discrete Algorithms (SODA)*, pages 570–579, 2005.

---

[4] If we define α as $\max_{v,u \in V}\{|c(v) - c(u)|\}$, it is convenient to assume that all costs are at least 1, otherwise a trivial bound of $\Theta(k)$ can be shown for the online version.

4. P. Berman and C. Coulston. Online algorithms for Steiner tree problems. In *Proceedings of the Twenty-Ninth Annual ACM Symposium on the Theory of Computing*, pages 344–353, 1997.

5. M. Bern and P. Plassmann. The Steiner problem with edge lengths 1 and 2. *Information Processing Letters*, 32:171–176, 1989.

6. M. Charikar, C. Chekuri, T. Cheung, Z. Dai, A. Goel, S. Guha, and M. Li. Approximation algorithms for directed steiner problems. *Journal of Algorithms*, 1(33):73–91, 1999.

7. M. Faloutsos. *The Greedy the Naive and the Optimal Multicast Routing–From Theory to Internet Protocols*. PhD thesis, University of Toronto, 1998.

8. M. Faloutsos, R.Pankaj, and K. C. Sevcik. The effect of asymmetry on the on-line multicast routing problem. *Int. J Found. Comput. Sci.*, 13(6).889–910, 2002.

9. U. Feige. A threshold of ln n for approximating set cover. *Journal of the ACM*, 45(4):634–652, 1998.

10. M.X. Goemans and D.P. Williamson. A general approximation technique for constrained forest problems. *SIAM Journal on Computing*, 6(24), 1995.

11. S. Guha and S. Khuller. Improved methods for approximating node weighted steiner trees and connected dominating sets. *Information and Computation*, (150):228–248, 1999.

12. M. Imase and B. Waxman. The dynamic Steiner tree problem. *SIAM Journal on Discrte Mathematics*, 4(3):369–384, 1991.

13. R. Karp. Reducibility among combinatorial problems. *Complexity of Computer Computations*, pages 85–103, 1972.

14. P. Klein and R. Ravi. A nearly best-possible approximation algorithm for node-weighted steiner trees. *Journal of Algorithms*, (19):104–115, 1995.

15. R. Raz and S. Safra. A sub-constant error-probability low-degree test, and a sub-constant error-probability PCP characterization of NP. In *Proceedings of the twenty-ninth annual ACM Symposium on Theory of Computing*, pages 475–484, 1997.

16. G. Robins and A. Zelikovsky. Improved Steiner tree approximation in graphs. In *Proceedings of the eleventh annual ACM-SIAM symposium on Discrete algorithms*, pages 770–779, 2000.

17. A. Segev. The node-weighted steiner tree problem. *Networks*, (17):1–17, 1987.

18. M. Thimm. On the approximability of the Steiner tree problem. *Theoretical Computer Science*, 295(1):387–402, 2003.

19. L. Trevisan. Non-approximability results for optimization problems on bounded degree instances. In *Proceedings of the thirty-third annual ACM Symposium on Theory of Computing*, pages 453–461, 2001.

20. V. Vazirani. *Approximation Algorithms*. Springer, 2001.

21. A. C.-C. Yao. Probabilistic computations:towards a unified measure of complexity. In *In Proceedings of the 17th Annual IEEE Symposium on Foundations of Computer Science*, pages 222–227, 1997.

On Guarding Rectilinear Domains

Matthew J. Katz* and Gabriel S. Roisman**

Department of Computer Science
Ben-Gurion University of the Negev, Beer-Sheva 84105, Israel
{matya, roismang}@cs.bgu.ac.il

Abstract. We prove that guarding the vertices of a rectilinear polygon P, whether by guards lying at vertices of P, or by guards lying on the boundary of P, or by guards lying anywhere in P, is NP-hard. For the first two proofs (i.e., vertex guards and boundary guards), we construct a reduction from minimum piercing of 2-intervals. The third proof is somewhat simpler; it is obtained by adapting a known reduction from minimum line cover.

We also consider the problem of guarding the vertices of a 1.5D rectilinear terrain by vertex guards. We establish an interesting connection between this problem and the problem of computing a minimum clique cover in chordal graphs. This connection yields a 2-approximation algorithm for the guarding problem.

1 Introduction

Problems dealing with visibility coverage are often called *art-gallery problems*. The "classical" art-gallery problem is to place guards in a polygonal region, such that every point in the region is visible to one (or more) of the guards. More formally, given a domain P, one needs to find a set $\mathcal{G}$ of points in P, of minimum cardinality, such that every point in P is seen by at least one of the points, called *guards*, in $\mathcal{G}$. Often there are some restrictions on the location of the guards; e.g., guards may lie only at vertices (in which case they are called *vertex guards*).

The classical art-gallery problem, where guards may lie anywhere in the polygon or only at vertices, is known to be NP-hard, even if the underlying domain is a simple polygon [24, 19, 1]. Moreover, Eidenbenz et al. [11, 12] have shown that these problems are APX-hard. Schuchardt and Hecker [26] proved that these problems remain NP-hard if we restrict our attention to (simple) rectilinear polygons. Their proof is based on a reduction from 3SAT.

In this paper we study two art-gallery problems. The first is the problem of guarding the *vertices* of a rectilinear polygon (GVRP) P. We consider three versions of this problem. In the first version guards may lie anywhere on the boundary of P but not in the interior of P, in the second version guards may lie only at vertices of P, and in the third version guards may lie anywhere in P. We prove

---

* Partially supported by grant No. 2000160 from the U.S.-Israel Binational Science Foundation.
** Partially supported by the Lynn and William Frankel Center for Computer Sciences.

L. Arge and R. Freivalds (Eds.): SWAT 2006, LNCS 4059, pp. 220–231, 2006.

that all three versions remain NP-hard. For the first two proofs (i.e., boundary guards and vertex guards), we construct a reduction from minimum piercing of 2-intervals, where a 2-interval is the union of two disjoint line-segments on the real line. For the third proof, we construct a reduction from minimum line cover. (Note that minimum line cover has been used previously in hardness proofs for art-gallery problems by, e.g., Brodén et al. [4] and Joseph Mitchell. However, in order to use it in our setting, one needs sophisticated gadgets.)

The second problem that we study is that of guarding the vertices of a 1.5D rectilinear terrain by vertex guards. (A 1.5D rectilinear terrain is defined by an x-monotone chain T of horizontal and vertical line segments; two vertices u, v of T see each other, if the line segment $\overline{uv}$ does not pass below T.) This problem arises when one wants, e.g., to place security cameras along a wall. We establish an interesting connection between this problem and the problem of computing a minimum clique cover in chordal graphs (see below for the definition of chordal graph). This connection yields a 2-approximation algorithm for the guarding problem.

Recently Ben-Moshe et al. [2] presented a constant-factor approximation algorithm for computing a set of vertex guards for a 1.5D terrain that is defined by a *strictly* x-monotone polygonal chain. Their algorithm, however, cannot be applied (at least not immediately) to a 1.5D rectilinear terrain, since strict x-monotonicity is necessary at several places in their work. Moreover, the constant of approximation of their algorithm, as well as of the subsequent, purely theoretical, algorithm of Clarkson and Varadarajan [8], is big. We also note that the idea of using perfect graph theory in the context of guarding is not new; see, e.g., [21].

More Related Work. Combinatorial art-gallery problems have been studied for three decades; see, e.g., [23, 17, 25, 27] for surveys. The classical combinatorial result, the "art gallery theorem", states that $\lfloor n/3 \rfloor$ guards are sufficient and sometimes necessary to guard an n-vertex simple polygon [7]. Combinatorial results on the number of guards needed for various forms of guarding on terrains are given in [3].

Researches have mostly concentrated on obtaining good approximations. Ghosh [15] gave an $O(\log n)$-approximation for optimal guarding of a polygon by vertex guards, based on standard set cover results. Recent work [10, 16] has focused on methods that efficiently apply the Brönnimann-Goodrich technique [5]. Efrat and Har-Peled [10] obtain an $O(\log k^*)$-approximation algorithm for simple polygon guarding with vertex guards, using time $O(n(k^*)^2 \log^4 n)$, where k^* is the optimal number of vertex guards. Their technique can be applied to non-vertex guards, lying at points of a dense grid, adding a factor polylogarithmic in the grid density to the time bound. (No approximation algorithm is known if the guards are completely unrestricted and every point in the polygon must be guarded.) Their results apply also to polygons with holes and to 2.5D terrains, still with polylogarithmic approximation factors.

Very recently, Nilsson [22] presented a constant-factor approximation algorithm for guarding a monotone polygon. Using this algorithm, he also obtains an $O((c^*)^2)$-algorithm for guarding a rectilinear polygon, where c^* is the size of an optimal guarding set.

Finally, for 1.5D terrains (i.e., for an x-monotone polygonal chain), Chen et al. [6] claim that by modifying the hardness proof of [24] one can show that the problem is NP-hard; details are omitted and are still to be verified.

2   GVRP Is NP-Hard

In this section we show that all three versions of GVRP are NP-Hard. We begin with the version where guards may lie only on the boundary of the polygon.

2.1   Guards May Lie Only on the Boundary

We show that if the guards are restricted to lie on the boundary of the polygon, then GVRP is NP-hard. We construct a reduction from minimum piercing of 2-intervals.

The 2-Interval Piercing Problem. A 2-interval o is the union of two line-segments t_a and t_b on the x-axis, that can be separated by a vertical slab of constant width c_0. The minimum 2-interval piercing problem is defined as follows. Let O be a set of n 2-intervals. Find a set $\mathcal{P}$ of points on the x-axis, such that (i) for each 2-interval $o \in O$ there exists a point $p \in \mathcal{P}$ that pierces o (i.e., that lies in o), and (ii) $\mathcal{P}$ is as small as possible. Let D2IP denote the corresponding decision problem, that is, given O and an integer $k > 0$, decide whether there exists a piercing set for O of cardinality k. In the full version of this paper we show that D2IP is NP-hard, although we suspect that it is well known.

Reduction from D2IP. We first present the gadget that we shall use. We call it *d-gadget* (short for double gadget), see Fig. 1. Any guard below the line l is *local*. Some of the vertices of a d-gadget can only be guarded by a local guard (e.g., vertices x, y, and z). It is easy to see that in order to guard all these vertices one needs at least 3 local guards. However, any 3 local guards that

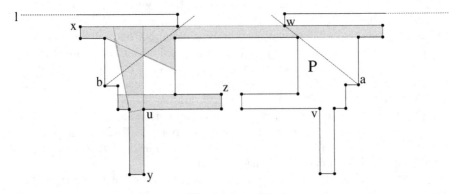

Fig. 1. A d-gadget

guard all these vertices cannot see both a and b. Moreover, one can locate 3 local guards on the boundary of a d-gadget, such that all the vertices of the d-gadget are guarded except for either a or b. (E.g., locate 3 guards at the vertices u, v, and w, respectively.) Thus, another guard is required in order to guard all the vertices of a d-gadget. This guard does not have to be local; it can lie anywhere on the portion of the boundary of the polygon that is seen from the unguarded vertex a or b.

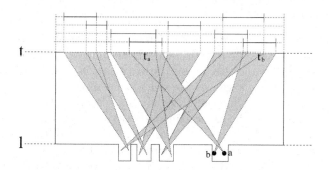

Fig. 2. The rectilinear polygon P. Each of the "holes" on the bottom represents a d-gadget.

We define a reduction function f from D2IP to GVRP with boundary guards. Given an instance $\{O, k\}$ of D2IP, f constructs a rectilinear polygon P, such that the vertices of P can be guarded by $3|O| + k$ boundary guards if and only if there is a piercing set for O of size k. In particular, f constructs a rectilinear polygon P with $|O|$ d-gadgets (see Fig. 2). The length of the top edge t of P is determined by the 2-intervals in O. For each 2-interval $o \in O$, $o = \{t_a, t_b\}$, f

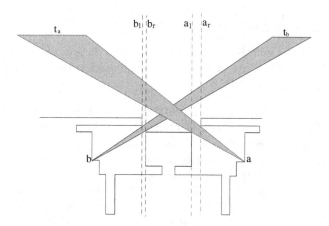

Fig. 3. The portion of t visible from a (resp. b) can be adjusted by setting the auxiliary lines a_l and a_r (resp. b_l and b_r)

constructs a d-gadget g below the line l. f locates g and adjusts it, so that the vertex a (resp. b) is (boundary) seen from outside of g by any point on t_a (resp. t_b) and only by these points. Figure 3 shows such a construction. The portion of t that is visible from vertex a (resp. b) can be controlled by setting the auxiliary lines a_l and a_r (resp. b_l and b_r). (Recall that the distance between t_a and t_b is at least c_0.)

Lemma 1. *The rectilinear polygon P that is obtained can be guarded by $3|O|+k$ boundary guards if and only if there exists a piercing set for O of size k.*

Proof. Assume first that there exists a piercing set for O of size k. We describe how to guard the vertices of P with $3|O| + k$ boundary guards. For each point p in the piercing set, we locate a guard at p (which is of course on t). By the construction above, these k guards see at least one of the vertices g_a, g_b in each of the $|O|$ d-gadgets. In addition, these guards see all vertices of P that are not below the line l. Finally, as explained above, one can locate, in each of the d-gadgets, 3 local (boundary) guards that see all the rest of the vertices of this d-gadget. Hence, the total number of guards is $3|O| + k$.

Assume now that the vertices of P can be guarded by $3|O| + k$ boundary guards. We show a piercing set for O of size k. As we argued above, each d-gadget requires at least 3 local guards. For each d-gadget g that is guarded by more than 3 local guards, we replace these local guards by 3 local boundary guards that see all the vertices of g except for the vertex g_a, and by a guard in t_a. Hence, we have at most k guards that are located on the top edge t. These guards constitute a piercing set for O, since, for each d-gadget g, at least one of the vertices g_a, g_b is seen by a guard on t. In other words, for each 2-interval $o \in O$, there is a guard on t that lies in o.

The following theorem summarizes the result of this subsection.

Theorem 1. GVRP *with boundary vertices is NP-hard.*

2.2   Guards May Lie Only at Vertices

We show that if the guards are restricted to lie at vertices of the polygon, then GVRP remains NP-hard. As in the previous subsection (boundary guards), we construct a reduction from D2IP.

Reduction form D2IP. In addition to d-gadgets, we shall also use *ear gadgets*, see Fig. 4. The vertices of an ear gadget can be guarded by a single guard that is located in the shaded rectangle (e.g., by a guard that lies at one of the vertices u or v). Moreover, any set of guards that sees all the vertices of an ear gadget must include a guard in the shaded rectangle.

We define a reduction function f from D2IP to GVRP with vertex guards. Given an instance $\{O, k\}$ of D2IP, f constructs a rectilinear polygon P, such that the vertices of P can be guarded by $m + 3|O| + k$ vertex guards if and only if there is a piercing set for O of size k, where $m \leq 2|O|$ is the number of different right endpoints of the line segments corresponding to the 2-intervals in O.

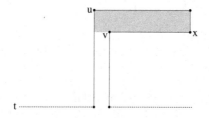

Fig. 4. An ear gadget

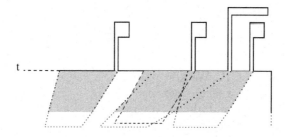

Fig. 5. An ear gadget is attached at each right endpoint of a segment on t

f constructs a rectilinear polygon with $|O|$ d-gadgets, as in Sect. 2.1. In addition (see Fig. 5), f attaches an ear gadget at each right endpoint of a line segment on t (i.e., at each right endpoint of a line segment of a 2-interval in O).

Observation. Let x be any point on t that pierces a subset of the line segments (corresponding to a subset of O). Then, we may move x to the first vertex to its right (which is a vertex of an ear gadget), without exiting any of the segments in the subset.

Lemma 2. *The rectilinear polygon P that is obtained can be guarded by $m + 3|O| + k$ vertex guards if and only if there exists a piercing set for O of size k.*

Proof. Assume first that there exists a piercing set for O of size k. One can locate $3|O| + k$ guards, as described in the proof of Lemma 1, such that these guards see all vertices of P except for two or more vertices in each of the ear gadgets. The $3|O|$ local guards can be placed at vertices. Let p be one of the k guards. According to the observation above, we can move p to the first vertex to its right without "losing" any of the vertices g_a, g_b that it sees. Thus, by placing m additional guards, one per ear gadget, we obtain a set of vertex guards that sees all the vertices of P.

Assume now that the vertices of P can be guarded by $m + 3|O| + k$ vertex guards. We have at least one guard in each ear gadget that cannot see any vertex below the line l. Hence, as explained in Lemma 1, we have a piercing set for O of size k.

The following theorem summarizes the result of this subsection.

Theorem 2. GVRP *with vertex guards is NP-hard.*

2.3 Guards May Lie Anywhere

We show that if the guards may lie anywhere in the polygon, i.e., both in the interior and on the boundary, then GVRP is NP-hard. We construct a reduction from the minimum line cover problem (MLCP).

The Minimum Line Cover Problem. The minimum line cover problem is defined as follows. Let $\mathcal{L} = \{l_1, ..., l_n\}$ be a set of n lines in the plane. Find a set $\mathcal{P}$ of points, such that for each line $l \in \mathcal{L}$ there is a point in $\mathcal{P}$ that lies on l, and $\mathcal{P}$ is as small as possible. Let DLCP denote the corresponding decision problem, that is, given $\mathcal{L}$ and an integer $k > 0$, decide whether there exists a cover of size k. DLCP is known to be NP-hard [20]. Moreover, MLCP was shown to be APX-hard [4, 18].

Reduction from DLCP. We first present the gadget that we shall use. We call it *s-gadget* (short for single gadget), see Fig. 6. Some of the vertices of a s-gadget (e.g., vertices x and y) can only be guarded by a local guard (i.e., by a guard below the line l through the two top vertices in Fig. 6). It is easy to see that in order to guard all these vertices one needs at least one local guard, and any single local guard that sees all these vertices cannot see a.

We define a reduction function f from DLCP to GVRP with guards anywhere. Given an instance $\{\mathcal{L}, k\}$ of DLCP, f constructs a rectilinear polygon P, such that the vertices of P can be guarded by $n + k$ guards if and only if there is a cover for $\mathcal{L}$ of size k. Let R be a large enough rectangle, such that all the vertices of the arrangement of $\mathcal{L}$ lie in the interior of R. For each line $l \in \mathcal{L}$, f constructs a s-gadget g at one of the endpoints of the line segment $l \cap R$, in such a way that the vertex a of g can be guarded from outside g only from points on $l \cap R$, see Fig. 7. Let P be the rectilinear polygon that is obtained.

Lemma 3. *The rectilinear polygon P that is obtained can be guarded by $n + k$ guards if and only if there is a cover for $\mathcal{L}$ of size k.*

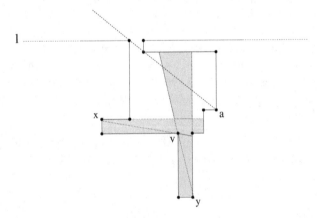

Fig. 6. A s-gadget

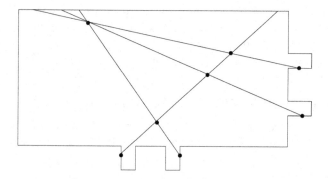

Fig. 7. The rectilinear polygon P. Each of the "holes" on the bottom and on the right represents a s-gadget.

Proof. Follows immediately from the construction above.

The following theorem summarizes the result of this subsection.

Theorem 3. GVRP *(with guards anywhere) is NP-hard.*

3   Guarding the Vertices of a 1.5D Rectilinear Terrain

A *1.5D terrain* (or simply, a *terrain*) T is a polygonal chain specified by n vertices $V(T) = \{v_1, \ldots, v_i = (x_i, y_i), \ldots, v_n\}$, such that $x_i \leq x_{i+1}$ (often strict monotonicity is assumed). The vertices induce $n-1$ edges $E(T) = \{e_1, \ldots, e_i = (v_i, v_{i+1}), \ldots, e_{n-1}\}$. Let $p = (p_x, p_y)$ and $q = (q_x, q_y)$ be two points on T. We say that p *sees* q (and q *sees* p) if the line segment $\overline{pq}$ lies above T, or, more precisely, does not intersect the open region that is bounded from above by T and from the left and right by the downwards vertical rays emanating from v_1 and v_n.

A terrain T is a *1.5D rectilinear terrain* (or in short, a *r-terrain*) if each edge $e \in E(T)$ is either horizontal or vertical. A vertex v_i of a r-terrain T is *convex* (resp. *reflex*) if the angle formed by the edges e_{i-1} and e_i above T is of 90 (resp. 270) degrees. In r-terrains, we distinguish between two types of convex vertices — left convex and right convex. A convex vertex is left (resp. right) convex if

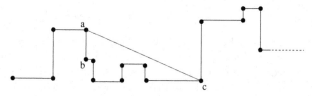

Fig. 8. A r-terrain; a is reflex, b is left convex, and c is right convex; a and c see each other, but b and c do not

e_{i-1} (resp. e_i) is vertical. We denote the set of left convex vertices by $V_{lc}(T)$ and the set of right convex vertices by $V_{rc}(T)$. For example, in Fig. 8 vertex a is reflex, vertex b is left convex, and c is right convex.

A subset of vertices $\mathcal{G} \subseteq V(T)$ *guards* a subset of vertices $V' \subseteq V(T)$ if each of the vertices in V' is seen by at least of one the vertices (guards) in $\mathcal{G}$.

3.1   Some Properties of Terrains and r-Terrains

In this section we explore some of the geometric properties of terrains and r-terrains. The following claim was stated and proved in [2].

Claim 1. Let a,b,c and d be four points on a terrain T, such that $a_x < b_x < c_x < d_x$, where q_x is the x-coordinate of point q. If a sees c and b sees d, then a sees d.

One of the main differences between terrains and r-terrains is presented in the following trivial claim.

Claim 2. Let T be a r-terrain, $v \in V_{rc}(T)$, and p a point on T. If p sees v, then $p_x \leq v_x$.

Clearly this is false for general terrains. Other unique properties of r-terrains are stated below.

Claim 3. If $\mathcal{G} \subseteq V(T)$ guards a subset $V' \subseteq V(T)$, then there exists a subset $\widehat{\mathcal{G}} \subseteq V(T)$ of reflex vertices, such that $\widehat{\mathcal{G}}$ guards V' and $|\widehat{\mathcal{G}}| \leq |\mathcal{G}|$.

Proof. Replace each left convex vertex in $\mathcal{G}$ by the vertex immediately preceding it in $V(T)$, and replace each right convex vertex in $\mathcal{G}$ by the vertex immediately succeeding it in $V(T)$.

Claim 4. If $\mathcal{G} \subseteq V(T)$ guards all the convex vertices of a r-terrain T (i.e., $\mathcal{G}$ guards the set $V_{lc}(T) \cup V_{rc}(T)$), then $\mathcal{G}$ guards all the vertices of T (and all the vertical edges of T).

Proof. Let $v \in V(T)$ be a reflex vertex. Then, at least one of its two neighboring vertices u must be convex. It is easy to see that the guard in $\mathcal{G}$ that sees u must also see v (and the vertical edge (u, v)).

Lemma 4. *Let u, v and w be three right convex vertices of a r-terrain T, such that $u_x < v_x < w_x$. If there exist two vertices $g_1, g_2 \in V(T)$, such that g_1 sees both u and v and g_2 sees both u and w, then there exists a vertex that sees all three vertices u,v and w. Moreover, the one between g_1 and g_2 that precedes the other in the sequence of vertices defining T is such a vertex.*

Proof. We first show that if g_1 lies to the left of g_2, then g_1 sees w. Consider the four vertices g_1, g_2, v and w. We know that $g_{1_x} < g_{2_x} < v_x < w_x$. Since g_1 sees v and g_2 sees w, we conclude by Claim 1 that g_1 also sees w. Assume now that g_2 lies to the left of g_1. If g_1 lies to the left of u, then we may conclude that g_2 also sees v, again by Claim 1. If, however, g_1 lies directly above u, then the vertices u, g_1, and v are necessarily consecutive in the sequence of vertices defining T, and it is easy to see that in this case g_2 must also see v. Finally, if $g_{1_x} = g_{2_x}$, then the higher of the two (that is also the one that precedes the other in the sequence of vertices defining T) also sees the third vertex.

3.2 Guarding the Vertices of a r-Terrain

Let T be a r-terrain. We present an algorithm that computes a set of (vertex) guards $\mathcal{G} \subseteq V(T)$ for $V(T)$ (i.e., each vertex in $V(T)$ is seen by a guard in $\mathcal{G}$), and prove that $\mathcal{G}$ is a 2-approximation, that is, $|\mathcal{G}| \leq 2m$, where m is the size of an optimal set of (vertex) guards for $V(T)$.

The algorithm computes optimal guard sets $\mathcal{G}_r$ for $V_{rc}(T)$ and $\mathcal{G}_l$ for $V_{lc}(T)$, and then outputs the set $\mathcal{G} = \mathcal{G}_r \cup \mathcal{G}_l$. According to Claim 4, $\mathcal{G}$ is a guard set for $V(T)$. Moreover, since $|\mathcal{G}_r|, |\mathcal{G}_l| \leq m$, $\mathcal{G}$ is a 2-approximation.

It remains to describe how to compute an optimal guard set for $V_{rc}(T)$ (alternatively, $V_{lc}(T)$). Although the final algorithm for computing such a guard set is simple and reminiscent of one of the base-case algorithms in [2], it is interesting, since it is the product of a connection that we discover between the problem of computing an optimal guard set for $V_{rc}(T)$ and the problem of computing a minimum clique cover for an appropriate chordal graph.

Several definitions are needed before we can proceed. A graph $G = (V, E)$ is *chordal* if every cycle of length four or more has a *chord*, that is, an edge that joins two non-consecutive vertices of the cycle. A *clique cover* of G is a collection $V_1, \ldots, V_k$ of subsets of V, such that each of them induces a complete subgraph of G (i.e., a clique) and $V_1 \cup \cdots \cup V_k = V$. In general, the minimum clique cover problem (i.e., compute a clique cover of minimum cardinality) is NP-hard [13]. However, if G is chordal, then a minimum clique cover can be found in polynomial time [14].

We now construct a graph G_r over the vertex set $V_{rc}(T)$. Draw an edge between two vertices $u, v \in V_{rc}(T)$ if and only if there exists a vertex $g \in V(T)$ that sees both u and v. Next we claim that G_r is chordal.

Lemma 5. G_r *is chordal.*

Proof. Let $C = \{v_{i_1}, \ldots, v_{i_k}\}$ be a cycle of length at least four in G_r. Let v be the leftmost vertex in C and let $v', v'' \in C$ be its two adjacent vertices in the cycle. We know that there exists a vertex guard g_1 that sees both v and v', and a vertex guard g_2 that sees both v and v''. Moreover, since v is right convex (see Claim 2), g_1 and g_2 cannot lie to the right of v. Therefore, by Lemma 4, there exists a vertex guard g that sees all three vertices v, v', v'', implying that C has a chord, namely, there exists an edge in G_r between v' and v''.

The following lemma, together with the fact that a minimum clique cover of G_r can be computed in polynomial time, implies a polynomial time algorithm for computing an optimal guard set for $V_{rc}(T)$.

Lemma 6. *A subset $\widehat{V}$ of $V_{rc}(T)$ induces a clique of G_r if and only if there exists a vertex guard in $V(T)$ that sees all the vertices in $\widehat{V}$.*

Proof. If $u \in V(T)$ sees all the vertices in a subset $\widehat{V}$ of $V_{rc}(T)$, then, by the definition of G_r, $\widehat{V}$ induces a clique of G_r. Assume now that $\widehat{V} \subseteq V_{rc}(T)$ induces a clique of G_r. If $|\widehat{V}| = 2$, then, by definition, there exists a vertex in $V(T)$ that sees both vertices in $\widehat{V}$. Assume therefore that $|\widehat{V}| \geq 3$. Let u be the leftmost vertex, in the sequence of vertices defining T, that sees both the leftmost vertex v_1 in $\widehat{V}$ and another vertex v_i in $\widehat{V}$. Let v_j be any other vertex in $\widehat{V}$. Then since $\widehat{V}$ induces a clique, there must be a vertex $u' \in V(T)$ that sees both v_1 and v_j. According to Lemma 4 u must also see v_j.

A Direct Algorithm for Computing $\mathcal{G}_r$. The algorithm for computing a minimum clique cover of a chordal graph [14], is based on the following two properties of chordal graphs. (i) Every chordal graph has a *simplicial vertex*; i.e., a vertex v whose set of adjacent vertices forms a clique in the graph [9]. (ii) An induced subgraph of a chordal graph is chordal. Thus, given a chordal graph G, one can compute a minimum clique cover by repeating the following step until done: Find a simplicial vertex in the current subgraph (initially G), and remove it and its adjacent vertices from the subgraph.

Let S be the set of simplicial vertices that were found during the execution of the algorithm. On the one hand, S is an independent set of vertices, hence, a minimum clique cover of G is of size at least $|S|$. On the other hand, each of the $|S|$ subsets that were removed during the execution of the algorithm forms a clique in G. Thus, these subsets constitute a minimum clique cover of G.

We now describe a direct algorithm for computing $\mathcal{G}_r$, an optimal guard set for $V_{rc}(T)$. Let v be the leftmost vertex in $V_{rc}(T)$, and let $C_v \subset V_{rc}(T)$ be the subset of vertices w for which there exists a guard in $V(T)$ that sees both v and w. It follows (similar to the proof of Lemma 6) that there exists a single guard in $V(T)$ that sees all the vertices in $\{v\} \cup C_v$. We thus find such a guard u, and repeat the above step for the remaining unguarded vertices in $V_{rc}(T)$. Let L be the set of left vertices that were found during the execution of the algorithm. Then, as in the algorithm for computing a minimum clique cover, at least $|L|$ guards are required to guard $V_{rc}(T)$, and since the algorithm finds exactly $|L|$ guards, it is optimal.

Finally, it is easy to implement the above guarding algorithm in $O(n^2)$ time. The following theorem summarizes the result of this subsection.

Theorem 4. *Let T be a 1.5D rectilinear terrain. One can compute in $O(n^2)$ time a set of vertex guards $\mathcal{G}$ for $V(T)$ (and for all vertical edges of T), such that $|\mathcal{G}| \leq 2m$, where m is the size of an optimal set of vertex guards for $V(T)$.*

References

1. A. Aggarwal. *The Art Gallery Theorem: Its Variations, Applications and Algorithmic Aspects.* Ph.D. thesis, Johns Hopkins University, 1984.
2. B. Ben-Moshe, M. J. Katz, and J. S. B. Mitchell. A constant-factor approximation algorithm for optimal terrain guarding. In *Proc. 16th ACM-SIAM Sympos. on Discrete Algorithms*, pages 515–524, 2005.
3. P. Bose, T. Shermer, G. Toussaint, and B. Zhu. Guarding polyhedral terrains. *Comput. Geom. Theory Appl.*, 7:173–185, 1997.
4. B. Brodén, M. Hammar, and B. J. Nilsson. Guarding lines and 2-link polygons is APX-hard. In *Proc. 13th Canad. Conf. Comput. Geom.*, pages 45–48, 2001.

5. H. Brönnimann and M. T. Goodrich. Almost optimal set covers in finite VC-dimension. *Discrete Comput. Geom.*, 14:263–279, 1995.
6. D. Z. Chen, V. Estivill-Castro, and J. Urrutia. Optimal guarding of polygons and monotone chains. In *Proc. 7th Canad. Conf. Comput. Geom.*, pages 133–138, 1995.
7. V. Chvátal. A combinatorial theorem in plane geometry. *Combinatorial Theory Series B* 18:39–41, 1975.
8. K. L. Clarkson and K. R. Varadarajan. Improved approximation algorithms for geometric set cover. In *Proc. 21st Annu. ACM Sympos. Comput. Geom.*, pages 135–141, 2005.
9. G. A. Dirac. On rigid circuit graphs. Abh. Math. Sem. Univ. Hamburg, 25:71–76, 1961.
10. A. Efrat and S. Har-Peled. Locating guards in art galleries. In *2nd IFIP International Conference on Theoretical Computer Science*, pages 181–192, 2002.
11. S. J. Eidenbenz. *(In-)Approximability of Visibility Problems on Polygons and Terrains*. Ph.D. thesis, ETH Zürich, Switzerland, 2000.
12. S. J. Eidenbenz, C. Stamm, and P. Widmayer. Inapproximability results for guarding polygons and terrains. *Algorithmica*, 31:79–113, 2001.
13. M. R. Garey and D. S. Johnson. *Computers and Intractability: A Guide to the Theory of NP-Completeness*. W. H. Freeman and Company, New York, 1979.
14. F. Gavril. Algorithms for minimum coloring, maximum clique, minimum covering by cliques, and maximum independent set of a chordal graph. *SIAM J. Computing*, 1(2):180–187, 1972.
15. S. K. Ghosh. Approximation algorithms for art gallery problems. In *Proc. Canadian Inform. Process. Soc. Congress*, pages 429–434, 1987.
16. H. González-Banos and J.-C. Latombe. A randomized art-gallery algorithm for sensor placement. In *Proc. 17th Annual Symposium on Computational Geometry*, pages 232–240, 2001.
17. J. M. Keil. Polygon decomposition. In J.-R. Sack and J. Urrutia, editors, *Handbook of Computational Geometry*, pages 491–518. Elsevier Science Publishers B.V. North-Holland, Amsterdam, 2000.
18. V. S. A. Kumar, S. Arya, and H. Ramesh. Hardness of a set cover with intersection 1. In *Proc. 27th International Colloquium, ICALP*, pages 624–635, 2000.
19. D. T. Lee and A. K. Lin. Computational complexity of art gallery problems. *IEEE Trans. Inform. Theory*, IT-32:276–282, 1986.
20. N. Megiddo and A. Tamir. On the complexity of locating linear facilities in the plane. Oper. Res. Lett., 1:194–197, 1982.
21. R. Motwani, A. Raghunathan, and H. Saran. Covering orthogonal polygons with star polygons: The perfect graph approach. *Comput. Syst. Sci.*, 40:19–48, 1990.
22. B. J. Nilsson. Approximate guarding of monotone and rectilinear polygons. In *Proc. 32nd International Colloquium, ICALP*, pages 1362–1373, 2005.
23. J. O'Rourke. *Art Gallery Theorems and Algorithms*. The International Series of Monographs on Computer Science. Oxford University Press, New York, NY, 1987.
24. J. O'Rourke and K. J. Supowit. Some NP-hard polygon decomposition problems. *IEEE Trans. Inform. Theory*, IT-30:181–190, 1983.
25. T. C. Shermer. Recent results in art galleries. *Proc. IEEE*, 80(9):1384–1399, 1992.
26. D. Schuchardt and H.-D. Hecker. Two NP-hard art-gallery problems for ortho-polygons. *Mathematical Logic Quarterly*, 41:261–267, 1995.
27. J. Urrutia. Art gallery and illumination problems. In J.-R. Sack and J. Urrutia, editors, *Handbook of Computational Geometry*, pages 973–1027. Elsevier Science Publishers B.V. North-Holland, Amsterdam, 2000.

Approximation Algorithms for the Minimum Convex Partition Problem

Christian Knauer[1] and Andreas Spillner[2]

[1] Institute of Computer Science, Freie Universität Berlin
christian.knauer@inf.fu-berlin.de
[2] Institute of Computer Science, Friedrich-Schiller-Universität Jena
spillner@minet.uni-jena.de

Abstract. We present two algorithms that compute constant factor approximations of a minimum convex partition of a set P of n points in the plane. The first algorithm is very simple and computes a 3-approximation in $O(n \log n)$ time. The second algorithm improves the approximation factor to $\frac{30}{11} < 2.7273$ but it is more complex and a straight forward implementation will run in $O(n^2)$ time. The claimed approximation factors are proved under the assumption that no three points in P are collinear. As a byproduct we obtain an improved combinatorial bound: there is always a convex partition of P with at most $\frac{15}{11}n - \frac{24}{11}$ convex regions.

1 Introduction

Let P denote a set of n points in the plane such that no three points in P are collinear. By $CH(P)$ we denote the convex hull of P. A *convex partition* of P is a set E of straight line segments with endpoints in P, called edges, such that edges do not cross each other and partition $CH(P)$ into a set $\mathcal{R}(E)$ of empty convex regions. A region is *empty* if it does not contain a point of P in its interior. An example of a convex partition is given in Figure 1. The points in the input point set are drawn as solid disks. Note that the edges of $CH(P)$ are contained in every convex partition of P.

The *Minimum Convex Partition* problem (MCP) is to compute a convex partition E of P such that the number of regions in $\mathcal{R}(E)$ is minimum. This problem appears in the following context. We want to set up a network connecting the points in P. If the edges used as links in the network form a convex partition of P then a simple randomized algorithm for oblivious routing can be used as shown by Bose et al. [2]. If we want to minimize the number of links used to form the network on P but still be able to use the simple routing algorithm we could build the network according to a minimum convex partition of P.

There are a number of exact algorithms for MCP. Fevens et al. have shown that this problem can be solved in $O(n^{3h+3})$ time if the points in P lie on h nested convex hulls [4]. Grantson and Levcopoulos presented a fixed-parameter algorithm for MCP running in $O(k^{6k-5} 2^{16k} n)$ time [5]. There is also a fixed-parameter algorithm for this problem running in $O(2^k k^3 n^3 + n \log n)$ time [9].

L. Arge and R. Freivalds (Eds.): SWAT 2006, LNCS 4059, pp. 232–241, 2006.

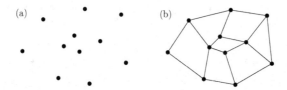

Fig. 1. A point set and a convex partition of this point set

The parameter k in both algorithms is the number of *inner points* in P, which are the points in P lying in the interior of $CH(P)$.

We are neither aware of a polynomial time exact algorithm nor of a proof of NP-hardness for MCP. However, Lingas [7] has shown that the related problem of partitioning a polygon with n vertices by diagonals into a minimum number of convex pieces is NP-hard for polygons with holes. For polygons without holes Keil and Snoeyink give an $O(n^3)$ time algorithm [6].

We are also not aware of any approximation algorithms for MCP. It seems the only result into this direction is a combinatorial bound derived by Neumann-Lara et al. [8]: there is always a convex partition of P with at most $\frac{10}{7}n - \frac{18}{7}$ convex regions. Their inductive proof can be turned into an $O(n^2)$ time algorithm to find such a convex partition but it is not analyzed with respect to its properties as a possible approximation of a minimum convex partition of P.

We will present two approximation algorithms for MCP. Note that the exact algorithms mentioned above can all deal with input point sets containing three or more collinear points. Our algorithms will also compute a convex partition for such point sets. But the number of convex regions in the resulting partition might fail to be within the claimed approximation factor. Our paper is structured as follows. In Section 2 we give a lower bound on the number of convex regions in a minimum convex partition. In Section 3 we show that MCP admits a simple 3-approximation algorithm running in $O(n \log n)$ time. In Section 4 we show that building on the ideas of Neumann-Lara et al. [8] it is possible to give a $\frac{30}{11}$-approximation algorithm running in $O(n^2)$ time. We conclude in Section 5.

2   A Lower Bound

In this section we want to bound the number of convex regions in a minimum convex partition of P from below. We will refer to the vertices of $CH(P)$ also as *outer points*. Note that there are $n - k$ outer points in P where k, as before, is the number of inner points in P. We will assume that $k \geq 3$. For $k \in \{0, 1, 2\}$ the problem is easy and the algorithm to be presented in Section 3 will compute a minimum convex partition of P.

Let $CH_{in}(P)$ denote the convex hull of the inner points in P. First we classify the vertices of $CH_{in}(P)$. Let v be a vertex of $CH_{in}(P)$ and u and w its neighbor-vertices. Let H_w be the half plane that is bounded by the straight line through u and v and that does not contain w. Similarly, let H_u be the half plane that

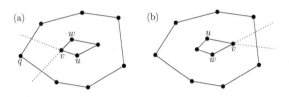

Fig. 2. Types of vertices of the convex hull of the inner points

is bounded by the straight line through v and w and that does not contain u. Vertex v is of *type* (a) iff $H_u \cap H_w$ contains an outer point q of P. An example is given in Figure 2(a). Vertex v is of type (b) iff $H_u \cap H_w$ does not contain any outer point of P. An example is given in Figure 2(b). Note that if vertex v is of type (a) then there is at least one edge connecting v to an outer point in every convex partition of P. Similarly, if vertex v is of type (b) then there are at least two edges connecting v to outer points in every convex partition of P. Let a and b denote the number of vertices of $CH_{in}(P)$ of type (a) and (b), respectively.

Lemma 1. *For every convex partition E of P holds $\frac{k}{2} + b + \frac{a}{2} + 1 \leq |\mathcal{R}(E)|$.*

Proof. Let E be a convex partition of P. We count the number of edges in E incident to each point in P. Every outer point has at least two incident edges in E that are part of the boundary of $CH(P)$. In addition there are in total at least $a + 2b$ edges incident to outer points that connect them to inner points that are vertices of $CH_{in}(P)$. Since every region in $\mathcal{R}(E)$ is convex and no three points in P are collinear every inner point has at least three incident edges in E. Thus, the sum of the number of incident edges in E over all points in P is at least $2(n - k) + 3k + a + 2b = 2n + k + a + 2b$. This gives us $|E| \geq n + a/2 + b + k/2$. Plugging this into the Eulerian relation $1 = |P| - |E| + |\mathcal{R}(E)|$ we obtain $1 + a/2 + b + k/2 \leq |\mathcal{R}(E)|$.   □

Note that in the proof we used the assumption that no three points in P are collinear. Collinear points can decrease the number of convex regions in a minimum convex partition below the bound in Lemma 1. In Figure 3(a) we give an example. Furthermore our lower bound is almost tight. This can be seen by arranging n points on a slightly perturbed (to avoid collinearities) hexagonal

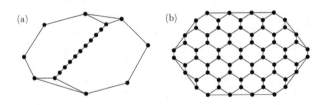

Fig. 3. Examples illustrating the remarks after Lemma 1

grid. This is indicated in Figure 3(b). As the number of points grows, k/n tends to 1 and every inner point has exactly three incident edges.

3   A Simple 3-Approximation Algorithm

We first give a tight combinatorial upper bound on the number of convex regions in a minimum convex partition of P. What we need in order to relate it to the lower bound in Lemma 1 is an upper bound in terms of k. The next lemma states such an upper bound.

Lemma 2. *There is a convex partition E of Γ such that $|\mathcal{R}(E)| \leq \frac{3}{2}k + \frac{3}{2}$.*

Proof. The proof is by induction on the number k of inner points in P.

If $k = 0$ then the set of edges of $CH(P)$ is a convex partition of P and it has one convex region. Thus the upper bound holds for $k = 0$.

If $k = 1$ let q denote the single inner point in P. We triangulate the set of outer points $P \setminus \{q\}$ arbitrarily. There is a unique triangle D in this triangulation that contains the point q in its interior. We obtain a convex partition E of P by adding to the set of edges of $CH(P)$ the three edges connecting q with the vertices of triangle D. This is indicated in Figure 4(a). Since $|\mathcal{R}(E)| = 3$ the upper bound holds for $k = 1$.

Now we consider the case $k \geq 2$ and suppose the upper bound holds for every set of points with less than k inner points. As before, let $CH_{in}(P)$ denote the convex hull of the inner points in P. We consider an arbitrary edge e of $CH_{in}(P)$. Without loss of generality we can assume that edge e is horizontal. Then let p_1 denote the left endpoint of e and p_2 the right endpoint of e. Let $L(e)$ denote the horizontal straight line that contains e. Again without loss of generality we can assume that $CH_{in}(P)$ is contained in the half plane above $L(e)$. The straight line $L(e)$ intersects one edge f_1 of $CH(P)$ to the left of point p_1 and another edge f_2 of $CH(P)$ to the right of p_2. The endpoint of edge f_i lying above $L(e)$ is denoted by q_i, the endpoint below $L(e)$ by r_i, $i \in \{1, 2\}$. The situation is shown in Figure 4(b) where $CH_{in}(P)$ is indicated by shading.

Now we consider a ray R emanating from point q_1 and containing edge f_1. We rotate R counterclockwise around q_1 until it is tangent to $CH_{in}(P)$ at a point t_1. Similarly we obtain a point t_2: A ray emanating from point q_2 and containing edge f_2 is rotated clockwise around q_2 until it is tangent to $CH_{in}(P)$. Now we have a convex chain C that goes from q_1 to t_1 then follows the boundary of

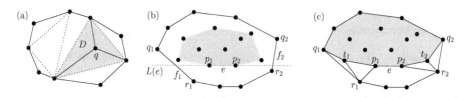

Fig. 4. Proof of the upper bound

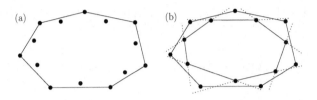

Fig. 5. Tightness of the upper bound

$CH_{in}(P)$ from t_1 over p_1 and p_2 to t_2 and from there it goes to q_2. This convex chain C together with the part of the boundary of $CH(P)$ that we traverse by walking from q_2 back to q_1 avoiding edges f_1 and f_2 forms a convex polygon Q. The situation is indicated in Figure 4(c). The convex polygon Q is drawn shaded.

Now suppose there are l inner points in P that lie on the convex chain C. Then there lie $k - l$ points of P in the interior of Q. Since $l \geq 2$ we can use the induction hypothesis and conclude that there exists a convex partition for the vertices of Q and the points of P in the interior of Q with at most $\frac{3}{2}(k - l) + \frac{3}{2}$ convex regions. By connecting every inner point on C from t_1 to p_1 with r_1 and similarly every inner point on C from p_2 to t_2 with r_2 we obtain a convex partition of P with at most $\frac{3}{2}(k-l)+\frac{3}{2}+l+1 = \frac{3}{2}k+\frac{3}{2} - \frac{1}{2}l+1$ convex regions. Since $l \geq 2$ this is at most $\frac{3}{2}k + \frac{3}{2}$.   □

Note that the upper bound in Lemma 2 is almost tight in the sense that there are point sets with k inner points such that every convex partition of this point set has at least $\frac{3}{2}k+1$ convex regions. To see this we consider a convex polygon T with k edges. Then we place one point very close to the midpoint of each edge of T in the interior of T. The construction is indicated in Figure 5(a). Now the vertices of T and the points placed close to the edges of T form a point set with $n = 2k$ points and k of these points are inner points. All the inner points are vertices of the convex hull of the inner points. If we place each inner point close enough to the corresponding edge of T then every vertex of the convex hull of the inner points will be of type (b). Hence, by Lemma 1 every convex partition of the constructed set of points will have at least $k/2 + k + 1 = \frac{3}{2}k + 1$ convex regions.

Note that Arkin et al. [1] used a similar construction to show that their combinatorial bound in terms of k on the so called reflexivity of P is tight. It seems that tight bounds in terms of n are much more difficult to prove both for convex partitions and the reflexivity of P. The inductive proof of Lemma 2 can easily be turned into an algorithm.

Corollary 1. *We can compute a 3-approximation of a minimum convex partition of P in $O(n \log n)$ time.*

Proof. By Lemma 1, since $a \geq 0$ and $b \geq 0$, we have that every convex partition of P induces at least $\frac{k}{2}+1$ convex regions. According to Lemma 2 there is always a convex partition E of P with $|\mathcal{R}(E)| \leq \frac{3}{2}k + \frac{3}{2}$. Now $3(\frac{k}{2} + 1) \geq \frac{3}{2}k + \frac{3}{2}$ and hence E is a 3-approximation.

In order to achieve $O(n \log n)$ running time we first compute the nested convex hulls of P. This can be done in $O(n \log n)$ time with an algorithm by Chazelle [3]. Having computed the nested convex hulls of P all the computation in one recursive step of the algorithm, except computing the edges of the partition found in this step, can be done in $O(\log n)$ time based on binary search. Since each edge of the convex partition computed is output only once, the total time needed for output is in $O(n)$. This yields a total running time in $O(n \log n)$.   □

4   Improving the Approximation Factor

As in Section 3 we start with a combinatorial upper bound on the number of convex regions in a minimum convex partition of P. To exploit the full power of the lower bound in Lemma 1 we tried to find an upper bound that in addition to k also brings a and b into play. We will assume $k \geq 3$.

Lemma 3. *There is a convex partition E of P such that $|\mathcal{R}(E)| \leq a + 2b + \frac{15}{11}k - \frac{24}{11}$.*

Proof. The basic idea is to first find a good convex partition for the inner points in P and then to partition the region A between the boundary of the convex hull of the inner points and the boundary of the convex hull of P. In Figure 6(a) we have a convex partition of the inner points. It is not hard to see that we can partition the region A into $a + 2b$ convex regions by adding one or two suitable edges incident to each vertex of type (a) and b, respectively. An example is given in Figure 6(b). Hence, it remains to find a convex partition for the inner points with at most $\frac{15}{11}k - \frac{24}{11}$ convex regions. This can be done according to Lemma 4. Plugging in the upper bound of Neumann-Lara et al. would yield at most $\frac{10}{7}k - \frac{18}{7}$ convex regions.   □

Lemma 4. *For every set P of n points in the plane there is a convex partition E such that $|\mathcal{R}(E)| \leq \frac{15}{11}n - \frac{24}{11}$.*

Proof. Our general approach is to show an upper bound of the form $\alpha n - \beta$ with $\alpha \geq 1$ and $\beta \geq 0$. We will use induction on n. The best values for α and β, namely $\alpha = \frac{15}{11}$ and $\beta = \frac{24}{11}$, are determined by a set of inequalities derived

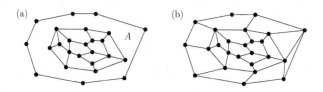

Fig. 6. The basic idea for the proof of the upper bound

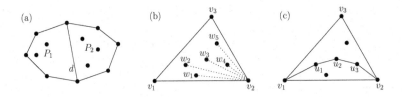

Fig. 7. The general idea of our construction

in the subsequent argumentation. For $3 \leq n \leq 8$ we use the upper bound of Neumann-Lara et al. and obtain the following inequalities. This forms the basis of our inductive proof.

$$3\alpha - \beta \geq 1 \qquad 4\alpha - \beta \geq 3 \qquad 5\alpha - \beta \geq 4$$
$$6\alpha - \beta \geq 6 \qquad 7\alpha - \beta \geq 7 \qquad 8\alpha - \beta \geq 8 \qquad (1)$$

Now suppose that $n \geq 9$. If the convex hull of P has more than three vertices we employ the same argument as Neumann-Lara et al. [8]: We split the point set by a diagonal d into two subsets P_1 and P_2 as shown in Figure 7(a). Let n_i denote the number of points in P_i. Then by the induction hypothesis we have a convex partition for P_i with at most $\alpha n_i - \beta$ convex regions. We glue them together along diagonal d and obtain a convex partition of P. Note that $n_1 + n_2 = n + 2$ and that d can be removed from the set of edges in the resulting partition. Hence, we have a convex partition of P with at most $\alpha(n_1 + n_2) - 2\beta - 1 = \alpha n - \beta + (2\alpha - \beta - 1)$ regions. Thus, α and β must satisfy:

$$2\alpha - \beta - 1 \leq 0 \qquad (2)$$

It remains to consider the case that the convex hull of P has exactly three vertices. So let v_1, v_2 and v_3 denote the vertices of the convex hull of P as shown in Figure 7(b). We sort the inner points in P according to its clockwise angular order around vertex v_2. Let $w_1, w_2, \ldots, w_{n-3}$ denote the resulting sorted sequence. We now consider the convex hull H of $\{v_1, w_1, w_2, \ldots, w_l, v_2\}$ for some $l \in \{1, \ldots, n-3\}$. Let $u_1, \ldots, u_{j-1}$ denote the subsequence of those points in $w_1, \ldots, w_l$ that are vertices of H. An example is shown in Figure 7(c). When we connect each of the vertices $u_1, \ldots, u_{j-1}$ with v_3 by an edge we obtain a partition of the convex hull of P into j triangles $D_1, \ldots, D_j$ and a convex polygon Q with $j + 1$ vertices. This is shown in Figure 8(a).

In order to describe our idea more precisely we introduce a little more notation. The edge that is shared by triangle D_i and Q we denote by e_i. Furthermore, the edge shared by D_i and D_{i+1} we denote by f_i. An example is given in Figure 8(b).

The idea is to partition each of the triangles $D_1, \ldots, D_j$ by using the induction hypothesis. The convex polygon Q contains only $l - j + 1$ points of P in its interior. If l is not too large (we will use $l = 6$) then we can partition Q directly, which might save us some convex regions compared to a call to the induction hypothesis for Q. Finally, we glue the partitions for the triangles and for Q together along

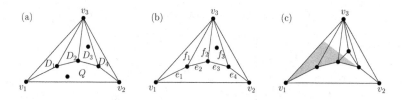

Fig. 8. Partition each piece and then glue them together

the edges $e_1, \ldots, e_j$ and $f_1, \ldots, f_{j-1}$. Depending on the structure of the convex partition for $D_1, \ldots, D_j$ and Q some of these edges might be removed when forming the resulting partition of P.

Our first goal is to give a general formula for an upper bound on the number of convex regions in the convex partition we obtain by following the strategy outlined above. We call a triangle D_i empty iff D_i does not contain a point of P in its interior. Let s denote the number of non-empty triangles among $D_1, \ldots, D_j$. Furthermore, let $\mu(Q)$ denote an upper bound on the number of convex regions in the partition of Q.

Claim. The resulting convex partition of P has at most

$$\rho(n, l, j, s) = [\alpha n - \beta] + [j - \alpha l + (3\alpha - \beta - 2)s - 3\alpha + \beta + 1 + \mu(Q)]$$

regions. For $s \in \{0, 1\}$ the bound can be improved to $\rho(n, l, j, s) - 1$.

Let n_i denote the number of points in P that are contained in triangle D_i including the three vertices of D_i. Suppose the triangles $D_{i_1}, \ldots, D_{i_s}$ are non-empty. Then by simply glueing together the convex partitions for the triangles $D_1, \ldots, D_j$ and for Q we obtain a convex partition of P with at most $\alpha(n_{i_1} + \cdots + n_{i_s}) - s\beta + (j - s) + \mu(Q)$ regions. By counting we obtain $n_{i_1} + \cdots + n_{i_s} + 3(j - s) = n + 2(j - 1) - (l - j + 1)$. This gives us $n_{i_1} + \cdots + n_{i_s} = n - l + 3s - 3$. We substitute this into the bound above and obtain $[\alpha n - \beta] + [j - \alpha l + (3\alpha - \beta - 1)s - 3\alpha + \beta + \mu(Q)]$.

Now by our construction if triangle D_i, $i \in \{1, \ldots, j - 1\}$, is non-empty then edge f_i can be removed. This is because the lower part of each such triangle is empty. This is indicated in the example in Figure 8(c) by shading. There triangle D_3 is non-empty and thus edge f_3 can be removed.

Hence, if there are s non-empty triangles we can remove at least $s - 1$ of the edges $f_1, \ldots, f_{j-1}$. So we obtain the desired bound $\rho(n, l, j, s)$. Finally, if $s = 0$ than subtracting $s - 1$ unnecessarily adds 1. And if $s = 1$ then we can remove at least one edge: If the single non-empty triangle is one of the D_i with $i \in \{1, \ldots, j - 1\}$ then we can remove f_i. If only D_j is non-empty then we can remove one of the edges f_{j-1} or e_j. This finishes the proof of Claim 4.

We can simplify (and thereby weaken) the bound $\rho(n, l, j, s)$ by requiring that

$$3\alpha - \beta - 2 \leq 0. \tag{3}$$

Note that the values of α and β we are aiming at satisfy this inequality with some slack. Hence, this requirement eases our argumentation but it is not the reason

Table 1. Overview of the values of $\rho_1(6, j)$ and $\rho_2(6, j)$

j	$\mu(Q)$	$\rho_1(6, j)$	$\rho_2(6, j)$
2	8	$-9\alpha + \beta + 10$	$-3\alpha - \beta + 7$
3	7	$-9\alpha + \beta + 10$	$-3\alpha - \beta + 7$
4	6	$-9\alpha + \beta + 10$	$-3\alpha - \beta + 7$
5	4	$-9\alpha + \beta + 9$	$-3\alpha - \beta + 6$
6	3	$-9\alpha + \beta + 9$	$-3\alpha - \beta + 6$
7	1	$-9\alpha + \beta + 8$	$-3\alpha - \beta + 5$

that we obtain those particular values for α and β. Now with (3), if $s \in \{0, 1\}$, we have

$$\rho(n, l, j, s) \leq [\alpha n - \beta] + [j - \alpha l - 3\alpha + \beta + \mu(Q)].$$

And if $s \in \{2, \ldots, j\}$, we have

$$\rho(n, l, j, s) \leq [\alpha n - \beta] + [j - \alpha l + (3\alpha - \beta - 2)2 - 3\alpha + \beta + 1 + \mu(Q)]$$
$$= [\alpha n - \beta] + [j - \alpha l + 3\alpha - \beta - 3 + \mu(Q)].$$

Thus, we need that

$$\rho_1(l, j) = [j - \alpha l - 3\alpha + \beta + \mu(Q)] \leq 0 \quad \text{and}$$
$$\rho_2(l, j) = [j - \alpha l + 3\alpha - \beta - 3 + \mu(Q)] \leq 0.$$

As mentioned above we will use the value $l = 6$. Smaller values do not seem to lead to the desired value of α. For larger values our argumentation becomes more complicated and it is not sure that it gives us a better bound. In Table 1 we have computed $\rho_1(6, j)$ and $\rho_2(6, j)$ for $j \in \{2, \ldots, l + 1\} = \{2, \ldots, 7\}$. The bound on $\mu(Q)$ is obtained as

$$min\left\{ \left\lfloor \frac{3}{2}(l - j + 1) + \frac{3}{2} \right\rfloor, \left\lfloor \frac{10}{7}(l + 2) - \frac{18}{7} \right\rfloor \right\}.$$

Recall that $l - j + 1$ is the number of points in the interior of Q. So the first term is the bound from Lemma 2. The second term is the bound of Neumann-Lara et al.

Hence, if $s \in \{0, 1\}$ then α and β must satisfy

$$-9\alpha + \beta + 10 \leq 0. \tag{4}$$

This is okay with respect to the desired values of α and β. If $s \in \{2, \ldots, j\}$ then this is only true for $j \in \{5, 6, 7\}$, i.e., the desired values of α and β will satisfy

$$-3\alpha - \beta + 6 \leq 0. \tag{5}$$

So the only cases we have to discuss in detail are $s \in \{2, \ldots, j\}$ and $j \in \{2, 3, 4\}$. The details can be found in the full version of the paper. The result of this case analysis is that α and β must also satisfy

$$\alpha - 2\beta + 3 \leq 0. \tag{6}$$

The inequalities (1)-(6) form the constraints of a linear program. Solving this linear program gives the desired constants $\alpha = \frac{15}{11}$ and $\beta = \frac{24}{11}$. $\qquad\square$

Corollary 2. *We can compute in $O(n^2)$ time a $\frac{30}{11}$-approximation of a minimum convex partition of P.*

Proof. According to Lemma 1 every convex partition of P induces at least $\frac{k}{2} + \frac{a}{2} + b + 1$ convex regions. In Lemma 3 we have shown that there is always a convex partition E of P with $|\mathcal{R}(E)| \leq a + 2b + \frac{15}{11}k - \frac{24}{11}$. Since $\frac{30}{11}(\frac{k}{2} + \frac{a}{2} + b + 1) \geq a + 2b + \frac{15}{11}k - \frac{24}{11}$ this partition E is a $\frac{30}{11}$-approximation. It is straight forward to turn the proof of Lemma 3 into an $O(n^2)$ time algorithm. $\qquad\square$

5   Concluding Remarks

It would be desirable to have an algorithm computing a constant-factor approximation even for point sets with many collinear points. Furthermore a substantially smaller approximation factor would also be nice. However, it seems to us that both goals can only be achieved by exploiting geometric properties more specific to the particular input point set P. How this can be done remains an open question.

Acknowledgments. We would like to thank Thomas Fevens for pointing out the application to oblivious routing mentioned in Section 1.

References

1. E. M. Arkin, S. P. Fekete, F. Hurtado, J. S. B. Mitchell, M. Noy, V. Sacristán, and S. Sethia. On the reflexivity of point sets. In *Discrete and Computational Geometry: The Goodman-Pollack Festschrift*, volume 25, pages 139–156. Springer, 2003.
2. P. Bose, A. Brodnik, S. Carlsson, E. D. Demaine, R. Fleischer, A. López-Ortiz, P. Morin, and J. I. Munro. Online routing in convex subdivisions. *Internat. J. Comput. Geom. Appl.*, 12:283–296, 2002.
3. B. Chazelle. On the convex layers of a planar set. *IEEE Trans. Inform. Theory*, 31:509–517, 1985.
4. T. Fevens, H. Meijer, and D. Rappaport. Minimum convex partition of a constrained point set. *Discrete Appl. Math.*, 109:95–107, 2001.
5. M. Grantson and C. Levcopoulos. A fixed-parameter algorithm for the minimum number convex partition problem. In *JCDCG*, volume 3742 of *LNCS*, pages 83–94. Springer, 2005.
6. J. M. Keil and J. Snoeyink. On the time bound for convex decomposition of simple polygons. *Internat. J. Comput. Geom. Appl.*, 12:181–192, 2002.
7. A. Lingas. The power of non-rectilinear holes. In *Internat. Colloq. Automata Lang. Program.*, volume 140 of *LNCS*, pages 369–383. Springer, 1982.
8. V. Neumann-Lara, E. Rivero-Campo, and J. Urrutia. A note on convex decompositions of a set of points in the plane. *Graphs and Combinatorics*, 20:223–231, 2004.
9. A. Spillner. Optimal convex partitions of point sets with few inner points. In *CCCG*, pages 34–37, 2005.

Approximation of Octilinear Steiner Trees Constrained by Hard and Soft Obstacles

Matthias Müller-Hannemann[1] and Anna Schulze[2]

[1] Technische Universität Darmstadt, Department of Computer Science,
Hochschulstraße 10, 64289 Darmstadt, Germany
`muellerh@algo.informatik.tu-darmstadt.de`
[2] Zentrum für Angewandte Informatik Köln, Weyertal 80, 50931 Köln, Germany
`schulze@zpr.uni-koeln.de`

Abstract. The novel octilinear routing paradigm (X-architecture) in VLSI design requires new approaches for the construction of Steiner trees. In this paper, we consider two versions of the shortest octilinear Steiner tree problem for a given point set K of terminals in the plane: (1) a version in the presence of hard octilinear obstacles, and (2) a version with rectangular soft obstacles.

The interior of hard obstacles has to be avoided completely by the Steiner tree. In contrast, the Steiner tree is allowed to run over soft obstacles. But if the Steiner tree intersects some soft obstacle, then no connected component of the induced subtree may be longer than a given fixed length L. This kind of length restriction is motivated by its application in VLSI design where a large Steiner tree requires the insertion of buffers (or inverters) which must not be placed on top of obstacles.

For both problem types, we provide reductions to the Steiner tree problem in graphs of polynomial size with the following approximation guarantees. Our main results are (1) a 2–approximation of the octilinear Steiner tree problem in the presence of hard rectilinear or octilinear obstacles which can be computed in $O(n \log^2 n)$ time, where n denotes the number of obstacle vertices plus the number of terminals, (2) a $(2 + \varepsilon)$–approximation of the octilinear Steiner tree problem in the presence of soft rectangular obstacles which runs in $O(n^3)$ time, and (3) a $(1.55+\varepsilon)$–approximation of the octilinear Steiner tree problem in the presence of soft rectangular obstacles.

Keywords: Approximation algorithms, Steiner trees, octilinear routing, obstacles, VLSI design.

1   Introduction

Background and Motivation. Octilinear routing is a novel routing paradigm in VLSI design, the so-called X-architecture [1], which has recently been introduced. In addition to vertical and horizontal wires, octilinear routing allows wiring in 45- and 135-degree directions. Compared to traditional and state-of-the-art rectilinear (Manhattan) routing, such a technology promises clear advantages in wire length but also in via reduction. As a consequence a significant

L. Arge and R. Freivalds (Eds.): SWAT 2006, LNCS 4059, pp. 242–254, 2006.

chip performance improvement and power reduction can be obtained (with estimations being in the range of 10% to 20% improvement) [2, 3, 4]. To enable such a technology, novel algorithmic approaches for the construction of octilinear Steiner trees are needed.

An *octilinear Steiner tree* is a tree that interconnects a set of points (*terminals*) in the plane with minimum length such that every line segment uses one of the four given orientations. Even more general routing architectures are obtained if a fixed set of uniformly oriented directions is allowed. For an integer parameter $\lambda \geq 2$, consecutive orientations are separated by a fixed angle of π/λ. A λ-*geometry* is a routing environment in which every line segment uses one of the given orientations. Manhattan routing can then be seen as the special case $\lambda = 2$ and the X-architecture as the case $\lambda = 4$. In this paper we focus on the octilinear case (although most of our results can be generalized to arbitrary $\lambda \geq 2$). We study approximation algorithms for the octilinear Steiner tree problem with different types of obstacles.

Hard and Soft Obstacles. In VLSI design preplaced macros or other circuits are obstacles. Throughout this paper, an *obstacle* is a connected region in the plane bounded by a simple polygon such that all obstacle edges lie within the 4-geometry (*octilinear obstacle*). If all boundary edges of an obstacle are rectilinear, we call such an obstacle a *rectilinear obstacle*. For a given set of obstacles $\mathcal{O}$ we require that the obstacles be disjoint, except for possibly a finite number of common points. In practice, obstacles can be assumed to be axis-parallel rectangles. An obstacle which prohibits wiring and therefore has to be avoided completely will be referred to as a *hard obstacle*. Due to the availability of several routing layers, most obstacles usually do not block wires, but it is impossible to place a buffer (or inverter) on top of an obstacle. A large Steiner tree requires the insertion of buffers (or inverters) in such a way that no induced subtree without any buffers becomes too large. This application in VLSI design motivates and translates into our model of *soft obstacles*. In this case the Steiner tree is allowed to run over obstacles; however, if we intersect the Steiner tree with some obstacle, then no connected component of the induced subtree may be longer than a given fixed length L.

Related Work. The rectilinear and the Euclidean Steiner tree problem have been shown to be NP-Hard in [5] and [6], respectively. Quite recently, we have been able to prove that the octilinear Steiner tree problem is also NP-hard in the strong sense [7]. Most previous work on the octilinear Steiner tree problem considered the problem *without obstacles*. Exact approaches to the octilinear Steiner tree problem have been developed by Nielsen, Winter and Zachariasen [8] and Coulston [9]. Nielsen et al. report the exact solution to a large instance with 10000 terminals within two days of computation time. An exact algorithm for obstacle-avoiding Steiner trees in the Euclidean metric has been developed by Zachariasen and Winter [10].

For rectilinear Steiner tree problems for point sets in the plane, the most successful approaches are based on transformations to the related Steiner tree problem in graphs. Given a connected graph $G = (V, E)$, a length function ℓ, and a set of terminals $K \subseteq V$, a *Steiner tree* is a tree which contains all vertices of K and is a subgraph of G. A Steiner tree T is a *Steiner minimum tree* of G if the length of T is minimum among all Steiner trees. An implementation by Althaus, Polzin and Daneshmand [11] is the currently strongest available exact approach for both the Steiner tree problem in graphs and the rectilinear Steiner tree problem. The best available approximation guarantee for the Steiner problem in general graphs is $\alpha = 1 + \frac{\ln 3}{2} \approx 1.55$, obtained by Robins and Zelikovsky [12].

Unfortunately, in the octilinear case, the only known transformation to the Steiner tree problem in graphs is based on a generalization of the Hanan-grid and requires $O(n^{2^{O(n)}})$ many vertices [13, 14, 15]. Hence, this transformation is not polynomial. Müller-Hannemann and Schulze [7] recently constructed a graph of size $O(n^2/\varepsilon^2)$ which contains for every $\varepsilon > 0$ a $(1 + \varepsilon)$-approximation for the case without obstacles and with hard obstacles.

We would like to point out that the well-known approximation schemes of Arora [16] and Mitchell [17] are only applicable to the octilinear Steiner tree problem without obstacles. For the octilinear Steiner tree problem *without obstacles* heuristics have been proposed by Kahng et al. [18] and Zhu et al. [19].

Müller-Hannemann and Peyer [20] showed that the rectilinear Steiner tree problem in the presence of soft obstacles can be 2-approximated in $O(n^2 \log n)$ time, where n denotes the number of terminals plus the number of obstacle vertices. They also presented a $(1.55 + \varepsilon)$-approximation for rectangular obstacles. In this paper we generalize these results to the octilinear Steiner tree problem. However, it will turn out that the problem becomes substantially more complicated and requires novel techniques both in design and analysis. We are not aware of any other exact approaches or heuristics in the presence of obstacles.

Our Methodology. We provide transformations from the octilinear Steiner tree problem in the plane with obstacles to the Steiner tree problem in graphs which contain approximate solutions. To achieve a 2–approximation our aim is to construct a *path preserving graph*, i.e., a graph which contains a shortest octilinear path between any pair of terminals. With respect to obstacles, the graph should only contain feasible paths and only feasible Steiner trees. (Note that for soft obstacles the latter does not follow from the feasibility of all paths.) These properties ensure that any approximation algorithm based on this graph for the Steiner tree problem will produce a feasible Steiner tree. In particular, we may use Mehlhorn's [21] implementation of a minimum spanning tree based approximation which runs in time $O(m + n \log n)$ on a graph with n nodes and m edges. This approach yields a 2–approximation, and we can show that the analysis is asymptotically tight.

Heading for a good running time, our secondary goal is to construct small path preserving graphs. Shortest paths in the presence of polygonal obstacles have already been studied intensively. See the surveys of Mitchell [22] and Lee, Yang, and Wong [23]. Our construction of small path preserving graphs generalizes techniques in previous work of Wu et al. [24] and Clarkson et al. [25].

To achieve a $(1 + \varepsilon)$-approximation we develop a different technique based on t-restricted Steiner trees. A Steiner tree is a *full* Steiner tree if all its terminals are leaves. Any Steiner tree can be decomposed into its full components. A t-restricted Steiner tree is a Steiner tree where all full components have at most t terminals. The boundary of each obstacle is discretized by auxiliary vertices with a distance of at most Δ between neighboring vertices. (Δ can be chosen so that we obtain a polynomial number of auxiliary vertices and still achieve the desired accuracy.) Inside obstacles, we approximate an optimal tree with the help of t-restricted Steiner trees for some constant t. Each of these trees respects the length restriction L for the obstacle. Outside obstacles, a grid-like graph through the terminals and obstacle vertices is refined by additional lines so that it contains a sufficiently close approximation. These ideas will be made precise in Section 4.

Our Contribution. We summarize the main results of this paper:

- There is a 2–approximation of the octilinear Steiner tree problem in the presence of hard octilinear obstacles which can be computed in $O(n \log^2 n)$ time, where n denotes the number of obstacle vertices plus the number of terminals.
- For any integer k, we obtain a $(2 + \frac{1}{k})$–approximation which runs in $O(k^2 n^3)$ time for the octilinear Steiner tree problem with soft rectangular obstacles.
- We construct a graph of polynomial size containing a $(1 + \varepsilon)$-approximation of the octilinear Steiner tree problem with rectangular soft obstacles. Hence, the currently strongest approximation guarantee by Robins and Zelikovsky for the Steiner tree problem in graphs implies a $(1.55 + \varepsilon)$-approximation for this problem. This matches the best known guarantees for the rectilinear case [20].

Overview. The remaining part of the paper is organized as follows. In Section 2, we describe how to construct shortest path preserving graphs for hard obstacles of size $O(n \log n)$. The more complicated construction of shortest paths for soft obstacles will be explained in Section 3. Finally, we show how to construct a graph of polynomial size which contains a $(1 + \varepsilon)$-approximation for rectangular soft obstacles. Due to strict page limitations, all proofs will appear only in the journal version which can be downloaded from http://www.algo.informatik. tu-darmstadt.de/muellerh/approx_octilinear.pdf.

2   Octilinear Shortest Paths Amidst Hard Obstacles

Throughout this section, let K be a set of points (terminals) in the plane and O be a set of octilinear obstacles. Denote by V_O the set of obstacle vertices. Let $n = |K| + |V_O|$. In this section we will show how to construct shortest path preserving graphs.

Octilinear Track Graphs. As a first step we construct a path preserving graph based on visibility. Our construction may be viewed as a generalization of that of Wu et al. [24], which was designed for rectilinear polygons and rectilinear paths. To simplify our discussion we add to our scene a bounding box containing all obstacles and all terminals. Clearly all desired paths will run within this bounding box. A *track tr* generated by a point t and an orientation is a line segment that starts at t and ends when it first hits an obstacle edge or the bounding box. The generated endpoints of tracks are called *track-induced Steiner points*. For each terminal t and each feasible orientation we construct a track in both directions from t. Similarly, we introduce tracks for each convex obstacle vertex v. More precisely, if $e_1 = (v_1, v)$ and $e_2 = (v, v_2)$ are polygon edges incident with v in clockwise order of the polygon, denote by r_1 the ray in direction from v_1 to v, and by r_2 the ray in direction from v_2 to v. We construct a track generated by v for all feasible directions which do neither go through the interior of the obstacle nor through the interior of the sector spanned by ray r_1 and r_2 in counter-clockwise order. See Fig. 1. The intersections among all tracks and their endpoints are made the vertices of the track graph. The edges are the track segments between the intersections. The construction is completed by adding edges connecting two consecutive track-induced Steiner points or polygon vertices along the boundary of each obstacle. The length of an edge in the track graph is simply the octilinear distance between its endpoints. See the middle part of Fig. 1 for a small example which illustrates this construction. The track graph consists of $O(n)$ many tracks which induce $O(n^2)$ many vertices and edges.

Sparser Path-Preserving Graphs. To improve upon the quadratic space bound of the track graph we use an idea of Clarkson et al. [25] and extend their approach to the octilinear case. We construct a sparser path-preserving graph $G = (V, E)$ as follows. The vertex set is constructed in two rounds. In the first round, we create V_1 as the union of

1. the set of all terminals K,
2. the set of all obstacle vertices V_O, and
3. the set of track-induced Steiner points for tracks induced by K and V_O.

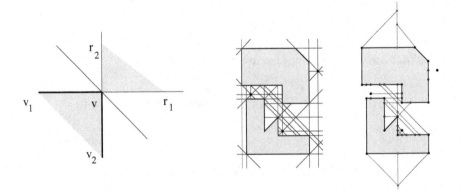

Fig. 1. Illustration of the graph construction. Left: The tracks around a convex vertex v of some obstacle. There is no track inside the shaded area. Middle: The track graph for an instance with three terminals (black dots) and two hard octilinear obstacles. Right: The first vertical cut line construction.

With respect to V_1 we create the set V_2 recursively by adding more Steiner points along vertical, horizontal and diagonal so-called *cut lines*. We explain the construction for vertical cut lines. A vertical cut line is placed at the median of the x-coordinates of all vertices. Vertices in V_1 generate projection points on the line. Projections are performed in all feasible orientations so that we may get up to three projections points on the line for each vertex in V_1 (in the rectilinear setting, Clarkson et al. need to project only orthogonally onto the cut line). Two points are mutually *visible* to each other if the straight line segment between them contains no obstacle point in its interior. All those projection points on a cut line which are visible from some inducing point in V_1 are put into a vertex set V_2. Moreover, we add the intersection points of the cut line with obstacle vertices or with non-parallel obstacle edges to V_2. The following edges are inserted into E. Two consecutive Steiner points on the line are connected by an edge if these points are visible to each other. We also add edges from each vertex in V_1 to its corresponding projection points.

This procedure is repeated recursively with the vertices respectively on the left and right sides of the cut line. The union of all these vertices yields $V = V_1 \cup V_2$. See Fig. 1 for a vertical cut line on the highest level. There are $O(\log n)$ many levels of recursion, and in each level we will create $O(n)$ many vertices and edges. This gives in total $O(n \log n)$ vertices and edges. Finally, for each obstacle we have edges between consecutive vertices from V on its boundary.

Correctness of the Construction. We now sketch the proof that the graph G has the desired properties.

Theorem 2.1. *For any two vertices from K the constructed graph G contains a shortest octilinear path.*

The validity of this theorem is based on the following three lemmas. A *segment S* of a path P is a subpath with the property that all its edges have the same orientation. Hence, any path can be thought of as composed by a sequence of *inclusion-maximal segments*, i.e. longest subpaths with the same orientation.

Lemma 2.2. *The track graph contains a shortest octilinear path for any two vertices from K.*

Proof. (Sketch) For arbitrary $s, t \in K$ choose a shortest octilinear path which has the fewest number k of inclusion-maximal segments which do not lie in the track graph. If $k = 0$, the path is completely contained in the track graph and the lemma holds. Otherwise, one obtains a contradiction by showing how to modify the path such that it remains length-minimal but contains fewer segments not lying in the track graph. $\square$

Lemma 2.3. *For any two vertices $p, q \in K$ there is a shortest octilinear path (in the plane, not restricted to G) which visits a sequence of vertices $p = v_0, v_1, v_2, \ldots, v_k = q$ from V_1 and for each two subsequent vertices v_i and v_{i+1}, $i = 0, \ldots, k-1$, these vertices are connected as short as possible in 4-geometry (i.e. with the same distance as if there were no obstacles).*

Lemma 2.4. *Let p and q be two vertices of V_1 such that a shortest octilinear path between p and q in the track graph does not contain any other vertex of V_1. Then the graph G contains a path from p to q of minimum length in 4-geometry.*

Our construction yields a graph with $O(n \log n)$ vertices and edges.

Lemma 2.5. *Given a set of terminals and a set of octilinear obstacles with n vertices in total, there is a graph with $O(n \log n)$ vertices and edges which contains for every pair of terminals a shortest octilinear path.*

By Lemma 2.5, we can apply Mehlhorn's implementation of the spanning tree heuristic to a graph with $O(n \log n)$ vertices and edges. This immediately implies the following theorem.

Theorem 2.6. *There is a 2-approximation of the octilinear Steiner tree problem with hard octilinear obstacles. Such an approximation can be computed in $O(n \log^2 n)$ time.*

3   Soft Obstacles

For soft obstacles we introduce length restrictions for those portions of a tree T which run over obstacles. Namely, for a given parameter $L \in \mathbb{R}_0^+$ we require the following for each obstacle $O \in \mathcal{O}$ and for each strictly interior connected component T_O of $(T \cap O) \setminus \partial O$: the length $\ell(T_O)$ of such a component must not be longer than the given length restriction L. Note that the intersection of a Steiner minimum tree with an obstacle may consist of more than one connected component and that our length restriction applies individually for each connected component. For ease of exposition, we restrict our presentation of soft obstacles to (axis-parallel) rectangular obstacles. Generalizations to rectilinear and octilinear soft obstacles are possible and do not change the asymptotic size of the resulting graphs.

Track Graph Construction. For soft obstacles, an analogous construction of the track graph is substantially more complicated than for hard obstacles. (This is in sharp contrast to the rectilinear case). We obtain the track graph by applying the following rules inductively. See Fig. 2 for an illustration of each rule.

1. We generate track lines for all terminals and all feasible orientations. But in contrast to hard obstacles, a track does not end as soon as it hits an obstacle. It only ends at an obstacle if the intersection of the track line with the obstacle exceeds the given length restriction L. Hence, we distinguish between Steiner points which are endpoints of a track due to a length restriction, called *L-Steiner points*, and all other Steiner points generated as intersections of a track line and obstacles. The latter type of Steiner points will still be called *track-induced Steiner points*.

2. Similarly, we introduce track lines through all edges of rectangular polygons. This yields $O(n)$ track lines and may cause $O(n^2)$ many track-induced Steiner points. This already implies that the number of track-induced Steiner points will be one order of magnitude larger than for hard obstacles.

3. Additional tracks are needed to make shortcuts when an obstacle causes a deviation due to the length restriction L.

   For each edge $e = (p_1, p_2)$ of an obstacle with length $\ell(e) > L/\sqrt{2}$ we do the following. At the points on e with distance $L/\sqrt{2}$ from the corners p_1 and p_2 we respectively generate tracks which have an angle of 45 and 135 degrees with e and run through the obstacle but do not exceed the length restriction. This yields another $O(n)$ track lines and $O(n^2)$ track-induced Steiner points.

4. Next suppose that a track tr ends at a point p of an edge e of some obstacle due to the length restriction and hits e with an angle of 45 degrees. If the edge f which is opposite to e in such a rectangle has a distance not exceeding L from e, we let the track continue inside the rectangle up to a certain point q. At q the track bends by an angle of 135 degrees and continues until it hits edge f, say at r. The point q is chosen in such a way that length of the two segments $\overline{pq}$ and $\overline{qr}$ together equals the length restriction L. Finally, at r a new track parallel to tr is created. Note that tracks generated for this item do not increase the asymptotic complexity.

5. Now consider the following situation. A track tr enters an obstacle O at some point p on edge a in an angle of 45 degrees and leaves the obstacle at some point q on an edge b of O which is adjacent to a. Furthermore, we assume that the length of b exceeds L. Then we start a new track tr_2 at q which runs orthogonally to tr through the obstacle, provided that $\ell(tr_2 \cap O) < L$ (i.e., the intersection of tr_2 with O does not exceed L; if equality holds this track has already been inserted). As a track may cross many obstacles each of which potentially induces a new track of the just described kind, and newly generated tracks in turn may induce further tracks of this kind, we have to be careful not to generate infinitely many new tracks. Therefore, the generation process is done in rounds for each track generated in Items 1-4. In each round, we create a tree of new tracks, called *track tree*. The root r of such a track tree is one of the tracks generated by Items 1-4. Every induced new track is made an immediate successor of its inducing track. A round ends if no new track is induced. To make each round finite, we add the following rule. Consider a fixed round and suppose that we have generated in step i of this round a track tr_i from a Steiner point on rectangle side e. If in a later step $j > i$ we would have to insert a further track tr_j from the very same rectangle side e due to Item 5 and this track would have track tr_i as a predecessor in the track tree, such a track is not necessary. This is because in such a scenario the generated tracks would form a full cycle around a rectangle, and clearly no cycle can be in a shortest path. Hence, our rule is not to generate a further track in such cases. By applying this rule, we have a finite number of tracks.

6. Suppose that a track tr ends at an obstacle O due to the length restriction and hits edge e of O orthogonally at some point q. Moreover, suppose q has

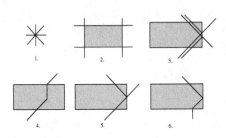

Fig. 2. The different types of tracks for soft obstacles

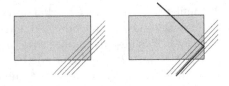

Fig. 3. The extra tracks inserted at a corner of some obstacle to approximate shortest paths (left). Clearly, the "thick" path is only slightly longer than an approximation using one of the extra tracks (right).

a distance of less than $L/\sqrt{2}$ from some obstacle corner v on e. Then we add a segment and a new track to shortcut the way around O (the latter only if its intersection with O does not exceed L). See again Fig. 2. We handle such tracks as in the previous item.

This completes the construction of our track graph. In the same way as for hard obstacles we can prove the correctness of this construction.

Lemma 3.1. *The constructed track graph contains a shortest length-restricted path between every pair of terminals.*

Approximate Shortest Paths. The track graph as described above may have exponential size. With a smarter construction one can bound the size of the track graph by $O(n^3)$ (but the proof then becomes quite complicated). In this paper, we therefore prefer a simpler construction which uses approximate shortest paths. For any integer k, we obtain $(1 + \frac{1}{k})$-approximate shortest paths. The idea is to leave out Items 5 and 6 of the track graph construction (which are responsible for the blow up in the graph size). Instead, we insert $k - 1$ additional tracks for each corner of an obstacle. These tracks "cut off" the corner and are placed in distance $\frac{j \cdot L}{\sqrt{2} \cdot k}$ from the corner for $j = 1, \ldots, k - 1$. See Fig. 3.

This construction induces $O(kn)$ many new tracks which are responsible for $O(kn)$ new track-induced Steiner points per obstacle. Next we apply the same sparsification technique as for hard obstacles and make sure that every path in our graph is feasible with respect to our length restriction. We do this in two steps. In the first step, we regard all obstacles as hard obstacles and use the modified cut line approach on the set of original vertices, terminals and all track-induced Steiner points. The overall number of Steiner points is $O(kn^2)$. Hence, the sparsification technique outside obstacles yields $O(kn^2 \log(kn))$ many vertices and edges.

In the second step, we add connections between vertices and Steiner points on the boundary of obstacles. In the previous discussion we observed that we may have $O(kn)$ many track-induced Steiner points lying on the boundary of an obstacle O. Locally these Steiner points can be regarded as terminals which have to be connected pairwise without violating the length bound L.

Lemma 3.2. *Let O be a rectilinear obstacle with t terminals on its boundary. Then we need $O(t^2)$ many edges for a graph which has (1) to represent shortest paths between any pair of terminals respecting the length restriction L, and (2) does not contain any path exceeding the length restriction L inside some obstacle.*

Thus, we can now apply Lemma 3.2 with $t = O(kn)$ and get $O(k^2n^2)$ edges inside a single obstacle, for a total of $O(k^2n^3)$ edges inside all obstacles. It is easy to see that shortest paths between terminals in this modified graph will be at most a factor of $(1 + \frac{1}{k})$ longer than shortest paths.

Lemma 3.3. *There is a graph for soft rectangular obstacles with $O(kn^2 \log(kn))$ many vertices and $O(k^2n^3)$ many edges which contains a $(1 + \frac{1}{k})$-approximative shortest path between any pair of terminals for any integer k. Moreover, all paths in this graph respect the length restriction L inside obstacles. The graph can be constructed in time proportional to its size.*

As the obtained graph contains only length-feasible paths, we can apply Mehlhorn's implementation of the minimum spanning tree heuristic to construct a Steiner tree. We finally obtain:

Theorem 3.4. *For any fixed k, we can find a $(2 + \frac{1}{k})$-approximation of the octilinear Steiner tree problem with soft rectangular obstacles in time $O(n^3)$.*

We conclude by mentioning that our analysis is tight. It is possible to construct a class of instances for which our approximation algorithm asymptotically achieves a performance guarantee of 2.

4   Improved Approximation Guarantee

As outlined in the Introduction, we can construct a graph of polynomial size which contains a $(1 + \varepsilon)$-approximation for the octilinear Steiner tree problem with soft rectangular obstacles. Next we give a detailed description of this construction. For the analysis, however, we have to refer to the full version of this paper. The graph construction requires the following five steps:

Step 1: The very first step is to compute an axis-parallel box which contains an optimal Steiner tree. Everything outside such a box can then be safely ignored in the subsequent steps. For the analysis it is important that the side length b of this box can be bounded by a constant times the length of an optimal Steiner tree T_{opt}. To achieve this goal, we can run the minimum spanning tree based approximation. Let us assume that this approximation yields a tree of length $\ell(T_{MST})$. Denote by $BB(K)$ the bounding box of the given terminal set, that is, the smallest axis-parallel rectangle which includes all terminals. Let bb be the maximal side length of $BB(K)$. Now we can define $b := bb + 2\ell(T_{MST})$. Clearly, an axis-parallel box B of side length b centered at the barycenter of $BB(K)$ is large enough to contain an optimal Steiner tree. Since the minimum spanning tree yields a 2-approximation and $bb \le \ell(T_{opt})$, we also have

$$b \le 5 \cdot \ell(T_{opt}). \tag{1}$$

Step 2: We build a refinement of a Hanan-like grid graph restricted to the area of B. This refinement is parameterized by some parameter k (to be determined later). More specifically, we subdivide the boundary of box B equidistantly with k points into $k + 1$ segments and add for each subdivision point additional lines in all four feasible orientations of the octilinear geometry. To this set of lines we add lines through each terminal and each vertex of an obstacle in all feasible directions. Let G be the graph induced by intersections of these lines restricted to the area inside B (including the boundary of B).

Step 3: The resulting graph may allow subtrees inside obstacles which violate the length restriction L. Therefore, we delete all nodes and edges which lie strictly inside some obstacle.

Step 4: Let $t \in \mathbb{N}$ be another parameter which will be chosen as a constant depending on ε but independent from the given instance. For each obstacle O and for each subset S of at most t vertices on the boundary of O compute an optimal Steiner tree for S which respects the length restriction L inside O. We add each such Steiner tree to the current graph and identify common boundary vertices. Since t is a constant, there is only a polynomial number of these small Steiner tree instances and each of these trees can be computed in constant time (basically by enumerating over all possible tree topologies).

Step 5: Finally, we want that our graph contains a feasible almost shortest octilinear path between any pair of vertices on the boundary of obstacles. More specifically, we require that these paths approximate the true shortest paths by a factor of $1 + 1/(k + 1)$. We can compute these paths and their lengths by the methods from Section 3 and add them to the graph. On the resulting graph $G = G(k, t)$, parameterized by k and t, we can then solve the Steiner tree problem for the given terminal set K.

Choosing $k = \lceil cn/\varepsilon \rceil$ for some constant $c = 374$, we obtain

Theorem 4.1. *Let α denote the approximation guarantee for an algorithm solving the Steiner tree problem in graphs. Given a terminal set K, a set of rectangular soft obstacles O with length restriction L, and some $\varepsilon > 0$, there is an $(\alpha + \varepsilon)$-approximation of the octilinear Steiner tree problem with length restriction L inside obstacles.*

References

1. http://www.xinitiative.org
2. Teig, S.L.: The X architecture: not your father's diagonal wiring. In: SLIP '02: Proceedings of the 2002 International Workshop on System-Level Interconnect Prediction, ACM Press (2002) 33–37
3. Chen, H., Cheng, C.K., Kahng, A.B., Măndoiu, I., Wang, Q.: Estimation of wirelength reduction for λ-geometry vs. Manhattan placement and routing. In: Proceedings of SLIP'03, ACM Press (2003) 71–76
4. Paluszewski, M., Winter, P., Zachariasen, M.: A new paradigm for general architecture routing. Proceedings of the 14th ACM Great Lakes Symposium on VLSI (GLSVLSI) (2004) 202–207

5. Garey, M.R., Johnson, D.S.: The rectilinear Steiner tree problem is NP-complete. SIAM Journal on Applied Mathematics **32** (1977) 826–834
6. Garey, M.R., Graham, R.L., Johnson, D.S.: The complexity of computing Steiner minimal trees. SIAM Journal on Applied Mathematics **32** (1977) 835–859
7. Müller-Hannemann, M., Schulze, A.: Hardness and approximation of octilinear Steiner trees. In: Proceedings of the 16th International Symposium on Algorithms and Computation (ISAAC 2005), Hainan, China. Volume 3827 of Lecture Notes in Computer Science. Springer (2005) 256–265
8. Nielsen, B.K., Winter, P., Zachariasen, M.: An exact algorithm for the uniformly-oriented Steiner tree problem. In: 10th Annual European Symposium on Algorithms (ESA 2002). Volume 2461 of Lecture Notes in Computer Science. Springer (2002) 760–772
9. Coulston, C.: Constructing exact octagonal Steiner minimal trees. In: ACM Great Lakes Symposium on VLSI. (2003) 1–6
10. Zachariasen, M., Winter, P.: Obstacle-avoiding Euclidean Steiner trees in the plane: An exact approach. In: Workshop on Algorithm Engineering and Experimentation. Volume 1619 of Lecture Notes in Computer Science. (1999) 282–295
11. Althaus, E., Polzin, T., Daneshmand, S.V.: Improving linear programming approaches for the Steiner tree problem. Research Report MPI-I-2003-1-004, Max-Planck-Institut für Informatik, Saarbrücken, Germany (2003)
12. Robins, G., Zelikovsky, A.: Improved Steiner tree approximation in graphs. Proceedings of the 11th Annual ACM-SIAM Symposium on Discrete Algorithms (2000) 770–779
13. Du, D.Z., Hwang, F.K.: Reducing the Steiner problem in a normed space. SIAM Journal on Computing **21** (1992) 1001–1007
14. Lee, D.T., Shen, C.F.: The Steiner minimal tree problem in the λ-geometry plane. In: Proceedings 7th International Symposium on Algorithms and Computations (ISAAC 1996). Volume 1178 of Lecture Notes in Computer Science., Springer (1996) 247–255
15. Lin, G.H., Xue, G.: Reducing the Steiner problem in four uniform orientations. Networks **35** (2000) 287–301
16. Arora, S.: Polynomial time approximation schemes for the Euclidean traveling salesman and other geometric problems. Journal of the ACM **45** (1998) 753–782
17. Mitchell, J.S.B.: Guillotine subdivisions approximate polygonal subdivisions: A simple polynomial-time approximation scheme for geometric TSP, k-MST, and related problems. SIAM Journal on Computing **28** (1999) 1298–1309
18. Kahng, A.B., Măndoiu, I.I., Zelikovsky, A.Z.: Highly scalable algorithms for rectilinear and octilinear Steiner trees. Proceedings 2003 Asia and South Pacific Design Automation Conference (ASP-DAC) (2003) 827–833
19. Zhu, Q., Zhou, H., Jing, T., Hong, X., Yang, Y.: Efficient octilinear Steiner tree construction based on spanning graphs. Proceedings 2004 Asia and South Pacific Design Automation Conference (ASP-DAC) (2004) 687–690
20. Müller-Hannemann, M., Peyer, S.: Approximation of rectilinear Steiner trees with length restrictions on obstacles. In: 8th Workshop on Algorithms and Data Structures (WADS 2003). Volume 2748 of Lecture Notes in Computer Science. Springer (2003) 207–218
21. Mehlhorn, K.: A faster approximation algorithm for the Steiner problem in graphs. Information Processing Letters **27** (1988) 125–128

22. Mitchell, J.S.B.: Geometric shortest paths and network optimization. In Sack, J.R., Urrutia, J., eds.: Handbook of Computational Geometry. Elsevier (2000) 633–701
23. Lee, D.T., Yang, C.D., Wong, C.K.: Rectilinear paths among rectilinear obstacles. Discrete Applied Mathematics **70** (1996) 185–215
24. Wu, Y.F., Widmayer, P., Schlag, M.D.F., Wong, C.K.: Rectilinear shortest paths and minimum spanning trees in the presence of rectilinear obstacles. IEEE Transactions on Computing (1987) 321–331
25. Clarkson, K.L., Kapoor, S., Vaidya, P.M.: Rectilinear shortest paths through polygonal obstacles in $O(n(\log n)^2)$ time. In: Proceedings of the 3rd Annual ACM Symposium on Computational Geometry. (1987) 251–257

Simultaneous Embedding with Two Bends per Edge in Polynomial Area

Frank Kammer

Institut für Informatik, Universität Augsburg, D-86135 Augsburg, Germany
kammer@informatik.uni-augsburg.de

Abstract. The simultaneous embedding problem is, given two planar graphs $G_1 = (V, E_1)$ and $G_2 = (V, E_2)$, to find planar embeddings $\varphi(G_1)$ and $\varphi(G_2)$ such that each vertex $v \in V$ is mapped to the same point in $\varphi(G_1)$ and in $\varphi(G_2)$. This article presents a linear-time algorithm for the simultaneous embedding problem such that edges are drawn as polygonal chains with at most two bends and all vertices and all bends of the edges are placed on a grid of polynomial size. An extension of this problem with so-called fixed edges is also considered.

A further linear-time algorithm of this article solves the following problem: Given a planar graph G and a set of distinct points, find a planar embedding for G that maps each vertex to one of the given points. The solution presented also uses at most two bends per edge and a grid whose size is polynomial in the size of the grid that includes all given points. An example shows two bends per edge to be optimal.

1 Introduction

The visualization of information has become very important in recent years. The information is often given in the form of graphs, which should at the same time aesthetically please and convey some meaning. Many aesthetic criteria exist, such as straight-line edges, few bends, a limited number of crossings, depiction of symmetry and a small area of the drawing given, e.g., a minimal distance between two vertices.

If graphs change over the course of time or if different relations among the same objects are presented in graphs, it is often useful to recognize the features of the graph that remain unchanged. If each graph is drawn in its own way, in other words if the graphs are embedded independently, there is probably only little correlation. Therefore, the embeddings of the graphs have to be constructed simultaneously to achieve that all or at least some features of the graph are fixed.

A viewer of a graph quickly develops a mental map consisting basically in the positions of the vertices. If k planar graphs with the same vertex set V are presented, it is desirable that the positions of all vertices in V remain fixed. This problem is called *simultaneous embedding*. An extension of the problem is the so-called *simultaneous embedding with fixed edges*: In addition to the k graphs, a set of edges F is given. A feasible solution is an embedding of the k graphs such that all vertices and all edges in F have fixed embeddings. An algorithm for the

L. Arge and R. Freivalds (Eds.): SWAT 2006, LNCS 4059, pp. 255–267, 2006.

simultaneous embedding problem for k planar graphs with few bends per edge helps to find an embedding with few bends per edge for graphs of *thickness* k. The thickness of a graph G is the minimum number of planar subgraphs into which the edges of G can be partitioned. Since a graph of thickness k can be embedded in k layers without any edge crossings, thickness is an important concept in VLSI design. Additionally, an algorithm for the simultaneous embedding of k planar graphs with fixed edges helps to find an embedding of a graph of thickness k such that certain sets of edges are drawn straight-line as well as identically in all layers.

Definition 1. *A* k-*bend embedding of* $G = (V, E)$ *is an embedding such that each edge in* E *is drawn as a polygonal chain with* $\leq k$ *bends. Thus, an edge with* l *bends consists of* $l + 1$ *straight-line segments.*

Unless stated otherwise, the following embeddings place all vertices and all bends on a grid of size polynomial in the number of vertices. According to results of Pach and Wenger [9], for any number of planar graphs on the same vertex set of size n, an $O(n)$-bend simultaneous embedding is possible. Erten and Kobourov [6] show with a small example that a 0-bend simultaneous embedding does not always exist for two planar graphs. They show that three bends suffice to embed two planar graphs and that one bend is enough in the case of two trees. By using a new algorithm presented in Section 3, this article shows in Section 2 that the number of bends per edge in a simultaneous embedding of two planar graphs can be reduced to two.

Erten and Kobourov also examine simultaneous embeddings with fixed edges in the special case where one input graph is a tree and the other is a path. For special kinds of graphs (caterpillar and outerplanar graphs), Brass et al. [2] show how to embed simultaneously two of the special graphs such that all edges are fixed. For general graphs, the simultaneous embedding problem with fixed edges is considered in Section 4. However, if all edges are fixed, this problem is already for almost all instances of two planar graphs not solvable (Section 5)—even if the number of bends per edge is unbounded. Therefore, the algorithm presented in Section 4 works only with sets of fixed edges with certain properties.

Another variation of the simultaneous embedding problem is described in [1] by Bern and Gilbert: Given a straight-line planar embedding of a planar graph with convex and 4-sided faces, find a suitable location for dual vertices such that the edges of the dual graph are also straight-line segments and cross only their corresponding primal edges.

Kaufmann and Wiese [7] present an algorithm for the *vertices-to-points* problem, which computes an embedding of a planar graph such that the vertices are drawn on a grid at given points. If all vertices and all bends are placed on a grid whose size is polynomial in the size of the grid that includes all given points, their embedding requires up to three bends per edge, but via a similar algorithm as for the simultaneous embedding problem, a 2-bend embedding can be constructed (Sections 2 and 3). If an outer face is specified, Kaufmann and Wiese show that an 1-bend embedding for the vertices-to-points problem is not possible in general. In Section 5, a very short proof of the same lower bound is presented, but now no outer face must be specified.

2   Finding an Embedding

Since the same ideas as already described in [7, 2, 6] are used, these will only be sketched. Many parts of these ideas help to find a 2-bend embedding for both of the two problems below. Assume for the time being that for all planar graphs $G = (V, E)$ considered in the following, a Hamilton cycle C exists and is known. Moreover, let f_G be a bijective function that maps each vertex to a number in $\{1, \ldots, |V|\}$ such that consecutive vertices in C have consecutive numbers modulo $|V|$. The knowledge of the Hamilton cycle C is useful because in a planar embedding of G, each edge not part of C is either completely inside or completely outside C. In the following two problems are defined and their solutions are presented subsequently.

Definition 2. *The simultaneous embedding problem is, given two planar graphs $G_1 = (V, E_1)$ and $G_2 = (V, E_2)$, to find planar embeddings $\varphi(G_1)$ and $\varphi(G_2)$ such that all vertices are fixed, i.e. $\forall v \in V : \varphi_1(v) = \varphi_2(v)$.*

As a first step to find a simultaneous embedding for G_1 and G_2, associate each vertex v with two numbers x, y, where $x = f_{G_1}(v)$ and $y = f_{G_2}(v)$. Use the two numbers of each vertex as its coordinates. Embed the edges in G_1 and G_2 by applying the procedure described below the following definition once for G_1 with *direction = horizontal* and once for G_2 with *direction = vertical*.

Definition 3. *Let $G = (V, E)$ be a planar graph and let P be a set of distinct points in the plane. The vertices-to-points problem is to find a planar embedding φ such that $\forall v \in V : \varphi(v) \in P$.*

For an embedding, sort the given points according to their x-coordinates. Map the vertex v with number $i = f_G(v)$ to the point with the i'th smallest x-coordinate. Continue the embedding of the edges with *direction = horizontal*.

In the following the procedure to embed the edges is described:

Denote the graph under consideration by $G = (V, E)$ and the edge $\{f_G^{-1}(1), f_G^{-1}(|V|)\}$ by $\hat{e}$. W.l.o.g. assume that *direction = horizontal*. Otherwise turn around the construction by 90 degree.

First, embed the edges of the Hamilton path $P = C \setminus \{\hat{e}\}$ as straight lines. For each edge $e \in P$ let x_e and y_e be the absolute values of the differences of the x- and y-coordinates of the endpoints of e. Set $\alpha = \min_{e \in P} \tan(x_e/y_e)$. For each vertex v, let l_v be the vertical line through v. Using a combinatorial embedding of G, partition the edges not part of C in linear time into two sets E_1 and E_2 such that each set can be embedded inside (or outside) the Hamilton cycle without edge intersections. Add the edge $\hat{e}$ to E_1, say. Embed each edge $\{u, v\}$ in E_1 below P and in E_2 above P as part of two rays starting from vertex u to the right of l_u and from vertex v to the left of l_v, if $f_G(u) < f_G(v)$. Draw each ray in such a way that the angle between the ray and the corresponding vertical line is α and cut off the two rays at their point of intersection. If a vertex has several incident edges embedded on the same side of P or if the point of intersection is not on the grid, modify the angle slightly such that planarity is preserved. This yields a 1-bend embedding of G.

However one problem remains: How to find a Hamilton cycle and what to do if no Hamilton cycle exists. The solution is to modify G. According to Chiba and Nishizeki [3], G can be made 4-connected preserving planarity by repeated applying

Operation 1: adding an auxiliary edge and

Operation 2: splitting an original edge of G once and adding a new vertex between the two parts of the split edge.

Denote this modified graph by G'. In [4], Chiba et al. show that every 4-connected graph has a Hamilton cycle that can be found in linear time. Use an embedding for G' to obtain an embedding for G by removing the new edges, merging the embeddings of the two parts of each split edge and replacing each new vertex by a bend for the corresponding edge.

Observe that an edge $e = \{v_1, v_2\}$ in G corresponds to at most two split edges $e_1 = \{v_1, v_{\text{new}}\}$ and $e_2 = \{v_{\text{new}}, v_2\}$ in G'. If both edges e_1, e_2 are embedded with one bend and there is a further bend between the edges e_1, e_2 at v_{new}, the edge e is embedded with three bends. As we see later, one part of the two split edges is inside and the other part is outside the Hamilton cycle used. Thus, this third bend at v_{new} exists only if v_{new} does not appear between v_1 and v_2 in the Hamilton path used for the embedding.

To see this, consider the next two examples.

If v_{new} is behind v_1 and v_2 on the Hamilton path, the edge e is drawn from v_1 rightwards to v_{new} and then leftwards to v_2. But if v_{new} is between v_1 and v_2, the two rays at v_{new} are drawn as one line from the bend point of e_1 through v_{new} to the bend point of e_2.

Using a shrinking angle during the process of embedding instead of an almost fixed angle α, Kaufmann and Wiese described in [7] how to remove the bend point at v_{new}, but this solution requires a grid of exponential size to place the bends of the edges.

Since it is essential where the numbering along the Hamilton cycle starts, let us consider the problem of finding a so-called *closable* Hamilton path. A Hamilton path is closable if it is contained in a Hamilton cycle. A closable Hamilton cycle makes it more explicit which part of the Hamilton cycle is used to number the vertices.

Definition 4. *An edge-extension of a planar graph G is a planar graph G^+ obtained from G by adding auxiliary edges or by splitting edges, i.e. replacing each such edge by a path of length two whose edges are split edges and whose midpoint is a so-called new vertex of degree 2. Thus, each edge in G corresponds to a unique path in G^+ of arbitrary length.*

Given a planar graph, an edge-extension is constructed in linear-time in the next section such that each edge in G corresponds to a path of length ≤ 2 in G^+. Moreover, a closable Hamilton path in G^+ is found at the same time that has the *between property*:

Definition 5. *Let G^+ be an edge-extension of $G = (V, E)$ and let P be a Hamilton path in G^+. P has the between property (in G^+ with respect to G) if each new vertex that was inserted between the two split parts of an edge $\{u, v\}$ is between u and v on the Hamilton path P.*

From the considerations, we can conclude the following.

Theorem 6. *Given two planar graphs G_1 and G_2 based on the same vertex set of size n, a 2-bend simultaneous embedding of G_1 and G_2 can be found in $O(n)$ time such that all vertices and all bends can be placed on a grid whose bounding box is of size $n^{O(1)}$.*

Theorem 7. *Given a planar graph $G = (V, E)$ and a set of at least $|V|$ distinct points P on a grid, a 2-bend embedding of G can be found in linear time such that each vertex is embedded on a point in P and such that the area of the embedding of G is polynomial in the size of the grid.*

3   Finding a Closable Hamilton Path

An extension H of G is first constructed. Although H will not be planar, a closable Hamilton path in H will help to construct a closable Hamilton path in a planar extension of G. Obtain $G' = (V, E)$ by triangulating G. Denote by $\varphi(G')$ a combinatorial embedding of G' and choose an arbitrary face of φ to be the outer face. Let $G'_D = (W, F)$ be the dual graph of G', but without a vertex (and its edges) for the outer face. For each vertex $w \in W$ representing a face A of $\varphi(G')$, denote by $\Delta(w)$ the set of the three vertices on the boundary of A. Define $D = \{(u, v) \mid u \in W \wedge v \in \Delta(u)\}$ and $H = (V \cup W, E \cup F \cup D)$. See Fig. 1 for an example, but for the time being ignore the distinction between vertices inside and outside the set A_{i-1}. Define an area as the union of some faces of $\varphi(G')$ and their boundaries. For an area A, let $V_A \subseteq V \cup W$ be the set of vertices in A, let $V_A^- \subseteq V \cap V_A$ be the set of vertices on the border of A adjacent to a vertex in $V \setminus V_A$, and let $E_A^- \subseteq E$ be the set of edges on the border of A. Choose $\hat{e} = \{u_1, u_2\} \in E$ as an arbitrary edge incident to the outer face of $\varphi(G')$. W.l.o.g. assume that u_1 is visited just before u_2 on a clockwise travel on the border of the outer face. Let $w \in W$ be the vertex of the dual graph that corresponds to the inner face of $\varphi(G')$ incident to $\hat{e}$. Moreover, denote the area of this inner face by A_0 and let u_3 be the third vertex incident to this inner face (i.e. $\Delta(w) = \{u_1, u_2, u_3\}$). Thus $V_{A_0} = \{u_1, u_2, u_3, w\}$, $V_{A_0}^- \subseteq \{u_1, u_2, u_3\}$ and $E_{A_0}^- = \{\{u_1, u_2\}, \{u_2, u_3\}, \{u_1, u_3\}\}$.

Using $P_0 = (\{u_2, u_3\}, \{u_3, w\}, \{w, u_1\}, \{u_1, u_2\})$ as a first simple path in H and A_0 as the processed area, the aim is to extend P_0 and A_0 stepwise such that the following invariants are true after each step i for the processed area A_i and the current path P_i:

Invariant 1: P_i is a simple path containing all vertices in V_{A_i}.

Invariant 2: For all edges $\{u, v\} \in E$ that are crossed by a dual edge e_D on P_i, the subpath of P_i between u and v contains e_D.

Invariant 3: The vertices in P_i occur in the same order in P_i and on the border of A_i, starting with u_2.

Invariant 4: For all edges $(u, v) \in E_{A_i}^-$ one of the following is true:

    Property a: (u, v) is part of the current path P_i.

    Property b: Let $w \in W$ be the dual vertex corresponding to the face of G' that is incident to (u, v) and inside the processed area A_i. Then either (u, w) or (v, w) is part of the current path P_i.

These invariants are all true for P_0 and A_0. Initially ($i = 0$) and in each step i, calculate the sets $V_{A_i}, V_{A_i}^-, E_{A_i}^-$ and for each vertex v the list $V_{A_i}^v = \{u \in V \mid \{v, u\} \in E \land |\{v, u\} \cap V_{A_i}| = 1\}$, ordered in counter clockwise order around v in $\varphi(G')$. This list contains all vertices adjacent to v that are relative to v on the opposite side of A_i. Begin each list with the vertex that is met first on a clockwise travel on the border of A_i starting with u_2. If step i adds a vertex $s \in V$ to the processed area, all these sets and lists can be updated in time $O(degree\ of\ s)$.

Step i is carried out as follows: Choose $s \in V_{A_{i-1}}^v$ for some vertex $v \in V_{A_{i-1}}^-$ on the border of A_{i-1}. While only one vertex of V is to be added to the processed area, test if the processed area together with the edges from s to vertices in $V_{A_{i-1}}^s$ encloses additional vertices $t \in V \setminus (V_{A_{i-1}} \cup \{s\})$. If such a vertex t exists, put s on a stack and process t first.

The test of whether such a vertex t exists is easy: Let $v_0, \ldots, v_k$ be the vertices of the ordered list $V_{A_{i-1}}^s$. Consider also Fig. 1. These vertices are all adjacent to s and they appear in clockwise order on the border of A_{i-1}. Consider in $\varphi(G')$ the vertices adjacent to s in counter clockwise order from v_0 to v_k. If these vertices are all in $V_{A_{i-1}}^s$, no such vertex t exists. Otherwise choose t as the first vertex found that does not belong to $V_{A_{i-1}}^s$. After processing t, continue this check for s.

If no such vertex t exists (any more), the $k+1$ vertices in $V_{A_{i-1}}^s$ together with s define k faces $W_s = \{w_1, \ldots w_k\}$. Number these faces such that w_j is incident

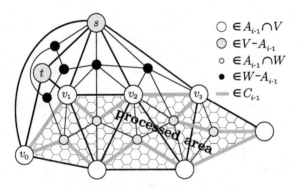

Fig. 1. Extended graph H of a graph $G' = (V, E)$

to v_{j-1} and v_j. In other words, each vertex $w \in W_s$ is adjacent in H to s and to two vertices in $V_{A_{i-1}}^-$. Extend the processed area A_{i-1} by the faces in W_s. For calculating the simple path P_i, two cases are considered. Figures 2 and 3 illustrate the cases 1 and 2, respectively.

Case 1. For some $j \in \{1, \ldots, k\}$, the edge $\{v_{j-1}, v_j\}$ lies on P_{i-1}. Set

$$P_i = (P_{i-1} \setminus \{v_{j-1}, v_j\})$$
$$\cup \{\{v_j, w_{j+1}\}, \{w_{j+1}, w_{j+2}\}, \ldots, \{w_{k-1}, w_k\}\{w_k, s\}\}$$
$$\cup \{\{s, w_1\}, \{w_1, w_2\}, \ldots, \{w_{j-1}, w_j\}, \{w_j, v_{j-1}\}\}.$$

Case 2. Otherwise. Let $\hat{w} \in W \cap V_{A_{i-1}}$ be the vertex inside the processed area A_{i-1} adjacent to v_0 and v_1. Since property a of Invariant 4 does not hold, we know that $\{v_0, \hat{w}\} \in P_{i-1}$ or $\{v_1, \hat{w}\} \in P_{i-1}$. In the first case set $\hat{v} = v_0$ and $\hat{P} = \{\{\hat{v}, s\}, \{\hat{w}, w_1\}, \{w_1, w_2\}\}$; in the other case set $\hat{v} = v_1$ and $\hat{P} = \{\{\hat{w}, w_1\}, \{w_1, s\}, \{\hat{v}, w_2\}\}$. Then

$$P_i = (P_{i-1} \setminus \{\hat{w}, \hat{v}\}) \cup \hat{P}$$
$$\cup \{\{w_2, w_3\}, \ldots, \{w_{k-1}, w_k\}, \{w_k, s\}\}.$$

By the construction of P_i and since Invariant 3 held before the i'th step, Invariants 1 and 2 are true after the i'th step. Since the border of A_i results from the border of A_{i-1} by a replacement of $v_1, \ldots, v_{k-1}$ by s and since the simple

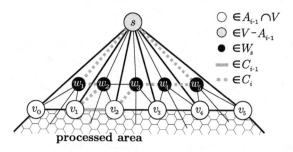

Fig. 2. Face w_2 is incident to an edge in P_{i-1}

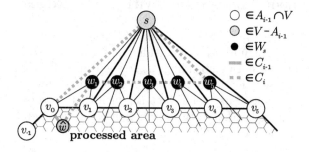

Fig. 3. No face in W_s is incident to an edge in P_{i-1}

path P_i is an extension of P_{i-1} such that s is inserted between some vertices in $\{v_0, \ldots, v_k\}$, Invariant 3 is preserved.

Observe that for each new edge e on the boarder of the processed area (i.e. $e \in E_{A_i}^- \setminus E_{A_{i-1}}^-$), either Property a or b of Invariant 4 is true. Furthermore, in Case 1, the edge $\{v_{j-1}, v_j\} \in P_{i-1} \setminus P_i$ is not in $E_{A_i}^-$ any more after step i. In Case 2, let $v_{-1} \in \Delta(\hat{w}) \setminus \{v_0, v_1\}$. If $\{v_{-1}, v_0\} \in E_{A_{i-1}}^-$, then v_0 is adjacent to only three vertices in A_{i-1} and thus $\{v_{-1}, v_0\} \in P_{i-1}$. Altogether, Invariant 4 is also true after the i'th step.

After $|V| - 3$ steps, $A_{|V|-3}$ equals to the whole internal area of G'. Because of Invariant 1, a closable Hamilton path $P_{|V|-3}$ in H is found. It remains to show how to use the knowledge of a closable Hamilton path in H to find a closable Hamilton path P in a planar extension of G' that is also a planar extension of G. Let $v_{\sigma_1}, \ldots, v_{\sigma_{|V|}}$ be the order of the vertices of V as they appear on $P_{|V|-3}$.

The closable Hamilton path P in an edge-extension of G is constructed by connecting the vertices v_{σ_i} and $v_{\sigma_{i+1}}$ $(1 \leq i < |V|)$. If $\{v_{\sigma_i}, v_{\sigma_{i+1}}\} \in E$, add $\{v_{\sigma_i}, v_{\sigma_{i+1}}\}$ to P. Otherwise draw an edge p from v_{σ_i} to $v_{\sigma_{i+1}}$ such that only the faces are visited that are also visited by $P_{|V|-3}$ and such that each edge in E crossed by p is also crossed by $P_{|V|-3}$. Each time p crosses an edge $e \in E$, break e into two split edges and add a new vertex between them. Also replace p by a path of auxiliary edges that traverses all these new vertices and thus connects v_{σ_i} and $v_{\sigma_{i+1}}$. Add all these newly inserted auxiliary edges to P. Since $P_{|V|-3}$ is a simple path and each edge in E is crossed by only one edge in F, the construction of P breaks each edge $\{u, v\}$ in E into at most two split edges $\{u, v_{\text{new}}\}$ and $\{v_{\text{new}}, v\}$. Additionally, because of Invariant 2, the new vertex v_{new} is between u and v on P. Therefore P has the between property.

Definition 8. *Call an edge-extension G^+ of G a good edge-extension if each new vertex is only incident to two auxiliary and to two split edges.*

Theorem 9. *A good edge-extension G^+ of a planar graph G and a closable Hamilton path P in G^+ can be found in linear time such that each edge in G corresponds to a path of length two in G^+ and P has the between property.*

As discussed in Section 2, this proves Theorems 6 and 7.

4   Simultaneous Embedding with Fixed Edges

Let $G_1 = (V, E_1)$ and $G_2 = (V, E_2)$ be two planar graphs and let $F \subset E_1 \cup E_2$. The goal is to find a simultaneous embedding of G_1 and G_2 such that the edges in F can be drawn in both embeddings as straight lines; in particular, edges in $F \cap E_1 \cap E_2$ are drawn identically in the two embeddings. However, F must have some special properties. First, let F be a set such that no vertex is incident to more than one fixed edge. Later, this restriction is relaxed. Iterate the following once for $G = G_1$ and once for $G = G_2$.

Find a Hamilton cycle C in a good edge-extension of G in which no fixed edge is split, i.e. no fixed edge is crossed by C. Using a more difficult case distinction we can use the algorithm of Section 3 to find a Hamilton cycle in such an edge-extension. However, since we can later handle paths of fixed edges that are crossed several times by the Hamilton cycle, in particular, since we can handle a fixed edge crossed by C, details are omitted. Let φ be the used combinatorial embedding of the algorithm in Section 3. The edges of F are added now successively to C.

Consider the situation shown in Fig. 4. Let $\{\hat{u}, \hat{v}\} \in F$ be an edge that is not part of the Hamilton cycle. Since a Hamilton cycle contains all vertices, two other edges incident to $\hat{u}$ and $\hat{v}$, respectively, are part of the Hamilton cycle. For each vertex v and an incident edge e, denote by E_v^e the sequence of edges incident to v in clockwise order around v in φ starting with e. We add the edge $\{\hat{u}, \hat{v}\}$ to C in two steps.

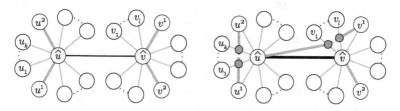

Fig. 4. A fixed edge f (black) and a part of H (bold)

Let $\{u^1, \hat{u}\}$ and $\{u^2, \hat{u}\}$ be the first and second edge in $E_{\hat{u}}^{\{\hat{u},\hat{v}\}}$, respectively, that is part of the Hamilton cycle. Replace successively each edge $\{u_i, \hat{u}\}$ in the list $E_{\hat{u}}^{\{u^1,\hat{u}\}}$ between $\{u^1, \hat{u}\}$ and $\{u^2, \hat{u}\}$—but not equal to one of these—by a new vertex u_i^{new} and the split edges $\{u_i, u_i^{\text{new}}\}$ and $\{u_i^{\text{new}}, \hat{u}\}$. Let $u_1^{\text{new}}, \ldots, u_k^{\text{new}}$ be the new vertices of this step. Replace the part $u^1, \hat{u}, u^2$ of the Hamilton cycle by $u^1, u_1^{\text{new}}, \ldots, u_k^{\text{new}}, u^2$ by the use of new auxiliary edges. Let $\{v^1, \hat{v}\}$ and $\{v^2, \hat{v}\}$ be the first and second edge in $E_{\hat{v}}^{(\hat{u},\hat{v})}$, respectively, that is part of the Hamilton cycle. Replace successively each edge $\{v_i, \hat{v}\}$ in the list $E_{\hat{v}}^{(\hat{u},\hat{v})}$ between $\{\hat{u}, \hat{v}\}$ and $\{v^1, \hat{v}\}$—but not equal to one of these—by a new vertex v_i^{new} and the split edges $\{v_i, v_i^{\text{new}}\}$ and $\{v_i^{\text{new}}, \hat{v}\}$. Let $v_1^{\text{new}}, \ldots, v_l^{\text{new}}$ be the new vertices of this step. Replace the part $v^1, \hat{v}, v^2$ of the Hamilton cycle by $v^1, v_1^{\text{new}}, \ldots, v_l^{\text{new}}, \hat{u}, \hat{v}, v^2$ by the use of new auxiliary edges.

Now, the edge $\{\hat{u}, \hat{v}\}$ is part of C. Observe that this edge is never removed by the subsequent steps. Moreover, no edge in F and no auxiliary edge is ever split. Calling the parts of a multiple split edge further on split edges, we can conclude the following.

Corollary 10. *Given a planar graph G and a set of fixed edges F such that no vertex is incident to ≥ 2 fixed edges, a good edge-extension G^+ of G and a Hamilton cycle C in G^+ can be found such that $F \subset C$.*

Property 11. We can always assume that both auxiliary edges of a new vertex v_{new} are part of the Hamilton cycle C. Otherwise remove v_{new}, its auxiliary edges and merge its split edges. Possibly reroute C.

We can use the ideas of Section 2 to obtain a simultaneous embedding and to draw all edges in F as straight lines. However, we do not know how many bends are necessary for an edge outside the Hamilton cycle. The following lemma helps us to limit the number of bends per edge. Let $V_1 = V$ and let V_2 be the set of new vertices of G^+. Use the following lemma iteratively for each path Q of length > 3 in G^+ corresponding to an edge in G. Observe that Q and C have no edges in common and all edges of Q are split edges. Since the edge-extension G^+ is good and because of Property 11, the application of the lemma below needs no edge splitting and the obtained edge-extension remains good.

Lemma 12. *Let $H = (V_1 \cup V_2, E)$ be a planar graph and let C be a cycle in H that visits all vertices of V_1. Additionally, let $Q = (v_1, v_2, \ldots, v_k)$ be a path in H whose endpoints belong to V_1 and whose remaining vertices all belong to V_2. H can be modified by adding edges and splitting some edges e neither part of C nor part of Q incident to an inner vertex at most two times such that a cycle $\hat{C}$ can be found that visits all vertices of V_1 and $\hat{C}$ crosses Q at most two times.*

Due to space limitations, a proof of Lemma 12 is omitted. Figure 5 sketches one iteration of the proof. Observe that a path Q that is crossed two times by C can be reduced to a path of length 3 (Property 11).

Corollary 13. *Let G be a planar graph, let F be a set of edges and let G^+ be a good edge-extension of G with a Hamilton cycle $C \supseteq F$. Another good edge-extension G^+_{new} of G with a Hamilton cycle C_{new} can be constructed such that C_{new} also contains all edges in F and each edge in G corresponds to a path of length ≤ 3 in G^+_{new}.*

In the following, we consider a generalized set of fixed edges. Moreover, the following algorithm works directly with the algorithm of Section 3.

Definition 14 (star-free). *For a given graph $G = (V, E)$, a set of edges $F \subseteq E$ is star-free if F does not contain three edges with a common endpoint.*

Definition 15 (cycle-free). *For a given graph $G = (V, E)$, a set of edges $F \subseteq E$ is cycle-free if each cycle spanned by F is a Hamilton cycle.*

Let $G_1 = (V, E_1)$ and $G_2 = (V, E_2)$ be two planar graphs and let F be a set of edges that is star- and cycle-free with respect to G_1 and G_2. These graphs are handled now one after another. The set F can contain several paths of fixed edges. For the graph under consideration, let $Q_1, \ldots, Q_r$ denote the paths in F that can not be extended. Again, using the ideas of Section 2, we need a Hamilton cycle C in an edge-extension G^+ that contains all fixed edges.

This can be done iteratively by adding complete paths Q_i for $i = 1, \ldots, r$ to the Hamilton cycle. Construct an arbitrary Hamilton cycle C_0 with the algorithm of Section 3 and let C_i be the Hamilton cycle after step i that contains $Q_1, \ldots, Q_i$.

It remains to show how to add one path Q_i to C_{i-1}. First, use Lemma 12 to reduce the crossings of Q_i and C_{i-1}.

As shown in Fig. 6 by the dashed edges, reroute the ≤ 2 crossings of Q_i and C_{i-1} around one of the endpoints of Q_i. At the same time, handle the complete path Q_i of fixed edges similarly to one fixed edge: Add Q_i to C_{i-1} as shown in Fig. 6 by the dotted edges.

Each edge incident to a vertex on Q_i is split ≤ 2 times by Lemma 12, ≤ 2 times by the rerouting and ≤ 1 time by the step that adds Q_i to C_{i-1}. Altogether, such an edge is split ≤ 5 times. Since an edge in G can be incident only to two inner vertices of paths $Q_1, \ldots, Q_r$, an edge can be split $\leq 2 \cdot 5 = 10$ times after iterating over all $Q_1, \ldots, Q_r$. Again, use Lemma 12 to reduce the crossings of each edge and C_r to two without removing an edge of F from C_r. Use the algorithm of Section 2 to find a 5-bend simultaneous embedding of G_1 and G_2. With a similar argument as for Lemma 3.2 in [7], the number of bends per edge can be reduced to 3 at the expense of exponential area for the embedding.

Fig. 5. Three crossings of Q and C can be reduced to one crossing

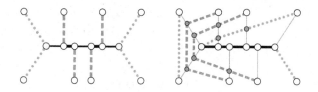

Fig. 6. A path of fixed edges Q (black) and some edges of a Hamilton path (dashed and dotted)

Corollary 16. *A 5-bend simultaneous embedding of two planar graphs with a star- and cycle-free set of fixed edges can be found in linear time. If the area may be arbitrary, three bends suffice.*

5 A Lower Bound and Other Restrictions

The graph shown in Fig. 7 clearly has no Hamilton path, since the white vertices outnumber the black ones by two, but form an independent set.

Lemma 17. *No 1-bend embedding for the vertices-to-points problem is possible in general.*

Proof. Let G be a planar, triangulated graph without a Hamilton path and let P be a set of vertices on a line. Since G has no Hamilton path, there must be two vertices embedded to consecutive points being not adjacent. Since G is triangulated, there is no face incident to these two vertices. Therefore, an edge $\{u, v\}$ with two bends has to exist that crosses the line between the two consecutive points. See Fig. 8(a).

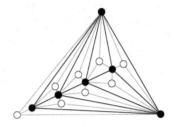

Fig. 7. A triangulated graph without a Hamilton path

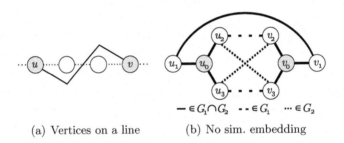

(a) Vertices on a line   (b) No sim. embedding

Fig. 8. Two counterexamples

The algorithm in the last section can only handle a star- and cycle-free set of fixed edges. The question arises whether this restriction is necessary or not. Consider first the case where two triangulated planar graphs and a not cycle-free set of fixed edges are given. Denote the cycle of fixed edges by $C \subseteq F$. If there are two vertices not part of C that are on the same side of the cycle in one of the two graphs and on different sides in the other graph, no simultaneous embedding is possible. Second, consider two triangulated planar graphs and two vertices u_0 and v_0 that are incident to at least three fixed edges $\{u_0, u_1\}, \{u_0, u_2\}, \{u_0, u_3\}$ and $\{v_0, v_1\}, \{v_0, v_2\}, \{u_0, v_3\}$, respectively. See Fig. 8(b). If in one graph the pairs of vertices $\{u_1, v_1\}, \{u_2, v_2\}$ and $\{u_3, v_3\}$, in the other graph the pairs of vertices $\{u_1, v_1\}, \{u_2, v_3\}$ and $\{u_3, v_2\}$ are connected by vertex-disjoint paths, respectively, again no simultaneous embedding is possible.

References

1. M. Bern and J. R. Gilbert. Drawing the planar dual. *Information Processing Letters 43(1)*: 7-13, 1992.
2. P. Brass, E. Cenek, C. Duncan, A. Efrat, C. Erten, D. Ismailescu, S. Kobourov, A. Lubiw, and J. Mitchell. On Simultaneous Planar Graph Embeddings. *LNCS 2748: Algorithms and Data Structures, 8th International Workshop (WADS 2003)*: 219-230.
3. N. Chiba and T. Nishizeki. Arboricity and subgraph listing algorithms. *SIAM J. Comput., 14*: 210-223, 1985.
4. N. Chiba and T. Nishizeki. The hamiltonian cycle problem is linear-time solvable for 4-connected planar graphs. *Journal of Algorithm, 10(2)*: 187-211, 1989.
5. M. B. Dillencourt, D. Eppstein, and D. S. Hirschberg. Geometric thickness of complete graphs. *Journal of Graph Algorithms and Applications, 4(3)*: 5-17, 2000.
6. C. Erten and S. G. Kobourov. Simultaneous embedding of planar graphs with few bends *LNCS 3383: Proc. 12th Int. Symp. Graph Drawing (GD 2004)*: 195-205.
7. M. Kaufmann and R. Wiese. Embedding Vertices at Points: Few Bends Suffice for Planar Graphs. *Journal of Graph Algorithms and Applications. 6(1)*: 115-129, 2002.
8. P. Mutzel, T. Odental, and M. Scharbrodt. The thickness of graphs: a survey. *Graphs Combin., 14(1)*: 59-73, 1998.
9. J. Pach and R. Wenger. Embedding planar graphs at fixed vertex locations. *Graphs and Combinatorics 17*: 717-728, 2001.

Acyclic Orientation of Drawings[*]

Eyal Ackerman[1], Kevin Buchin[2], Christian Knauer[2], and Günter Rote[2]

[1] Department of Computer Science, Technion—Israel Institute of Technology,
Haifa 32000, Israel
ackerman@cs.technion.ac.il
[2] Institute of Computer Science, Freie Universität Berlin, Takustr.
9, 14195 Berlin, Germany
{buchin, knauer, rote}@inf.fu-berlin.de

Abstract. Given a set of curves in the plane or a topological graph, we ask for an orientation of the curves or edges which induces an acyclic orientation on the corresponding planar map. Depending on the maximum number of crossings on a curve or an edge, we provide algorithms and hardness proofs for this problem.

1 Introduction

Let G be a *topological graph*, that is, a graph drawn in the plane such that its vertices are distinct points and its edges are Jordan arcs, each connecting two vertices and containing no other vertex. In this work we further assume that G is a *simple* topological graph, i.e., every pair of edges intersects at most once, either at a common vertex or at a crossing point.

An *orientation* of (the edges of) a graph is an assignment of a direction to every edge in the graph. For a given undirected (abstract) graph an orientation with no directed cycle can be easily computed in linear time by performing a depth-first search on the graph and then orienting every edge from the ancestor to the descendent. However, is it always possible to find an orientation of the edges of a topological graph, such that a traveller on that graph will not be able to return to his starting position even if allowed to move from one edge to the other at their crossing point? Such an orientation is called an *acyclic orientation*. Rephrasing it in a more formal way, let $M(G)$ be the planar map induced by G, that is, the map obtained by adding the crossing points of G as vertices, and subdividing the edges of G accordingly. Then we ask for an orientation of the edges of G such that the induced directed planar map $M(G)$ is acyclic.

Clearly, if the topological graph is *x-monotone*, that is, every vertical line crosses every edge at most once, then one can orient each edge from left to right. Travelling on the graph under such an orientation, one always increases

[*] Work by the first author was done while he was visiting the Freie Universität Berlin, and was partly supported by a Marie Curie scholarship. Research by the second author was supported by the Deutsche Forschungsgemeinschaft within the European graduate program "Combinatorics, Geometry, and Computation" (No. GRK 588/2).

L. Arge and R. Freivalds (Eds.): SWAT 2006, LNCS 4059, pp. 268–279, 2006.

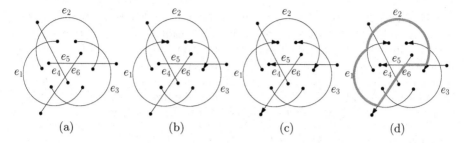

Fig. 1. A non-orientable topological graph

the x-coordinate and therefore there can be no directed cycle. Fig. 1(a) provides an example for a topological graph which has no acyclic orientation: The edges e_1, e_2, e_3 cannot be all oriented clockwise or counter-clockwise, so assume w.l.o.g. that e_1 and e_2 are oriented clockwise, while e_3 is oriented counter-clockwise as in Fig. 1(b). To prevent a cycle with e_1 and e_3, e_6 must be oriented downwards. Similarly, e_5 must be directed leftwards to prevent a cycle with e_2 and e_3 (see Fig. 1(c)). However, this yields the cycle shown in Fig. 1(d). The degree of every vertex in this example is one, i.e. it consists of a set of *curves* embedded in the plane with distinct endpoints. We will consider the case of curves separately.

It turns out that determining whether a topological graph (resp., a set of curves) has an acyclic orientation depends crucially on the maximum number of times an edge in the graph (resp., a curve) can be crossed. Given a (simple) topological graph G on n vertices, such that each edge in G is crossed at most once, we show that one can find an acyclic orientation of G in $O(n)$ time. When four crossings per edge are allowed, deciding whether there exists an acyclic orientation becomes NP-complete. Topological graphs with few crossings per edge were considered in several works in the literature [1, 2, 3]. For a set of n curves with distinct endpoints in which each pair of curves intersects at most once and every curve is crossed at most k times, we describe an $O(n)$-time orientation algorithm for the case $k \leq 3$. When $k \geq 5$ finding an acyclic orientation of the set of curves is NP-complete.

The rest of this paper is organized as follows. In Sect. 2 we study the problem of finding an acyclic orientation for a set of curves. Then, in Sect. 3 we consider the more general case where the input is a topological graph. Finally, we give some concluding remarks in Sect. 4, and mention a few related open problems.

2   Acyclic Orientation of a Set of Curves in the Plane

Throughout this paper we assume the intersections between the curves are known in advance. Given a set of curves $\mathcal{C}$, the vertices of the planar map $M(\mathcal{C})$ induced by $\mathcal{C}$ are the crossing points between the curves, while the edges of $M(\mathcal{C})$ are segments of the curves that connect two consecutive crossing points on a curve. Note that in contrast to the "classical" planar map, we do not consider the pieces that terminate at the endpoints because they cannot contribute to a cycle. As we have mentioned above, the maximum number of crossings per curve plays an important role when we ask for an acyclic orientation of a set of curves. If every curve is crossed at most once, then $M(\mathcal{C})$ contains no edges, and therefore any orientation of $\mathcal{C}$ is acyclic. If $\mathcal{C}$ is a set of curves with at most two crossing points per curve, then $M(\mathcal{C})$ is a union of cycles and paths and thus finding an acyclic orientation of $\mathcal{C}$ is also easy in this case. Hence, the first non-trivial case is where each curve is crossed at most three times. In this case we have:

Theorem 1. *Let $\mathcal{C}$ be a set of n curves in the plane, such that every pair of curves intersects at most once and each curve has at most three crossings. Then one can find an acyclic orientation of $\mathcal{C}$ in $O(n)$ time.*

This result is proved in Sect. 2.1, while in Sect. 2.2 we show:

Theorem 2. *Consider the class of sets of curves in the plane with the following properties: every pair of curves intersects at most once and each curve has at most five crossings. Deciding whether a set of curves from this class has an acyclic orientation is NP-complete.*

2.1 Curves with at Most Three Crossings per Curve

Let $\mathcal{C}$ be a set of n curves in the plane, such that every pair of curves intersect at most once and each curve has at most three crossings. In this section we describe an algorithm for obtaining an acyclic orientation of $\mathcal{C}$. We start by constructing $M(\mathcal{C})$, the planar map induced by $\mathcal{C}$. Clearly, an (acyclic) orientation of $\mathcal{C}$ induces an (acyclic) orientation of the edges in $M(\mathcal{C})$.

Every connected component of $M(\mathcal{C})$ can be oriented independently, therefore we describe the algorithm assuming $M(\mathcal{C})$ is connected. Suppose $\mathcal{C}$ contains a curve c which is crossed less than 3 times. By removing c we obtain a set of $n-1$ curves in which there must be at least two curves (the ones crossed by c) which are crossed at most twice. We continue removing the curves, until none is left. Then we reinsert the curves in reverse order. During the insertion process we reconstruct $M(\mathcal{C})$ and define a total order of its vertices. For this we store the vertices of $M(\mathcal{C})$ in a data structure suggested by Dietz and Sleator [4]. This data structure supports the following operations, both in $O(1)$ worst-case time:

1. INSERT(X,Y): Insert a new element Y immediately after the element X.
2. ORDER(X,Y): Compare X and Y.

By inserting Y after X and then switching their labels we can also use this data structure to insert a new element immediately *before* an existing element in constant time. We also keep a record of the maximal element in the order, MAX (that is, we update MAX when a new element is added after it).

We now describe the way a curve c is reinserted. For every curve c' that has already been added and is crossed by c (there are at most two such curves) we take the following actions. Let x be the crossing point of c and c'. If c' has no other crossing points, then x is inserted after MAX. In case c' has exactly one crossing point x', we insert x after x' when c' is oriented from x' to x, and before x' otherwise. Otherwise, suppose c' has two crossing points x'_1 and x'_2, such that $x'_1 < x'_2$. Then we insert x before x'_1 if x'_1 is the middle point on c' among the three points; after x'_1 if x is the middle point; and after x'_2 if x'_2 is the middle point. In this way, if c' has now three crossing points, they are ordered consistently. Finally we orient c arbitrarily if it has less than two crossings, or from the smaller crossing to the larger one, in case it has two crossings. We refer to the algorithm described above as Algorithm 1.

Lemma 1. *Let C be a set of n curves such that every curve is crossed at most three times and there is a curve that is crossed at most twice. Then Algorithm 1 finds an acyclic orientation of C in $O(n)$ time.*

Proof. Since a total order is defined on the vertices of $M(C)$ and it is easy to verify that every edge is oriented from its smaller vertex to its larger one, it follows that there is no directed cycle in $M(C)$. Computing the connected components of $M(C)$ requires $O(n)$ time. Removing and adding a curve is performed in constant time, therefore the overall time complexity is $O(n)$. □

The more complicated case is when all the curves in C are crossed exactly three times. The general idea in this case is to:

1. find a set of curves S that form an undirected cycle in $M(C)$;
2. orient $C \setminus S$ using Algorithm 1;
3. orient S such that:
   (a) the curves in S do not form a directed cycle; and
   (b) it is impossible to 'hop' on S from $C \setminus S$, 'travel' on S, and 'hop' off back to $C \setminus S$.

Henceforth, we assume that every curve in C is crossed three times. We tackle the orientation problem based on whether or not there is a crossing point x whose degree in $M(C)$ is 3. Suppose x_0 is a crossing point of degree 3, that is, it is the crossing point of two curves, c_0 and c, such that x_0 is an extreme crossing point on c and the middle crossing point on c_0. Denote by a_0 and a_1 (resp., a and b) the other crossing points on c_0 (resp., c) (see Fig. 2(a), the edges of the planar map are drawn in thick).

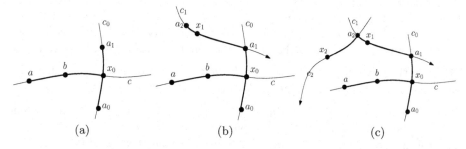

(a) (b) (c)

Fig. 2. Orienting the curves where there is a degree 3 crossing point

We proceed by temporarily removing x_0 and looking for an undirected path in $M(\mathcal{C})$ between a_0 and a_1. If there is no such path, then suppose that there is no path from a_1 to a (the case there is no path from a_0 to a is handled in a similar way). We can solve the orientation problem of $\mathcal{C}$ by solving two sub-problems: first orienting the curves reachable from a_1 (without going through x_0), then orienting the rest of the curves. Both sub-problems can be solved using Algorithm 1.

Now assume we have found a simple path p between a_1 and a_0. Denote by $c_1, c_2, \ldots, c_k$ the curves on this path from a_1 to a_0. Let C be the cycle formed by p, (a_0, x_0) and (x_0, a_1), and let $S = \{c_0, c_1, \ldots, c_k\}$. Our algorithm proceeds by finding an acyclic orientation of $\mathcal{C} \setminus S$ (using Algorithm 1) and then orienting the curves in S. If $c \notin S$ and the direction of c in the acyclic orientation of $\mathcal{C} \setminus S$ is from x_0 to a, then we switch the direction of every curve in $\mathcal{C} \setminus S$. It is easy to see that the curves in $\mathcal{C} \setminus S$ still do not form a directed cycle.

Next we provide the details of the orientation of the curves in S. Denote by a_i the crossing point of c_{i-1} and c_i, $i = 1, \ldots, k-1$, and let $a_0 = a_k$ be the crossing point of c_0 and c_k. If we traverse C starting at x_0 along the curves $c_0, c_1, \ldots, c_k$, then the curve c_i is traversed from the point a_i to the point a_{i+1}, $i = 1, \ldots, k$. Every curve has a third crossing point, that we denote by x_i, and refer to as the *connection point* of c_i. Note that x_i may or may not be on C, and that it is possible that $x_i = x_j$ for $i \neq j$. We say that x_i is a *before*-connection point if a_i is between x_i and a_{i+1} on c_i; x_i is an *after*-connection point if a_{i+1} is between x_i and a_i on c_i; and we say that x_i is a *middle*-connection point if it is not an extreme crossing point on c_i.

We will orient the curves in S such that there will be no directed path between two connection points through S (apart from some cases that will be discussed later on). We start by orienting c_1 from x_1 to a_1 (see Fig. 2(b)), thus making it impossible to "walk" from x_0 to x_1 using c_0 and c_1. Next, we assign an orientation to c_2. The orientation of c_1 already prevents walking on c_1 and c_2 either from x_1 to x_2, or the other direction. We assign an orientation to c_2 such that both directions are impossible (see Fig. 2(c) for an example). We continue orienting the curves $c_3, \ldots, c_k$ in a similar way, making it impossible to reach x_i from x_{i+1} and the other way around, by using the curves c_i and c_{i+1}, for $i = 1, 2, \ldots, k-1$. Finally, we set an orientation to c_0 as follows: In case x_k is already unreachable from x_0 (using the curves c_0 and c_k), we set the orientation of c_0 such that x_0 is unreachable from x_k. Otherwise, we make sure x_k is unreachable from x_0. Orienting S this way guarantees that for $i = 1, \ldots, k-1$ one cannot go from x_i to x_{i+1} or vice versa, using c_i and c_{i+1}. It also guarantees that x_1 is unreachable from x_0 using c_0 and c_1, and that x_k is unreachable from x_0 using c_0 and c_k.

Observation 1. *For every* $i = 1, \ldots, k$ *there is no connection point* $x_j \neq x_i$ *such that there is a directed path on the curves in S from* x_j *to* x_i.

Proof. We prove the claim by induction on number of curves in the path between the two connection points. If the two connection points are on two crossing (that is, adjacent) curves, then the algorithm guarantees that there is no path from x_j to x_i (note that $x_i \neq x_0$). Suppose that there are x_i and x_j such that there is a directed path from x_j to x_i on S, and assume w.l.o.g. that c_{j+1} is the next curve on that path. Note that $c_{j+1} \neq c_0$, for otherwise there is path from x_0 to x_i which is shorter than the path from x_j to x_i. x_{j+1} must be a before-connection point, since otherwise there is a path from x_j to x_{j+1} using c_j and c_{j+1}. However, it follows that there is a path from x_{j+1} to x_i which is shorter than the path from x_j to x_i. $\qquad \square$

Observation 2. C *is not a directed cycle.*

Proof. Assume our orientation results in a directed cycle $a_0 \rightarrow a_1 \rightarrow \cdots \rightarrow a_k = a_0$ (for a directed cycle in the other direction the proof is similar). According to the rules by which c_0 is oriented, it follows that x_k is an after-connection point (otherwise one can walk from x_k to x_0 on c_k and c_0). Considering the orientation of c_k and c_{k-1} and the fact there is no path using them from x_{k-1} to x_k or vice versa, one concludes that x_{k-1} is also an after-connection point. Proceeding in a similar manner implies that x_1 is also an after-connection point. However, in this case c_1 should have been oriented in the reverse direction in order to prevent a path from x_0 to x_1 using c_0 and c_1. $\qquad \square$

Since C is not a directed cycle and Algorithm 1 finds an acyclic orientation of the curves in $C \setminus S$, it remains to verify that there is no cycle that involves curves from S and from $C \setminus S$. If there is such a cycle then when traversing it one must 'hop' on S at some connection point, 'travel' on S for a while, and then 'hop' off S. However, it follows from Observation 1 that one can 'hop' off S only at x_0. Since c is directed from a to x_0 and x_0 is an extreme crossing point on c, the way from x_0 on c leads to a "dead-end" and cannot be part of a directed cycle.

Finally, we have to consider the case where every curve in C is crossed exactly three times, but there is no crossing point whose degree is three (see Fig. 3 for an example). This means that all degrees are 2 and 4, and the endpoints of each edge have different degrees. In this case we first look for an undirected cycle (there must be one as the degree of every vertex in $M(C)$ is at least 2). During our search, after arriving at a vertex v through one of the two curves defining v, we leave through the other curve. Let C be the undirected cycle found, and denote by $c_1, c_2, \ldots, c_k$ the curves forming C (in that order). Again, we first orient the curves in $C \setminus \{c_1, c_2, \ldots, c_k\}$ using Algorithm 1, and then assign orientations to the curves $c_1, c_2, \ldots, c_k$. Let $a_1, a_2, \ldots, a_k$ be the vertices of C, and let $x_1, x_2, \ldots, x_k$ be the connection point of $c_1, c_2, \ldots, c_k$, respectively. Thus, the degree of the points $a_1, a_2, \ldots, a_k$ must alternate between 2 and 4, and k must be even. It also follows from the way we search for a cycle, that the connection points $x_1, x_2, \ldots, x_k$ are alternating 'before' and 'after'-connection points. By orienting the curves $c_1, c_2, \ldots, c_k$ in an alternating manner (see Fig. 3) we make sure that C is not a directed cycle, and that it is impossible to 'hop' on C at some connection point and then 'hop' off at another connection point. Therefore, the resulting orientation is acyclic. Let us refer to the algorithm describe above for the case every curve is crossed exactly three times as Algorithm 2.

Lemma 2. *Let C be a set of n curves such that every curve is crossed exactly three times. Then Algorithm 2 finds an acyclic orientation of C in $O(n)$ time.*

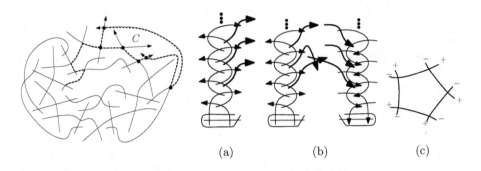

Fig. 3. Orienting the curves when there is no degree 3 crossing point

Fig. 4. A reduction from NAE-k-SAT to orientation of curves with at most 5 crossings per curve. (a) a variable, (b) a gadget for handling negation and extra-crossings, (c) a clause.

Proof. It follows from the correctness of Algorithm 1 and from the discussion above that Algorithm 2 finds an acyclic orientation of C. A cycle is removed only once, and then it is guarantied that there will be some curves with less than three crossings (in each connected component), and therefore we can apply Algorithm 1 on the remaining curves. All the operations concerning finding the cycle, removing it, and orienting the involved curves can be performed in $O(n)$. Thus the overall time complexity is linear in the number of curves. □

Combining lemmata 1 and 2 completes the proof of Theorem 1.

2.2 Curves with at Most Five Crossings per Curve

In this section we show that deciding whether there exists an acyclic orientation of a set of curves with at most 5 crossings per curve is intractable. We will reduce this problem from the following NP-complete variant of the satisfiability problem [5]:

Definition 1. *An instance of* NOT-ALL-EQUAL-k-SAT ($k \geq 3$) *is given by a collection of clauses, each containing exactly k literals. The problem is to determine whether there exists a truth assignment such that each clause has at least one true and one false literal.*

Proof (Theorem 2). The problem is clearly in NP. The problem is shown to be NP-hard by reduction from NOT-ALL-EQUAL-k-SAT to the acyclic orientation problem for $k \geq 3$.

We will have gadgets to represent variables and clauses and we will connect variables to clauses in which they appear by *wires*. Drawing the NAE problem in the plane introduces crossings between the wires. We call these crossings *extra-crossings* in order to distinguish them from the crossings between the curves.

A variable is encoded as shown in Fig. 4(a) where orientations correspond to Boolean signals. In any acyclic orientation all the curves drawn as arrows either have the orientation depicted or the opposite. The thick curves are used as wires and can have three further crossings. The construction uses $4+3k$ curves to generate k wires. Fig. 4(b) shows the encoding of a NOT gate. It uses two wires from one variable and one from the other. The latter is used to propagate the signal across an extra-crossing without introducing a sixth crossing on a curve (this wire has only four crossings). Using this gadget a signal is negated before and after the extra-crossing, thus preventing the introduction of new cycles through the extra-crossing point. The encoding of a clause with k literals is done by k curves forming a k-gon, as shown in Fig. 4(c) for $k - 5$. A wire enters at the plus or at the minus sign depending on whether its corresponding literal in the clause is negated. The edges of the k-gon form a directed cycle if and only if all the literals of the clause are true or all are false. A solution to the NOT-ALL-EQUAL-k-SAT problem will therefore yield an acyclic orientation of the curves. Conversely, if there is no solution, any orientation will either have a cycle at a clause encoding, or have outgoing edges at a variable encoding with different orientations, forcing a cycle within the variable.   □

3 Acyclic Orientation of Topological Graphs

Given a topological graph in which no edge is crossed, one can use the simple algorithm for abstract graphs described in the Introduction to find an acyclic orientation. Thus, the first non-trivial case is when every edge is crossed at most once. In Sect. 3.1 we show that in this case we have:

Theorem 3. *Let G be a simple topological graph on n vertices in which every edge is crossed at most once. Then G has an acyclic orientation. Moreover, such an orientation can be found in $O(n)$ time.*

In Sect. 3.2 we show:

Theorem 4. *Consider the class of simple topological graphs on n vertices in which every pair of edges crosses at most once and each edge has at most four crossings. Deciding whether a graph of this class has an acyclic orientation is NP-complete.*

3.1 Topological Graphs with at Most One Crossing per Edge

Before proving Theorem 3 we recall some basic terms and facts from graph theory.

Definition 2. *An undirected graph $G = (V, E)$ is biconnected if there is no vertex $v \in V$ such that $G \setminus \{v\}$ is not connected. A biconnected component of a connected graph G is a maximal set of vertices that induce a biconnected subgraph.*

Observation 3. *Let G be an undirected graph and let C be a simple cycle in G. Denote by B the biconnected component containing C, and let $e \notin C$ be an edge connecting two vertices of C. Then B is a biconnected component of $G \setminus \{e\}$.*

Definition 3. *Given a biconnected graph $G = (V, E)$ and an edge $\{s, t\} \in E$, an st-numbering (or st-ordering) of G is a bijection $\ell : V \to \{1, 2, \ldots, |V|\}$ such that: (a) $\ell(s) = 1$; (b) $\ell(t) = |V|$; and (c) for every vertex $v \in V \setminus \{s, t\}$ there are two edges $\{v, u\}, \{v, w\} \in E$ such that $\ell(v) < \ell(u)$ and $\ell(v) > \ell(w)$.*

Given an st-numbering we will not make a distinction between a vertex and its st-number. An st-numbering of a graph G naturally defines an orientation of the edges of G: direct every edge $\{u, v\}$ from u to v if $u < v$ and from v to u otherwise.

Lemma 3 (Tamassia, Tollis [6]). *Let $G = (V, E)$ be a plane biconnected multi-graph such that $|V| > 2$, and denote by G' the directed plane graph induced by some st-numbering of G. Let f be a face of G, and denote by G'_f the graph induced by the edges of G' bounding f. Then G'_f has exactly one source and one sink and consists of two directed paths from the source to the sink.*

Algorithm 3. Acyclic orientation of a topological graph with at most one crossing per edge

---

Input: A topological graph G with at most one crossing per edge.
Output: An acyclic orientation of G.

1: **for** each pair of crossing edges $\{a, b\}$ and $\{c, d\}$ **do**
2:     add each of the edges $\{a, c\}, \{a, d\}, \{b, c\}$, and $\{b, d\}$;
3: **end for**
4: compute the biconnected components of the new (multi-)graph;
5: **for** each biconnected component C **do**
6:     temporarily delete all pairs of crossing edges in C;
7:     compute an st-numbering of the remaining subgraph;
8:     reinsert all pairs of crossing edges in C;
9:     orient each edge of C according to the st-numbering;
10: **end for**
11: remove the edges added in line 2;

Proof (Theorem 3). Let G be a simple topological graph with n vertices and m edges in which every edge is crossed at most once. We will show that Algorithm 3 computes an acyclic orientation of G. Denote by G' the graph obtained after adding the edges in lines 1–3. It is always possible to add the edges in line 2 without introducing new crossings. We add them close to the edges $\{a, b\}$ and $\{c, d\}$, such that when these edges are deleted in line 6, the new edges form a face. It may happen that some new edges are parallel to existing edges, and thus the graph may become a multigraph. After this step the vertices of each crossing pair of edges lie on a simple 4-cycle. It is enough to verify that each biconnected

component of G' is acyclically oriented, since (a) every simple cycle in the underlying abstract graph is contained entirely in some biconnected component, and
(b) the crossings do not introduce any interaction between different biconnected
components, as all the vertices of a crossing pair of edges lie on a simple 4-cycle
in G' and therefore are in the same biconnected component. Thus, let us look
at a biconnected component C of G'. We denote by C'' the graph obtained by
removing all pairs of crossing edges from C. Since we only remove edges which
are chords of a cycle C'' is biconnected, therefore, in line (7) an st-numbering of
C'' is indeed computed.

Clearly, one can obtain an acyclic orientation of an abstract graph by numbering the vertices of the graph and directing every edge from its endpoint with the
smaller number to its endpoint with the larger number. Therefore, it is enough
to verify that the crossing points do not introduce a *bad* "shortcut", that is a
path from a vertex u to a vertex v such that $v < u$. Let $(a, b), (c, d)$ be a pair
of crossing edges. Denote by f the 4-face $a - c - b - d - a$ of C''. According to
Lemma 3 the digraph induced by f and the computed st-numbering has only one
source and sink. Therefore, we have to consider only two cases based on whether
the sink and the source are adjacent in f. One can easily verify by inspection
that in both cases no bad shortcut is formed. Thus Algorithm 3 produces an
acyclic orientation.

Algorithm 3 can be implemented to run in time linear in the number of vertices: Finding the biconnected components of a graph takes $O(n + m)$ time [7],
as does the computation of an st-numbering [8]. Therefore the overall running
time is $O(n+m)$. However the maximum number of edges in a topological graph
in which every edge is crossed at most once is $4n - 8$ [3], thus the time and space
complexity of Algorithm 3 is $O(n)$.   □

3.2   Topological Graphs with at Most Four Crossings per Edge

In this section we show that deciding whether there exists an acyclic orientation
of a topological graph with at most four crossings per edge is NP-complete.

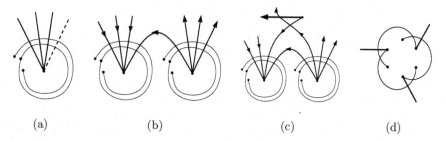

Fig. 5. A reduction from NAE-k-SAT to acyclic orientation of a topological graph with
at most 4 crossings per edge. (a) a variable, (b) a NOT gate, (c) an edge (wire) with
only one crossing so far, (d) a clause gate.

Proof (Theorem 4). An acyclic orientation can be verified in polynomial time, therefore the problem is in NP. As for the case of a set of curves, the reduction is done from NOT-ALL-EQUAL-k-SAT.

The encoding of a variable is shown in Fig. 5(a). A clause can be encoded as in the case of curves, however the slightly more complex encoding shown in Fig. 5(d) uses only one crossing per ingoing wire instead of two. The NOT gate shown in Fig. 5(b) can again be used to split wires. It is not suitable for handling extra-crossings, since the arc connecting a literal and its negation uses all its four crossings for negation. Instead, an edge with only one crossing can be constructed as shown in Fig. 5(c). This edge can then be used to propagate the signal across an extra-crossing, ending in the same construction on the other side. Extra-crossings cannot introduce new cycles, since the arcs at a variable or its negation are again all ingoing or outgoing.   □

4   Discussion

For topological graphs with at most one crossing per edge we showed that an acyclic orientation always exists (and can be found in linear time). When the maximum number of crossings per edge is at least four, deciding whether an acyclic orientation of the graph exists is NP-complete. An obvious open question is what happens when the maximum number of crossings per edge is two or three.

Observation 4. *There is a simple topological graph G with at most three crossings on each curve, which has no acyclic orientation.*

Proof. Such a graph can be constructed with the gadgets of the NP-hardness proof in the case of at most four crossings per edge. The encoding of a variable (Fig. 5(a)) with only three outgoing wires uses at most three crossings per edge and two crossings for the outgoing edges. The outgoing edges all have the same orientation. The encoding of a clause (Fig. 5(d)) with three literals has at most three crossings per edge, and at most one crossing per outgoing wire. In an acyclic orientation of the clause gadget it is impossible for all three outgoing wires to be oriented all inwards or all outwards. Combining a variable and a clause yields a graph with at most three crossings per edge and no acyclic orientation.   □

We do not know whether it can be decided in polynomial time whether a topological graph with at most three crossings per edge has an acyclic orientation. The situation is worse for topological graphs with at most two crossings per edge: So far we were unable to find an example which has no acyclic orientation, or to prove that every such graph is acyclic-orientable.

A special case is where all the vertices in the graph have degree 1. This case corresponds to asking the acyclic orientation question for a set of curves. Clearly, if the problem can be solved (or decided) for topological graphs with at most k crossings per edge, then it can be solved for curves with at most k crossings per curve. It would be interesting to determine whether there is a construction that provides a reduction from topological graphs with at most k crossings per edge to a set of curves with at most k' crossings per curve for some k.

For curves with at most three crossings per curve we provided a linear time algorithm that finds an acyclic orientation. For five crossings per curve we showed that the problem becomes NP-complete. A set of curves with at most four crossings per curve does not always have an acyclic orientation, as Fig. 1 implies. However, the complexity status of the decision problem for such sets of curves is also open. Two other interesting open questions are: (1) What happens if we only require acyclic *faces*? and (2) What happens if we look for an orientation such that for every pair of vertices, u, v, in the induced planar map there is a directed path from u to v or vice versa?

Our original motivation for considering the acyclic orientation of curves does not concern curves in the plane but in surfaces. An important property [9, Theorem 1] of arrangements of double pseudolines in the Möbius band can be formulated in terms of the acyclicity of the 1-skeleton of a certain arrangement of oriented curves in a cylinder. It would be interesting to obtain results concerning acyclic orientations of curves on other surfaces. Also it would be interesting to study acyclic orientations of graphs with more general dependencies between or constraints on the orientation of edges.

Acknowledgments. We thank Michel Pocchiola for suggesting the problem concerning the acyclicity of curves in the plane. We also thank Scot Drysdale, Frank Hoffmann, and Klaus Kriegel for helpful discussions.

References

1. Grigoriev, A., Bodlaender, H.L.: Algorithms for graphs embeddable with few crossings per edge. In: Proc. 15th Int. Symp. on Fundamentals of Computation Theory (FCT). Vol. 3623 of Lecture Notes in Computer Science. Springer (2005) 378–387
2. Pach, J., Radoicic, R., Tardos, G., Tóth, G.: Improving the crossing lemma by finding more crossings in sparse graphs. In: Proc. 20th ACM Symp. on Computational Geometry (SoCG), Brooklyn, NY. (2004) 68–75
3. Pach, J., Tóth, G.: Graphs drawn with few crossings per edge. Combinatorica **17**(3) (1997) 427–439
4. Dietz, P., Sleator, D.: Two algorithms for maintaining order in a list. In: Proc. 19th Ann. ACM Symp. on Theory of Computing (STOC), NYC, NY. (1987) 365–372
5. Schaefer, T.J.: The complexity of satisfiability problems. In: Proc. 10th Ann. ACM Symp. on Theory of Computing (STOC), San Diego, CA. (1978) 216–226
6. Tamassia, R., Tollis, I.G.: Algorithms for visibility representations of planar graphs. In: Proc. 3rd Ann. Symp. on Theoretical Aspects of Computer Science, Orsay, France. Vol. 210 of Lecture Notes in Computer Science. Springer (1986) 130–141
7. Even, S.: Graph Algorithms. Computer Science Press (1979)
8. Even, S., Tarjan, R.E.: Computing an st-numbering. Theoretical Computer Science **2**(3) (1976) 339–344
9. Habert, L., Pocchiola, M.: A homotopy theorem for arrangements of double pseudolines. In: Proc. 20th European Workshop on Comput. Geom. (EWCG), Delphi, Greece. (2006) 211–214

Improved Algorithms for Quantum Identification of Boolean Oracles

Andris Ambainis[1,*], Kazuo Iwama[2,**], Akinori Kawachi[3,***],
Rudy Raymond[2,†], and Shigeru Yamashita[4,‡]

[1] Department of Combinatorics and Optimization, University of Waterloo
ambainis@math.uwaterloo.ca
[2] Graduate School of Informatics, Kyoto University
{iwama, raymond}@kuis.kyoto-u.ac.jp
[3] Graduate School of Information Science and Engineering,
Tokyo Institute of Technology
kawachi@is.titech.ac.jp
[4] Graduate School of Information Science, Nara Institute of Science and Technology
ger@is.naist.jp

Abstract. The oracle identification problem (OIP) was introduced by Ambainis et al. [3]. It is given as a set S of M oracles and a blackbox oracle f. Our task is to figure out which oracle in S is equal to the blackbox f by making queries to f. OIP includes several problems such as the Grover Search as special cases. In this paper, we improve the algorithms in [3] by providing a mostly optimal upper bound of query complexity for this problem: (*i*) For any oracle set S such that $|S| \leq 2^{N^d}$ ($d < 1$), we design an algorithm whose query complexity is $O(\sqrt{N \log M / \log N})$, matching the lower bound proved in [3]. (*ii*) Our algorithm also works for the range between 2^{N^d} and $2^{N/\log N}$ (where the bound becomes $O(N)$), but the gap between the upper and lower bounds worsens gradually. (*iii*) Our algorithm is robust, namely, it exhibits the same performance (up to a constant factor) against the noisy oracles as also shown in the literatures [2, 11, 18] for special cases of OIP.

1 Introduction

We study the following problem, called the Oracle Identification Problem (OIP): Given a hidden N-bit vector $f = (a_1, \ldots, a_N) \in \{0,1\}^N$, called an *oracle*, and a *candidate set* $S \subseteq \{0,1\}^N$, OIP requires us to find which oracle in S is equal to f. OIP has been especially popular since the emergence of quantum computation, e.g., [7,8,9,11,13,18]. For example, suppose that we set

  * Supported by CIAR, NSERC, ARO and IQC University Professorship.
 ** Supported in part by Scientific Research Grant, Ministry of Japan, 16092101.
*** Supported by Scientific Research Grant, Ministry of Japan, 17700007 and 16092206.
 † Currently at IBM Tokyo Research Laboratory.
 ‡ Supported in part by Scientific Research Grant, Ministry of Japan, 16092218.

L. Arge and R. Freivalds (Eds.): SWAT 2006, LNCS 4059, pp. 280–291, 2006.

$S = \{(a_1, \ldots, a_N) |$ exactly one $a_i = 1\}$. Then this OIP is essentially the same as Grover search [17]. In [3], Ambainis et al. extended the problem to a general S. They proved that the total cost of *any* OIP with $|S| = N$ is $O(\sqrt{N})$, which is optimal within a constant factor since this includes the Grover search as a special case and for the latter an $\Omega(\sqrt{N})$ lower bound is known (e.g., [9]). For a larger S, they obtain nontrivial upper and lower bounds, $O(\sqrt{N \log M \log N} \log \log M)$ and $\Omega(\sqrt{N \log M / \log N})$, respectively, but unfortunately, there is a fairly large gap between them.

Our Result. Let $M = |S|$. (*i*) If $M \leq 2^{N^d}$ for a constant d (< 1), then the cost of our new algorithm is $O(\sqrt{N \log M / \log N})$ which matches the lower bound obtained in [3]. (Previously we have an optimal upper bound only for $M = N$). (*ii*) For the range between 2^{N^d} and $2^{N / \log N}$, our algorithm works without any modification and the (gradually growing) gap to the lower bound is at most a factor of $O(\sqrt{\log N \log \log N})$. (*iii*) Our algorithm is robust, namely, it exhibits the same performance (up to a constant factor) against the *noisy* oracles as shown in the literatures [2, 11, 18] for special cases of OIP.

Our algorithms use two operations: (*i*) The first one is a simple query (*S-query*) to the hidden oracle, i.e., to obtain the value (0 or 1) of a_i by specifying the $\log N$-bit index i. The cost for this query is one per each. (*ii*) The second one is called a *G-query* to the oracle: By specifying a set $T = \{i_1, \ldots, i_r\}$ of indices, we can obtain, if any, an index $i_j \in T$ s.t. $a_{i_j} = 1$ and nill otherwise. If there are two or more such i_j's then one of them is chosen at random. The cost for this query is $O(\sqrt{|T|/K})$ where $K = \left| \{i_j \mid i_j \in T \text{ and } a_{i_j} = 1\} \right| + 1$. This query is stochastic, i.e., the answer is correct with a constant probability. Obviously our goal is to minimize the cost for solving the OIP with a constant success probability. Note that we incur the cost for only S- and G-queries (i.e., the cost for any other computation is zero), and it turns out that our query model is equivalent to the standard query complexity one, e.g., [6].

If we use the two queries as blackbox subroutines together with their cost rule, then any knowledge about quantum computation is not needed in the design and analysis of our algorithms. Since S is a set of M 0/1-vectors of length N, it is naturally given as a 0/1 matrix Z of N columns and M rows. For a given Z, our basic strategy is quite simple: if there is a column which includes a balanced number of 0's and 1's, then we ask the value of the oracle at that position by using an S-query. This reduces the number of candidates by a constant factor. Otherwise, i.e., if every column has, say, a small fraction of 1's, then S-queries may seldom reduce the candidates. In such a situation, the idea is that it is better to use a G-query by selecting a certain number of columns in T than repeating S-queries. In order to optimize this strategy, our new algorithm controls the size of T very carefully. This contrasts with the previous method [3] that uses G-queries always with $T = \{1, \ldots, N\}$.

Previous Work. Suppose that we wish to solve some problem over input data of N bits. Presumably, we need all the values of these N bits to obtain a correct answer, which in turn requires N (simple) queries to the data. In a certain situation, we do not need all the values, which allows us to design a variety of sublinear-time (classical) algorithms, e.g., [12, 16, 20]. This is also true when the input is given with some premise, for which giving a candidate set as in this paper is the most general method. Quickly approaching to the hidden data using the premise information is the basis of algorithmic learning theory. In fact, Atici et al. in [5] independently use techniques similar to ours in the context of quantum learning theory. One of their results, which states the existence of a quantum algorithm for learning a concept class S whose parameter is γ_S with $O(\log|S|\log\log|S|/\sqrt{\gamma_S})$ queries, almost establishes a conjecture of $O(\log|S|/\sqrt{\gamma_S})$ queries in [19].

2   S-Queries, G-Queries and Robustness

Recall that an instance of OIP is given as a set $S = \{f_1, \ldots, f_M\}$ of oracles, each $f_i = (f_i(1), \ldots, f_i(N)) \in \{0, 1\}^N$, and a hidden oracle $f \in S$ which is not known in advance. We are asked to find the index i such that $f = f_i$. We can access the hidden oracle f through a unitary transformation U_f, which is referred to as an *oracle call*, such that $U_f|x\rangle|0\rangle = |x\rangle|f(x)\rangle$, where $1 \leq x \leq N$ denotes the bit-position of f whose value (0 or 1) we wish to know. This bit-position might be a superposition of two or more bit-positions, i.e., $\sum_i \alpha_i|x_i\rangle$. Then the result of the oracle call is also a superposition, i.e., $\sum_i \alpha_i|x_i\rangle|f(x_i)\rangle$. The query complexity counts the number of oracle calls being necessary to obtain a correct answer i with a constant probability.

In this paper we will not use oracle calls directly but through two subroutines, S-queries and G-queries. (Both can be viewed as classical subroutines when used.) An S-query, $SQ(i)$, is simply a single oracle call with the index i plus observation. It returns $f(i)$ with probability one and its query complexity is obviously one. A G-query, $GQ(T)$, where $T \subseteq \{1, \ldots, N\}$, returns $1 \leq i \leq N$ such that $i \in T$ and $f(i) = 1$ if such i exists and nill otherwise. We admit an error, namely, the answer may be incorrect but should be correct with a constant probability, say, $2/3$. Although details are omitted, it is easy to see that $GQ(T)$ can be implemented by applying Grover Search only to the selected positions T. Its query complexity is given by the following lemma.

Lemma 1 ([10]). $GQ(T)$ *needs* $O(\sqrt{|T|/K})$ *oracle calls, where*
$K = |\{j|\ j \in T \ and \ f(j) = 1\}| + 1.$

If f is a *noisy oracle*, then its unitary transformation is given as follows [2]: $\tilde{U}_f|x\rangle|0\rangle|0\rangle = \sqrt{p_x}|x\rangle|\phi_x\rangle|f(x)\rangle + \sqrt{1-p_x}|x\rangle|\psi_x\rangle|\neg f(x)\rangle$, where $2/3 \leq p_x \leq 1$, $|\phi_x\rangle$ and $|\psi_x\rangle$ (the states of working registers) may depend on x. As before $|x\rangle$ (and hence the result also) may be a superposition of bit-positions. Since an oracle call itself includes an error, an S-query should also be stochastic. $\tilde{SQ}(i)$ returns $f(i)$ with probability at least $2/3$ (and $\neg f(i)$ with at most $1/3$). G-queries, $\tilde{GQ}(T)$, are already stochastic, i.e., succeed to find an answer with probability at least $2/3$ if there exists one, and they do not need modification.

Lemma 2 ([18]). *Let K and T be as before. Then $\tilde{G}Q(T)$ needs $O(\sqrt{|T|/K})$ noisy oracle calls.*

In this paper our oracle mode is almost always noisy. Therefore we simply use the notation SQ and GQ instead of $\tilde{S}Q$ and $\tilde{G}Q$.

3   Algorithms for Small Candidate Sets

3.1   Overview of the Algorithm

Recall that the candidate set S ($|S| = M$) is given as an $M \times N$ matrix Z. Before we give our main result in the next section, we discuss the case that Z is small, i.e., $M = \mathrm{poly}(N)$ in this section, which we need in the main algorithm and also will be nice to understand the basic idea. Since our goal is to find a single row from the M ones, a natural strategy is to reduce the number of candidate rows (a subset of rows denoted by S) step by step. This can be done easily if there is a column, say, j which is "balanced," i.e., which has an approximately equal number of 0's and 1's in $Z(S)$, where $Z(S)$ denotes the matrix obtained from Z by deleting all rows not in S. Then by asking the value of $f(j)$ by an SQ(j), we can reduce the *size* of S (i.e., the number of oracle candidates) by a constant factor. Suppose otherwise, that there are no such good columns in $Z(S)$. Then we gather a certain number of columns such that the set T of these columns is "balanced," namely, such that the number of rows which has 1 somewhere in T is a constant fraction of $|S|$. (See Fig. 1 where the columns in T are shifted to the left.) Now we execute GQ(T) and we can reduce the size of S by a constant fraction according to whether GQ(T) returns nill (S is reduced to S_2 in Fig. 1) or not (S is reduced to S_1 in Fig. 1). Then we move to the next iteration until $|S|$ becomes one.

The merit of using GQ(T) is obvious since it needs at most $O(\sqrt{|T|})$ queries while we may need roughly $|T|$ queries if asking each position by S-queries. Even so, if $|T|$ is too large, we cannot tolerate the cost for GQ(T). So, the key issue here is to set a carefully chosen upper bound for the size of T. If we can select T within this upper bound, then we are happy. Otherwise, we just give up constructing T and use another strategy which takes advantage of the sparseness of the current matrix $Z(S)$. (Obviously $Z(S)$ is sparse since we could not select a T of small size.)

It should be also noted that in each iteration the matrix $Z(S)$ should be *one-sensitive*, namely the number of 1's is less than or equal to the number of 0's in every column. (The reason is obvious since it does not make sense to try to find 1 if almost all entries are 1.) For this purpose we implicitly apply the *column-flipping* procedure in each iteration. Suppose that some column, say j, of $Z(S)$ has more 1's than 0's. Then this procedure "flips" the value of $f(j)$ by adding an extra circuit to the oracle (but without any oracle call). Let this oracle be $\overline{f(j)}$ and $\overline{Z(S(j))}$ be the matrix obtained by flipping the column j of $Z(S)$. Then obviously $f \in S$ iff the matrix $\overline{Z(S(j))}$ contains the row $\overline{f(j)}$, i.e., the problem does not change essentially. Note that the column-flipping is the same as that in [3], where the OIP matrix was written as a $N \times M$ (number of columns $\times$ number of rows) 0-1 matrix instead of the more common $M \times N$ one.

3.2 Procedure RowReduction(• ••) for Reducing Oracles Candidates

This procedure narrows S in each iteration, where T is a set of columns and l is an integer ≥ 1 necessary for error control. See Procedure 1 for its pseudocode. Case 1: If f has one or more 1's in T like f_1 in Fig. 1, then $k = \mathrm{GQ}(T)$ gives us one of the positions of these 1's, say the circled one in the figure. The procedure returns with the set S_1' of rows in the figure, i.e., the rows having a 1 in the position selected by the $\mathrm{GQ}(T)$. Case 2: If f has no 1's in T like f_2 in the figure, then $k = $ nill (i.e., $\mathrm{GQ}(T)$ correctly answered). Even if $k \neq$ nill ($\mathrm{GQ}(T)$ failed) then Majority(k, l, f), i.e., the majority of $60l$ samples of $f(k)$, is 0 with high probability regardless of the value of k. Therefore the procedure returns with the set S_2 of rows, i.e., the rows having no 1's in T. The parameter l guarantees the success probability of this procedure as follows.

Lemma 3. *The success probability and the number of oracle calls in the procedure RowReduction(T, l) are $1 - O(l/3^l)$ and $l(O(\sqrt{|T|}) + l)$, respectively.*

The success probability can be derived from an ordinary argument by Chernoff bound, and the rigorous proof can be found in [4].

3.3 Procedure RowCover(• ••) for Collecting Position of Queries

As mentioned in Sec. 3.1, we need to make a set T of columns being balanced as a whole. This procedure is used for this purpose where $Z(S)$ is the current matrix and $0 < r \leq 1$ controls the size of T. See Procedure 2 for its pseudocode. As shown in Fig. 2, the procedure adds columns $t_1, t_2, \ldots,$ to T as long as a new addition t_i increases the number of covered rows ($= |\mathrm{PositiveRow}(T, Z)|$) by a factor of r or until the number of covered rows becomes $|S|/4$. We say that RowCover *succeeds* if it finishes with S' such that $|S'| \leq \frac{3|S|}{4}$ and *fails* otherwise. Suppose that we choose a smaller r. Then this guarantees that the resulting $Z(S)$ when RowCover fails is more sparse, which is desirable for us as described later. However since $|T| \leq 1/r$, a smaller r means a larger T when the procedure succeeds, which costs more for G-queries in RowReduction. Thus, we should choose the minimum r such that the query complexity for the case that RowCover keeps succeeding as long as the total cost does not exceed the total limit ($= O(\sqrt{N})$).

3.4 Analysis of the Whole Algorithm

Now we are ready to prove our first theorem:

Theorem 1. *The $M \times N$ OIP can be solved with a constant success probability by querying the blackbox oracle $O(\sqrt{N})$ times if $M = poly(N)$.*

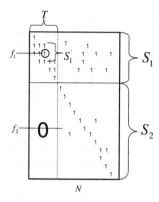

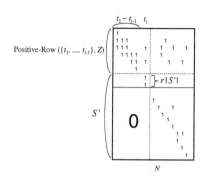

Fig. 1. Reducing the candidate set by G-queries $GQ(T)$ on the column set T

Fig. 2. Constructing the column set T by RowCover(S, r)

Proof. See Procedure 5 for the pseudocode of the algorithm ROIPS(S, Z) (Robust OIP algorithm for Small Z). We call this procedure with $S = \{1, \ldots, M\}$ (we need this parameter since ROIPS is also used in the later algorithm) and the given matrix Z. As described in Sec. 3.1, we narrow the candidate set S at lines 2 and 3. If RowCover at line 2 succeeds, then $|S|$ is sufficiently reduced. Even if RowCover fails, $|S|$ is also reduced similarly if RowReduction at line 3 can find a 1 by G-queries. Otherwise line 7 is executed where the current oracle looks like f_2 in Fig. 1. In this case, by finding a 1 in the positions $\{1, \ldots, N\} \setminus T$ by the G-query at line 7, $|S|$ is reduced to $|S| \log^4 N / N$, because we set $r = \log^4 N / N$ at line 2. Since the original size of S is N^c for a constant c, line 7 is executed at most $c + 1$ times.

Note that the selection of the value of r at line 2 follows the rule described in Sec. 3.3: Since $r = \log^4 N / N$, the size of T at line 3 is at most $N / \log^4 N$. This implies that the number of oracle calls at line 3 is $O(\log N \cdot \sqrt{N} / \log^2 N) = O(\sqrt{N} / \log N)$. Since line 3 is repeated at most $O(\log N)$ times, the total number of oracle calls at line 3 is at most $O(\sqrt{N})$. Line 7 needs $O(\sqrt{N})$ oracle calls, but the number of its repetitions is $O(1)$ as mentioned above. Thus the total number of oracle calls is $O(\sqrt{N})$.

Also by Lemma 1, the error probability of line 3 is at most $O(\log N / N)$. Since the number of repetitions is $O(\log N)$, this error probability is obviously small enough. The error probability of line 7 is constant but again this is not harmful since it is repeated only $O(1)$ times, and thus the error probability can be made as small as it is needed at constant cost.

4   Algorithms for Large Candidate Sets

4.1   Overview of the Algorithm

In this section, our $M \times N$ input matrix Z is large, i.e., M is superpolynomial. We first observe how the previous algorithm, ROIPS, would work for such a large Z. Due to the rule given in Sec. 3.3, the value of r at line 2 should be $\beta = \log M (\log \log M)^2 \log N/(2N)$. The calculation is not hard: Since we need $\log M$ repetitions for the main loop, we should assign roughly $\log \log M$ to l of RowReduction for a sufficiently small error in each round. Then the cost of RowReduction will be $\sqrt{1/\beta} \cdot \log \log M$. Furthermore, we have to multiply the number of repetitions by $\log M$ factor, which gives us $\sqrt{N \log M/\log N}$, the desired complexity. Thus it would be nice if RowCover keeps succeeding. However, once RowCover fails, each column can still include as many as $M\beta$ 1's which obviously needs too many repetitions of RowReduction at line 7 of ROIPS.

Recall that the basic idea of ROIPS is to reduce the number of candidates in the candidate set S by halving (the first phase) while the matrix is dense and to use the more direct method (the second phase) after the matrix becomes sufficiently sparse. If the original matrix is large, this strategy fails because, as mentioned above, the matrix does not become sufficiently sparse after the first phase. Now our idea is to introduce an "intermediate" procedure which reduces the number of the candidates more efficiently than the first phase. For this purpose, we use RowReductionExpire_MTGS, which tries to find a position of "1" in the oracle with multi-target Grover Search ($K > 1$ in Lemma 5) by assuming that the portion of such position, K/N, is sufficiently larger than $1/\beta$. If the assumption is indeed true then we apply RowReduction as before and moreover the number of G-queries in the main loop of RowReduction is repeated for a constant time of $\sqrt{N/K}$ on average.

However, it is of course possible that the actual number of repetitions is far different from the expected value. That is why we limit the maximum number of oracle calls spent in G-queries by MAX_QUERIES(N, M), a properly adjusted number which depends on the size of the OIP matrix, and will be referred in the hereafter without its arguments for simplicity. If the value of COUNT gets this value, then the procedure expires (just stops) with no answer, but this probability is negligibly small by selecting MAX_QUERIES appropriately. Notice also that because of the failure of phase 1, it is guaranteed that the number of 1's in each column is "fairly" small, which in turn guarantees that the degree of row reduction is satisfactory for us. See Procedure 8 for our new algorithm ROIPL.

Finally, when the assumption is false, RowReductionExpire_MTGS finishes after $\log \log(\log M/\log N)$ iterations of its main loop. In this case, we can prove that the matrix of the remaining candidates is very sparse and the number of its rows decreases exponentially by a single execution of RowReductionExpire_MTGS. Thus one can achieve our upper bound also (details are given in the next section).

4.2   Justification of the Algorithm

One can see that in ROIPL, oracle calls take place only at lines 6 and 11. As described in the previous overview, the total number of oracle calls in RowReduction at line 6 is $O(\sqrt{N \log M/\log N})$, and the whole execution of this part successfully ends up with high probability. For the cost of line 11, we can prove the following lemma.

Lemma 4. *The main loop (line 4 to 13) of ROIPL finishes with high probability before the value of COUNT reaches MAX_QUERIES(N, M).*

Proof. Note that there are two types of oracle calls in RowReductionExpire _MTGS at lines 11. The first type, Type A, is when portion of "1" in the hidden oracle is at least $1/4(\log|S|/(N\log N))$, and the other type, Type B, is when the portion of "1" is smaller. Let $W = W_A + W_B$ be the expected number of oracle calls, where W_A is the expected number of Type A calls and W_B, that of Type B calls. It is enough to prove that $W_A \leq \frac{2}{3}$MAX_QUERIES and $W_B < \frac{1}{3}$MAX_QUERIES. Here we give the following more simple averaging argument on the bounds of W_A and W_B. (The rigorous proof can be found in [4].)

We first prove that $W_A \leq \frac{2}{3}$MAX_QUERIES. First, note that RowReduction-Expire_MTGS for Type A should require an $O(1)$ expected number of iterations of GQ, each of which requires $O(\sqrt{N\log N/\log|S|})$ queries. Now, since phase 1 has failed, the number of rows having a "1" at some position in $T = \{1..N\}$ is at most $\beta|S|$. Thus, after the above $O(\sqrt{N\log N/\log|S|})$ queries the number of candidates is reduced by a factor of $\beta = (\frac{1}{2})^{\log(1/\beta)}$. Therefore, intuitively, to reduce the number of candidates by half, the number of queries spent in $GQ(T)$ is $O(\frac{1}{\log(1/\beta)}\sqrt{N\log N/\log|S|})$.

Thus we have the following recurrence relation:

$$W_A(|S|) \leq max(W_A(1), W_A(2), \cdots, W_A(|S|/2)) + O(\tfrac{1}{\log(1/\beta)}\sqrt{N\log N/\log|S|}),$$

where $W_A(|S|)$ is the number of Type A queries to distinguish the candidate set S. Since ROIPL starts with $|S| = M$ and ends with $|S| \approx N^{10}$ (note that $\beta|S| > 2$ if $|S| \approx N^{10}$), the above recurrence relation resolves to the following:

$$W_A(M) \leq W_A(M/2) + \tfrac{\sigma}{\log(1/\beta)}\sqrt{N\log N/\log M}$$

$$\leq \sigma\tfrac{\sqrt{N\log N}}{\log(1/\beta)}\left(\tfrac{1}{\sqrt{\log M}} + \tfrac{1}{\sqrt{\log(M/2)}} + \cdots + \tfrac{1}{\sqrt{10\log N}}\right)$$

$$\leq \sigma\tfrac{\sqrt{N\log N}}{\log(1/\beta)}\left(\tfrac{1}{\sqrt{\log M}} + \tfrac{1}{\sqrt{\log M-1}} + \cdots + 1\right)$$

$$\leq 2\sigma \cdot \tfrac{\sqrt{N\log M\log N}}{\log(1/\beta)},$$

where σ is a sufficiently large constant. Therefore, the total number of queries is $O(\sqrt{N\log M/\log N})$ since $\log(1/\beta) = \Omega(\log N)$ if $M \leq 2^{N^d}$. Note that if the above averaging argument is correct then $|S|$ can be reduced into a constant by just repeating line 11. However, this is not exactly true for ROIPL since $|S|$ can only be reduced until becoming poly(N) in order to obtain the desired number of query complexity (see the proof of Lemma 6 in [4]). Fortunately, in this case we can resort to ROIPS for identifying the hidden oracle out of poly(N) candidates with just $O(\sqrt{N})$ queries as in line 16, and thus achieve a similar result with the averaging argument.

For technical details of ROIPL, note that 1/3MAX_QUERIES is ten times the expected total number of queries supposing all queries are at line 11, i.e., the case with the biggest number of Type A queries. By Markov bound, the probability that the number of queries exceeds this amount is negligible (at most 1/10). We summarize the property of RowReductionExpire_MTGS in the following lemma which can be proven similarly as Lemma 3.

Lemma 5. *The success probability and the number of oracle calls of the procedure RowReductionExpire_MTGS($T, l, COUNT, r$) are $1 - O(l/3^l)$ and $l(O(\sqrt{1/r}) + l)$, respectively. Moreover, if there are more than r fraction of 1's in the current oracle, then the average number of queries is $O(\sqrt{1/r} + l)$.*

We next prove that $W_B < \frac{1}{3}$MAX_QUERIES. In this case, MultiTargetGQ fails and therefore the density of "1" at every row of the candidates is less than $\gamma = \frac{1}{4}\log|S|/(N\log N)$. Note that any two rows in S'' (the new S at the left-hand side of line 11) must be different, i.e., we have to generate $|S''|$ different rows by using at most γN 1's for each row. Let W be the number of rows in S'' which include at most $2\gamma N$ 1's. Then $|S''| - W$ rows include at least $2\gamma N$ 1's, and hence the number of such rows must be at most $|S|/2$. Thus we have $|S''| - W \leq |S|/2$ and it follows that $|S''| \leq 2W \leq 2\sum_{k=0}^{\lambda=\lceil 2\gamma N\rceil} \binom{N}{k}$. The right-hand side is at most $2 \cdot 2^{NH(\lambda/N)}$ (see e.g., [14], page 33), which is then bounded by $2|S|^{1/2}$ since $H(x) \approx x\log(1/x)$ for a small x. Thus, we have $|S''| \leq 2|S|^{1/2}$. Hence, the number of candidates decreases doubly exponentially, which means we need only $O(\log(\log M/\log N))$ iterations of RowReductionExpire_MTGS to reduce the number of the candidates from M to N^{10}. Note that we let $l = \log\log(\log M/\log N)$ at line 11 and therefore the error probability of its single iteration is at most $O(1/\log(\log M/\log N))$. Considering the number of iterations mentioned above, this is enough to claim that $W_B < \frac{1}{3}$MAX_QUERIES (see [4] for the proof in detail, where the actual bound of W_B is shown to be much smaller).

Now here is our main theorem in this paper.

Theorem 2. *The $M \times N$ OIP can be solved with a constant success probability by querying the blackbox oracle $O(\sqrt{N\frac{\log M}{\log N}})$ times if poly(N) $\leq M \leq 2^{N^d}$ for some constant d ($0 < d < 1$).*

Proof. The total number of oracle calls at line 6 is within the bound as described in Sec. 4.1 and the total number of oracle calls at line 11 is bounded by Lemma 4. As for the success probability, we have already proved that there is no problem for the total success probability of line 6 (Sec. 4.1) and lines 11 (Lemma 4). Thus the theorem has been proved.

4.3   OIP with •(•) Queries

Next, we consider the case when $M > 2^{N^d}$. Note that when $M = 2^{d'N}$, for a constant $d' \leq 1$, the lower bound of the number of queries is $\Omega(N)$ instead of $\Omega(\sqrt{N\log M/\log N})$. Therefore, it is natural to expect that the number of queries exceeds our bound as M approaches 2^N. Indeed, when $2^{N^d} < M < 2^{N/\log N}$, the number of queries of ROIPL is bigger than $O(\sqrt{N\log M/\log N})$ but still better than $O(N)$, as shown in the following theorem.

Theorem 3. *For $2^{N^d} \leq M < 2^{N/\log N}$, the $M \times N$ OIP can be solved with a constant success probability by querying the blackbox oracle $O(\frac{\sqrt{N}\log N \log M}{\log(1/\beta)})$ times for $\beta = min(\frac{\log M(\log\log M)^2 \log N}{2N}, \frac{1}{4})$.*

Proof. The algorithm is the same as ROIPL excepting the following: At line 1, we set β as before if $M < 2^{N/\log^3 N}$. Otherwise, i.e., if $2^{N/\log^3 N} \leq M \leq 2^{N/\log N}$, we set $\beta = 1/4$. Then, we can use almost the same argument to prove the theorem, which may be omitted.

5 Concluding Remarks

As mentioned above, our upper bound becomes trivial $O(N)$ when $M = 2^{N/\log N}$, while for bigger M [11] has already given a nice robust algorithm which can be used for OIP with $O(N)$ queries. A challenging question is whether or not there exists an OIP algorithm whose upper bound is $o(N)$ for $M > 2^{N/\log N}$, say, for $M = 2^{N/\log\log N}$. Even more challenging is to design an OIP algorithm which is optimal in the whole range of M. There are two possible scenarios: The one is that the lower bound becomes $\Omega(N)$ for some $M = 2^{o(N)}$. The other is that there is no such case, i.e., the bound is always $o(N)$ if $M = 2^{o(N)}$. At this moment, we do not have any conjecture about which scenario is more likely.

Procedure 1. RowReduction(T, l)

Require: $T \subseteq \{1, \ldots, N\}$ and $l \in \mathcal{N}$
1: **for** $j \leftarrow 1$ **to** l **do**
2:    $k \leftarrow$ GQ(T)
3:    **if** Majority$(k, \min(l, \log N), f) = 1$ **then**
4:       **return** PositiveRow$(\{k\}, Z)$
5:    **end if**
6: **end for**
7: **return** $\{1, \ldots, M\} \setminus$ PositiveRow(T, Z)

Procedure 2. RowCover(S, r)

Require: $S \subseteq \{1, \ldots, M\}$ and $0 < r < 1$
1: $T \leftarrow \{\}$
2: $S' \leftarrow S$
3: **while** $\exists i$ s.t. $|$PositiveRow$(\{i\}, Z(S'))| \geq r|S|$ and $|$PositiveRow$(T, Z(S))| < |S|/4$ **do**
4:    $T \leftarrow T \cup \{i\}$
5:    $S' \leftarrow S \setminus$ PositiveRow$(T, Z(S))$
6: **end while**
7: **return** T *//by one-sensitivity $|PositiveRow(T, Z(S))| < 3|S|/4$*

Procedure 3. PositiveRow(T, Z)

---

   return $\{i|\ j \in T$ and $Z(i,j) = 1\}$

Procedure 4. Majority(k, l, f)

---

   return the majority of $60l$ samples of $f(k)$ if $k \neq$ nill, else 0.

Procedure 5. ROIPS(S, Z)

1: **repeat**
2:    $T \leftarrow$ RowCover$(S, \log^4 N/N)$
3:    $S' \leftarrow S \cap$ RowReduction$(T, \log N)$
4:    **if** $|S'| \leq \frac{3}{4}|S|$ **then**
5:       $S \leftarrow S'$
6:    **else**
7:       $S \leftarrow S' \cap$ RowReduction$(\{1, \ldots, N\} \setminus T, 1)$
8:    **end if**
9: **until** $|S| \leq 1$
10: **return** S

Procedure 6. RowReductionExpire_MTGS(T, l, COUNT, r)

---

   the same as RowReduction(T, l) except that we add the folowing two: (i) the number
   of queries is added to COUNT and the empty set is returned when COUNT exceeds
   $MAX\_QUERIES(N, M)$ (defined in ROIPL) (ii) For $r > 0$: GQ(T) is replaced by
   MultiTargetGQ(T, r), a G-query on T assuming that there are more than r fraction
   of 1's in the current oracle, and at line 7 the set of all rows that have at most r
   fraction of 1's is returned instead.

Procedure 7. ROIPL(Z)

Require: $Z : M \times N$ 0-1 matrix and poly$(N) \leq M \leq 2^{N/\log N}$
1: $\beta \leftarrow \frac{\log M (\log \log M)^2 \log N}{2N}$; $S = \{1, \ldots, M\}$
2: MAX_QUERIES(N,M) $\leftarrow 45\sigma \frac{\sqrt{N \log M \log N}}{\log 1/\beta}$ $//\sigma$: a constant factor of Robust
   Quantum Search in [18]
3: COUNT $\leftarrow 0$ $//$Increased in RowReductionExpire
4: **repeat**
5:    $T \leftarrow$ RowCover(S, β)
6:    $S' \leftarrow S \cap$ RowReduction$(T, \log \log M)$
7:    **if** $|S'| \leq 3/4|S|$ **then**
8:       $S \leftarrow S'$
9:    **else**
10:       $S \leftarrow S'$
11:       $S \leftarrow S \cap$ RowReductionExpire_MTGS$(\{1 \ldots N\}, \log \log(\frac{\log M}{\log N}), \text{COUNT}, \frac{\log |S|}{4(N \log N)}))$
12:    **end if**
13: **until** $|S| \leq N^{10}$
14: $Z' \leftarrow Z(S)$
15: relabel S and Z' so that the answer to OIP of Z can be deduced from that of Z'
16: **return** ROIPS(S, Z')

References

1. S. Aaronson and A. Ambainis. Quantum search of spatial regions. In *Proc. of STOC '03*, pages 200–209, 2003.
2. M. Adcock and R. Cleve. A quantum Goldreich-Levin theorem with cryptographic applications. In *Proc. of STACS '02*, LNCS 2285, pages 323–334, 2002.
3. A. Ambainis, K. Iwama, A. Kawachi, H. Masuda, R. H. Putra, and S. Yamashita. Quantum identification of boolean oracles. In *Proc. of STACS '04*, LNCS 2996, pages 105–116, 2004.
4. A. Ambainis, K. Iwama, A. Kawachi, R. H. Putra, and S. Yamashita. Robust quantum algorithm for oracle identification. Preprint available at quant-ph/0411204.
5. A. Atici and R. A. Servedio. Improved bounds on quantum learning algorithms. Quantum Information Processing, pages 1–32, Jan. 2006.
6. R. Beals, H. Buhrman, R. Cleve, M. Mosca, and R. de Wolf. Quantum lower bounds by polynomials. In *IEEE Symposium on Foundations of Computer Science*, pages 352–361, 1998.
7. E. Bernstein and U. Vazirani. Quantum complexity theory. *SIAM J. Comput.*, 26(5):1411–1473, 1997.
8. D. Biron, O. Biham, E. Biham, M. Grassl, and D. A. Lidar. Generalized Grover search algorithm for arbitrary initial amplitude distribution. In *Proc. of QCQC '98*, LNCS 1509, pages 140–147, 1998.
9. M. Boyer, G. Brassard, P. Høyer, and A. Tapp. Tight bounds on quantum searching. *Fortschritte der Physik*, vol. 46(4-5), 493–505, 1998.
10. G. Brassard, P. Høyer, M. Mosca, A. Tapp. Quantum amplitude amplification and estimation. In *AMS Contemporary Mathematics Series Millennium Volume entitled "Quantum Computation & Information"*, vol 305, pages 53–74, 2002.
11. H. Buhrman, I. Newman, H. Röhrig, and R. de Wolf. Robust quantum algorithms and polynomials. In *Proc. of STACS '05*, LNCS 3404, pages 593–604, 2005.
12. B. Chazelle, D. Liu and A. Magen. Sublinear geometric algorithms. In *Proc. of STOC '03*, 531-540, pages 531–540, 2003.
13. D. P. Chi and J. Kim. Quantum database searching by a single query. In *Proc. of QCQC '98*, LNCS 1509, pages 148–151, 1998.
14. G. D. Cohen, I. Honkala, S. N. Litsyn and A. Lobstein. Covering codes, North Holland, Amsterdam, The Netherlands, 1997.
15. E. Farhi, J. Goldstone, S. Gutmann, and M. Sipser. How many functions can be distinguished with k quantum queries? In *Phys. Rev. A 60*, 6, 4331–4333, 1999 (quant-ph/9901012).
16. O. Goldreich, S. Goldwasser, and D. Ron. Property testing and its connection to learning and approximation. In *Proc. of FOCS '96*, pages 339–348, 1996.
17. L. K. Grover. A fast quantum mechanical algorithm for database search. In *Proc. of STOC '96*, pages 212–218, 1996.
18. P. Høyer, M. Mosca, and R. de Wolf. Quantum search on bounded-error inputs. In *Proc. of ICALP '03*, LNCS 2719, pages 291–299, 2003.
19. M. Hunziker, D. A. Meyer, J. Park, J. Pommersheim and M. Rothstein The geometry of quantum learning. arXiv:quant-ph/0309059, to appear in Quantum Information Processing.
20. R. Krauthgamer and O. Sasson. Property testing of data dimensionality. In *Proc. of SODA '03*, pages 18–27, 2003.

Approximability of Minimum AND-Circuits[*]

Jan Arpe[1,**] and Bodo Manthey[2,***]

[1] Universität zu Lübeck, Institut für Theoretische Informatik
Ratzeburger Allee 160, 23538 Lübeck, Germany
arpe@tcs.uni-luebeck.de
[2] Universität des Saarlandes, Informatik
Postfach 151150, 66041 Saarbrücken, Germany
manthey@cs.uni-sb.de

Abstract. Given a set of monomials, the Minimum-AND-Circuit problem asks for a circuit that computes these monomials using AND-gates of fan-in two and being of minimum size. We prove that the problem is not polynomial time approximable within a factor of less than 1.0051 unless P = NP, even if the monomials are restricted to be of degree at most three. For the latter case, we devise several efficient approximation algorithms, yielding an approximation ratio of 1.278. For the general problem, we achieve an approximation ratio of $d - 3/2$, where d is the degree of the largest monomial. In addition, we prove that the problem is fixed parameter tractable with the number of monomials as parameter. Finally, we reveal connections between the MINIMUM AND-CIRCUIT problem and several problems from different areas.

1   Introduction

Given a set of Boolean monomials, the Minimum-AND-Circuit problem asks for a circuit that consists solely of logical AND-gates with fan-in two and that computes these monomials. The monomials may for example arise in the DNF-representation of a Boolean function or in some decomposed or factored form. Thus, the Minimum-AND-Circuit problem is of fundamental interest for automated circuit design, see Charikar et al. [3, Sect. VII.B] and references therein. In this paper, we assume that all variables always occur positively; no negations are permitted. The investigation of minimum AND-circuits from a complexity theoretic standpoint was proposed by Charikar et al. [3]. According to them, no approximation guarantees have been proved at all yet.

We give the first positive and negative approximability results for the Minimum-AND-Circuit problem. Specifically, we show that the problem is not approximable within a factor of less than $\frac{983}{978}$ unless P = NP, even if the monomials are

---

[*] A full version of this work with all proofs is available as Report 06-045 of the Electronic Colloquium on Computational Complexity (ECCC).

[**] Supported by DFG research grant RE 672/4.

[***] Work done as a member of the Institut für Theoretische Informatik of the Universität zu Lübeck and supported by DFG research grant RE 672/3.

L. Arge and R. Freivalds (Eds.): SWAT 2006, LNCS 4059, pp. 292–303, 2006.

restricted to be of maximum degree three (Sect. 3). For the latter variant, we present several algorithms and prove an upper bound of 1.278 on its approximation ratio (Sect. 4). If the number of occurrences of each submonomial of size two in the input instance, called the *multiplicity*, is bounded from above by a constant $\mu \geq 3$, similar hardness results are achieved (Sect. 3) and the upper bounds are slightly improved (Sect. 4.4). For $\mu = 2$, the problem is even in P (Sect. 4.2). However, if we allow the monomials to be of degree four, it remains open whether the case $\mu = 2$ is solvable in polynomial time. We prove that the general problem with multiplicity bounded by μ is approximable within a factor of μ (Sect. 6.2).

In general, restricting the monomials to be of degree at most d admits a straightforward approximation within a factor of $d - 1$, which we improve to $d - 3/2$ (Sect. 6.1). If the degrees are required to be exactly d and in addition, the multiplicity is bounded by μ, we prove an upper bound on the approximation ratio of $\mu(d - 1)/(\mu + d - 2)$ (Sect. 6.2).

Besides from fixing the maximum degree or the multiplicity of the input monomials, we consider fixing the number of monomials (Sect. 5). We show that Minimum-AND-Circuit instances have small problem kernels, yielding a fixed parameter tractable algorithm (for terminology, see Downey and Fellows [6]). In other words, the Minimum-AND-Circuit problem restricted to instances with a fixed number of monomials is in P.

There are two evident generalizations of AND-circuits. The first one is to ask for a minimum *Boolean* circuit (with AND-, OR-, and NOT-gates) that computes a given function. This problem has, for example, been investigated by Kabanets and Cai [7]; its complexity is still open. Even if the functions to be computed consist solely of positive monomials, allowing the circuit to contain AND- and OR-gates can reduce the circuit size, as has been shown by Tarjan [11] (see also Wegener [13]).

The second one is to consider monomials over other structures such as the additive group of integers or the monoid of finite words over some alphabet (see also Sect. 6.3). While the former structure leads to *addition chains* [9, Sect. 4.6.3], the latter yields the *smallest grammar problem* which has attracted much attention in the past few years; a summary of recent results has been provided by Charikar et al. [3, Sect. I and II]. In fact, Charikar et al.'s suggestion to investigate minimum AND-circuits was motivated by the lack of understanding the hierarchical structure of grammar-based compression. In particular, there is a bunch of so-called *global algorithms* for the smallest grammar problem which are believed to achieve quite good approximation ratios, but no one has yet managed to prove this.

2   Preliminaries

2.1   Monomials and Circuits

We study the design of small circuits that simultaneously compute given monomials $M_1, \ldots, M_k$ over a set of Boolean variables $X = \{x_1, \ldots, x_n\}$. More precisely, a *(Boolean) monomial* is an AND-product of variables of a subset of X, and by an *AND-circuit*, we mean a circuit consisting solely of AND-gates with fan-in two. We identify a monomial $M = x_{i_1} \wedge \ldots \wedge x_{i_d}$ with the subset $\{x_{i_1}, \ldots, x_{i_d}\}$, which we denote by M again. Since we only use one type of operation, we often omit the $\wedge$ signs and simply write $x_{i_1} \ldots x_{i_d}$. The *degree* of M is $|M|$.

An *(AND-)circuit* $\mathcal{C}$ over X is a directed acyclic graph with node set $G(\mathcal{C})$ *(gates)* and edge set $W(\mathcal{C})$ *(wires)* satisfying the following properties:

1. To each input variable $x \in X$ is associated exactly one *input gate* $g_x \in G(\mathcal{C})$ that has indegree zero and arbitrary outdegree.
2. All nodes that are not input nodes have indegree exactly two and arbitrary outdegree. These nodes are called *computation gates*.

We denote the set of computation gates of $\mathcal{C}$ by $G^*(\mathcal{C})$, i.e., $G^*(\mathcal{C}) = G(\mathcal{C}) \setminus \{g_x \mid x \in X\}$. The *circuit size* of $\mathcal{C}$ is equal to the number of computation gates of $\mathcal{C}$, i.e., $\text{size}(\mathcal{C}) = |G^*(\mathcal{C})|$. A gate g *computes* the monomial $\text{val}(g)$, which is defined as follows:

1. $\text{val}(g_x) = x$.
2. For a computation gate g with predecessors g_1 and g_2, $\text{val}(g) = \text{val}(g_1) \wedge \text{val}(g_2)$.

The circuit $\mathcal{C}$ *computes* a Boolean monomial M if some gate in $\mathcal{C}$ computes M. It computes a set $\mathcal{M}$ of monomials if it computes all monomials in $\mathcal{M}$. Such a circuit is called *a circuit for* $\mathcal{M}$. The gates that compute the monomials in $\mathcal{M}$ are referred to as the *output gates*. Output gates, unless they are input gates at the same time, are computation gates, too, and hence contribute to the circuit size. This makes sense since in a physical realization of the circuit, such gates have to perform an AND-operation—in the same way as all non-output computation gates.

A *subcircuit* $\mathcal{C}'$ of a circuit $\mathcal{C}$ is a subgraph of $\mathcal{C}$ that is again a circuit. In particular, $\mathcal{C}'$ contains all "induced" input gates. For $g \in G(\mathcal{C})$, let $\mathcal{C}_g$ be the minimal subcircuit of $\mathcal{C}$ containing g. Since $\mathcal{C}_g$ is a circuit, it contains all input gates g_x with $x \in \text{val}(g)$. Moreover, $\mathcal{C}_g$ contains at least $|\text{val}(g)| - 1$ computation gates. Let $\mathcal{M}$ be a set of monomials and $\mathcal{C}$ be a circuit for $\mathcal{M}$. For each $M \in \mathcal{M}$, denote the gate that computes M by g_M and write $\mathcal{C}_M$ for $\mathcal{C}_{g_M}$.

A gate is called *strict* if its predecessors compute disjoint monomials. A circuit is called *strict* if all of its gates are strict. Any non-strict circuit for a Min-AC instance $\mathcal{M}$ of maximum degree at most four can be turned into a strict circuit for $\mathcal{M}$ of the same size. This is not true if the monomials are allowed to be of degree five or more (Sect. 6.1).

Let $S \subseteq X$. The *multiplicity of* S in $\mathcal{M}$ is the number of occurrences of S in $\mathcal{M}$ as a submonomial, i.e.,

$$\text{mult}_{\mathcal{M}}(S) = |\{M \in \mathcal{M} \mid S \subseteq M\}|.$$

The *maximum multiplicity of* $\mathcal{M}$ is defined by

$$\text{mult}(\mathcal{M}) = \max_{|S| \geq 2} \text{mult}_{\mathcal{M}}(S) .$$

It is equal to the number of occurrences of the most frequent pair of variables in $\mathcal{M}$.

2.2 Optimization Problems

For an introduction to the approximation theory of combinatorial optimization problems, we refer to Ausiello et al. [2]. For an optimization problem P and an instance I for P, we write $\text{opt}_P(I)$ for the measure of an optimum solution for I.

Let $\mathcal{A}$ be an approximation algorithm for P, i.e., an algorithm, that on an instance I of P, outputs an admissible solution $\mathcal{A}(I)$. The *approximation ratio* $\rho_{\mathcal{A}}(I)$ *of* $\mathcal{A}$ *at* I is the ratio between the measure $m(\mathcal{A}(I))$ of a solution $\mathcal{A}(I)$ output by $\mathcal{A}$ and the size of an optimal solution, i.e., $\rho_{\mathcal{A}}(I) = \frac{m(\mathcal{A}(I))}{\text{opt}_P(I)}$. The *approximation ratio* $\rho_{\mathcal{A}}$ *of* $\mathcal{A}$ is the worst-case ratio of all ratios $\rho_{\mathcal{A}}(I)$, i.e., $\rho_{\mathcal{A}} = \max_I \rho_{\mathcal{A}}(I)$.

The Minimum-AND-Circuit problem, abbreviated Min-AC, is defined as follows: Given a set of monomials $\mathcal{M} = \{M_1, \ldots, M_k\}$ over a set of Boolean input variables $X = \{x_1, \ldots, x_n\}$, find a circuit $\mathcal{C}$ of minimum size that computes $\mathcal{M}$.

Throughout the paper, k denotes the number of monomials, n denotes the number of input variables, and $N = \sum_{M \in \mathcal{M}} |M|$ denotes the total input size. In addition, we always assume that $X = \bigcup_{M \in \mathcal{M}} M$.

We denote by Min-d-AC the Minimum-AND-Circuit problem with instances restricted to monomials of degree *at most* d. The problem where the degrees are required to be *exactly* d is denoted by Min-Ed-AC.

A *vertex cover* of a graph G is a subset $\hat{V} \subseteq V$ such that every edge has at least one endpoint in $\tilde{V}$. This definition also applies to hypergraphs. Aside from Min-AC, we will encounter the following optimization problems: The vertex cover problem, denoted by Min-VC, is defined as follows: Given an undirected graph G, find a vertex cover of G of minimum size.

The restriction of Min-VC to graphs of maximum degree d is denoted by Min-d-VC. A hypergraph is called *r-uniform* if all of its edges have size exactly r. The vertex cover problem for r-uniform hypergraphs, denoted by Min-r-UVC, is: Given an r-uniform hypergraph G, find a vertex cover of G of minimum size.

Finally, Maximum-Coverage is the following optimization problem: Given a hypergraph G and a number $r \in \mathbb{N}$, find r edges $e_1, \ldots, e_r \in E$ such that $\bigcup_{i=1}^{r} e_i$ is of maximum cardinality.

3 Hardness

We show that Minimum-AND-Circuit is NP-complete and that there is no polynomial-time approximation algorithm that achieves an approximation ratio of less than $\frac{983}{978}$ unless $P = NP$. To do this, we reduce Min-VC to Min-AC.

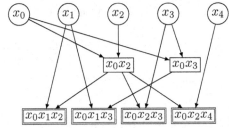

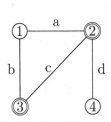

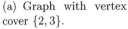

(a) Graph with vertex cover $\{2,3\}$.

(b) Circuit for the Min-3-AC instance $\{M_a, M_b, M_c, M_d\}$ with $M_a = x_0x_1x_2$, $M_b = x_0x_1x_3$, $M_c = x_0x_2x_3$, and $M_d = x_0x_2x_4$.

Fig. 1. A graph with a vertex cover and the corresponding circuit as constructed in Section 3

Let $G = (V, E)$ be an undirected graph with $n = |V|$ vertices and $m = |E|$ edges. We construct an instance of Min-AC as follows. For each node $v \in V$, we have a variable x_v. In addition, there is an extra variable x_0. For each edge $e = \{v, w\} \in E$, we construct the monomial $M_e = x_0x_vx_w$. Our instance of Min-AC is then $\mathcal{M}_G = \{M_e \mid e \in E\}$. Note that $|M| = 3$ for all $M \in \mathcal{M}_G$. Moreover, if G has maximum degree Δ, then $\mathcal{M}_G$ has maximum multiplicity Δ. Clearly, $\mathcal{M}_G$ can be constructed in polynomial time. An example is shown in Figure 1.

There is a one-to-one correspondence between the sizes of the vertex cover and the circuit: We have $\mathrm{opt}_{\mathsf{Min\text{-}AC}}(\mathcal{M}_G) = |E| + \ell$, where $\ell = \mathrm{opt}_{\mathsf{Min\text{-}VC}}(G)$. Furthermore, given a circuit $\mathcal{C}$ of size $|E| + \ell'$ for $\mathcal{M}_G$, we can compute a vertex cover $\tilde{V}$ of G with $|\tilde{V}| \leq \ell'$ in polynomial time. This together with recent inapproximability results by Chlebík and Chlebíková [4] yields the following theorems.

Theorem 1. Min-AC *is NP-complete,* APX-*hard and cannot be approximated in polynomial time within a factor of less than* $\frac{983}{978} > 1.0051$ *unless* P = NP. *This holds even for* Min-3-AC *restricted to instances with maximum multiplicity six.*

Theorem 2. Min-3-AC *restricted to instances of maximum multiplicity three is NP-complete,* APX-*complete, and cannot be approximated in polynomial time within a factor of less than* $\frac{269}{268} > 1.0037$ *unless* P = NP.

4   Approximation Algorithms for Min-3-AC

In this section, we provide several polynomial-time approximation algorithms for Min-3-AC, the problem of computing minimum AND-circuits for monomials of degree at most three. Note that the lower bounds proved in Section 3 hold already for Min-E3-AC.

Without loss of generality, we may assume that all monomials have degree exactly three for the following reasons. Firstly, we do not need any computation gates to compute monomials of degree one, so we can delete such monomials from the input. Secondly, for each input monomial of size two, we are forced to construct an output gate. On the other hand, we should use this gate wherever we can for other input monomials, so we can delete all monomials of degree two from the input and substitute all occurrences of such monomials in the other monomials by extra variables. We repeat this process until no more monomials of size two are in the input. As we have already mentioned in Section 2, we can assume without loss of generality that circuits for Min-3-AC instances are strict. Moreover, if all monomials are of degree exactly three, then a circuit can be assumed to consist of two layers of computation gates. The gates of the first layer compute monomials of size two, and the gates of the second layer are the output gates.

Since each monomial M of degree at most three can be computed by a circuit of size two, we can construct a *trivial circuit* C_{triv} for a Min-3-AC instance $\mathcal{M}$ of size $2k$, where k is the number of monomials. On the other hand, the computation of k monomials obviously requires at least k gates. Thus, we obtain an upper bound of 2 on the polynomial-time approximation ratio for Min-3-AC. In the following, we show how to improve this bound.

4.1 Algorithm "Cover"

We first reduce Min-3-AC to Min-3-UVC, the problem of finding a vertex cover in three-uniform hypergraphs. Subsequently, we will present our algorithms.

Let $\mathcal{M}$ be a Min-3-AC instance. We introduce some notation that will be used throughout this paper. For $M \in \mathcal{M}$, let

$$\text{pairs}(M) = \{S \subseteq X \mid |S| = 2 \wedge S \subseteq M\}$$

be the set of pairs contained in M. Note that $|\text{pairs}(M)| = 3$. Furthermore, let $\text{pairs}(\mathcal{M}) = \bigcup_{M \in \mathcal{M}} \text{pairs}(M)$ be the set of all pairs of variables appearing in $\mathcal{M}$.

Let $\mathcal{C}$ be a circuit for $\mathcal{M}$. Then $\mathcal{C}$ consists of two layers, the second one containing the $k = |\mathcal{M}|$ output gates. In the first layer, certain monomials of size two are computed: for each monomial $M \in \mathcal{M}$, one of the pairs $S \in \text{pairs}(M)$ has to be computed at the first level of $\mathcal{C}$. The task is thus to find a minimum set of pairs $S \in \text{pairs}(M)$ such that each monomial $M \in \mathcal{M}$ contains one such pair. This corresponds to finding a minimum vertex cover of the three-uniform hypergraph $H(\mathcal{M}) = (V, E)$ described in the following. The node set is the set of pairs appearing in $\mathcal{M}$, i.e., $V = \text{pairs}(\mathcal{M})$, and for each monomial $M \in \mathcal{M}$, there is a hyperedge containing the pairs that appear in M, i.e., $E = \{\text{pairs}(M) \mid M \in \mathcal{M}\}$. A circuit $\mathcal{C}$ for $\mathcal{M}$ with gates computing the pairs $S_1, \ldots, S_\ell$ at its first level corresponds to the vertex cover of $H(\mathcal{M})$ given by $\{S_i \mid 1 \leq i \leq \ell\}$ and vice versa. We denote the circuit corresponding to a vertex cover $\tilde{V}$ by $C_{\tilde{V}}$.

Our first polynomial-time approximation algorithm for Min-3-AC is based on the reduction we have just presented (Algorithm 1). The set $\tilde{V}$ consists of all nodes that are incident with the matching $\tilde{E}$. Thus the size of $\tilde{V}$ equals $3 \cdot |\tilde{E}|$. $\tilde{V}$ is a vertex cover since $\tilde{E}$ cannot be enlarged. On the other hand, any vertex cover of $H(\mathcal{M})$ must include at least one vertex from each hyperedge of the maximum matching $\tilde{E}$, so any vertex cover of $IG(\mathcal{M})$ must be of size at least $|\tilde{E}|$. In conclusion, we have $|\tilde{V}| \leq 3 \cdot \text{opt}_{\text{Min-3-UVC}}(H(\mathcal{M}))$.

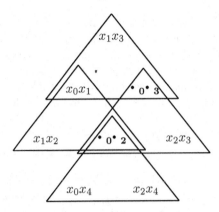

Fig. 2. The hypergraph $H(\mathcal{M})$ associated with the Min-AC instance $\mathcal{M}$ introduced in Figure 1. Each triangle represents a hyperedge. The two bold monomials constitute a vertex cover.

Algorithm 1. Algorithm COVER for Min-3-AC.

    Input $\mathcal{M} = \{M_1, \ldots, M_k\}$.
1: Compute the hypergraph $H = H(\mathcal{M})$.
2: Compute greedily an inclusion-maximal matching $\tilde{E}$ in H, i.e., a collection of disjoint hyperedges that cannot be enlarged.
3: Let $\tilde{V} = \bigcup_{e \in \tilde{E}} e$.
4: Compute $\mathcal{C} = \mathcal{C}_{\tilde{V}}$.
5: Output $\mathcal{C}$.

Overall, COVER achieves the following approximation performance.

Lemma 1. *Let* $\mathrm{opt}_{\mathsf{Min\text{-}3\text{-}AC}}(\mathcal{M}) = k + \ell$. *Then* COVER *outputs a circuit* $\mathcal{C}_{\mathrm{COVER}}$ *for* $\mathcal{M}$ *of size at most* $k + 3 \cdot \ell$.

In case that $\ell \geq \frac{1}{3}k$, COVER outputs a circuit that is larger than the trivial one. Choosing to output the trivial circuit instead, yields an algorithm with an approximation ratio of 3/2. Thus, we have already found an algorithm that achieves a non-trivial approximation ratio. In the course of this paper, we will improve this ratio to below 1.3.

4.2   Algorithm "Match"

Before we present our next algorithm, we introduce another technical utility. Associate with $\mathcal{M}$ the *intersection graph* $IG(\mathcal{M})$ defined as follows: the nodes of $IG(\mathcal{M})$ are the monomials of $\mathcal{M}$, and two monomials $M, M' \in \mathcal{M}$ are connected by an edge iff $|M \cap M'| = 2$. An example is shown in Figure 3.

    MATCH (Algorithm 2) is a polynomial-time algorithm; in particular, a maximum matching in $IG(\mathcal{M})$ can be computed in time $O(n^{2.5})$ [1]. The approximation performance of MATCH is stated in the following

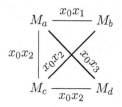

Fig. 3. Intersection graph $IG(\mathcal{M})$ associated with the Min-AC instance $\mathcal{M}$ introduced in Figure 1. The edges are labeled by the pairs that their endpoints have in common. The bold edges constitute a maximal matching.

Algorithm 2. Algorithm MATCH for Min-3-AC.

Input $\mathcal{M} = \{M_1, \ldots, M_k\}$.
1: Compute $G = IG(\mathcal{M})$.
2: Compute a matching $\tilde{E}$ of G of maximum cardinality.
3: For each $\{M, M'\} \in \tilde{E}$:
4:    Add a gate computing $M \cap M'$ to $\mathcal{C}$.
5:    Add subcircuits computing M and M' to $\mathcal{C}$, using two additional gates.
6: For each $M \in \mathcal{M} \setminus \bigcup_{e \in \tilde{E}} e$ (not incident with $\tilde{E}$):
7:    Add a subcircuit computing M, using $|M| - 1$ gates.
8: Output $\mathcal{C}$.

Lemma 2. *Let* $\mathrm{opt}_{\mathsf{Min\text{-}3\text{-}AC}}(\mathcal{M}) = k + \ell$. *Then* MATCH *outputs a circuit* $\mathcal{C}_{\mathrm{MATCH}}$ *for* $\mathcal{M}$ *of size at most* $\frac{3}{2} \cdot k + \frac{1}{2} \cdot \ell$.

Although the analysis of MATCH is not needed for our best upper bound result for Min-3-AC, the algorithm is the only one for which we can prove a non-trivial approximation ratio for Min-d-AC in case that $d \geq 4$.

For Min-3-AC with instances restricted to a multiplicity of at most two, MATCH computes an optimum solution. Thus, Min-3-AC restricted to instances with a maximum multiplicity of at most two can be solved in polynomial time.

4.3 Algorithm "Greedy"

Our last algorithm GREEDY (Algorithm 3) greedily constructs gates for pairs that occur most frequently in the input instance $\mathcal{M}$ until each remaining pair is shared by at most two monomials. At that point, instead of proceeding in an arbitrary order, an optimal solution is computed for the remaining monomials. The latter task is achieved by MATCH, as we have stated in Section 4.2.

Lemma 3. *Let* $\mathcal{M} = \{M_1, M_2, \ldots, M_k\}$ *be an instance for* Min-3-AC *such that* $\mathrm{opt}_{\mathsf{Min\text{-}3\text{-}AC}}(\mathcal{M}) = k + \ell$. *Then* GREEDY *outputs a circuit* $\mathcal{C}_{\mathrm{GREEDY}}$ *for* $\mathcal{M}$ *of size at most*

$$\min\left\{ \frac{4}{3} \cdot k + \ell, \left(1 + \frac{1}{e^2}\right) k + 2\ell \right\}.$$

It does not make much sense to reiterate the last step of the analysis since this would give us a circuit of size larger than $k + 3\ell$, the size achieved by COVER.

Algorithm 3. Algorithm GREEDY for Min-3-AC.

Input $\mathcal{M} = \{M_1, \ldots, M_k\}$.

1: While there exists an $S \in \binom{X}{2}$ such that $|\{M \in \mathcal{M} \mid S \subseteq M\}| \geq 3$:
2:    Arbitrarily select $S \in \binom{X}{2}$ with maximum $|\{M \in \mathcal{M} \mid S \subseteq M\}|$.
3:    Add a gate computing S to $\mathcal{C}$.
4:    For each $M \in \mathcal{M}$ with $S \subseteq M$:
5:       Add subcircuit computing M to $\mathcal{C}$, using at most $|M| - 2$ additional gates.
6:    $\mathcal{M} \leftarrow \mathcal{M} \setminus \{M\}$.
7: $\mathcal{C}' \leftarrow$ MATCH$(\mathcal{M})$.
8: $\mathcal{C} \leftarrow \mathcal{C} \cup \mathcal{C}'$.
9: Output $\mathcal{C}$.

Corollary 1. *The approximation ratio achieved by* GREEDY *for* Min-3-AC *is at most* $\frac{5e^2-3}{4e^2-3} \approx 1.278$.

The best lower bound that we are able to show for the approximation ratio of GREEDY is $10/9$.

4.4 Summary of Approximation Ratios

In this subsection, we summarize the approximation ratios of the algorithms presented in the preceding subsections and present some improvements for Min-3-AC instances with bounded multiplicity. So far, we have found the following bounds for the approximation ratios of the Min-3-AC algorithms:

$$
\begin{aligned}
\rho_{\text{COVER}} &\leq \frac{k+3\ell}{k+\ell} & \text{increasing in } \ell\,, \\
\rho_{\text{GREEDY}} &\leq \frac{(1+e^{-2})k+2\ell}{k+\ell} & \text{increasing in } \ell\,, \\
\rho_{\text{GREEDY}} &\leq \frac{\frac{4}{3}k+\ell}{k+\ell} & \text{decreasing in } \ell\,, \\
\rho_{\text{MATCH}} &\leq \frac{\frac{3}{2}k+\frac{1}{2}\ell}{k+\ell} & \text{decreasing in } \ell\,.
\end{aligned}
$$

These approximation ratios are presented in Figure 4. Concerning restricted multiplicity, we can show the following result.

Theorem 3. *The* Min-3-AC *problem restricted to instances of maximum multiplicity* μ, $\mu \in \{3, 4, 5\}$, *is approximable within a factor of*

- $5/4 = 1.25$ *if* $\mu = 3$,
- $19/15 = 1.2\overline{6}$ *if* $\mu = 4$, *and*
- $23/18 = 1.2\overline{7}$ *if* $\mu = 5$.

5   Fixing the Number of Monomials

Min-AC is fixed parameter tractable with respect to the number k of monomials in the input instance. For more details on fixed parameter tractability, we refer to Downey and Fellows [6].

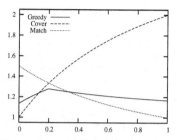

(a) Upper bounds for GREEDY, COVER, and MATCH.

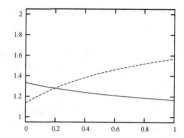

(b) Upper bounds for GREEDY given by Lemma 3.

Fig. 4. Approximation ratios of the Min-3-AC algorithms dependent on the ratio ℓ/k

Theorem 4. Min-AC, *parameterized by the number of input monomials, is fixed parameter tractable. This means that there are a function $f : \mathbb{N} \to \mathbb{N}$ and a polynomial $p : \mathbb{N} \to \mathbb{N}$ such that* Min-AC *can be solved deterministically in time* $f(k) + p(N)$.

6 Concluding Remarks and Future Research

6.1 Approximation Algorithms for Min-•-AC, • ≥ 4

Obviously, the approximation ratio of Min-d-AC is at most $d - 1$ since on the one hand, every monomial of degree at most d can be computed by at most $d - 1$ separate gates and on the other hand, any circuit contains at least one gate per monomial of the input instance. It is easy to see that MATCH achieves the slightly better approximation ratio $d - \frac{3}{2}$ (which is tight).

We are particularly curious about whether Min-d-AC is approximable within a factor of $o(d)$ or whether it is possible to show an $\Omega(d)$ hardness result.

For $d \geq 4$, there are several possibilities of generalizing the greedy algorithm, which coincide for $d = 3$.

The algorithms GREEDY and MATCH produce strict circuits. Already for $d = 5$, we can construct Min-AC instances $\mathcal{M}$ of maximum degree d such that any strict circuit for $\mathcal{M}$ is roughly $4/3$ times larger than a minimum non-strict circuit.

Corollary 2. *Any approximation algorithm for* Min-AC *(or even* Min-5-AC*) that produces only strict circuits does not achieve an approximation ratio better than $4/3$.*

6.2 Approximation of Instances with Bounded Multiplicity

In Section 4.2, we showed that Min-3-AC instances with maximum multiplicity two are optimally solvable in polynomial time. In contrast, Min-3-AC with instances restricted to maximum multiplicity three is hard to solve, as we saw in Section 3. We leave it as an open problem whether Min-d-AC instances with

$d \geq 4$ are polynomial time solvable. Nonetheless we can provide a positive approximability result for general Min-AC instances with bounded multiplicity.

Theorem 5. Min-AC *with instances restricted to be of maximum multiplicity μ is polynomial-time approximable within a factor of μ.*

Theorem 5 also follows from a more general result by Wegener [13, Sect. 6.6] about *Boolean sums*, which are collections of disjunctions of (positive) Boolean variables, and thus are dual to collections of monomials. Wegener [13, Def. 6.1] defines such a collection to be (h, k)-disjoint if $h + 1$ disjunctions have at most k common summands. In particular, sets of monomials of multiplicity μ correspond to $(\mu, 1)$-disjoint collections. The claim then follows from [13, Lem. 6.1] by plugging in $h = \mu$ and $k = 1$. Although the lemma is only stated for collections in which the number of input variables equals the number of disjunctions, it also holds if these numbers differ.

We can improve the result of Theorem 5 for Min-Ed-AC restricted to instances with bounded multiplicity using the fact that for these instances, all output gates have frequency one.

Theorem 6. *The* Min-Ed-AC *problem with instances restricted to be of maximum multiplicity μ is polynomial-time approximable within a factor of $\frac{\mu(d-1)}{\mu+d-2}$.*

This implies an improved approximation ratio of $3/2$ compared to the ratio of $5/2$ achieved by MATCH for general Min-4-AC instances.

6.3   Generalizations and Related Problems

Let us first mention some applications that arise as alternative interpretations of the problem in this paper. Viewing monomials M over X as subsets of X (see also Sect. 2), an AND-gate computes the *union* of the sets computed by its predecessors. Thus, AND-circuits may be interpreted as compact representations of set systems. Since each gate has to be evaluated only once, the circuit may be considered as a straight-line program that generates the set system. Furthermore, in a Boolean matrix-vector product, each entry of the result is a disjunction (or a parity, depending on which type of "sum" is considered) of the vector entries corresponding to the positions of 1s in the matrix rows. Thus, if many vectors have to be multiplied by the same matrix, it may be useful to preprocess the matrix by constructing a circuit that computes all disjunctions (with indeterminates) first.

Table 1. The circuit problem for several semigroup structures and parameters

S	$\circ$	k	n	Description	Remark
$\{0,1\}$	$\wedge$	arb.	arb.	Boolean monomials, Min-AC	
$\mathbb{Z}$	$+$	1	1	Addition chains [9, 12]	complexity unknown
$\mathbb{Z}$	$+$	arb.	1	Extended addition chains	NP-complete [5]
Σ^*	concat.	1	arb.	Grammar-based compression [8] of strings over alphabets of size n	NP-complete for $n \geq 3$ [10], complexity unknown for $n \leq 2$

Beside Boolean variables and monomials, it is natural to consider monomials over other structures. In general, the variables $x \in X$ take values from some semigroup $(S, \circ)$ (note that we assume the structure to be associative since otherwise it makes no sense to design small circuits). In case that S is non-commutative, the predecessors of a gate have to be ordered. Table 1 shows several examples of semigroups and other parameters with their corresponding circuit problem. As one can see, many seemingly different problems turn out to be instantiations of a general *semigroup circuit problem*.

References

1. Ravindra K. Ahuja, Thomas L. Magnanti, and James B. Orlin. *Network Flows: Theory, Algorithms, and Applications*. Prentice-Hall, 1993.
2. Giorgio Ausiello, Pierluigi Crescenzi, Giorgio Gambosi, Viggo Kann, Alberto Marchetti-Spaccamela, and Marco Protasi. *Complexity and Approximation: Combinatorial Optimization Problems and Their Approximability Properties*. Springer, 1999.
3. Moses Charikar, Eric Lehman, Ding Liu, Rina Panigrahy, Manoj Prabhakaran, Amit Sahai, and Abbi Shelat. The smallest grammar problem. *IEEE Transactions on Information Theory*, 51(7):2554–2576, 2005.
4. Miroslav Chlebík and Janka Chlebíková. Complexity of approximating bounded variants of optimization problems. *Theoretical Computer Science*, 354(3):320–338, 2006.
5. Peter J. Downey, Benton L. Leong, and Ravi Sethi. Computing sequences with addition chains. *SIAM Journal on Computing*, 10(3):638–646, 1981.
6. Rodney G. Downey and Michael R. Fellows. *Parameterized Complexity*. Monographs in Computer Science. Springer, 1999.
7. Valentine Kabanets and Jin-Yi Cai. Circuit minimization problems. In *Proc. of the 32nd Ann. ACM Symp. on Theory of Computing (STOC)*, pages 73–79. ACM Press, 2000.
8. John C. Kieffer and En-hui Yang. Grammar based codes: A new class of universal lossless source codes. *IEEE Transactions on Information Theory*, 46(3):737–754, 2000.
9. Donald E. Knuth. *Seminumerical Algorithms*, volume 2 of *The Art of Computer Programming*. Addison-Wesley, 2nd edition, 1981.
10. James A. Storer and Thomas G. Szymanski. The macro model for data compression. In *Proc. of the 10th Ann. ACM Symp. on Theory of Computing (STOC)*, pages 30–39. ACM Press, 1978.
11. Robert E. Tarjan. Complexity of monotone networks for computing conjunctions. *Annals of Discrete Mathematics*, 2:121–133, 1978.
12. Edward G. Thurber. Efficient generation of minimal length addition chains. *SIAM Journal on Computing*, 28(4):1247–1263, 1999.
13. Ingo Wegener. *The Complexity of Boolean Functions*. Wiley-Teubner, 1987.

Triangles, 4-Cycles and Parameterized (In-)Tractability

Venkatesh Raman and Saket Saurabh

The Institute of Mathematical Sciences,
Chennai 600 113
{vraman, saket}@imsc.res.in

Abstract. We show that several problems that are hard for various parameterized complexity classes on general graphs, become fixed parameter tractable on graphs with no small cycles.

More specifically, we give fixed parameter algorithms for DOMINATING SET, t-VERTEX COVER (where we need to cover at least t edges) and several of their variants on graphs that have no triangles or cycles of length 4. These problems are known to be $W[i]$-hard for some i in general graphs. We also show that the Dominating Set problem is $W[2]$-hard in bipartite graphs and hence on triangle free graphs.

In the case of INDEPENDENT SET and several of its variants, we show them fixed parameter tractable even in triangle free graphs. In contrast, we show that the DENSE SUBGRAPH problem (related to the CLIQUE problem) is $W[1]$-hard on graphs with no cycles of length at most 5.

1   Introduction

Parameterized Complexity is a recent approach to deal with intractable computational problems. For decision problems with input size n, and a parameter k (which typically, and in all the problems we consider in this paper, is the solution size), the goal here is to design an algorithm with runtime $f(k)n^{O(1)}$ where f is a function of k alone, against a trivial $n^{k+O(1)}$ algorithm. Problems having such an algorithm is said to be fixed parameter tractable (FPT), and such algorithms are practical when small parameters cover practical ranges. There is also a theory of parameterized intractability using which one can identify parameterized problems that are unlikely to admit fixed parameter tractable algorithms. The book by Downey and Fellows [3] provides a good introduction to the topic of parameterized complexity. For recent developments see books by Flum and Grohe [4] and Niedermeier [6].

VERTEX COVER is a celebrated fixed parameter problem where the problem is well solved for parameter size up to 100. Similarly INDEPENDENT SET and DOMINATING SET are problems that are known to be hard for different levels of W-hierarchy. Our main contribution in the paper is to show that these two problems and several of their variants become fixed parameter tractable if the input graph has no short cycles.

L. Arge and R. Freivalds (Eds.): SWAT 2006, LNCS 4059, pp. 304–315, 2006.
© Springer-Verlag Berlin Heidelberg 2006

In Section 2, we look at the DOMINATING SET problem and show that the problem is $W[2]$-complete even in bipartite graphs. As far as we know this wasn't known before. Our observation means that the problem is $W[2]$-complete in triangle free graphs. Then we show that the problem is FPT if the input graph has no 3 or 4-cycles. It turns out that this result can be generalized to several (more than half a dozen) variants of the DOMINATING SET problem.

In Section 3, we look at t-VERTEX COVER and t-DOMINATING SET problems. These are generalizations of VERTEX COVER and DOMINATING SET problems: In the t-VERTEX COVER problem, we are interested in finding a set of at most k vertices covering at least k edges and in the t-DOMINATING SET problem the objective is to find a set of at most k vertices that dominates at least l vertices. Both these problems are parameterized in two different ways: by k alone and by both k and t. Both these problems are fixed parameter tractable when parameterized by both k and t. Blašer[1] gave $O(2^{O(t)}n^{O(1)})$ algorithm for both the problems using color coding technique. Recently, Guo et. al. [7] have shown that t-VERTEX COVER is $W[1]$-hard when parameterized by k alone. It is easy to see that the t-DOMINATING SET is $W[2]$-hard by a reduction from dominating set. We show that both these problems are fixed parameter tractable in graphs with no cycles of length less than 5, when parameterized by k alone.

In Section 4, we look at INDEPENDENT SET problem and several of its variants. We show that these problems are fixed parameter tractable even in triangle free graphs.

While Sections 2, 3 and 4 show that several problems that are $W[i]$ hard for some i, become fixed parameter tractable in graphs with no small cycles, in Section 5, we exhibit a problem that is $W[1]$-hard in graphs with no small cycles. This is the 'dense subgraph problem'. Here, given a graph $G = (V, E)$ and positive integers k and l, the problem is to determine whether there exists a set of at most k vertices $C \subseteq V$ such that the induced subgraph on C has at least l edges; here k is the parameter.

Section 6 gives some concluding remarks and open problems.

In the rest of the paper, we assume that all our graphs are simple and undirected. Given a graph $G = (V, E)$, n represents number of vertices, and m represents the number of edges. For a subset $V' \subseteq V$, by $G[V']$ we mean the subgraph of G induced on V'. By $N(u)$ we represent all vertices (excluding u) that are adjacent to u, and by $N[u]$, we refer to $N(u) \cup \{u\}$. Similarly, for a subset $D \subseteq V$, we define $N[D] = \cup_{v \in D} N[v]$. By the girth of a graph, we mean the length of the shortest cycle in the graph. We say that a graph is a G_i graph if the girth of the graph is at least i. A vertex is said to dominate all its neighbors.

2   Dominating Set and Its Variants

In this section we look at the DOMINATING SET problem and its variants.

DOMINATING SET: Given a graph $G = (V, E)$ and an integer $k \geq 0$, determine whether there exists a set of at most k vertices $D \subseteq V$ such that for every vertex $u \in V$, $N[u] \cap D \neq \emptyset$.

We say that the set D 'dominates' the vertices of G. We first show that DOMI-NATING SET problem is $W[2]$-complete in bipartite graphs by a reduction from the same problem in general undirected graphs. Then we give a fixed parameter tractable algorithm for the problem in graphs without cycles of length 3 or 4.

2.1   W-Hardness of Dominating Set Problem in Bipartite Graphs

Theorem 1. DOMINATING SET *problem is $W[2]$-complete in bipartite graphs.*

Proof. We prove this by giving a reduction from the DOMINATING SET problem in general undirected graphs. Given an instance $(G = (V, E), k)$ of DOMINATING SET, we construct a bipartite graph $H = (V', E')$. We create two copies of V namely $V_1 = \{u_1 \mid u \in V\}$ and $V_2 = \{u_2 \mid u \in V\}$. If there is an edge (u, v) in E then we draw the edges (u_1, v_2) and (v_1, u_2). We also draw edges of the form (u_1, u_2) for every $u \in V$. We create two new vertices z_1 and z_2, with z_1 in V_1 and z_2 in V_2. We add an edge from every vertex in V_1 to z_2. This completes the construction of H.

We show that G has a dominating set of size k if and only if H has a dominating set of size $k + 1$. Let D be a dominating set of size k in G. Then clearly $D' = \{u_1 \mid u \in D\} \cup \{z_2\}$ is a dominating set of size $k + 1$ in H. Conversely, let K be a dominating set in H of size $k + 1$. Observe that either z_1 or z_2 must be part of K as z_2 is the unique neighbor of z_1. Without loss of generality, we can assume that $z_2 \in K$, as otherwise we could delete z_1 and include z_2 in K and still have a dominating set of size at most $k + 1$ in H. Now take $D = \{u \mid u \in V, u_1 \text{ or } u_2 \in K\}$. Clearly D is of size k. We show that D is a dominating set in G. For any $u \notin D$, $u_2 \notin K$ and hence there exists some $v_1 \in K$ such that v_1 dominates u_2 in H. But this implies $v \in D$ and $(v, u) \in E$, which shows that v dominates u. This proves that D is a dominating set of size k for G and establishes the theorem.   □

2.2   FPT Algorithm for Dominating Set in G_5 Graphs

We give a fixed parameter tractable algorithm for the DOMINATING SET problem in graphs with girth at least 5 (G_5 graphs) and also observe that various other W-hard problems become tractable in G_5 graphs.

Our algorithm follows a branching strategy where at every iteration we find a vertex that needs to be included in the Dominating Set which we are trying to construct. Once a vertex is included, we can at best delete that vertex. Though the neighbors of the vertex are dominated, we can not remove these vertices from further consideration as they can be useful to dominate other vertices.

Hence we resort to a coloring scheme for the vertices. At any point of time of the algorithm, the vertices are colored as below:

1. Red - The vertex is included in the the dominating set D which we are trying to construct.
2. White - The vertex is not in the set D, but it is dominated by some vertex in D.
3. Black - The vertex is not dominated by any vertex of D.

Now we define the dominating set problem on the graph with vertices colored with White, Black or Red as above. We call a graph colored red, white and black as above, as a rwb-graph.

RWB-DOMINATING SET: Let G be a G_5 graph (graph with girth at least 5) with vertices colored with Red, White or Black satisfying the following conditions, and let k be a positive integer parameter. Let R, W and B be the set of vertices colored red, white and black respectively.
1. Every white vertex is a neighbor of a red vertex.
2. Black vertices have no red neighbors.
3. $|R| \leq k$

Does G have at most $k - |R|$ vertices that dominate all the black vertices?

It is easy to verify that if we start with a general G_5 graph with all vertices colored black, and color all vertices we want to include in the dominating set as red, and their neighbors as white, the graph we obtain at every intermediate step is a rwb-graph, and the problem we will have at the intermediate steps is the RWB-DOMINATING SET problem.

The following lemma essentially shows that if the rwb-graph has a black or white vertex dominating more than k black vertices, then such a vertex must be part of every solution of size at most k, if exists.

Lemma 1. *Let $(G(R \cup W \cup B, E), k)$ be an instance of the RWB-DOMINATING SET problem where G is a G_5 graph and k a positive integer. Let v be a black or white vertex with more than $k - |R|$ black neighbors. Then if G has a set of size at most $k - |R|$ that dominates all black vertices, then v must be part of every such set.*

Proof. Let D be a set of size $k - |R|$ that dominates all black vertices in G, and suppose $v \notin D$. Let X be the set of black neighbors of v which are not in D and Y be the set of black neighbors of v in D. So $|X| + |Y| > k - |R|$. Observe that for every $v_x \in X$ we have a neighbor $u_x \in D$ which is not in Y (otherwise v, v_x, u_x is a 3 length cycle). Similarly, for x, $y \in X$, $x \neq y \Rightarrow u_x \neq u_y$. Otherwise v, x, u_x, y will form a cycle of length 4. This means that $|D| \geq |X| + |Y| > k - |R|$ which is a contradiction. $\square$

Given a rwb-graph, Lemma 1 suggests the following simple reduction rules.

(R1) If there is a white or a black vertex v having more than $k - |R|$ black neighbors, then color v red and color its neighbors white.
(R2) If a white vertex is not adjacent to a black vertex, delete the white vertex.
(R3) If there is an edge between two white vertices, delete the edge.
(R4) If $|R| > k$, then stop and return NO.

The following Lemma follows from Lemma 1.

Lemma 2. *Let $G = (R \cup W \cup B, E)$ be an instance of RWB-DOMINATING SET and let $G' = (R' \cup W' \cup B', E')$ be the reduced instance after applying rules $(R1)$ to $(R4)$ once. Let k be an integer parameter. Then G is an yes instance if and* *only if G' is an yes instance. I.e. G has a set of size at most $k - |R|$ dominating all vertices in B if and only G' has a set of size at most $k - |R'|$ dominating all vertices in B'.*

Let G be an instance of RWB-DOMINATING SET and let G' be the reduced instance after applying the reduction rules $(R1) - (R4)$ until no longer possible. Then we show that if G' is an yes instance (and hence G is an yes instance), the number of vertices in G' is bounded by polynomial in k. More precisely we show the following lemma.

Lemma 3. *Let (G, k) be an yes instance of RWB-DOMINATING SET and (G', k') be the reduced instance of (G, k) after applying the rules $(R1) - R(4)$ until no longer possible. Then, the number of vertices in G' is bounded by a polynomial function of k.*

Proof. Let R', B' and W' be the set of vertices colored red, black and white respectively in G'. We argue that each of $|R'|$, $|B'|$ and $|W'|$ is bounded by a function of k.

Because of $(R4)$ (and the fact that G' is an yes instance), $|R'| \leq k$.

Because of $(R1)$, every vertex colored white or black has at most $k - |R'|$ black neighbors. Also we know that no red vertex has a black neighbor. Since G' is an yes instance, there are at most k ($k - |R'|$ to be more precise) black or white vertices dominating all black vertices. Since each of them can dominate at most k black vertices, we conclude that $|B'|$ can be at most k^2.

We argue that $|W'| \leq k^3$. Towards this end, we just show that every black vertex has at most k white neighbors. Since $|B'| \leq k^2$, and every white vertex is adjacent to some black neighbor (because of $(R2)$ and $(R3)$), the conclusion will follow.

Note that every white vertex has a red neighbor. Observe that the white neighbors of any black vertex (any vertex for that matter) will have all distinct red neighbors. I.e. if w_1 and w_2 are white neighbors of a black vertex b, then there is no overlap between the red neighbors of w_1 and the red neighbors of w_2. This is because if w_1 and w_2 have a common red neighbor r, then we will have a 4-cycle b, w_1, r, w_2, b. Since $|R'| \leq k$, it follows that a black vertex can have at most k white neighbors.

This proves the required claim.   □

Thus we have the following theorem.

Theorem 2. RWB-DOMINATING SET *problem can be solved in $O(k^{k+O(1)} + n^{O(1)})$ time in G_5 graphs.*

Proof. It is easy to see that the reduction rules $(R1)$ to $(R4)$ take polynomial time to execute. When none of these rules can be executed, by Lemma 3, we have that the number of vertices in the resulting graph is $O(k^3)$, and each vertex has at most k black neighbors. We can just try all possible subsets of size at most k of the vertex set of the reduced graph, to see whether that subset dominates all the black vertices. If any of them does, then we say YES and NO otherwise. This will take $O(k^{3k+O(1)})$ time.

Alternatively, we can apply a branching technique on the black vertices, by selecting a black vertex or any of its neighbors in the dominating set. More precisely, let v be a black vertex. Then we branch on $N[v]$ by including $w \in N[v]$ in the possible dominating set D we are constructing and look for a solution of size $k - 1$ in $G - \{w\}$ where w is colored red and all its neighbors are colored white for every $w \in N[v]$. Details are easy and omitted. This will result in an $O((k + 1)^{k+O(1)})$ time algorithm. □

Now to solve the general Dominating Set problem in G_5 graphs, simply color all vertices black and solve the resulting RWB-DOMINATING SET problem using Theorem 2. Thus we have

Theorem 3. *Parameterized* DOMINATING SET *problem can be solved in* $O(k^{k+O(1)} + n^{O(1)})$ *time in* G_5 *graphs.*

Other variants of dominating set problem which are W[2]-hard can be shown to be fixed parameter tractable in a similar way. We state the theorem without giving the proof. The reader can refer [3, 2] for the definitions of these problems.

Theorem 4. *Parameterized versions of* CONSTRAINT DOMINATING SET, IRRE-DUNDANT SET, EXACT EVEN SET, EXACT ODD SET, ODD SET, DOMINATING THRESHOLD SET, INDEPENDENT DOMINATING SET, MAXIMAL IRREDUNDANT SET, RED BLUE DOMINATING SET *problems are fixed parameter tractable in* G_5 *graphs.*

3   •-Vertex Cover and •-Dominating Set Problems

t-VERTEX COVER and t-DOMINATING SET problems are respectively, generalizations of classical VERTEX COVER and DOMINATING SET problems. Here the objective is not to cover all the edges or to dominate all the vertices but to cover at least t edges or to dominate at least t vertices with at most k vertices. More precisely they are defined as follows:

> t-VERTEX COVER: Given a graph $G = (V, E)$ and positive integers k and t, does there exist a set of at most k vertices $C \subseteq V$ such that $|\{e = (u, v) \in E \mid C \cap \{u, v\} \neq \emptyset\}| \geq t$.
> t-DOMINATING SET: Given a graph $G = (V, E)$ and positive integers k and t, does there exist a set of at most k vertices $D \subseteq V$ such that $|N[D]| \geq t$.

The t-VERTEX COVER and t-DOMINATING SET problems are parameterized in two ways. They are either parameterized by k or by t and k. Both these problems are FPT when parameterized by both k and t [1] and are hard for different level of W-hierarchy when parameterized by k alone [7].

Here, we first give a simple algorithm for t-VERTEX COVER when parameterized by both t and k and then show that this problem is FPT even when parameterized by k alone in G_5 graphs. We then extend this result for t-DOMINATING SET problem in G_5 graphs when parameterized by k alone.

Our algorithms for t-VERTEX COVER depend on the following lemma.

Lemma 4. *Let $(G = (V, E), k, t)$ be an yes instance of t-VERTEX COVER and v be a vertex of maximum degree in G. Then there exists a t-vertex cover C whose intersection with $N[v]$ is non empty, i.e. $N[v] \cap C \neq \emptyset$.*

Proof. Since G is an yes instance of the problem there exists a t-vertex cover C of size at most k and covering at least t edges. If $N[v] \cap C = \emptyset$ then choose $C' = C - \{u\} + \{v\}$ where u is any vertex in C. Since v is a vertex of highest degree and none of its neighbors is in C, C' also covers at least t edges and is of size at most k.   □

Suppose that the given graph has maximum degree bounded by d. Since there exists a t-vertex cover with some neighbour of the maximum degree vertex, we can branch on one of the (at most) d neighbours of the maximum degree vertex giving raise to a d-way branching. The following theorem is immediate from this.

Theorem 5. *Let G be a graph with maximum degree d. Then t-VERTEX COVER can be solved in $O(d^k n)$ time.*

Given a graph $G = (V, E)$ and positive integer parameters t and k, if there exists a vertex of degree at least t then we get a t-vertex cover by chosing the vertex. So without loss of generality, we can assume that every vertex has degree at most $t - 1$. Then from Theorem 5, we have

Corollary 1. *t-VERTEX COVER can be solved in $O(t^k n)$ in general graphs.*

Suppose, instead of trying to cover at least t edges, we want to cover all but t edges (where t is a parameter). That is, we want an induced subgraph on $n - k$ vertices with at most t edges. We call it as the $(m - t)$-VERTEX COVER problem. Such a parameterization is known as dual parameterization and dual problems are, in general, natural and equally interesting[3, 10]. For example VERTEX COVER is fixed parameter tractable whereas the dual of VERTEX COVER is the INDEPENDENT SET problem (which is the same as choosing $n - k$ vertices to cover all edges) and is W[1] complete.

The $(m - t)$-VERTEX COVER problem can also be parameterized in two ways, by k alone and by k and t. When we have both t and k as parameters then we solve this problem by branching on an edge $e = (u, v)$. Here we branch by choosing either the vertex u or the vertex v or e which means that we are looking for a solution which contains either u or v or does not cover e. So we get the following branching recurrence:

$$T(k, t) \leq 2T(k - 1, t) + T(k, t - 1).$$

This immediately gives us the following theorem

Theorem 6. *$(m - t)$-VERTEX COVER can be solved in $O(3^{t+k} n + m)$ time. I.e. $(m - t)$ VERTEX COVER is fixed parameter tractable if parameterized by t and k.*

Now we show that the t-VERTEX COVER problem is FPT in G_5 graphs when parameterized by k alone. We will see that this result also applies to $(m - t)$-VERTEX COVER problem when parameterized by k alone.

Theorem 7. t-VERTEX COVER *is fixed parameter tractable in* G_5 *graphs when parameterized by* k *alone. The algorithm runs in* $O(2^{O(k \log k)} n + m)$ *time.*

Proof. Without loss of generality we can assume that the maximum degree of this graph is not bounded by a function of k, otherwise the problem is fixed parameter tractable by Theorem 5. Let v_0 be a vertex of highest degree and let $v_1, v_2, \ldots, v_r$ be its neighbors. Further assume that

$$deg(v_1) \geq deg(v_2) \geq \cdots deg(v_k) \geq \cdots \geq deg(v_r).$$

Let $A = \{v_0, v_1, \cdots, v_k\}$. We show that v_0 or one of its k highest degree neighbors must be in some t-vertex cover. More precisely, we prove the following claim:

Claim: There exists a t-vertex cover C such that $A \cap C \neq \emptyset$.

The claim says that there exist a t-vertex cover C containing at least one vertex of A. We then branch on the vertices of the set A, and look for a solution of size $k - 1$, covering $t - deg(v_i)$ edges in $G - \{v_i\}$, where $0 \leq i \leq k$ and recursively use this claim on the respective subgraphs. Hence the claim proves that t-vertex cover is fixed parameter tractable.

Now we are left with proving the claim. We show the claim by contradiction. Assume to the contrary that no t-vertex cover intersects A. By Lemma 4 we know that there exists a t-vertex cover C containing one of v_0's neighbors. Let v_l be a neighbor of v_0 in C. Because of our assumption $l > k$. Suppose for some v_i, $1 \leq i \leq k$, $N(v_i) \cap C = \emptyset$. Then we can obtain a t-vertex cover $C' = C - \{v_l\} + \{v_i\}$ of size at most k and covering at least t edges as $deg(v_i) \geq deg(v_l)$. So we now assume that for each v_i, $1 \leq i \leq k$, $N(v_i) \cap C \neq \emptyset$. Let $B_i = N(v_i) \cap C$. Observe that for each i, B_i does not contain v_l otherwise that will imply v_0, v_i, v_l is a triangle. Suppose for some $i \neq j$, $u \in B_i \cap B_j$ then v_0, v_i, u, v_j is a cycle of length 4. Hence $B_i \cap B_j = \emptyset$ for all i, j such that $i \neq j$. So this implies that $\sum_{i=1}^{k} |B_i| \geq k$. So we have $B_i \neq \emptyset$, $B_i \subseteq C - \{v_l\}$ and their pairwise intersections are empty. But this implies $\sum_{i=1}^{k} |B_i| \leq |C - \{v_l\}| \leq k - 1$ which contradicts that $\sum_{i=1}^{k} |B_i| \geq k$. This in turn proves the claim.

Since we branch on the vertices of A whose size is bounded by $k + 1$, we get an algorithm of time complexity $O((k+1)^k n)$.   □

Since the runtime in Theorem 7 was independent of t, we get

Theorem 8. $(m - t)$-VERTEX COVER *can be solved in* $O(2^{O(k \log k)} n + m)$ *time in* G_5 *graphs when parameterized by only* k.

By arguments similar to those used in Theorem 7, we can show the following:

Theorem 9. t-DOMINATING SET *can be solved in* $O(2^{O(k \log k)} n^{O(1)})$ *time in* G_5 *graphs when parameterized by only* k.

4   Independent Set and Its Variants in G_4 Graphs

INDEPENDENT SET problem asks for an induced subgraph on k vertices which only contains isolated vertices. More precisely:

INDEPENDENT SET : Given a graph $G = (V, E)$ and an integer $k \geq 0$, determine whether there exists a set of at most k vertices $I \subseteq V$ such that the subgraph induced by I does not contain any edges.

INDEPENDENT SET problem is W[1]-hard in general graphs. We show that INDEPENDENT SET and some of its variants are fixed parameter tractable in *triangle free* graphs. We use *Ramsey theory* to get a kernel of size $O(k^2)$ for these problems.

Theorem 10. *Parameterized Independent Set problem can be solved in $O(kn + 2^{O(k \log k)})$ in G_4 graphs (triangle free graphs).*

Proof. Given any two integers p and q, there exists a number $R(p, q)$ such that any graph on at least $R(p, q)$ vertices contains an independent set of size p or a clique of size q. $R(p, q)$, for various values of p and q are known as *Ramsey Numbers*. It is well known that $R(p, q) \leq \binom{p+q-2}{q-1}$ [8]. And if $n > R(p, q)$ then either an independent set of size p or a clique of size q can be found in $O((p+q)n)$ time by transforming the inductive arguments used in the proof of Theorem 27.3 in [9] for the upper bound of $R(p, q)$ to a constructive algorithm.

If $k \leq 2$, then we can check in linear time whether the graph has an independent set of size 2 or not. So let us assume that $k \geq 3$. If the number of vertices $n > k^2 \geq R(k, 3)$ then we know that this graph has either an independent set of size k or a clique of size 3. But since the input graph is triangle free, we know it must have an independent set of size k and can be found in $O(kn)$ time. Otherwise we know that $n \leq k^2$. In this case, we try all possible subsets to see whether the graph has an independent set of size k or not. If any of them does, then we answer YES and answer NO otherwise. This will take $O(k^{O(k)})$ or $O(2^{O(k \log k)})$ time. This completes the proof. □

Theorem 10 can be extended to a larger class of problems where one is interested in finding a subset inducing a "hereditary property". A graph property Π is a collection of graphs. A graph property Π is non-trivial if Π has at least one graph and does not include all graphs. A non-trivial property is said to be *hereditary* if a graph G is in property Π implies that every *induced subgraph* of G is also in Π. Given any property Π, let $P(G, k, \Pi)$ be the problem defined below:

   $P(G, k, \Pi)$: Given a graph $G = (V, E)$ and a positive integer k, determine whether there exists a set of k vertices $V' \subseteq V$ such that $G[V']$ is in Π.

Khot et al. [10] studied this problem and showed the following theorem.

Theorem 11. *[10](Khot and Raman) Let Π be a hereditary property that includes all independent sets but not all cliques (or vice versa). Then the problem $P(G, k, \Pi)$ is W[1] hard.*

We state fixed parameter tractable version of this theorem in G_4 graphs. The proof of this theorem follows along the same line as of Theorem 10.

Theorem 12. *Let Π be a hereditary property that includes all independent sets but not all cliques. Then the problem $P(G, k, \Pi)$ restricted to G_4 graphs is fixed parameter tractable and can be solved in $O(kn + 2^{O(k \log k)} n^{O(1)})$ time.*

Given a graph $G = (V, E)$ and a positive integer $k \geq 0$, ACYCLIC SUBGRAPH, BIPARTITE SUBGRAPH and PLANAR SUBGRAPH problems ask whether there exists a subset $V' \subseteq V$, such that $|V'| \geq k$ and $G[V']$ is acyclic, bipartite or planar respectively. We refer [8] for definitions of these terms. All these problems are known to be W[1]-hard [3, 10] in general graphs. As a corollary to Theorem 12 we have following:

Corollary 2. ACYCLIC, BIPARTITE *and* PLANAR SUBGRAPH *problems are fixed parameter tractable with time complexity* $O(kn + 2^{O(k \log k)} n^{O(1)})$ *in* G_4 *graphs.*

Corollary 2 shows that ACYCLIC and PLANAR SUBGRAPH problems are fixed parameter tractable in bipartite graphs. In fact we can easily get much improved FPT algorithms for these problems in bipartite graphs. Observe that bipartite graph has an independent set (and hence a planar subgraph) on $n/2$ vertices. So, if $k \leq n/2$ then for both these problems the answer is YES and otherwise $k > n/2$ or $n < 2k$ and hence we get a kernel of size at most $2k$ for both ACYCLIC and PLANAR SUBGRAPH problems in bipartite graphs. Now we check all k sized subsets of the vertex set to see whether the subset induces an acyclic subgraph or planar subgraph. Since $\binom{n}{k} \leq \binom{2k}{k} \leq 2^{2k} = 4^k$, we get $O(4^k n^{O(1)})$ time algorithm for both these problems in bipartite graphs.

Fomin et al. [5] has shown that thee minimum feedback vertex set can be found in $O(1.8621^n n^{O(1)})$ time in bipartite graphs. Minimum feedback vertex set is a complement of the vertex set of the maximum ACYCLIC SUBGRAPH problem, where the objective is to find the minimum number of vertices whose removal results in an acyclic subgraph. So together with this result and kernel of size $2k$ we get $O(1.8621^{2k} n^{O(1)}) = O(3.47^k n^{O(1)})$ time algorithm for the ACYCLIC SUBGRAPH problem. Putting together everything we get the following theorem.

Theorem 13. *Parameterized* ACYCLIC SUBGRAPH *and* PLANAR SUBGRAPH *problems can be solved in* $O(3.47^k k^{O(1)} + n^3)$ *and* $O(4^k k^{O(1)} + n^3)$ *time respectively in bipartite graphs.*

5   Is Everything Easy ?

In contrast to the results presented in the previous sections, here we show a problem to be $W[1]$-hard even in bipartite graphs with girth at least 6 (G_6 graphs). Observe that in graphs with large girth the CLIQUE problem is trivial. We look at DENSE SUBGRAPH problem [11] which is a generalization of the CLIQUE problem.

> DENSE SUBGRAPH: Given a graph $G = (V, E)$ and positive integers k and l, determine whether there exists a set of at most k vertices $C \subseteq V$ such that $G[C]$ has at least l edges, i.e. the induced subgraph on C has at least l edges. (Note that l is at most $\binom{k}{2}$.)

DENSE SUBGRAPH is clearly W[1]-hard when parameterized by k. But we give a reduction from CLIQUE to DENSE SUBGRAPH problem parameterized by k which shows that the problem is W[1]-hard even in bipartite graphs with girth at least 6.

Theorem 14. DENSE SUBGRAPH *is* $W[1]$-*hard in bipartite graphs with girth at least 6 when parameterized by* k.

Proof. We give a reduction from CLIQUE. Let (G, k) be an instance of CLIQUE. We make graph $G = (V, E)$ bipartite by subdividing every edge. Let $G' = (V', E')$ be the resulting subgraph. Here, $V' = V \cup W$ where $W = \{w_{uv} \mid (u, v) \in E\}$ and E', the set of edges, consists of (u, w_{uv}) and (v, w_{uv}) for every $w_{uv} \in W$. Take $k' = k + \binom{k}{2}$ and $l = 2\binom{k}{2}$. We claim that G has a clique of size k if and only if G' has a subgraph on k' vertices with at least l edges.

Observe that G' is a bipartite graph as every cycle is of even length and the girth is at least 6 as girth of G is at least 3. Also note that every vertex in W has degree 2 as they represent edges in the original graph. Now suppose G has a clique of size k on vertex set $C = \{v_1, v_2, \cdots, v_k\}$. Then $C' = C \cup \{w_{uv} \mid u, v \in C\}$ is a vertex set of dense subgraph in G' having k' vertices and l edges as $G[C]$ has at least $\binom{k}{2}$ edges.

Conversely, let C' be a set of k' vertices such that $G'[C']$ has at least l edges. Since G' is a bipartite graph, $G'[C']$ is also bipartite. Let $O = V(G) \cap C'$. Suppose $|O| > k$. Now $G'[C']$ is a bipartite graph with O in one part and $N = C' - O$ in other part and every vertex in N has degree at most 2. Since the number of edges is strictly less than l, $|O| \leq k$. Now we show that $|O| \geq k$. We show this by counting the degree of vertices in N. Let n_1 and n_2 be the degree 1 and degree 2 vertices in N respectively. Then the number of edges in $G[C']$ is :

$$|E(G[C'])| = 2n_2 + n_1 \ \leq \ 2\binom{|O|}{2} + \left(k + \binom{k}{2} - \binom{|O|}{2} - |O|\right) \ < \ l = 2\binom{k}{2}.$$

This is because $n_2 \leq \binom{|O|}{2}$ (as there are n_2 edges in $G[O]$) and $|O| \leq k$. This implies that $|O| \geq k$ and hence $|O| = k$. As a result of this, $|N| = \binom{k}{2}$ and every vertex in N has degree 2. Every vertex of degree 2 in N represents an edge in $G[O]$. This shows that the vertices of O form a clique in the original graph. □

6   Conclusion and Discussions

In this paper we showed that if the input graphs do not possess short cycles then the neighborhood problems like dominating set, independent set and their variants are fixed parameter tractable. We have also shown that the restriction on girth is optimal if we do not put further restriction on the graph classes. This is the first time, to our knowledge, the complexity of graph problems are classified by girth.

Most of the algorithms given here are just parameterized complexity classification algorithms. We believe that the vast literature known for these problems can be applied to obtain more efficient FPT algorithms. Obtaining a $O(c^k n^{O(1)})$, c a constant, algorithm for all these problems remain an open problem. We conclude with the following conjecture:

Conjecture 1. t-Vertex Cover problem is W[2]-hard in undirected graphs and remains hard even for bipartite graphs when parameterized by k alone.

Currently, it is known to be $W[1]$-hard [7] when parameterized by k.

References

1. M. Bläser. *Computing small partial coverings.* Information Processing Letters **85(6)** (2003) 327-331.
2. M. Cesati. http://bravo.ce.uniroma2.it/home/cesati/research/compendium/
3. R. G. Downey and M. R. Fellows. *Parameterized Complexity.* Springer Verlag (1999).
4. J. Flum and M. Grohe. *Parameterized Complexity Theory.* Springer-Verlag, (2006).
5. F. V. Fomin and A. V. Pyatkin. *Finding Feedback Vertex Set in a Bipartite Graph.* Reports in Informatics, University of Bergen. Report No. 291, (2005).
6. R. Niedermeier. *Invitation to Fixed-Parameter Algorithms.* Oxford Lecture Series in Mathematics and Its Applications, Oxford University Press, (2006).
7. J. Guo, R. Niedermeier and S. Wernicke. *Parameterized complexity of generalized Vertex Cover problems..* In the Proceeding of 9th WADS, LNCS **3608** (2005) 36-48.
8. F. Harary. *Graph Theory.* Addison-Wesley Publishing Company (1969).
9. S. Jukna. *Extremal Combinatorics.* Springer-Verlag (2001).
10. S. Khot and V. Raman. *Parameterized complexity of finding Subgraphs with hereditary properties.* Theoretical Computer Science **289(2)** (2002) 997-1008.
11. G. Kortsarz and D. Peleg. *On Choosing a Dense Subgraph .* In the Proceeding of 34th FOCS, (1993) 692-701.

Better Approximation Schemes for Disk Graphs[*]

Erik Jan van Leeuwen

CWI, Kruislaan 413, 1098 SJ Amsterdam, the Netherlands
E.J.van.Leeuwen@cwi.nl

Abstract. We consider Maximum Independent Set and Minimum Vertex Cover on disk graphs. We propose an asymptotic FPTAS for Minimum Vertex Cover on disk graphs of bounded ply. This scheme can be extended to an EPTAS on arbitrary disk graphs, improving on the previously known PTAS [8]. We introduce the notion of level density for disk graphs, which is a generalization of the notion of ply. We give an asymptotic FPTAS for Maximum Independent Set on disk graphs of bounded level density, which is also a PTAS on arbitrary disk graphs. The schemes are a geometric generalization of Baker's EPTASs for planar graphs [3].

1 Introduction

With the ever decreasing size of communication and computing devices, mobility is a key word at the start of the 21st century. Using wireless connections, mobile devices can join local or global communication networks. In trying to understand the properties of such wireless networks, the (unit) disk graph model is frequently used. A *disk graph* is the intersection graph of a set of disks in the plane. This means that each disk corresponds to a vertex of the graph and there is an edge between two vertices if the corresponding disks intersect. The set of disks is called a *disk representation* of the graph. A *unit disk graph* has a disk representation where all disks have the same radius. Besides their practical purposes, (unit) disk graphs have interesting theoretical properties as well.

In a previous paper [21], we considered unit disk graphs of bounded density, leading to new approximation schemes for several optimization problems. Here we extend these ideas to disk graphs and introduce the notion of bounded level density. We give an asymptotic FPTAS for Maximum Independent Set on disk graphs of bounded level density, which is also a PTAS on arbitrary disk graphs. Furthermore, we show there exists an EPTAS for Minimum Vertex Cover on arbitrary disk graphs, improving results of Erlebach, Jansen, and Seidel [8]. As each planar graph is a 1-ply disk graph [11], these results are a geometric generalization of the EPTASs for planar graphs by Baker [3].

2 Preliminaries

An *independent set* $I \subseteq V$ of a graph $G = (V, E)$ contains only non-adjacent vertices (i.e. $u, v \in I \Rightarrow (u, v) \notin E$). A set $C \subseteq V$ *covers* $V' \subseteq V$ if $u \in C$ or

---

[*] This research was supported by the Bsik project BRICKS.

L. Arge and R. Freivalds (Eds.): SWAT 2006, LNCS 4059, pp. 316–327, 2006.

$v \in C$ for each $(u, v) \in E \cap (V' \times V')$. If C covers V, then S is a *vertex cover*. We seek independent sets of maximum size and vertex covers of minimum size.

For each instance x of a maximization (minimization) problem and any $\epsilon > 0$, a *polynomial-time approximation scheme (PTAS)* delivers in time polynomial in $|x|$ (for fixed ϵ) a feasible solution of value within a factor $(1 - \epsilon)$ (respectively $(1 + \epsilon)$) of the optimum. An *efficient PTAS (EPTAS)* delivers such a solution in time polynomial in $|x|$ and $f(\frac{1}{\epsilon})$ for some function f only dependent on $\frac{1}{\epsilon}$, while a *fully polynomial-time approximation scheme (FPTAS)* delivers such a solution in time polynomial in $|x|$ and $\frac{1}{\epsilon}$. An *asymptotic FPTAS (FPTAS$^\omega$)* gives a feasible solution in time $|x|$ and $\frac{1}{\epsilon}$ and attains the approximation factor if $|x| > c_\epsilon$, for some constant c_ϵ only dependent on ϵ.

A *path decomposition* of a graph $G = (V, E)$ is a sequence $(X_1, X_2, \ldots, X_p)$ of subsets of V (called *bags*) such that 1) $\bigcup_{1 \leq i \leq p} X_i = V$, 2) for all $(v, w) \in E$, there is an i $(1 \leq i \leq p)$ such that $v, w \in X_i$, and 3) $X_i \cap X_k \subseteq X_j$ for all i, j, k with $1 \leq i < j < k \leq p$. The *width* of a path decomposition $(X_1, X_2, \ldots, X_p)$ is $\max_{1 \leq i \leq p} |X_i| - 1$. The *pathwidth* of a graph $G = (V, E)$ is the minimum width of any path decomposition of G [18].

3   Previous Work

Clark, Colbourn, and Johnson [6] showed that Maximum Independent Set and Minimum Vertex Cover are NP-hard for (unit) disk graphs. The problems remain NP-hard under the assumption of bounded (level) density [20].

On unit disk graphs, Marathe et al. [13] give constant factor approximation algorithms. Different PTASs are presented by Hunt et al. [10] and Matsui [15] and PTASs exist even if no disk representation is known [17]. On λ-precision disk graphs of bounded radius ratio, Hunt et al. [10] show FPTAS$^\omega$s exist. In a λ-precision disk graph, the distance between any two disk centers is at least λ. Marx [14] gives an EPTAS for Minimum Vertex Cover on unit disk graphs.

Alber and Fiala [2] show that Maximum Independent Set is fixed-parameter tractable for λ-precision disk graphs of bounded radius ratio. Marx [14] recently showed that Maximum Independent Set is W[1]-hard for general (unit) disk graphs. However, $O(n^{O(\sqrt{k})})$-time fixed-parameter algorithms are known [1, 2].

On disk graphs, Malesińska [12] and Marathe et al. [13] give constant factor approximation algorithms. Erlebach, Jansen, and Seidel [8] give a PTAS for Maximum Independent Set and Minimum Vertex Cover. Chan [4] proposes a different PTAS for Maximum Independent Set on intersection graphs of fat objects (a set of disks is considered to be fat). Chan [5] also gives a PTAS for Maximum Independent Set on unit-height rectangle intersection graphs of bounded ply.

4   The Ply of Disk Graphs

Let $D = \{D_i \mid i = 1, \ldots, n\}$ be a set of disks in the plane and $G = (V, E)$ the corresponding disk graph. Scale the disks by a factor 2^w for some integer w, such that each disk has radius at least $\frac{1}{2}$. In the following, we will not distinguish between a set of disks and the graph they induce.

Previously [21], we showed that an FPTAS$^\omega$ exists for Maximum Independent Set, Minimum Vertex Cover, and Minimum (Connected) Dominating Set on unit disk graphs of bounded density. The density of a unit disk graph is the maximum number of disk centers in any 1×1 square. A careful examination of the proof of these schemes shows that they can be extended to disk graphs of bounded density and constant maximum radius. Observe that while scaling can make the maximum radius arbitrarily small, this can increase the density quadratically. Therefore these schemes do not generalize to disk graphs of arbitrary radius. Hence another approach is needed.

The *ply* of a point p in the plane with respect to D is the number of disks of D containing p (i.e. having p inside the disk). Then the *ply* of D is the maximum ply of any point in the plane [16]. Observe that disk graphs of bounded ply are more general than disk graphs of bounded density and bounded maximum radius. Hence an FPTAS$^\omega$ for disk graphs of bounded ply would generalize previous results. Below we give such an approximation scheme for Minimum Vertex Cover. This scheme uses the following properties of disk graphs of bounded ply.

Lemma 1. *Given a set D of disks of ply γ, the number of disks of radius at least r intersecting*

- *a line of length k is at most $\frac{4}{r\pi}(k + 4r)\gamma$,*
- *the boundary of a $k \times k$ square $(k \geq 4r)$ is at most $\frac{16}{r\pi}k\gamma$,*
- *a $k \times k$ square is at most $\frac{(k+4r)^2}{r^2\pi}\gamma$,*
- *two perpendicular, intersecting lines of length k is at most $\frac{8}{r\pi}(k + 2r)\gamma$.*

This lemma implies the following pathwidth upper bound.

Lemma 2. *Given a set D of disks of ply γ, there exists a path decomposition of the disks of radius at least r intersecting a $k \times k$ square of width at most $\frac{4}{r\pi}(k + 4r)\gamma - 1$ and consisting of at most $\frac{(k+4r)^2}{r^2\pi}\gamma$ bags.*

5   Approximating Minimum Vertex Cover

To approximate the minimum vertex cover problem, we use the *shifting technique* introduced by Baker [3] and Hochbaum and Maass [9]. To apply this technique, a decomposition of the minimum vertex cover problem into smaller subproblems is needed. Here we use a decomposition of the disks similar to the ones proposed by Hochbaum and Maass [9], Erlebach, Jansen, and Seidel [8], and Chan [4]. Combining the shifting technique with this decomposition yields the desired approximation factor (see Section 7).

First partition the disks into levels. A disk D_i has *level* $j \in \mathbb{Z}_{\geq 0}$ if its radius r_i satisfies $2^{j-1} \leq r_i < 2^j$. Since each disk has radius at least $\frac{1}{2}$, each disk is indeed assigned a level. The level of the largest disk is denoted by l. For a set of disks D, let $D_{=j}$ denote those disks in D of level j. Similarly, we define $D_{\geq j}$

as the set of disks of level at least j, $D_{>j,<j'}$ as the set of disks of level greater than j, but smaller than j', and so on.

Now let $k \geq 5$ be an odd positive integer. For each level j, we decompose the plane into $k2^j \times k2^j$ squares such that these squares induce a quadtree. Formally, for each level j, we consider the horizontal lines $y = hk2^j$ and vertical lines $x = vk2^j$ ($h, v \in \mathbb{Z}$). The squares induced by these lines are called *level j squares*, or simply j-squares.

Note that each j-square is completely contained in some $(j+1)$-square. Conversely, each $(j+1)$-square S contains exactly 4 j-squares, denoted by S_1 through S_4. The squares $S_1, \dots, S_4$ are *siblings* of each other. We let D^S denote the set of disks intersecting S and $D^{b(S)}$ denotes the set of disks which intersect the boundary of S. Furthermore, we define $D^{i(S)} = D^S - D^{b(S)}$ (i.e. the set of disks fully contained in the interior of S) and let $D^{+(S)} = \bigcup_{i=1}^4 D^{b(S_i)} - D^{b(S)}$ (i.e. the set of disks intersecting the boundary of at least one of the four children of S, but not the boundary of S itself). The meaning of combinations like $D^{b(S)}_{\leq j}$ should be self-explaining. We use $j(S)$ to denote the level of a square S.

6   A Close to Optimal Vertex Cover

We prove the following theorem, which will be auxiliary to our main theorem.

Theorem 1. *Let D be a set of disks of ply γ, $k \geq 5$ an odd positive integer, and OPT a minimum vertex cover for D. Then in time $O(k^2 n^2 \, 2^{\frac{48}{\pi} k \gamma})$, we can find a vertex cover VC for D such that $|VC| \leq \sum_S \left| OPT^S_{=j(S)} \right|$, where the sum is over all squares S.*

We can obtain a vertex cover with the required size by applying bottom-up dynamic programming to the j-squares. Roughly speaking, for each j-square S, we consider all subsets of $D^{b(S)}_{>j}$ (the disks of level greater than j intersecting the boundary of S). For each such subset, we compute a close to optimal vertex cover for D^S containing this subset. Formally, we define for each j-square S and each $W \subseteq D^{b(S)}_{>j}$ a function $\mathsf{size}(S, W)$. This function is defined recursively on j.

$$
\mathsf{size}(S, W) = \begin{cases} \min \left\{ |T| \mid T \subseteq D^S_{=j} \cup D^{i(S)}_{>j}; \; T \cup W \text{ covers } D^S \right\} & \text{if } j = 0; \\ \min\limits_{U \subseteq D^{+(S)}_{>j-1} \cup D^{b(S)}_{=j}} \left\{ |U| + \sum_{i=1}^4 \mathsf{size}\left(S_i, (U \cup W)^{b(S_i)}\right) \right\} & \text{if } j > 0. \end{cases}
$$

Let $\mathsf{sol}(S, W)$ be the subfamily of D attaining $\mathsf{size}(S, W)$ or $\emptyset$ if $\mathsf{size}(S, W)$ is ∞.

6.1   Properties of the size- and sol-Functions

We first show that the sum of $\mathsf{size}(S, \emptyset)$ over all level l squares S attains the value mentioned in Thm. 1. Let OPT again be a minimum vertex cover for D.

Lemma 3. $\sum_{S; \, j(S)=l} \mathsf{size}(S, \emptyset) \leq \sum_S \left| OPT^S_{=j(S)} \right|.$

Proof. Apply induction on j. We prove that the following invariant holds:

$$\text{size}(S, OPT^{b(S)}_{>j}) \le \sum_{S' \subseteq S} \left| OPT^{S'}_{=j(S')} \right| + \left| OPT^{i(S)}_{>j} \right|.$$

Here S is some j-square. For $j = 0$, the correctness of the invariant follows from the definition of size. So assume $j > 0$ and the invariant holds for all j'-squares with $j' < j$. Then from the description of size and by applying induction,

$$\text{size}(S, OPT^{b(S)}_{>j}) \le \sum_{i=1}^{4} \text{size}(S_i, OPT^{b(S_i)}_{>j-1}) + \left| OPT^{+(S)}_{>j-1} \right| + \left| OPT^{b(S)}_{=j} \right|$$

$$\le \sum_{i=1}^{4} \sum_{S'_i \subseteq S_i} \left| OPT^{S'_i}_{=j(S'_i)} \right| + \sum_{i=1}^{4} \left| OPT^{i(S_i)}_{>j-1} \right|$$

$$+ \left| OPT^{+(S)}_{>j-1} \right| + \left| OPT^{b(S)}_{=j} \right|$$

$$= \sum_{i=1}^{4} \sum_{S'_i \subseteq S_i} \left| OPT^{S'_i}_{=j(S'_i)} \right| + \left| OPT^{i(S)}_{>j-1} \right| - \left| OPT^{+(S)}_{>j-1} \right|$$

$$+ \left| OPT^{+(S)}_{>j-1} \right| + \left| OPT^{b(S)}_{=j} \right|$$

$$= \sum_{i=1}^{4} \sum_{S'_i \subseteq S_i} \left| OPT^{S'_i}_{=j(S'_i)} \right| + \left| OPT^{S}_{=j} \right| + \left| OPT^{i(S)}_{>j} \right|$$

$$= \sum_{S' \subseteq S} \left| OPT^{S'}_{=j(S')} \right| + \left| OPT^{i(S)}_{>j} \right|.$$

Since l is the level of the largest disk, $OPT^{i(S)}_{>j} = \emptyset$ and $OPT^{b(S)}_{>j} = \emptyset$ for all j-squares S with $j \ge l$. Hence

$$\sum_{S; j(S)=l} \text{size}(S, \emptyset) \le \sum_{S; j(S)=l} \sum_{S' \subseteq S} \left| OPT^{S'}_{=j(S')} \right| \le \sum_{S} \left| OPT^{S}_{=j(S)} \right|.$$

This proves the lemma. □

We prove the union of $\text{sol}(S, \emptyset)$ over all level l squares S is a vertex cover of D.

Lemma 4. $\bigcup_{S; j(S)=l} \text{sol}(S, \emptyset)$ *is a vertex cover of* D.

Proof. We again apply induction on j. For each j-square S and $W \subseteq D^{b(S)}_{>j}$, we claim that $\text{size}(S, W) \ne \infty$ if and only if there exists a subset of $D^{S}_{\le j} \cup D^{i(S)}_{>j}$ such that this subset and W cover D^S. For $j = 0$, this follows from the definition of size and sol. So assume $j > 0$ and the claim holds for all j'-squares with $j' < j$.

Let S be an arbitrary j-square and let W be an arbitrary subset of $D^{b(S)}_{>j}$. Suppose there is no subset of $D^{S}_{\le j} \cup D^{i(S)}_{>j}$ such that this subset and W cover

D^S. Then W does not cover $D_{>j}^{\mathrm{b}(S)}$, or $D_{\leq j}^S \cup D_{>j}^{\mathrm{i}(S)} \cup W$ would cover D^S. This implies that, for any $U \subseteq D_{>j-1}^{+(S)} \cup D_{=j}^{\mathrm{b}(S)}$ and for any $i = 1, \ldots, 4$, there is no subset of $D_{\leq j-1}^{S_i} \cup D_{>j-1}^{\mathrm{i}(S_i)}$ such that this subset and $U \cup W$ cover D^{S_i}. As the claim holds for the $(j-1)$-square S_i, $\mathsf{size}(S_i, U \cup W) = \infty$. Hence $\mathsf{size}(S, W) = \min_{U \subseteq D_{>j-1}^{+(S)} \cup D_{=j}^{\mathrm{b}(S)}} \left\{ |U| + \sum_{i=1}^4 \mathsf{size}\left(S_i, (U \cup W)^{\mathrm{b}(S_i)}\right) \right\} = \infty$.

Conversely, suppose there exists a subset C of $D_{\leq j}^S \cup D_{>j}^{\mathrm{i}(S)}$ such that $C \cup W$ covers D^S. Let $U = C \cap (D_{>j-1}^{+(S)} \cup D_{=j}^{\mathrm{b}(S)})$. Then for each $i = 1, \ldots, 4$, there exists a subset of $D_{\leq j-1}^{S_i} \cup D_{>j-1}^{\mathrm{i}(S_i)}$ such that this subset and $U \cup W$ cover D^{S_i} (simply take $(C - U) \cap D^{S_i}$). As the claim holds for the $(j-1)$-square S_i, $\mathsf{size}(S_i, U \cup W) \neq \infty$. Hence $\mathsf{size}(S, W) = \min_{U \subseteq D_{>j-1}^{+(S)} \cup D_{=j}^{\mathrm{b}(S)}} \left\{ |U| + \sum_{i=1}^4 \mathsf{size}\left(S_i, (U \cup W)^{\mathrm{b}(S_i)}\right) \right\} \neq \infty$. This proves the claim.

Since l is the level of the largest disk, $D_{>j}^{\mathrm{b}(S)} = \emptyset$ for all j-squares S with $j \geq l$. Hence for each l-square S, $D_{\leq j}^S \cup D_{>j}^{\mathrm{i}(S)} = D^S$. As D^S covers D^S, it follows from the claim that $\mathsf{size}(S, \emptyset) \neq \infty$. By the definition of sol, $\mathsf{sol}(S, \emptyset)$ covers D^S. Since each edge is contained in D^S for some l-square S, $\bigcup_{S;\, j(S)=l} \mathsf{sol}(S, \emptyset)$ is indeed a vertex cover of D. $\square$

6.2   Computing the size- and sol-Functions

We show that it is sufficient to compute size and sol for a limited number of j-squares. This can be done in the time stated in Thm. 1.

Call a j-square *non-empty* if it is intersected by a level j disk and *empty* otherwise. A j-square S is *relevant* if one of its three siblings is non-empty or there is a non-empty square S' containing S, such that S' has level at most $j + \lceil \log k \rceil$ (so each non-empty j-square is relevant). Note that this definition induces $O(k^2 n)$ relevant squares. A relevant square S is said to be a *relevant child* of another relevant square S' if $S \subset S'$ and there is no third relevant square S'', such that $S \subset S'' \subset S'$. Conversely, if S is a relevant child of S', S' is a *relevant parent* of S.

Lemma 5. *For each relevant 0-square S, all size- and sol-values for S can be computed in $O(k^2 \gamma\, 2^{(\frac{24}{\pi}k + \frac{16}{\pi})\gamma})$ time.*

Proof. From Lem. 1, $\left| D_{>0}^{\mathrm{b}(S)} \right|$ is bounded by $\frac{16}{\pi} k\gamma$. Hence all subsets W of $D_{>0}^{\mathrm{b}(S)}$ can be enumerated in $O(2^{\frac{16}{\pi} k\gamma})$ time. For a fixed set W, $\mathsf{size}(S, W)$ is defined as the size of a minimum size subset of $D_{=0}^S \cup D_{>0}^{\mathrm{i}(S)}$, such that this subset and W cover D^S. We may assume W covers $D_{>0}^{\mathrm{b}(S)}$, otherwise such a subset does not exist and $\mathsf{size}(S, W)$ is ∞. Then the requested subset is simply a minimum vertex cover for $D^S - W$. Following Lem. 2, there exists a path decomposition of D^S of width at most $\frac{8}{\pi}(k + 2)\gamma$ and using $O(k^2 \gamma)$ bags. By adapting the algorithm by Telle and Proskurowski [19], the requested cover can be computed in $O(k^2 \gamma\, 2^{\frac{8}{\pi}(k+2)\gamma})$ time. Therefore all size- and sol-values for S can be computed in $O(k^2 \gamma\, 2^{(\frac{24}{\pi}k + \frac{16}{\pi})\gamma})$ time. $\square$

Assume that the size- and sol-values of all relevant children of S are known.

Lemma 6. *For each relevant j-square S $(j > 0)$ with relevant $(j-1)$-square children, all size- and sol-values for S can be computed in $O(2^{\frac{48}{\pi}k\gamma})$ time.*

Proof. If one of the children $S_1, \ldots, S_4$ of S is relevant, then, by the definition of relevant, all children of S must be relevant. Following the definition of size, we need to enumerate all subsets W of $D_{>j}^{b(S)}$ and for each such W all subsets U of $D_{>j-1}^{+(S)} \cup D_{=j}^{b(S)}$. This is equivalent to enumerating all subsets of $D_{>j-1}^{b(S)} \cup D_{>j-1}^{+(S)}$. Following Lem. 1, the number of disks in $D_{>j-1}^{b(S)}$ is bounded by $\frac{32}{\pi}k\gamma$. Since disks intersecting the boundary of S are not in $D_{>j-1}^{+(S)}$, we can adapt the proof of Lem. 1 to show that $\left|D_{>j-1}^{+(S)}\right|$ is at most $\frac{16}{\pi}k\gamma$. But then enumerating all subsets of $D_{>j-1}^{b(S)} \cup D_{>j-1}^{+(S)}$ can be done in $O(2^{\frac{48}{\pi}k\gamma})$ time. Since size and sol of all relevant children of S are known and assuming that for a given W and U we can compute the sum in constant time, the running time of $O(2^{\frac{48}{\pi}k\gamma})$ follows. □

Lemma 7. *For each relevant j-square S $(j > 0)$ with no relevant children of level $j - 1$, all size- and sol-values for S can be computed in $O(n2^{\frac{32}{\pi}\gamma})$ time.*

Proof. Since S has no relevant children of level $j - 1$, S must be empty. By the definition of relevant, the nearest non-empty ancestor of S (if it exists) has level at least $j + \lceil \log k \rceil$. Hence $D_{>j-1}^S = D_{\geq j+\lceil \log k \rceil}^S$. As any disk of level at least $j + \lceil \log k \rceil$ has radius at least $\frac{1}{2}k$, $D_{\geq j+\lceil \log k \rceil}^S = D_{\geq j+\lceil \log k \rceil}^{b(S)}$ and thus $D_{>j-1}^S = D_{\geq j+\lceil \log k \rceil}^{b(S)}$. In particular, $D_{>j}^{b(S)} = D_{\geq j+\lceil \log k \rceil}^{b(S)}$. Using Lem. 1, $\left|D_{>j}^{b(S)}\right| = \left|D_{\geq j+\lceil \log k \rceil}^{b(S)}\right| \leq \frac{32}{\pi}\gamma$ and thus all subsets W of $D_{>j}^{b(S)}$ can be enumerated in $O(2^{\frac{32}{\pi}\gamma})$ time.

Consider a j'-square $S' \subset S$ for which there is no relevant j''-square S'' such that $S' \subseteq S''$. Using similar arguments as above, we can show that $D_{>j'-1}^{S'} = D_{\geq j+\lceil \log k \rceil}^{S'} = D_{\geq j+\lceil \log k \rceil}^{b(S')}$. In particular, this implies that $D_{>j'-1}^{i(S')}$ and $D_{=j'}^{b(S')}$ are empty. If $j' > 0$, then $D_{>j'-1}^{+(S')}$ is also empty. This simplifies the definition of size(S', W') for some $W' \subseteq D_{>j'}^{b(S')}$ to

$$
\text{size}(S', W') = \begin{cases} 0 & \text{if } j' = 0 \text{ and } W' \text{ covers } D_{>j'}^{b(S')}; \\ \infty & \text{if } j' = 0 \text{ and } W' \text{ does not cover } D_{>j'}^{b(S')}; \\ \sum_{i=1}^{4} \text{size}\left(S_i', W'^{b(S_i')}\right) & \text{if } j' > 0. \end{cases}
$$

Now let W be a subset of $D_{>j}^{b(S)}$. We may assume W covers $D_{>j}^{b(S)}$, otherwise size$(S, W) = \infty$. Note that $D_{>j'}^{b(S')} \subseteq D_{>j}^{b(S)}$ for any j'-square S' as above. By repeatedly applying the above simplification of the definition of size, it follows that size(S, W) is simply $\sum_{S''} \text{size}\left(S'', W^{b(S'')}\right)$, where the sum is over all relevant children S'' of S. As $j(S'') < j(S) - 1$ for any such relevant child S'', S'' must be non-empty or the sibling of a non-empty square. Since the number of non-empty squares is $O(n)$ and a square has precisely three siblings, the number of relevant children of S is $O(n)$. Hence this sum can be computed in $O(n)$ time.   □

Lemma 8. $\sum_{S;\,j(S)=l} \text{size}(S,\emptyset)$ can be computed in $O(k^2 n^2 2^{\frac{48}{\pi}k\gamma})$ time.

Proof. As observed before, there are $O(k^2 n)$ relevant squares. Let S be a relevant j-square without a relevant parent. Following Lemmas 5, 6, and 7, we can compute $\text{size}(S,\emptyset)$ for all such squares S in $O(k^2 n^2 2^{\frac{48}{\pi}k\gamma})$ time.

Now consider any level l square S. If S is relevant, then it cannot have a relevant parent. Hence by the preceding argument, $\text{size}(S,\emptyset)$ is known. If S is not relevant, then we can use the same arguments as in Lem. 7 to show that $\text{size}(S,\emptyset) = \sum_{S''} \text{size}\left(S'', W^{b(S'')}\right)$, where the sum is over all relevant j''-squares $S'' \subset S$ without a relevant parent. It follows that $\sum_{S;\,j(S)=l} \text{size}(S,\emptyset)$ can be computed in $O(k^2 n^2 2^{\frac{48}{\pi}k\gamma})$ time.   □

Proof (of Thm. 1). Follows directly from Lemmas 3, 4, and 8.   □

7   An EPTAS for Minimum Vertex Cover

We now apply the shifting technique to obtain a $(1 + \epsilon)$ approximation of the optimum. For some integer a $(0 \le a \le k - 1)$, define the decomposition as follows. We call a line of level j *active* if it is of the form $y = (hk + a2^{l-j})2^j$ or $x = (vk + a2^{l-j})2^j$ $(h, v \in \mathbb{Z})$. The active lines partition the plane into j-squares as before, except that they are now shifted by the shifting parameter a. The structure however remains the same, and thus we can apply Thm. 1 to compute a close to optimal vertex cover.

Let VC_a denote the set returned by the algorithm for some value of a $(0 \le a \le k - 1)$ and let $VC_{\min}$ be a smallest such set.

Lemma 9. $|VC_{\min}| \le (1 + \frac{12}{k})|OPT|$.

Proof. We claim a line of level j (i.e. of the form $y = h'2^j$ or $x = v'2^j$) is active for precisely one value of a. A horizontal line $y = h'2^j$ is active if $h' = hk + a2^{l-j}$ for some h and a, i.e. if $h' \equiv a2^{l-j} \mod k$. As $\gcd(k, 2^{l-j}) = 1$, such a value of a exists. Hence the line is active for at least one value of a.

Suppose a horizontal line of level j is active for two values of a. Then $hk + a2^{l-j} = h'k + a'2^{l-j}$ for some choice of $h, h', a,$ and a'. Simplifying gives $(h - h')k = (a'-a)2^{l-j}$, or $k|(a'-a)2^{l-j}$. Since k is odd, $k|(a'-a)$, which is impossible as $1 \le |a' - a| \le k - 1$. Therefore each horizontal line of level j is active for precisely one value of a. The same proof holds for vertical lines of level j.

Define the set $\overline{D}_a$ as those disks intersecting the boundary of a j-square S at their level, i.e. $\overline{D}_a = \bigcup_S D_{=j(S)}^{b(S)}$. A level j disk is in $\overline{D}_a$ if and only if it intersects an active line of level j. Since each line of level j is active for precisely one value of a and disks of level j have radius strictly less than 2^j, a level j disk can be in

$\overline{D_a}$ for at most 4 different values of a. Hence there exists a value of a (say a^*) for which $\left| OPT \cap \overline{D}_{a^*} \right| \leq \frac{4}{k} |OPT|$.

From Lem. 3, we know that $|VC_{a^*}| \leq \sum_S \left| OPT^S_{=j(S)} \right|$. Observe that for a fixed value of a, any disk can intersect at most 4 squares at its level. Then

$$
\begin{aligned}
|VC_{a^*}| &\leq \sum_S \left| OPT^S_{=j(S)} \right| \\
&= \sum_S \left| OPT^S_{=j(S)} - OPT^{b(S)}_{=j(S)} \right| + \sum_S \left| OPT^{b(S)}_{=j(S)} \right| \\
&\leq |OPT| - \left| OPT \cap \overline{D}_{a^*} \right| + 4 \left| OPT \cap \overline{D}_{a^*} \right| \\
&\leq |OPT| + \frac{12}{k} |OPT|.
\end{aligned}
$$

Hence $|VC_{\min}| \leq |VC_{a^*}| \leq (1 + \frac{12}{k}) |OPT|$ and the lemma follows.   □

Combining Thm. 1 and Lem. 9, we obtain the following result.

Theorem 2. *There exists an $FPTAS^\omega$ for Minimum Vertex Cover on disk graphs of bounded ply, i.e. of ply $\gamma = \gamma(n) = o(\log n)$.*

Proof. Consider any $\epsilon > 0$. Choose k as the largest odd integer such that $\frac{48}{\pi} k \gamma \leq \delta \log n$ for some constant $\delta > 0$. If $k < 5$, output D. Otherwise, using Thm. 1 and the choice of k, compute and output $VC_{\min}$ in $O(n^{2+\delta} \log^3 n)$ time. Hence in time polynomial in n and $\frac{1}{\epsilon}$, a feasible solution is computed. Furthermore, there exists an n_ϵ such that $k \geq \frac{12}{\epsilon}$ and $k \geq 5$ for all $n > n_\epsilon$. Therefore if $n > n_\epsilon$, it follows from Lem. 9 that $VC_{\min}$ is a $(1 + \epsilon)$ approximation of the optimum.   □

This approximation scheme can actually be extended to a scheme for Minimum Vertex Cover on arbitrary disk graphs.

Theorem 3. *There exists an EPTAS for Minimum Vertex Cover on disk graphs.*

Proof. Consider a point p in the plane of ply more than $\frac{1}{\epsilon}$. Note that the set of disks D_p containing p form a clique. Marx [14] observed that D_p is actually a $(1 + \epsilon)$-approximation of a minimum vertex cover for D_p. Hence we remove D_p from D and repeat until the ply is bounded by $\frac{1}{\epsilon}$. Using the algorithm by Eppstein, Miller, and Teng [7] to determine the ply of a set of disks, this can be done in $O(n^3 \log n)$ time.

Let D_0 denote the remaining set of disks. We use the above approximation scheme to compute a $(1 + \epsilon)$-approximation of a minimum vertex cover of D_0 in $O(\epsilon^{-3} n^2 2^{\frac{576}{\pi} \epsilon^{-2}})$ time. Combining the different approximations gives a $(1 + \epsilon)$-approximation of a minimum vertex cover of D.   □

This result improves the $n^{O(\epsilon^{-2})}$-time PTAS for Minimum Vertex Cover on disk graphs by Erlebach, Jansen, and Seidel [8].

8   Approximating Maximum Independent Set

We introduce the notion of level density. Partition the disks into levels as before. For each level j, let d_j denote the maximum number of level j disks in any $2^j \times 2^j$ square. Then the *level density*, denoted by d, is the maximum d_j over all levels j. Observe that disk graphs of bounded level density are more general than disk graphs of bounded ply, as they can contain arbitrary size cliques.

To simplify the notation ahead, we use D_o to refer to the original set of disks. Assume the level density of D_o is bounded by d. Let $k \geq 5$ be an odd positive integer to be determined later and let the plane be partitioned into j-squares as before. For each j, remove all level j disks intersecting the boundary of a j-square and denote the set of remaining disks by D. Now compute a maximum independent set of D. For each j-square S and each $W \subseteq D^{b(S)}_{>j}$, define $\mathsf{size}(S, W)$ as

$$
\begin{cases}
\max\left\{ |T| \;\middle|\; T \subseteq D^S_{=j} \cup D^{\mathsf{i}(S)}_{>j};\ T \cup W \text{ is an independent set} \right\} & \text{if } j = 0; \\[2ex]
\max_{U \subseteq D^{+(S)}_{>j-1}} \left\{ |U| + \sum_{i=1}^{4} \mathsf{size}\left(S_i, (U \cup W)^{b(S_i)}\right) \right\} & \text{if } j > 0.
\end{cases}
$$

Let $\mathsf{sol}(S, W)$ be the subset of D attaining $\mathsf{size}(S, W)$ or $\emptyset$ if $\mathsf{size}(S, W)$ is $-\infty$.

Lemma 10. $\sum_{S;\, j(S)=l} \mathsf{size}(S, \emptyset) \leq \sum_S \left| OPT^{\mathsf{i}(S)}_{=j(S)} \right|$, *where OPT is a maximum independent set of D_o.*

Lemma 11. $\bigcup_{S;\, j(S)=l} \mathsf{sol}(S, \emptyset)$ *is a maximum independent set of D_o.*

These two lemmas follow straightforwardly from the definition of size, sol, and D. To compute $\sum_{S;\, j(S)=l} \mathsf{size}(S, \emptyset)$, it is again sufficient to consider only relevant j-squares, where the definition of relevant is the same as before. In the analysis, we will apply the following theorem.

Theorem 4. *The maximum number of disjoint unit disks intersecting a unit square is 7.*

For lack of space, a proof is omitted. We also use the following proposition.

Proposition 1. *A set of size $c \cdot s$ has at most $c^s e^s$ distinct subsets of size s.*

Lemma 12. *For each relevant 0-square S, all size- and sol-values can be computed in $O(n^7 k^2 d \, (d\frac{3}{4}\pi e)^{\frac{24}{\pi} k})$ time.*

Proof. As disks in $D^{b(S)}_{\geq \lceil \log k \rceil}$ have radius at least $\frac{1}{2}k$, we can use Thm. 4 to bound the maximum size of any independent set in $D^{b(S)}_{\geq \lceil \log k \rceil}$ by 7. Hence all possible independent sets in $D^{b(S)}_{\geq \lceil \log k \rceil}$ can be enumerated in $O(n^7)$ time.

Using the assumption of bounded level density, we can bound $\left| D^{b(S)}_{>0,\, <\lceil \log k \rceil} \right|$ by $12kd$. As an independent set of disks has ply 1, it follows from Lem. 1 that any independent subset of $D^{b(S)}_{>0,<\lceil \log k \rceil}$ has size at most $\frac{16}{\pi}k$. Then, following Prop. 1, all independent sets of disks in $D^{b(S)}_{>0,<\lceil \log k \rceil}$ can be enumerated in $O((d\frac{3}{4}\pi e)^{\frac{16}{\pi}k})$ time. Hence all independent subsets W of $D^S_{>0}$ can be enumerated in $O(n^7(d\frac{3}{4}\pi e)^{\frac{16}{\pi}k})$ time. Using a path decomposition, $\mathsf{size}(S,W)$ for a fixed W can be computed in $O(k^2 d\, (d\frac{3}{4}\pi e)^{\frac{8}{\pi}k})$ time.   □

Lemma 13. *For each relevant j-square S with relevant $(j-1)$-square children, all* size- *and* sol-*values can be computed in $O(n^7(d\frac{3}{4}\pi e)^{\frac{32}{\pi}k})$ time.*

For relevant j-squares with relevant children of level less than $j-1$, we can show that the size- and sol-values can be computed in $O(n^8)$ time.

Lemma 14. $\sum_{S;j(S)=l} \mathsf{size}(S,\emptyset)$ *can be computed in $O(k^2 n^9 (d\frac{3}{4}\pi e)^{\frac{32}{\pi}k})$ time.*

Let a $(0 \le a \le k-1)$ be an integer. Shift the decomposition as before. Let IS_a be the independent set returned by the algorithm for some value of a $(0 \le a \le k-1)$ and let $IS_{\max}$ be a largest such set. Using similar ideas as in Lem. 9 and Thm. 2, we obtain the following.

Lemma 15. $|IS_{\max}| \ge (1 - \frac{4}{k})|OPT|$.

Theorem 5. *There exists an FPTAS$^\omega$ for Maximum Independent Set on disk graphs of bounded level density, i.e. of level density $d = d(n) = O(\text{polylog } n)$.*

Here polylog n is any polynomial in $\log n$. Now observe that d is bounded by n. Hence the worst case running time of the scheme is $O(k^3 n^9 (n\frac{3}{4}\pi e)^{\frac{32}{\pi}k})$.

Theorem 6. *There exists a PTAS for Maximum Independent Set on disk graphs.*

This PTAS improves on the $n^{O(k^2)}$-time PTAS by Erlebach, Jansen, and Seidel [8] and matches the $n^{O(k)}$-time PTAS by Chan [4].

9   Minimum Vertex Cover Revisited

In Lemmas 5, 6, and 7, instead of for instance enumerating all subsets of $D^{b(S)}_{>j}$, it suffices to enumerate those subsets covering $D^{b(S)}_{>j}$. Observe that the complement of a cover for $D^{b(S)}_{>j}$ is an independent set. Applying upper bounds on the size of independent sets in this way, we can improve the running time of the EP-TAS for Minimum Vertex Cover on disk graphs to $O(\epsilon^{-3} n^2 (e/\epsilon)^{\frac{576}{\pi}\epsilon^{-1}} + n^3 \log n)$.

Acknowledgements. The author thanks Lex Schrijver for many helpful suggestions and discussions.

References

1. Agarwal, P.K., Overmars, M., Sharir, M., "Computing maximally separated sets in the plane and independent sets in the intersection graph of unit disks" in *SODA 2004*, SIAM, 2004, pp. 516–525.
2. Alber, J., Fiala, J., "Geometric Separation and Exact Solutions for the Parameterized Independent Set Problem on Disk Graphs", *J. Algorithms* **52** 2 (2003), pp. 134–151.
3. Baker, B.S., "Approximation Algorithms for NP-Complete Problems on Planar Graphs", *JACM* **41** 1 (1994), pp. 153-180.
4. Chan, T.M., "Polynomial-time Approximation Schemes for Packing and Piercing Fat Objects", *J. Algorithms* **46** 2 (2003), pp. 178–189.
5. Chan, T.M., "A Note on Maximum Independent Sets in Rectangle Intersection Graphs", *Inf. Proc. Let.* **89** 1 (2004), pp. 19–23.
6. Clark, B.N., Colbourn, C.J., Johnson, D.S., "Unit Disk Graphs", *Discr. Math.* **86** 1–3 (1990), pp. 165–177.
7. Eppstein, D., Miller, G.L., Teng, S.-H., "A Deterministic Linear Time Algorithm for Geometric Separators and its Applications", *Fund. Inform.* **22** 4 (1995), pp. 309–329.
8. Erlebach, T., Jansen, K., Seidel, E., "Polynomial-time Approximation Schemes for Geometric Intersection Graphs" *SIAM J. Computing* **34** 6 (2005), pp. 1302–1323.
9. Hochbaum, D.S., Maass, W., "Approximation Schemes for Covering and Packing Problems in Image Processing and VLSI", *JACM* **32** 1 (1985), pp. 130–136.
10. Hunt III, D.B., Marathe, M.V., Radhakrishnan, V., Ravi, S.S., Rosenkrantz, D.J., Stearns, R.E., "NC-Approximation Schemes for NP- and PSPACE-Hard Problems for Geometric Graphs", *J. Algorithms* **26** 2 (1998), pp. 238–274.
11. Koebe, P., "Kontaktprobleme der konformen Abbildung", *Ber. Ver. Sächs. Ak. Wiss. Leipzig, Math.-Phys. Kl.*, **88** (1936), pp. 141–164.
12. Malesińska, E., *Graph-Theoretical Models for Frequency Assignment Problems*, PhD Thesis, Technical University of Berlin, Berlin, 1997.
13. Marathe, M.V., Breu, H., Hunt III, H.B., Ravi, S.S., Rosenkrantz, D.J., "Simple Heuristics for Unit Disk Graphs", *Networks* **25** (1995), pp. 59–68.
14. Marx, D., "Efficient Approximation Schemes for Geometric Problems?" in *Proc. ESA 2005*, LNCS **3669**, Springer-Verlag, Berlin, 2005, pp. 448-459.
15. Matsui, T., "Approximation Algorithms for Maximum Independent Set Problems and Fractional Coloring Problems on Unit Disk Graphs", in *Proc. JCDCG*, LNCS **1763**, Springer-Verlag, Berlin, 1998, pp. 194–200.
16. Miller, G.L., Teng, S.-H., Thurston, W., Vavasis, S.A., "Separators for Sphere-Packings and Nearest Neighbor Graphs", *JACM* **44** 1 (1997), pp. 1–29.
17. Nieberg, T., Hurink, J.L., Kern, W., "A Robust PTAS for Maximum Weight Independent Sets in Unit Disk Graphs" in *Proc. WG 2004*, LNCS **3353**, Springer-Verlag, Berlin, 2004, pp. 214–221.
18. Robertson, N., Seymour, P.D., "Graph Minors. I. Excluding a Forest", *J. Comb. Th. B* **35** (1983), pp. 39–61.
19. Telle, J.A., Proskurowski, A., "Algorithms for Vertex Partitioning Problems on Partial k-Trees", *SIAM J. Disc. Math.* **10** 4 (1997), pp. 529–550.
20. van Leeuwen, E.J., *Optimization Problems on Mobile Ad Hoc Networks – Algorithms for Disk Graphs*, Master's Thesis INF/SCR-04-32, Inst. of Information and Computing Sciences, Utrecht Univ., 2004.
21. van Leeuwen, E.J., "Approximation Algorithms for Unit Disk Graphs" in *Proc. WG 2005*, LNCS **3787**, Springer-Verlag, Berlin, 2005, pp. 351–361.

An Approximation Algorithm for the Wireless Gathering Problem[*]

Vincenzo Bonifaci[1,2], Peter Korteweg[1], Alberto Marchetti-Spaccamela[2], and Leen Stougie[1,3]

[1] Eindhoven University of Technology
Department of Mathematics and Computer Science
Den Dolech 2, 5600 MB Eindhoven, The Netherlands
v.bonifaci@tue.nl, p.korteweg@tue.nl, l.stougie@tue.nl
[2] University of Rome "La Sapienza"
Department of Computer and Systems Science
Via Salaria 113, 00198 Rome, Italy
bonifaci@dis.uniroma1.it, alberto@dis.uniroma1.it
[3] CWI, Kruislaan 413, 1098 SJ Amsterdam, The Netherlands
stougie@cwi.nl

Abstract. The Wireless Gathering Problem is to find a schedule for data gathering in a wireless static network. The problem is to gather a set of messages from the nodes in the network at which they originate to a central node, representing a powerful base station. The objective is to minimize the time to gather all messages. The sending pattern or schedule should avoid interference of radio signals, which distinguishes the problem from wired networks.

We study the Wireless Gathering Problem from a combinatorial optimization point of view in a centralized setting. This problem is known to be **NP**-hard when messages have no release time. We consider the more general case in which messages may be released over time. For this problem we present a polynomial-time on-line algorithm which gives a 4-approximation. We also show that within the class of shortest path following algorithms no algorithm can have approximation ratio better than 4. We also formulate some challenging open problems concerning complexity and approximability for variations of the problem.

1 Introduction

The last decade has seen a broad research focus on wireless networks [9, 10]. Mobile phones, Bluetooth data communication and ad hoc laptop networks are testimony of the wide range of applications for wireless networks. Current interest on wireless sensor networks further emphasizes the importance of wireless

[*] Work of the first author is partly supported by Dutch Ministry of Education, Culture and Science through a Huygens scholarship. Work of the first and third author is partly supported by EU FET Integrated Project AEOLUS, IST- 15964. Work of the authors is partly supported by EU COST-action 293, Graal.

L. Arge and R. Freivalds (Eds.): SWAT 2006, LNCS 4059, pp. 328–338, 2006.
© Springer-Verlag Berlin Heidelberg 2006

networks in the future [8]. One of the main issues concerning wireless networks is data communication. Much research is focused on finding efficient communication protocols, i.e. protocols which minimize energy consumption or maximize throughput of the network.

In wireless networks stations communicate with each other through radio signals. A radio signal has a *transmission radius*, the distance over which the signal is strong enough to send data and an *interference radius*, the distance over which the radio signal is strong enough to interfere with other radio signals.

Interference, also called collision, and *fading* of radio signals are severe problems in wireless networks and the main features that distinguish communication in wireless networks from communication in wired networks. Interference is the effect of radio signal loss due to the fact that multiple stations located within interference radius of each other try to communicate simultaneously. Fading is the effect of radio signal loss due to physical circumstances. Interference and fading cause data loss which results in lower throughput or equivalently higher completion times [12]. Furthermore, it causes higher energy consumption as data has to be resent.

Communication protocols can be divided into five different layers [11]: the application layer, the transport layer, the network layer, the data link layer and the physical layer. The transport layer is used to provide a data delivery protocol between the stations. The network layer is concerned with routing of data supplied by the transport layer. In order to route data, stations need to have information on the structure of the wireless network. See [10] for an overview of routing protocols which provide such information. The data link layer is concerned with finding medium access (MAC) schemes; such a scheme determines which nodes send data at a certain time.

MAC schemes can be characterized as either fixed assignment schemes or contention schemes [9]. A fixed assignment scheme provides an allocation of medium access resources; these resources are time and frequency. This leads to Time Division Multiple Access (TDMA) schemes where each station is allocated a time slot to send data. Similarly there are Frequency Division Multiple Access (FDMA) schemes where each station is allocated a frequency range to send data and Code Division Multiple Access (CDMA) schemes which is a combination of both time and frequency allocation. As resource allocation requires coordination fixed assignment schemes are common in a centralized setting.

Contention schemes are schemes where stations compete for medium access. In a contention scheme stations access the medium independently of other stations. The most common contention schemes are Carrier Sense Multiple Access (CSMA) schemes. Because stations compete to access the medium there is the possibility of collision. There are several techniques for collision detection (CD) or collision avoidance (CA). The 802.11 protocol [7], a widely used standard for Wireless Local Area Networks describes both a fixed assignment scheme and a contention scheme: the point coordination function (PCF) in a centralized setting and the distributed coordination function (DCF).

In this paper we consider the Wireless Gathering Problem (WGP). The WGP is to find a schedule for data gathering at a base station of a wireless static network which minimizes gathering time of messages. Gathering data at some base station is a well known process in wireless networks and is sometimes referred to as *many-to-one* communication. Typically, base stations are powerful stations which are able to process the data and act as gateways to other networks [6].

The Wireless Gathering Problem is the following: given is a static wireless network which consists of several stations (*nodes*) and one base station (the *sink*). Some stations contain a set of packets (*messages*) which have to be gathered at the base station. Wireless communication enables stations to communicate data packets to each other. We assume that time is discrete and that stations have a common clock, hence time can be divided into rounds. In our model a node may either send or receive a single packet during a round. Typically, not all nodes in the network can communicate with each other, hence packets have to be sent through several nodes before they can be gathered at the sink; this is called *multi-hop* routing. The problem consists of constructing a schedule without interference which determines for each packet both route and times at which it is sent. The objective of the problem is to find a schedule which minimizes a function of the completion times of the packets, i.e., the times at which the packets arrive at the sink. We consider minimization of maximum completion time.

The WGP combines the problem of data routing at the network layer and scheduling at the data link layer. We assume that the stations are provided with the necessary information on the graph structure, such as path distances. Thus we study the problem in a *centralized* setting. We also assume perfect radio signals, i.e. no signal loss due to fading. We consider both a version where messages have no release time, i.e. are released at time zero, and a version in which messages are released over time.

Much research has been devoted to the design of communication protocols, but most models do not consider interference. See [5] for a general overview of communication protocols. In the centralized setting without release times, the many-to-one gathering problem is closely related to the one-to-all personalized broadcast, where the base station wants to communicate a unique message to each station in the network. In [1] and [2] the authors study the one-to-all (non-personalized) broadcast on arbitrary graphs in a model similar to ours, for the case where both transmission radius and interference radius are unit distance. In the second paper the authors consider the WGP as a subproblem. For this problem they give a distributed randomized algorithm with expected completion time of $O((n+\delta)\log d)$, where n is the number of messages, δ the diameter of the graph and d the maximum degree. In [4] the authors give an optimal centralized greedy algorithm for the gathering problem without interference on a tree.

The Wireless Gathering Problem with arbitrary integral transmission and interference radius was formulated as a combinatorial optimization problem by Bermond et al. [3]. In this paper the authors prove that WGP is **NP**-hard even without release times (or equivalently, when all messages are released at time zero). For this case they present an approximation algorithm and sketch a proof that it gives an asymptotic 4-approximation and an asymptotic 3-approximation if transmission radius is equal to interference radius. Their algorithm partitions the nodes into layers and areas such that any set of nodes whose elements are in the same area but distinct layers may send simultaneously. Using this result and by choosing layers in a round-robin fashion the algorithm may pipeline messages over some shortest path towards the sink. The authors do not consider the case in which messages are released over time. In Section 2 we formulate the problem precisely.

In Section 3 we present an on-line polynomial-time greedy algorithm which for arbitrary release times gives a 4-approximation in general and a 3-approximation when transmission radius and interference radius are equal. Our results improve over those of Bermond et al. [3] only slightly in case all messages are released at time zero. However, our algorithm is simple, and the approximation ratio indeed holds for arbitrary release times. Both our algorithm and that of Bermond et al. send all messages along the shortest path to the sink. We prove that within the class of shortest path following algorithms no algorithm can have approximation ratio better than 4. Thus, within this class our algorithm is best possible. Furthermore, we prove that our algorithm is optimal on a chain with unit transmission radius and without release times when the sink is at one end of the chain. The complexity of all other variations of the problem on a chain and on a tree is to the best of our knowledge still open. These and some other challenging research opportunities on the Wireless Gathering Problem conclude the paper in Section 4.

2   Mathematical Formulation

We formulate the WGP as a graph optimization problem. Given is a graph $G = (V, E)$ with $|V| = n$, sink $s \in V$, and a set of messages (data packets) $M = \{1, 2, \ldots, m\}$. We assume that each edge has unit length. For each pair of nodes $u, v \in V$ we define the *distance* between u and v, denoted by $d(u, v)$, as the length of a *shortest path* from u to v in G. We introduce d_I as the interference radius and d_T as the transmission radius. Each message $j \in M$ has an *origin* $v_j \in V$.

We assume that time is discrete, say $\{0, 1, \ldots\}$; we call a time unit a *round*. The rounds are numbered $0, 1, \ldots$. During each round a node may either be *sending* a message, be *receiving* a message or be *inactive*. If $d(u, v) \leq d_T$ then u can send some message j to v during a round. If node u sends a message j to v in some round, then the pair (u, v) is called a *call* of message j during that round. Two calls (u, v) and (u', v') *interfere* if $d(u', v) \leq d_I$ or $d(u, v') \leq d_I$; otherwise the calls are *compatible*.

The solution of the WGP is a schedule of compatible calls such that all messages are sent to the sink. Given a schedule, let v_j^t be the node of message j at time t. $C_j = \min\{t : v_j^t = s\}$ is called the *completion time* of message j. We consider the minimization of $\max_j C_j$ (*makespan*). We assume that messages cannot be aggregated. Every message j has a release time $r_j \in \mathbb{Z}_+$ at which it enters the network. We consider the case in which all messages are released at the same time, i.e. $r_j = 0$ for all j, as a special case. In the off-line version all message information is known at time 0, in the on-line version information about a message becomes known only at its release time. The off-line WGP is equivalent to a one-to-many personalized broadcast problem: a time reverse gathering schedule provides a one-to-many personalized broadcast schedule.

3   A Greedy Algorithm

We present a greedy algorithm which assigns messages to calls according to some priority ordering. Each message is assigned to a call of maximum distance without causing interference. Let $\phi(j) = \lceil \frac{d(v_j,s)}{d_T} \rceil$, the minimum number of calls required for j to reach s.

PRIORITY GREEDY (PG). Given messages j and k, we say that j has higher priority than k if $r_j + \phi(j)$ is smaller than $r_k + \phi(k)$, ties broken arbitrarily. In every round, consider the available messages in order of decreasing priority, and send each next message as far as possible along a (possibly prefixed) shortest path from its current node to s, without creating interference with any higher-priority message.

Notice that PG is an on-line, polynomial-time algorithm. To analyze the worst-case approximation ratio of PG we first derive upper bounds on the completion time of each message in a PG solution.

Given a message j, we say that j is *blocked* in round t if, in round t, j cannot be sent over a shortest path towards s over distance d_T (or it cannot be sent to s if $d(v_j^t, s) \leq d_T$) because of interference with some higher-priority message in PG.

We define the following *blocking relation* on a PG schedule: $k \prec j$ if in the last round in which j is blocked, k is the message closest to j that is sent in that round and has a priority higher than j (ties broken arbitrarily). The blocking relation induces a directed graph $F = (M, A)$ on the message set M with an arc (k, j) for each $k, j \in M$ such that $k \prec j$. Observe that for any PG schedule F is a directed forest and the root of each tree of F is a message which is never blocked. For each j let $T(j) \subseteq F$ be the tree of F containing j, $b(j) \in M$ be the root of $T(j)$, and $P(j)$ the path in F from $b(j)$ to j. Let $h(j)$ be the length of $P(j)$. Finally, define $\gamma = 1 + \lceil \frac{d_I+1}{d_T} \rceil$ and let C_j denote the completion time of message j in a PG schedule.

Lemma 1. *For each message $j \in M$,*

$$C_j \leq r_{b(j)} + \phi(b(j)) + \sum_{i \in P(j), i \neq b(j)} \min\{\phi(i), \gamma\}.$$

Proof. The proof is by induction on $h(j)$. Any message j with $h(j) = 0$ is never blocked, hence $b(j) = j$, and the lemma is obviously true.

Otherwise, let t be the last round in which j is blocked by some message k, $k \prec j$. By definition of the blocking relation we have $d(v_j^t, v_k^t) \leq d_T + d_I$ and if $d(v_j^t, v_k^t) > d_I + 1$ then j, although blocked, is sent to v_j^{t+1} with $d(v_j^{t+1}, v_k^t) = d_I + 1$. Also, $d(v_k^t, s) \leq (C_k - t)d_T$, otherwise k would not reach s by time C_k. From time $t + 1$ on, j is forwarded to s over distance d_T each round, reaching s at

$$
\begin{aligned}
C_j &\leq t + 1 + \left\lceil \frac{d(v_k^t, s) + d(v_j^{t+1}, v_k^t)}{d_T} \right\rceil \\
&\leq t + 1 + C_k - t + \left\lceil \frac{d_I + 1}{d_T} \right\rceil \\
&= C_k + 1 + \left\lceil \frac{d_I + 1}{d_T} \right\rceil = C_k + \gamma.
\end{aligned}
$$

Also, $C_j \leq C_k + \phi(j)$, since after k reaches s, j will need no more than $\phi(j)$ rounds to reach s. Thus $C_j \leq C_k + \min\{\phi(j), \gamma\}$ and the lemma follows by applying the induction hypothesis to C_k.   □

Now we derive lower bounds on the optimal cost. We define the *critical radius* R^* as the greatest integer R such that no two nodes at distance at most R from s can receive a message in the same round. Notice that $R^* \geq \lfloor \frac{d_I - d_T}{2} \rfloor$. The *critical region* is the set of nodes at distance at most R^* from s. Define $\gamma^* = \lceil \frac{R^* + 1}{d_T} \rceil$. Let C_j^* denote the completion time of message j in an optimal solution.

Lemma 2. *Let $S \subseteq M$ be a nonempty set of messages. Then there is $k \in S$ such that*

$$
\max_{j \in S} C_j^* \geq r_k + \phi(k) + \sum_{j \in S, j \neq k} \min\{\phi(j), \gamma^*\}.
$$

Proof. Define $p_j = \min\{\phi(j), \gamma^*\}$ and $r_j' = r_j + \phi(j) - p_j$. Since in every round at most one message can move inside the critical region, any feasible solution to the Wireless Gathering Problem gives a feasible solution to a preemptive single machine scheduling problem in which the release time of job j (corresponding to message j) is r_j' and its processing time is p_j. By ignoring interference outside the critical region we can only decrease the optimum cost, thus a lower bound on the scheduling cost is also a lower bound on the gathering cost.

Now let k be the first message in S entering or being released in the critical region in the optimal schedule. In the scheduling relaxation, the makespan is at least the time at which the first job starts processing plus the sum of the processing times:

$$
\begin{aligned}
\max_{j \in S} C_j^* &\geq r_k' + \sum_{j \in S} p_j \\
&= r_k + \phi(k) + \sum_{j \in S, j \neq k} p_j.
\end{aligned}
\qquad \square
$$

Theorem 1. PRIORITY GREEDY *gives a* $\max\{2, \gamma/\gamma^*\}$*-approximation to the Wireless Gathering Problem with release times. When* $r_j = 0$ *for all* $j \in M$, PRIORITY GREEDY *gives a* γ/γ^**-approximation.*

Proof. Let j be the message having maximum C_j, and consider $T(j)$, the tree containing j in the forest induced by the blocking relation. We can apply Lemma 2 with $S = T(j)$ to obtain

$$\max_{i \in T(j)} C_i^* \geq r_k + \phi(k) + \sum_{i \in T(j), i \neq k} \min\{\phi(i), \gamma^*\} \tag{1}$$

where k is some message in $T(j)$. On the other hand, by using Lemma 1,

$$C_j \leq r_{b(j)} + \phi(b(j)) + \sum_{i \in P(j), i \neq b(j)} \min\{\phi(i), \gamma\} \tag{2}$$

$$\leq r_{b(j)} + \phi(b(j)) + \min\{\phi(k), \gamma\} + \sum_{i \in P(j), i \neq b(j), i \neq k} \min\{\phi(i), \gamma\}$$

$$\leq 2(r_k + \phi(k)) + \frac{\gamma}{\gamma^*} \sum_{i \in T(j), i \neq k} \min\{\phi(i), \gamma^*\}$$

where we used the fact that $b(j)$, being the root of $T(j)$, minimizes $r_i + \phi(i)$ in $T(j)$. Now let $\alpha \geq 0$ be such that $\sum_{i \in T(j), i \neq k} \min\{\phi(i), \gamma^*\} = \alpha(r_k + \phi(k))$. Then

$$\frac{C_j}{\max_{i \in T(j)} C_i^*} \leq \frac{2 + \frac{\gamma}{\gamma^*}\alpha}{1 + \alpha} \leq \max\{2, \gamma/\gamma^*\}.$$

For the case in which $r_j = 0$ for all $j \in M$, we proceed similarly, but also notice that since now $b(j) \in \operatorname{argmin}_{i \in T(j)}\{\phi(i)\}$, we have $\phi(b(j)) \leq \phi(k)$ and the claim follows by directly comparing (1) with (2).   □

Corollary 1. PRIORITY GREEDY *is 3-approximate if* $d_I = d_T$ *and 4-approximate in general for WGP with release times.*

Proof. We distinguish several cases:
Case 1 If $d_I \leq 2d_T - 1$ then $\gamma = 3$, which in particular proves the 3-approximation in case $d_I = d_T$.
Case 2: $d_I \leq 3d_T - 1$. Then $\gamma \leq 4$.
Case 3: $ld_T \leq d_I \leq (l+2)d_T - 1$ for any odd integer $l \geq 3$. Then $\gamma \leq l+3$ and as $r^* \geq \lfloor \frac{d_I - d_T}{2} \rfloor$ it follows that $\gamma^* \geq (l+1)/2$, hence $\gamma/\gamma^* \leq 2(l+3)/(l+1) \leq 3$.   □

The analysis shows that the ratio $\gamma/\gamma^* = 4$ only if $d_I/d_T \in [2, 3)$ and the ratio approaches 2 if d_I/d_T tends to infinity.

Corollary 2. *When* $r_j = 0$ *for all* $j \in M$, PRIORITY GREEDY *is optimal if* G *is a chain with* s *as an extreme and* $d_T = 1$.

Proof. If G is a chain and s an extreme, then the critical radius is $d_I + 1$. Thus, for $d_T = 1$ we have $\gamma^* = d_I + 2$. The claim follows since $\gamma = d_I + 2$ for $d_T = 1$ and without release times the approximation ratio is γ/γ^*.   □

PG sends messages over shortest paths. We show that no algorithm which sends each message j over a shortest path from v_j to s can be better than 3-approximate if $d_I = d_T$ and 4-approximate if $d_I > d_T$. This means that to find algorithms with lower approximation ratios than PG, messages need to be diverged from their shortest path to the sink if this path becomes congested.

First, consider the example of Figure 1 with $d_I = d_T = 1$. Nodes u_1, u_2, u_3 have $m/3$ messages each. Any shortest paths following algorithm sends all messages via u, yielding $\max_j C_j = 3m$. On the other hand there is a solution with no message passing u that implies $\max_j C_j^* \leq 3 + m$. The example can easily be extended for arbitrary $d_I = d_T$ such that PRIORITY GREEDY is a 3-approximation.

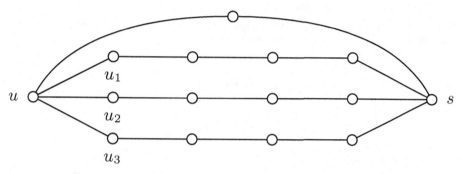

Fig. 1. Any shortest paths following algorithm is no better than a 3-approximation for $d_I = d_T = 1$

In case $d_I > d_T$ consider Figure 2. The nodes $u_1, \ldots, u_m$ each have 1 message. Let $d_I = 2$ and $d_T = 1$. Any shortest paths following algorithm sends all messages via u, yielding $\max_j C_j = 4m$. There is a solution with no message passing u that implies $\max_j C_j^* \leq 4 + m$. The example can easily be extended for arbitrary $d_I = 2d_T$ such that PRIORITY GREEDY is a 4-approximation.

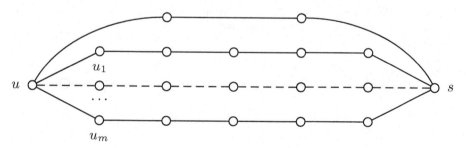

Fig. 2. Any shortest paths following algorithm is no better than a 4-approximation for $d_I = 2, d_T = 1$

In these examples the optimal schedule sends each message over a path of length exceeding the length of its shortest path by at most 1. This may suggest to consider algorithms which send messages over paths whose length does not exceed their shortest path length by some constant k. However, as can easily be verified, for each constant k we could change the length of the paths in the examples above, such that the optimal schedule sends each message over a path whose length exceeds the shortest path length by $k + 1$.

Improvement on the approximation ratio should come from algorithms that avoid congested paths. One such an idea is to use not only the shortest path but the k shortest paths, whichever of them is least congested. By adapting Figure 1, the example in Figure 3 is obtained, showing for $k = 2$ that even if we choose the 2 shortest internally vertex disjoint paths, the lower bound on the approximation ratio for $d_I = d_T$ remains unchanged. The example can be extended to any fixed k. Similarly the example of Figure 2 can be adapted to show that in case $d_I > d_T$ choosing any of k shortest paths, for fixed k, leaves the lower bound of 4 on the ratio unchanged.

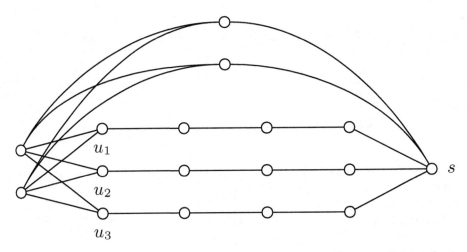

Fig. 3. Any shortest paths following algorithm using 2 internally vertex disjoint shortest paths is no better than a 3-approximation for $d_I = d_T = 1$

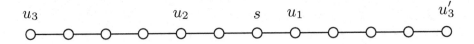

Fig. 4. PG is not optimal on a chain if the sink is not at an extreme

PG can be non-optimal on a chain if the sink is not at an extreme node of the chain. Consider the instance given by the graph in Figure 4. Let $d_T = 1$ and $d_I = 2$ and assume messages are released in u_1, u_2 and u_3 at time zero. PG would first send the message in u_1 to the sink, and then send the message in u_2 to the sink, resulting in a makespan of 7, while in an optimal solution the messages in u_2 and u_3 are forwarded until the message of u_2 reach the sink at time 2, then the messages of u_3 and u_1 are forwarded simultaneously, yielding a cost of 6. However, if the third message would have been released at node u'_3 instead of at node u_3 then PG would have been optimal. This shows that any optimal algorithm for the problem on a chain should take into account the position of the other messages when deciding which of the two messages nearest to the sink to send first. The complexity of the problem on a chain remains open.

4   Challenging Future Research Problems

In this paper we designed and analyzed algorithms for a basic centralized gathering problem (WGP) on a wireless static network with unit edge lengths. Specifically, we proved that a greedy algorithm has approximation ratio 4 for this WGP when minimizing maximum completion time. We also showed that our greedy algorithm yields the best possible approximation ratio within the class of algorithms in which each message is sent over a shortest path to the sink. It is a beautiful challenge to design algorithms that avoid congested paths, which have approximation ratios strictly less than 4 (or 3 if $d_I = d_T$). Our examples at the end of the previous section show that this is not so trivial. For instance selecting among k shortest paths (even disjoint) will not give lower ratios. On the other hand, from the proof of Corollary 1, one could concentrate on the subclass of problems with $d_I/d_T \in [2, 3)$ to improve on the ratio of 4, since in all other cases greedy has ratio at most 3.

All our results apply to general graphs; we have not considered specific graphs in depth. Specifically, the complexity of WGP on chains or trees is open, apart from the restricted case in Corollary 2. Our suspicion is that these problems are easy as well, but the greedy algorithm is not optimal on a chain for $d_T = 1$ and $d_I = 2$ as we have shown at the end of the previous section. The example can easily be extended to any combination of d_I and d_T, except $d_I = d_T$, for which greedy might be optimal. Also, our attempts to formulate the problem on a chain as a dynamic programming problem have failed so far.

It is interesting to study variations of the problem. An evident generalization would be to use an arbitrary distance function, instead of unit distances. We strongly believe that our algorithm is a constant approximation algorithm for arbitrary distance functions, though the approximation ratio may be worse. Although we did not prove it here, we have shown by similar arguments as used in this paper that for the problem without release times our algorithm also gives a 4-approximation when the objective function is the *sum* of completion times, and that this ratio is best possible among shortest path following algorithms. We will report on this result in a research report version of the paper.

A real challenge is to study a much more realistic version of the problem: the real problem is *on-line* with *distributed algorithms* at each of the nodes. In on-line WGP a message becomes known only at its release time and the sending schedules have to be adapted to newly released messages in an on-line fashion. In fact, our greedy algorithm is an on-line algorithm, though *centralized*, and thus gives a *competitive ratio* of 4. However, there is a significant gap between this upper bound and a simple lower bound on the competitive ratio of any deterministic centralized algorithm: we have constructed rather simple examples which give lower bounds of $7/5$ for $d_I > d_T$ and $4/3$ for $d_I = d_T$.

In distributed WGP coordination of communication is very limited. As interference can not always be detected a priori in this model, algorithms should be able to accommodate for retransmissions of lost data. Bar-Yehuda et al. [2] designed distributed randomized algorithms that do this for WGP with $d_T = d_I = 1$ without release times (thus in an off-line setting). They derive bounds on the expected number of rounds required to gather all messages at the sink. It would be interesting to exploit ideas in this paper to design (randomized) distributed on-line algorithms and obtain satisfactory competitive ratios. It may very well be that constant competitive ratios are not achievable.

References

1. R. Bar-Yehuda, O. Goldreich, and A. Itai. On the time-complexity of broadcast in multi-hop radio networks: an exponential gap between determinism and randomization. *Journal of Computer and System Sciences*, 45(1):104–126, 1992.
2. R. Bar-Yehuda, A. Israeli, and A. Itai. Multiple communication in multihop radio networks. *SIAM Journal on Computing*, 22(4):875–887, 1993.
3. J. Bermond, J. Galtier, R. Klasing, N. Morales, and S. Pérennes. Hardness and approximation of gathering in static radio networks. In *FAWN06, Pisa, Italy*, 2006.
4. S. N. Bhatt, G. Pucci, A. Ranade, and A. L. Rosenberg. Scattering and gathering messages in networks of processors. *IEEE Trans. Comput.*, 42(8):938–949, 1993.
5. P. Fraigniaud and E. Lazard. Methods and problems of communication in usual networks. *Discrete Appl. Math.*, 53:79–133, 1994.
6. M. Haenggi. Opportunities and challenges in wireless sensor networks. In *Handbook of Sensor Networks: Compact Wireless and Wired Sensing Systems*. CRC Press, Boca Raton, 2004.
7. IEEE Standard 802.11, Information technology–Telecommunications and information exchange between systems–Local and metropolitan area networks–Specific requirements–Part 11: Wireless LAN Medium Access Control (MAC) and Physical Layer (PHY) Specifications, 1999.
8. M. Ilyas and I. Mahgoub. *Handbook of Sensor Networks: Compact Wireless and Wired Sensing Systems*. CRC Press, Boca Raton, 2004.
9. K. Pahlavan and A. H. Levesque. *Wireless information networks*. Wiley-Interscience, New York, NY, USA, 1995.
10. C. E. Perkins. *Ad hoc networking*. Addison-Wesley Longman Publishing Co., Inc., Boston, MA, USA, 2001.
11. W. Su, E. Cayirci, and O. Akan. Overview of communication protocols for sensor networks. In *Handbook of Sensor Networks: Compact Wireless and Wired Sensing Systems*. CRC Press, Boca Raton, 2004.
12. D. Tse and P. Viswanath. *Fundamentals of Wireless Communication*. Cambridge University Press, Cambridge, 2005.

Minimum Membership Set Covering
and the Consecutive Ones Property

Michael Dom, Jiong Guo[*], Rolf Niedermeier, and Sebastian Wernicke[**]

Institut für Informatik, Friedrich-Schiller-Universität Jena,
Ernst-Abbe-Platz 2, D-07743 Jena, Fed. Rep. of Germany
{dom, guo, niedermr, wernicke}@minet.uni-jena.de

Abstract. The MINIMUM MEMBERSHIP SET COVER problem has recently been introduced and studied in the context of interference reduction in cellular networks. It has been proven to be notoriously hard in several aspects. Here, we investigate how natural generalizations and variations of this problem behave in terms of the consecutive ones property: While it is well-known that classical set covering problems become polynomial-time solvable when restricted to instances obeying the consecutive ones property, we experience a significantly more intricate complexity behavior in the case of MINIMUM MEMBERSHIP SET COVER. We provide polynomial-time solvability, NP-completeness, and approximability results for various cases here. In addition, a number of interesting challenges for future research is exhibited.

1 Introduction

SET COVER (and, equivalently, HITTING SET [1]) is a core problem of algorithmics and combinatorial optimization [2, 3]. The basic task is, given a collection $\mathcal{C}$ of subsets of a base set S, to select as few sets in $\mathcal{C}$ as possible such that their union is the base set. This models many resource allocation problems and generalizes fundamental graph problems such as VERTEX COVER and DOMINATING SET. SET COVER is NP-complete and only allows for a logarithmic-factor polynomial-time approximation [7]. It is parameterized intractable (that is, W[2]-complete) with respect to the parameter "solution size" [5, 14].

Numerous variants of set covering are known and have been studied [2, 4, 8, 9, 11, 16]. Motivated by applications concerning interference reduction in cellular networks, Kuhn et al. [10] very recently introduced and investigated the MINIMUM MEMBERSHIP SET COVER problem.

MINIMUM MEMBERSHIP SET COVER (MMSC)
Input: A set S, a collection $\mathcal{C}$ of subsets of S, and a nonnegative integer k.
Task: Determine if there exists a subset $\mathcal{C}' \subseteq \mathcal{C}$ such that $\bigcup_{C \in \mathcal{C}'} = S$ and $\max_{s \in S} |\{C \in \mathcal{C}' \mid s \in C\}| \leq k$.

---

[*] Supported by the Deutsche Forschungsgemeinschaft, Emmy Noether research group PIAF (fixed-parameter algorithms), NI 369/4.
[**] Supported by the Deutsche Telekom Stiftung.

L. Arge and R. Freivalds (Eds.): SWAT 2006, LNCS 4059, pp. 339–350, 2006.

In this natural variant again a base set S has to be covered with sets from a collection $\mathcal{C}$. By way of contrast to the classical SET COVER problem, however, the goal is not to minimize the number of sets from $\mathcal{C}$ required to do this, but the maximum number of occurrences each element from S has in the cover. Kuhn et al. [10] showed that MMSC is NP-complete and has similar approximation properties as the classical SET COVER problem.

A well-known line of attack against the hardness of SET COVER is to study special cases of practical interest. Perhaps the most famous one of these cases is SET COVER obeying the *consecutive ones property (c.o.p.)* [11, 12, 13, 16, 17]. Herein, the elements of S have the property that they can be ordered in a linear arrangement such that each set in the collection $\mathcal{C}$ contains only whole "chunks" of that arrangement, that is, without any gaps.[1] SET COVER instances with c.o.p. are solvable in polynomial time, a fact which is made use of in many practical applications [11, 13, 16, 17]. Thus, the question naturally arises whether such results can be transferred to MMSC. This is what we study here, arriving at a much more colorful scenario than in the classical case.

In order to thoroughly study MMSC, in particular with respect to the c.o.p., it is fruitful to consider the following generalization.

RED-BLUE HITTING SET (RBHS)
Input: An n-element set S, two collections $\mathcal{C}_{red}$ and $\mathcal{C}_{blue}$ of subsets of S, and a nonnegative integer k.
Task: Determine if there exists a subset $S' \subseteq S$ such that each set in $\mathcal{C}_{red}$ contains *at least one* element from S' and each set in $\mathcal{C}_{blue}$ contains *at most k* elements from S'.

MMSC is the same as RBHS for the case $\mathcal{C}_{red} = \mathcal{C}_{blue}$. However, the RBHS formulation now opens a wide field of natural investigations concerning the c.o.p., the point being that the c.o.p. may apply to either $\mathcal{C}_{red}$, $\mathcal{C}_{blue}$, both, or none of them. The c.o.p. in connection with RBHS leads to a number of different results concerning the computational complexity. This is what we explore here, Table 1 providing a general overview of known and new results.

The main messages from Table 1 (and this work) are:

- In case of only "partial" or even no consecutive ones properties (first three columns), the problem mostly remains NP-complete.
- In the case that both $\mathcal{C}_{red}$ and $\mathcal{C}_{blue}$ obey the c.o.p., only simple cases are known to be polynomial-time solvable but the general case remains open.
- The case that both $\mathcal{C}_{red}$ and $\mathcal{C}_{blue}$ obey the c.o.p. allows for a simple and efficient approximation which is only by additive term one worse than an optimal solution. Surprisingly, an optimal solution seems harder to achieve.

---

[1] The name "consecutive ones" refers to the fact that one may think of a SET COVER instance as a coefficient matrix M where the elements of S correspond to columns and the sets in $\mathcal{C}$ correspond to rows; An entry is 1 if the respective element is contained in the respective set and 0 otherwise. If the SET COVER instance has the c.o.p., then the columns of M can be permuted in such a way that the ones in each row appear consecutively.

Table 1. An overview of previously known results, new results presented in this paper, and open questions for future research regarding the computational complexity of the RED-BLUE HITTING SET problem

	no c.o.p. requirement	$\mathcal{C}_{red}$ has c.o.p.	$\mathcal{C}_{blue}$ has c.o.p.	$\mathcal{C}_{red}$ and $\mathcal{C}_{blue}$ have c.o.p.
No restrictions	NP-c [10]			+1-approx. (Thm. 9)
Fixed max. overlap k	NP-c (Thm. 5)	NP-c (Thm. 5)	NP-c (Thm. 7)	poly.-time (Thm. 11)
Fixed elem. occ. in $\mathcal{C}_{blue}$				
Fixed elem. occ. in $\mathcal{C}_{red}$				?
Card.-2 sets $\mathcal{C}_{red}$ or $\mathcal{C}_{blue}$	NP-c (Thms. 5, 6, 7, and 8)			trivial
Card.-2 sets $\mathcal{C}_{red}$ and $\mathcal{C}_{blue}$	linear-time (Cor. 4)			
Fix. card. sets $\mathcal{C}_{red}$				?
Fix. card. sets $\mathcal{C}_{blue}$	NP-c [10]	NP-c (Thms. 5, 7, and 8)		poly.-time (Thm. 11)
$\mathcal{C}_{blue}$ contains one set	NP-c (Thm. 2)	poly.-time (Cor. 12)	NP-c (Thm. 2)	poly.-time (Cor. 12)
Fix. num. sets in $\mathcal{C}_{blue}$		?		

Preliminaries. Formally, the consecutive ones property is defined as follows.

Definition 1. *Given a set $S = \{s_1, \ldots, s_n\}$ and a collection $\mathcal{C}$ of subsets of S, the collection $\mathcal{C}$ is said to have the* consecutive ones property *(c.o.p.) if there exists an order $\prec$ on S such that for every set $C \in \mathcal{C}$ and $s_i \prec s_k \prec s_j$, it holds that $s_i \in C \wedge s_j \in C \Rightarrow s_k \in C$.*

The following simple observation is useful for our NP-completeness proofs.

Observation 1. *Given a set $S = \{s_1, \ldots, s_n\}$ and a collection $\mathcal{C}$ of subsets of S such that all sets in $\mathcal{C}$ are mutually disjoint, the collection $\mathcal{C}$ has the c.o.p..*

In order to simplify the study of RED-BLUE HITTING SET, for a given instance $(S, \mathcal{C}_{red}, \mathcal{C}_{blue}, k)$ we will call k the *maximum overlap* and say that a set S' has the *minimum overlap property* if each set in $\mathcal{C}_{red}$ contains at least one element from S'. The set S' has the *maximum overlap property* if each set in $\mathcal{C}_{blue}$ contains at most k elements from S'. Thus, a set S' that has both the minimum and maximum overlap property constitutes a valid solution to the given instance of RED-BLUE HITTING SET.

2   Red-Blue Hitting Set Without C.O.P.

This section deals with the general RBHS problem, meaning that we make no requirement for $\mathcal{C}_{red}$ and $\mathcal{C}_{blue}$ concerning the c.o.p.. Being a generalization of

MMSC, RBHS is of course NP-complete in general. This even holds for some rather strongly restricted variants, as the next theorem shows.

Theorem 2. RBHS *is* NP-*complete even if the following restrictions apply:*

1. *The collection $\mathcal{C}_{blue}$ contains exactly one set, and*
2. *each set in $\mathcal{C}_{red}$ has cardinality 2.*

Proof. We show the theorem by a reduction from the NP-complete VERTEX COVER problem. Given a graph $G = (V, E)$ and a nonnegative integer k, this problem asks to find a size-k subset $V' \subseteq V$ such that for every edge in E, at least one of its endpoints is in V'. Given an instance (G, k) of VERTEX COVER, construct an instance of RBHS by setting $S := V$, $\mathcal{C}_{red} := E$, $\mathcal{C}_{blue} := \{V\}$ (that is, the collection $\mathcal{C}_{blue}$ consists of one set containing all elements of S), and setting the maximum overlap equal to k. It is easy to see that this instance of RBHS directly corresponds to the original vertex cover instance: We may choose at most k elements from S to be in the solution set S' such that at least one element from every set in $\mathcal{C}_{red}$ is contained in S'. □

As shown in the next theorem, polynomial-time solvable instances of RBHS arise when the cardinalities of all sets in the collection $\mathcal{C}_{red}$ are restricted to 2 and the maximum overlap $k = 1$.

Theorem 3. RBHS *can be solved in polynomial time if the maximum overlap $k = 1$ and all sets in $\mathcal{C}_{red}$ have cardinality at most 2.*

Proof. We prove the theorem by showing how the restricted RBHS instance can equivalently be stated as a 2-SAT problem; 2-SAT is well-known to be solvable in linear time [6].

For our reduction, we construct the following instance F of 2-SAT for a given instance $(S, \mathcal{C}_{red}, \mathcal{C}_{blue}, 1)$ of RBHS:

- For each element $s_i \in S$, where $1 \leq i \leq n$, F contains the variable x_i.
- For each set $\{s_{i_1}, s_{i_2}\} \in \mathcal{C}_{red}$, F contains the clause $(x_{i_1} \vee x_{i_2})$.
- For each set $\{s_{i_1}, \ldots, s_{i_d}\} \in \mathcal{C}_{blue}$, F contains $d(d-1)/2$ clauses $(\neg x_{i_a} \vee \neg x_{i_b})$ with $1 \leq a < b \leq d$.

If the resulting Boolean formula F has a satisfying truth assignment T, then $S' := \{s_i \in S : T(x_i) = true\}$ is a solution to the RBHS instance $(S, \mathcal{C}_{red}, \mathcal{C}_{blue}, 1)$: By construction, for each set in $\mathcal{C}_{red}$ at least one element must have been chosen in order to satisfy the corresponding clause. Hence, S' has the minimum overlap property. Also, no two elements s_{i_a}, s_{i_b} from a set in $\mathcal{C}_{blue}$ can have been chosen because this would imply that the corresponding clause $(\neg x_{i_a} \vee \neg x_{i_b})$ in F is not satisfied by T. Hence, S' also has the maximum overlap property and thus is a valid solution to the RBHS instance.

Omitting a formal proof here, it is easy to see that a solution set S' for $(S, \mathcal{C}_{red}, \mathcal{C}_{blue}, 1)$ can be used to construct a satisfying truth assignment T for F: For all $1 \leq i \leq n$, set $T(x_i) = true$ if $s_i \in S'$ and $T(x_i) = false$ otherwise. □

Corollary 4. RBHS *can be solved in linear time if all sets in $\mathcal{C}_{red}$ and $\mathcal{C}_{blue}$ have cardinality at most 2.* □

3   Red-Blue Hitting Set with Partial C.O.P.

In this section, we prove that RBHS remains NP-complete even under the requirement that either $\mathcal{C}_{red}$ or $\mathcal{C}_{blue}$ is to have the c.o.p.. To this end, we give reductions from the following restricted variant of the SATISFIABILITY problem:

RESTRICTED 3-SAT (R3-SAT)
Input: An n-variable Boolean formula F in conjunctive normal form where each variable x_i, $1 \leq i \leq n$, appears at most three times, each literal appears at most twice, and each clause contains at most three literals.
Task: Determine if there exists a satisfying truth assignment T for F.

It is well-known that R3-SAT is NP-complete (e.g., see [15, p. 183]).[2]

3.1   Consecutive Ones Property for $\mathcal{C}_{red}$

The following two theorems (Theorems 5 and 6) show that the requirement of $\mathcal{C}_{red}$ obeying the c.o.p. does not make RBHS tractable. The theorems complement each other in the sense that they impose different restrictions on the cardinalities of the sets $\mathcal{C}_{red}$ and $\mathcal{C}_{blue}$; Theorem 5 allows for size-3 sets in $\mathcal{C}_{red}$ and size-2 sets in $\mathcal{C}_{blue}$ (the reduction encodes clauses of a given R3-SAT instance in $\mathcal{C}_{red}$) while the converse holds true for Theorem 6 (the reduction encodes variables in $\mathcal{C}_{red}$).

Theorem 5. *RBHS is NP-complete even if all the following restrictions apply:*

1. *The collection $\mathcal{C}_{red}$ has the consecutive ones property.*
2. *The maximum overlap k is equal to one.*
3. *Each set in $\mathcal{C}_{red}$ has cardinality 3 and each set in $\mathcal{C}_{blue}$ has cardinality 2.*
4. *Each element from S occurs in exactly one set in $\mathcal{C}_{red}$ and each element from S occurs in at most two sets in $\mathcal{C}_{blue}$.*

Proof. We prove the theorem by a reduction from R3-SAT. Given an m-clause Boolean formula F that is an instance of R3-SAT, construct the following instance $(S, \mathcal{C}_{red}, \mathcal{C}_{blue}, k)$ of RBHS:

- The set S consists of elements $s_1^1, s_1^2, s_1^3, \ldots, s_m^1, s_m^2, s_m^3$. The element s_j^i corresponds to the i-th literal in the j-th clause of F. If the j-th clause has only two literals, then S contains only s_j^1 and s_j^2.
- Each set in $\mathcal{C}_{red}$ corresponds to a clause in F, that is, for the i-th clause in F, we add $\{s_i^1, s_i^2, s_i^3\}$ to $\mathcal{C}_{red}$ if it contains three literals and $\{s_i^1, s_i^2\}$ if it contains two literals.
- For each variable x in F and for all pairs of literals $l_1 = x, l_2 = \neg x$ in F: If l_1 is the i-th literal in the j-th clause and l_2 is the p-th literal in the q-th clause of F, $\mathcal{C}_{blue}$ contains the set $\{s_j^i, s_q^p\}$.
- The maximum overlap k is set to one.

---

[2] Note that it is essential for the NP-completeness of R3-SAT that the Boolean formula F may contain size-2 clauses, otherwise the problem is in P [15].

The construction is illustrated in Figure 1.

It is easy to see that, by the definition of R3-SAT, the constructed instance satisfies the restrictions claimed in the theorem; note that $\mathcal{C}_{red}$ has the consecutive ones property due to Observation 1. It remains to show that the constructed instance of RBHS has a solution iff F has a satisfying truth assignment T.

"$\Rightarrow$" Assume that the constructed instance of RBHS has a solution set S'. Let T be a truth assignment such that, for every $s_j^i \in S'$, the variable represented by s_j^i is set to *true* if the literal represented by s_j^i is positive, and *false* otherwise. This truth assignment is well defined because S' must have the maximum overlap property—it therefore cannot happen that two elements $s_j^i, s_q^p \in S'$ correspond to different literals of the same variable.

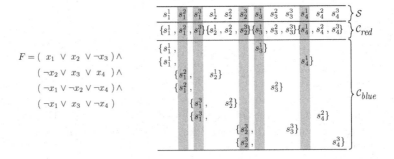

$$F = (\; x_1 \; \vee \; x_2 \; \vee \neg x_3 \;) \wedge$$
$$(\; \neg x_2 \vee \; x_3 \; \vee \; x_4 \;) \wedge$$
$$(\; \neg x_1 \vee \neg x_2 \vee \neg x_4 \;) \wedge$$
$$(\; \neg x_1 \vee \; x_3 \; \vee \neg x_4 \;)$$

Fig. 1. Example of encoding an instance of R3-SAT into an instance of RBHS (proof of Theorem 5). Each clause of the Boolean formula F is represented by a three-element set in $\mathcal{C}_{red}$. The sets in $\mathcal{C}_{blue}$ and the maximum overlap $k = 1$ ensure that no two elements from S can be chosen into a solution that correspond to conflicting truth assignments of the same variable. Observe how $S' = \{s_1^2, s_1^3, s_2^3, s_3^1, s_4^1\}$ (grey columns) constitutes a valid solution to the RBHS instance; accordingly, a truth assignment T which makes all the corresponding literals evaluate to *true* satisfies F.

To show that T constitutes a satisfying truth assignment for F, observe that, for each clause of F, at least one element from S' must correspond to a literal in this clause because S' has the minimum overlap property. On the one hand, if this element corresponds to a positive literal x_i, then $T(x_i) = true$, satisfying the clause. On the other hand, if the element corresponds to a negative literal $\neg x_i$, then $T(x_i) = false$, satisfying the clause.

"$\Leftarrow$" Let T be a satisfying truth assignment for F. Let S' be the set of elements in S that correspond to literals that evaluate to *true* under T. Then, S' has the minimum overlap property because at least one literal in every clause of F must evaluate to true under T and each set in $\mathcal{C}_{red}$ represents exactly one clause of F. Also, S' has the maximum overlap property because T is uniquely defined for every variable that occurs in F. Since S' has both the minimum and maximum overlap property, it is a valid solution to the RBHS instance.   □

The following theorem can be proven in a similar way as Theorem 5.

Theorem 6. RBHS *is* NP-*complete even if all the following restrictions apply:*

1. *The collection $\mathcal{C}_{red}$ has the consecutive ones property.*
2. *The maximum overlap k is equal to two.*
3. *Each set in $\mathcal{C}_{red}$ has cardinality 2 and each set in $\mathcal{C}_{blue}$ has cardinality 3.*
4. *Each element from S occurs in exactly one set in $\mathcal{C}_{red}$ and each element from S occurs in at most two sets in $\mathcal{C}_{blue}$.* □

3.2   Consecutive Ones Property for $\mathcal{C}_{blue}$

Note that by the proof of Theorem 2, RBHS is NP-complete already if $\mathcal{C}_{blue}$ contains just a single set and has the c.o.p.. However, this requires a non-fixed maximum overlap k and unrestricted cardinality of the sets contained in $\mathcal{C}_{blue}$. Therefore, if we want to show the NP-hardness of RBHS with the additional restriction that the maximum overlap k is fixed and the sets in $\mathcal{C}_{red}$ and $\mathcal{C}_{blue}$ have small cardinality, another reduction is needed. Analogously to Theorems 5 and 6, the following two theorems impose different restrictions on the cardinalities of the sets in $\mathcal{C}_{red}$ and $\mathcal{C}_{blue}$.

Theorem 7. RBHS *is* NP-*complete even if all the following restrictions apply:*

1. *The collection $\mathcal{C}_{blue}$ has the consecutive ones property.*
2. *The maximum overlap k is equal to one.*
3. *Each set in $\mathcal{C}_{red}$ has cardinality 3 and each set in $\mathcal{C}_{blue}$ has cardinality 2.*
4. *Each element from S occurs in at most two sets in $\mathcal{C}_{red}$ and each element from S occurs in exactly one set in $\mathcal{C}_{blue}$.*

Proof. Again, we give a reduction from R3-SAT. For a given n-variable Boolean formula F that is an instance of R3-SAT, construct the following instance $(S, \mathcal{C}_{red}, \mathcal{C}_{blue}, k)$ of RBHS:

- The set S consists of $2n$ elements $s_1, \bar{s}_1, \ldots, s_n, \bar{s}_n$, that is, for each variable x_i in F, S contains an element s_i representing the literal x and an element $\bar{s}_i$ representing the literal $\neg x$.
- For each clause in F, $\mathcal{C}_{red}$ contains a set of those elements from S that represent the literals of that clause.
- $\mathcal{C}_{blue} = \bigcup_{1 \le i \le n} \{\{s_i, \bar{s}_i\}\}$.
- The maximum overlap k is one.

Observe that this instance satisfies all restrictions claimed in the theorem; $\mathcal{C}_{blue}$ has the c.o.p. due to Observation 1. The reduction is illustrated by an example in Figure 2. It remains to show that the constructed instance has a solution iff F has a satisfying truth assignment T. We omit the details. □

The proof of the following theorem is similar to the one of Theorem 7.

$$
\begin{array}{c}
\begin{array}{cccccccc}
s_1 & \bar{s}_1 & s_2 & \bar{s}_2 & s_3 & \bar{s}_3 & s_4 & \bar{s}_4
\end{array} \left. \right\} \mathcal{S} \\
\end{array}
$$

$$
\begin{aligned}
F = (\ x_1 \ \lor\ x_2 \ \lor \neg x_3\ ) \land \quad &\Longrightarrow \\
(\ \neg x_2 \lor\ x_3 \ \lor\ x_4 \ ) \land \quad &\Longrightarrow \\
(\ \neg x_1 \lor \neg x_2 \lor \neg x_4\ ) \land \quad &\Longrightarrow \\
(\ \neg x_1 \lor\ x_3 \ \lor \neg x_4\ ) \quad &\Longrightarrow
\end{aligned}
\qquad
\left.
\begin{array}{l}
\{s_1 , \quad s_2 , \quad\quad \bar{s}_3\} \\
\{\bar{s}_2 , s_3 , \quad s_4\} \\
\{\bar{s}_1 , \quad \bar{s}_2 , \quad\quad \bar{s}_4\} \\
\{\bar{s}_1 , \quad s_3 , \quad\quad \bar{s}_4\}
\end{array}
\right\} \mathcal{C}_{red}
$$

$$
\{s_1 , \bar{s}_1\}\{s_2 , \bar{s}_2\}\{s_3 , \bar{s}_3\}\{s_4 , \bar{s}_4\} \left. \right\} \mathcal{C}_{blue}
$$

Fig. 2. Example of encoding an instance of R3-SAT into an instance of RBHS (proof of Theorem 7). Each clause of the Boolean Formula F is encoded into one set of $\mathcal{C}_{red}$. The sets in $\mathcal{C}_{blue}$ and the maximum overlap $k = 1$ ensure that no two elements from S can be chosen into a solution that correspond to conflicting truth assignments of the same variable. Observe how $S' = \{\bar{s}_1, s_2, s_3, s_4\}$ (grey columns) constitutes a valid solution to the RBHS instance; accordingly, a truth assignment T with $T(x_i) = true$ iff $s_i \in S'$ satisfies F.

Theorem 8. *RBHS is NP-complete even if all the following restrictions apply:*

1. *The collection $\mathcal{C}_{blue}$ has the consecutive ones property.*
2. *The maximum overlap k is equal to two.*
3. *Each set in $\mathcal{C}_{red}$ has cardinality 2 and each set in $\mathcal{C}_{blue}$ has cardinality 3.*
4. *Each element from S occurs in at most two sets in $\mathcal{C}_{red}$ and each element from S occurs in exactly one set in $\mathcal{C}_{blue}$.*   □

Note that if k is restricted to $k = 1$ instead of $k = 2$ in the instances discussed in the above theorem, they become polynomial-time solvable according to Theorem 3.

4   Red-Blue Hitting Set with C.O.P.

In this section, we make the requirement that both $\mathcal{C}_{red}$ and $\mathcal{C}_{blue}$ in a given instance $(S, \mathcal{C}_{red}, \mathcal{C}_{blue}, k)$ of RBHS obey the c.o.p. and call the resulting problem "RBHS with c.o.p." We present an approximation algorithm (Section 4.1) and show, among others, that for fixed k RBHS with c.o.p. is solvable in polynomial time (Section 4.2). This leads to the following observation.

Observation 2. RBHS *with c.o.p. is equivalent to* MMSC *with c.o.p..*

To see this observation, note that on the one hand, MMSC is obviously the special case of RBHS with identical red and blue subset collections. On the other hand, an RBHS instance $(S, \mathcal{C}_{red}, \mathcal{C}_{blue}, k)$ with c.o.p. can be transformed into an MMSC instance with c.o.p. by observing that for an optimal solution S' of RBHS, for any set $C_b \in \mathcal{C}_{blue}$ that contains no set from $\mathcal{C}_{red}$ as a subset we have $|C_b \cap S'| \le 2$. Thus, if $k > 2$, then we can safely remove such blue subsets from $\mathcal{C}_{blue}$. Then, solving RBHS on the resulting instance is equivalent to solving MMSC on the instance $(S, \mathcal{C}_{red} \cup \mathcal{C}_{blue}, k)$. If $k \le 2$ then both RBHS and MMSC with c.o.p. are solvable in polynomial time as it will be shown in Section 4.2.

To simplify the discussion in this section, we assume that the elements in $S = \{s_1, \ldots, s_n\}$ are sorted such that all subsets in $\mathcal{C}_{red}$ and $\mathcal{C}_{blue}$ have the c.o.p., that is, for every $1 \leq i \leq k \leq j \leq n$ and every set $C \in \mathcal{C}_{blue} \cup \mathcal{C}_{red}$ it holds that $s_i \in C \wedge s_j \in C \Rightarrow s_k \in C$. For each subset $C \subseteq S$, its *left index* $l(C)$ is defined as $\min\{i \mid s_i \in C\}$ and its *right index* $r(C)$ is defined as $\max\{i \mid s_i \in C\}$.

4.1 Approximation Algorithm

Here, we describe a polynomial-time approximation algorithm for RBHS with c.o.p. that has a guaranteed additive term of one compared to an optimal solution. To this end, we rephrase RBHS as an optimization problem:

Input: A set S and two collections $\mathcal{C}_{red}$ and $\mathcal{C}_{blue}$ of subsets of S.
Task: Find a subset $S' \subseteq S$ with $S' \cap C \neq \emptyset$ for all $C \in \mathcal{C}_{red}$ which minimizes $\max_{C' \in \mathcal{C}_{blue}}\{|C' \cap S'|\}$.

Our greedy approximation algorithm works as follows:

```
01    S' ← ∅, C'_red ← C_red
02    while C'_red ≠ ∅
03        C ← set from C'_red with minimum right index
04        S' ← S' ∪ s_{r(C)}, C'_red ← C'_red \ {C ∈ C'_red : C ∩ S' ≠ ∅}
05    return S'
```

Theorem 9. *For RBHS with c.o.p., the greedy algorithm polynomial-time approximates an optimum solution within an additive term of one.*

Proof. Obviously, the output S' of the greedy algorithm has the minimum overlap property since $S' \cap C \neq \emptyset$ for all $C \in \mathcal{C}_{red}$. It is also clear that the algorithms runs in $O(|S| \cdot |\mathcal{C}_{red}|)$ time. It remains to show the additive term.

Let C denote one subset in $\mathcal{C}_{blue}$ with $|C \cap S'| = \max_{C' \in \mathcal{C}_{blue}}\{|C' \cap S'|\}$. It is easy to observe that C contains at least $|C \cap S'| - 1$ mutually disjoint sets from $\mathcal{C}_{red}$ as subsets, implying that *any* solution for this instance has to contain at least $|C \cap S'| - 1$ elements from C in order to satisfy the minimum overlap property for these sets. Therefore, $|C \cap S'_{opt}| \geq |C \cap S'| - 1$ for any optimal solution S'_{opt}. ☐

4.2 Dynamic Programming

We now present a dynamic programming algorithm that solves RED-BLUE HITTING SET with c.o.p. in polynomial time provided that either the maximum overlap is a fixed constant k, the maximum cardinality of the sets in $\mathcal{C}_{blue}$ is a fixed constant c_{card}, or the maximum number of occurrences of an element in $\mathcal{C}_{blue}$ is a fixed constant c_{occ}.

We assume that the sets $C \in \mathcal{C}_{red}$ are ordered according to $l(C)$ and denote them with $R_1, \ldots, R_{|\mathcal{C}_{red}|}$; the sets of $\mathcal{C}_{blue}$ are ordered analogously and denoted with $B_1, \ldots, B_{|\mathcal{C}_{blue}|}$. If a set in $\mathcal{C}_{red}$ is a superset of another set in $\mathcal{C}_{red}$ it can be removed, and, therefore, for any two sets $R_i, R_j \in \mathcal{C}_{red}$ it holds that $l(R_i) < l(R_j) \Leftrightarrow r(R_i) < r(R_j)$. For an analogous reason we have $l(B_i) < l(B_j) \Leftrightarrow r(B_i) < r(B_j)$ for all $B_i, B_j \in \mathcal{C}_{blue}$. We call this property the *monotonicity* of the sets in $\mathcal{C}_{red}$ and $\mathcal{C}_{blue}$.

The idea of the dynamic programming algorithm is to build collections $D(i, j)$ of so-called partial solutions; each partial solution in a collection $D(i, j)$ covers all sets $R_1, \ldots, R_j$ with a minimal subset of $\{s_1, \ldots, s_i\}$ that fulfills the maximum overlap property. To this end, the algorithm uses a two-dimensional table $D(i, j)$ with $1 \leq i \leq n$ (where $n := |S|$) and $1 \leq j \leq |\mathcal{C}_{red}|$; to fill this table, two nested loops are used, the outer one iterating over i and the inner one iterating over j. Every entry of $D(i, j)$ that already has been processed contains a collection of tuples $(S_h, v_h), 1 \leq h \leq |D(i, j)|$, where each tuple (S_h, v_h) consists of a set $S_h \subseteq \{s_1, \ldots, s_i\}$ and a vector $v_h = (v_h^1, \ldots, v_h^{|\mathcal{C}_{blue}|})$. The sets S_h are called *partial solutions* and have the following properties:

1. Each S_h contains at least one element of every set $R_1, \ldots, R_j \in \mathcal{C}_{red}$.
2. No proper subset of a partial solution S_h covers all sets $R_1, \ldots, R_j$.
3. For $1 \leq q \leq |\mathcal{C}_{blue}|$, we have $v_h^q = |S_h \cap B_q| \leq k$.

It is obvious that if the entry $D(n, |\mathcal{C}_{red}|)$ is not empty, each partial solution in $D(n, |\mathcal{C}_{red}|)$ is a solution for the RBHS instance.

The first step for filling the table is to compute all entries $D(i, j)$ with $i = 1$, a trivial task. All other entries $D(i, j)$ are computed as follows: If $l(R_j) > i$ then $D(i, j)$ is empty. Otherwise, the partial solutions that have to be generated can be divided in two categories: Partial solutions not containing the element s_i and partial solutions containing s_i. The partial solutions not containing s_i can only contain elements from $\{s_1, \ldots, s_{i-1}\}$ and, therefore, are exactly the partial solutions of $D(i - 1, j)$.

The partial solutions in $D(i, j)$ that do contain s_i are computed as follows: By selecting s_i to be member of such a partial solution in $D(i, j)$, all sets in $\mathcal{C}_{red}$ that contain s_i are covered. Therefore, the other elements in the partial solution only have to cover those sets $R_p \in \{R_1, \ldots, R_j\}$ with $r(R_p) < i$. Hence, these elements form a partial solution in a collection $D(i - 1, j')$ where j' is the maximum possible index such that $r(R_{j'}) < i$. More formally, an entry $D(i, j)$ with $i > 1$ and $l(R_j) \leq i$ is computed as follows:

```
01    if  D(i − 1, j) ≠ ∅ then D(i, j) ← D(i − 1, j).
02    j' ← max{p ∈ {1, . . . , j} | r(Rp) < i}
03    if  D(i − 1, j') ≠ ∅ then for each (Sh, vh) ∈ D(i − 1, j') do
04        Insert a copy (S̃h, ṽh) of (Sh, vh) into D(i, j)
05        S̃h ← S̃h ∪ {si}
06        for each q ∈ {1, . . . , |Cblue| : si ∈ Bq} do
07            ṽhq ← ṽhq + 1
08            if  ṽhq > k then delete (S̃h, ṽh) from D(i, j)
                 and continue with the next tuple in D(i − 1, j')
```

In order to shrink the table size by eliminating redundant tuples, we perform the following *data reduction step* directly after the computation of an entry $D(i,j)$:

09 $\quad q' \leftarrow \max\{q \in \{1, \ldots, |\mathcal{C}_{blue}|\} : r(B_q) \leq i\}$
10 $\quad$ **if** $D(i,j)$ contains tuples (S_{h_1}, v_{h_1}) and (S_{h_2}, v_{h_2}) such that
$\qquad\quad r(S_{h_1}) \geq r(S_{h_2})$ and $\forall q > q' : v_{h_1}^q \leq v_{h_2}^q$ **then**
11 $\qquad$ Delete (S_{h_2}, v_{h_2}) from $D(i,j)$

The data reduction step step does not affect the correctness of the algorithm: If there is a solution S' for the RBHS instance with $S' = S_{h_2} \cup \hat{S}$ such that $l(\hat{S}) > i$, then $\hat{S} \cup S_{h_1}$ is obviously also a solution.

The following lemma helps us to give an upper bound for the number of tuples (S_h, v_h) in a collection $D(i,j)$. We omit the proof.

Lemma 10. *For a collection $D(i,j) \neq \emptyset$, let $D_x(i,j)$ be the tuples $(S_h, v_h) \in D(i,j)$ with $r(S_h) = x$. Then the number of tuples in $D_x(i,j)$ is bounded from above by $\min\{(k+1)^{c_{occ}}, \binom{c_{occ}+k}{k}\}$.* $\qquad\square$

Theorem 11. *RBHS with c.o.p. can be solved in polynomial time provided that either the maximum overlap k is a fixed constant, the maximum cardinality of the sets in $\mathcal{C}_{blue}$ is a fixed constant c_{card}, or the maximum number of occurrences of an element in $\mathcal{C}_{blue}$ is a fixed constant c_{occ}. More precisely, RBHS is solvable either in $|S|^{O(k)}$ time or in $|S|^{O(c_{occ})}$ time or in $|S|^{O(1)} \cdot c_{card}^{O(c_{card})}$ time.*

Proof. The correctness of the algorithm follows from its above description. It remains to show the running time, which basically depends on the size of the table (that is, the number of collections $D(i,j)$) and the number of tuples that have to be compared during the data reduction step.

The table size is $|S| \cdot |\mathcal{C}_{red}| \leq |S|^2$. An upper bound for the number of tuples in each collection can be derived from Lemma 10. The claimed running times follow, because c_{card} is bounded from above by $|S|$, k is bounded from above by $c_{card} - 1$, and c_{occ} is bounded from above by $|\mathcal{C}_{blue}|$ and by c_{card}. $\qquad\square$

Corollary 12. *RBHS with c.o.p. can be solved in polynomial time if $|\mathcal{C}_{blue}|$ is a constant.* $\qquad\square$

5 Conclusion

In this work, we initiated a study of MINIMUM MEMBERSHIP SET COVER and, more generally, RED-BLUE HITTING SET with respect to instances (partially) obeying the consecutive ones property. Many natural challenges for future work arise from our results. For instance, it is desirable to find out more about the polynomial-time approximability and the parameterized complexity [5, 14] of the variants of RED-BLUE HITTING SET proven to be NP-complete (see Table 1). Moreover, in three cases Table 1 exhibits unsettled questions concerning the computational complexity of the respective problems.

References

1. G. Ausiello, A. D'Atri, and M. Protasi. Structure preserving reductions among convex optimization problems. *Journal of Computer and System Sciences*, 21(1):136–153, 1980.
2. A. Caprara, P. Toth, and M. Fischetti. Algorithms for the set covering problem. *Annals of Operations Research*, 98:353–371, 2000.
3. T. H. Cormen, C. E. Leiserson, R. L. Rivest, and C. Stein. *Introduction to Algorithms*. MIT Press, 2001.
4. E. D. Demaine, M. T. Hajiaghayi, U. Feige, and M. R. Salavatipour. Combination can be hard: approximability of the unique coverage problem. In *Proc. 17th SODA*, pages 162–171. ACM Press, 2006.
5. R. G. Downey and M. R. Fellows. *Parameterized Complexity*. Springer, 1999.
6. S. Even, A. Itai, and A. Shamir. On the complexity of timetable and multicommodity flow problems. *SIAM J. Comput.*, 5(4):691–703, 1976.
7. U. Feige. A threshold of ln n for approximating set cover. *J. ACM*, 45(4):634–652, 1998.
8. N. Garg, V. V. Vazirani, and M. Yannakakis. Primal-dual approximation algorithms for integral flow and multicut in trees. *Algorithmica*, 18(1):3–20, 1997.
9. J. Guo and R. Niedermeier. Exact algorithms and applications for Tree-like Weighted Set Cover. *Journal of Discrete Algorithms*, 2006. To appear.
10. F. Kuhn, P. von Rickenbach, R. Wattenhofer, E. Welzl, and A. Zollinger. Interference in cellular networks: The minimum membership set cover problem. In *Proc. 11th COCOON*, volume 3595 of *LNCS*, pages 188–198. Springer, 2005.
11. S. Mecke and D. Wagner. Solving geometric covering problems by data reduction. In *Proc. 12th ESA*, volume 3221 of *LNCS*, pages 760–771. Springer, 2004.
12. J. Meidanis, O. Porto, and G. Telles. On the consecutive ones property. *Discrete Applied Mathematics*, 88:325–354, 1998.
13. G. L. Nemhauser and L. A. Wolsey. *Integer and Combinatorial Optimization*. Wiley, 1988.
14. R. Niedermeier. *Invitation to Fixed-Parameter Algorithms*. Oxford University Press, 2006.
15. C. H. Papadimitriou. *Computational Complexity*. Addison-Wesley, 1994.
16. N. Ruf and A. Schöbel. Set covering with almost consecutive ones property. *Discrete Optimization*, 1(2):215–228, 2004.
17. A. F. Veinott and H. M. Wagner. Optimal capacity scheduling. *Operations Research*, 10:518–532, 1962.

Approximating Rational Objectives Is as Easy as Approximating Linear Ones[*]

José R. Correa[1,**], Cristina G. Fernandes[2,***], and Yoshiko Wakabayashi[2,***]

[1] School of Business, Universidad Adolfo Ibáñez, Santiago, Chile
`correa@uai.cl`
[2] Department of Computer Science, Universidade de São Paulo, Brazil
{`cris, yw`}`@ime.usp.br`

Abstract. In the late seventies, Megiddo proposed a way to use an algorithm for the problem of minimizing a linear function $a_0 + a_1 x_1 + \cdots + a_n x_n$ subject to certain constraints to solve the problem of minimizing a rational function of the form $(a_0 + a_1 x_1 + \cdots + a_n x_n)/(b_0 + b_1 x_1 + \cdots + b_n x_n)$ subject to the same set of constraints, assuming that the denominator is always positive. Using a rather strong assumption, Hashizume et al. extended Megiddo's result to include approximation algorithms. Their assumption essentially asks for the existence of good approximation algorithms for optimization problems with possibly negative coefficients in the (linear) objective function, which is rather unusual for most combinatorial problems. In this paper, we present an alternative extension of Megiddo's result for approximations that avoids this issue and applies to a large class of optimization problems. Specifically, we show that, if there is an α-approximation for the problem of minimizing a nonnegative linear function subject to constraints satisfying a certain *increasing* property then there is an α-approximation ($1/\alpha$-approximation) for the problem of minimizing (maximizing) a nonnegative rational function subject to the same constraints. Our framework applies to covering problems and network design problems, among others.

1 Introduction

We address the problem of finding approximate solutions for a class of combinatorial optimization problems with rational objectives. Our starting point is the seminal work of Megiddo [19] who showed that, for a large class of problems, optimizing a rational objective can be done in polynomial time, as long as there is an efficient algorithm to optimize a linear objective. The class of problems we address has a natural motivation in particular in network design problems. Suppose we want to build a network where each link has a construction cost, as well as a profit. The profit could measure some overall social benefit

---

[*] Research partially supported by CNPq Prosul Proc. no. 490333/04-4 (Brazil).

[**] Research partially supported by CONICYT (Chile) through grants Anillo en Redes ACT08 and FONDECYT 1060035.

[***] Research partially supported by ProNEx - FAPESP/CNPq Proc. No. 2003/09925-5 (Brazil).

associated to the corresponding link or, for example, could be inversely related to the environmental damage its construction causes. The goal then would be to find a network (satisfying some connectivity requirements) that minimizes the cost-benefit ratio. In general, whenever we are faced with problems where cost-to-profit relations have to be optimized, we are in this situation.

Fractional programming has attracted attention since the sixties in the continuous optimization community [2, 8, 12]. Optimizing rational objectives is a particular case of fractional programming in which both the denominator and the numerator of the fraction to be optimized are linear functions, and several combinatorial optimization problems have been studied in this context. For instance, minimum or maximum average cost problems [3, 6, 10, 20], minimum cost-to-time ratio problems [5, 7, 18], minimum cost-reliability problems [15], fractional knapsack problems [1], fractional assignment problems [22], among others [17, 21]. A concrete example of this class of problems, that was considered recently, is the one studied by Gubbala and Raghavachari [10]: given a weighted k-connected graph G, find a k-connected spanning subgraph of G with minimum average weight. They looked at the edge and vertex connectivity versions of the problem. Besides proving that these problems are NP-hard, they presented a 3-approximation algorithm for the edge-connectivity version, and an $O(\log k)$-approximation algorithm for the vertex-connectivity version.

Although Megiddo's paper has motivated mostly works on exact algorithms, some works on approximation algorithms have also been carried out. The latter include the work of Hashizume, Fukushima, Katoh and Ibaraki [11] who extended Megiddo's approach to take into account approximation algorithms for a class of optimization problems under some assumptions. The result of Hashizume et al. [11] is of similar flavor to what we do here. They proved that an α-approximation to a combinatorial problem with a rational objective can be derived from an α-approximation for its linear counterpart. However, they need a rather strong assumption, namely, the existence of a polynomial-time algorithm for the linear version of the combinatorial problem that gives an α-approximate solution even if negative weights are allowed. As they show, this holds for the knapsack problem, allowing them to deduce approximation results for the fractional knapsack problem. Note however that this does not hold for most optimization problems, in particular, for the ones we consider here. For instance, for the problems considered by Gubbala and Raghavachari [10] with $k = 2$, there is no constant factor approximation algorithm, unless P = NP, if we allow the weights to be arbitrary (there is a reduction from Hamiltonian path using negative weights). Therefore, the results by Hashizume et al. cannot be applied to those problems (among others).

In this paper, we build on Megiddo's work to derive approximation algorithms for a class of combinatorial problems that contains many covering and network connectivity problems, including the ones considered by Gubbala and Raghavachari. Our approach guarantees that, if there is an α-approximation for a problem with a linear objective function with nonnegative coefficients, then there is an α-approximation for its (nonnegative) "rational" version.

A natural approach to solve a problem of the form minimize ax/bx subject to $x \in X$, is to repeatedly solve the problem of minimizing ax subject to $x \in X$ and $bx \geq B$, where B is a guess for the denominator. Unfortunately, it is not clear that this idea works in general because adding one more linear constraint may turn the linear problem (minimize ax subject to $x \in X$) into a harder problem. For instance if $X = \{0,1\}^n$ then minimize ax subject to $x \in X$ is a trivial problem. However, minimize ax subject to $x \in X$ and $\sum_{i=1}^{n} a_i x_i \geq \sum_{i=1}^{n} a_i/2$ is equivalent to the partition problem [9, SP12]. In terms of approximability, adding a linear constraint to a combinatorial problem may also turn it into a harder problem. An example is given by the standard IP formulation of minimum 2-edge connected spanning subgraph. Adding a linear constraint ($\sum_i x_i \leq n$, where n is the number of vertices in the given graph) turns it into the TSP (although, in this case the constraint is not a "covering" one).

1.1 The Setting

Our goal is to derive a general approximation technique for a class of NP-hard combinatorial problems that, besides the connectivity problems just mentioned, includes many well-known graph problems such as minimum vertex cover, minimum feedback arc and vertex set, minimum dominating set, minimum multiway cut, and also some general covering problems such as minimum set cover and minimum hitting set. Our results apply to rational functions of the form $(a_0 + a_1 x_1 + \cdots + a_n x_n)/(b_0 + b_1 x_1 + \cdots + b_n x_n)$, where the a_i's and b_i's are nonnegative integers and $x_i \in \{0,1\}$ for each i.

The class of problems to which our framework applies can be described as follows. Let $U = \{e_1, \ldots, e_n\}$ be a finite ground set, and f be a binary (i.e., $\{0,1\}$-valued) function defined on the subsets of U. We say f is *increasing* if $f(U) = 1$ and $f(S) \leq f(S')$ for all supersets S' of S. Our framework applies to any problem that seeks for a set $S \subseteq U$ satisfying $f(S) = 1$ such that its corresponding characteristic vector minimizes a rational objective as above. It is straightforward to verify that this class of problems contains all rational versions of the problems mentioned in the previous paragraph. Moreover, it contains the following class of integer programming problems (which are covering problems):

$$\min \{a^T x / b^T x : Ax \geq d, \ x \in \{0,1\}^n\}$$

where all entries of A, a, b and d are nonnegative rationals. Indeed, in this setting, the ground set will be $U = \{1, \ldots, n\}$ and the binary function $f(S)$ will be 1 if and only if the characteristic vector of S is a feasible solution for $Ax \geq d$. Since the entries of A and d are nonnegative, f is increasing (as long as the problem is not infeasible, that is, as long as $f(U) = 1$).

1.2 Main Result

Our main result states that if f is increasing and there is an α-approximation algorithm for the problem of finding a set $S \subseteq U$ such that $f(S) = 1$ minimizing a linear function with nonnegative coefficients, then there is an α-approximation algorithm to find a set $S \subseteq U$ such that $f(S) = 1$ minimizing a nonnegative rational function. Thus, this framework allows the "rational" version of several problems to inherit the approximation results for their standard versions.

For instance, we can improve upon the results obtained by Gubbala and Raghavachari [10]. They showed a 3-approximation (resp. a $(1 + 2\sqrt{2H_k} + 2H_k)$-approximation) for the problem of finding a minimum average weight k-edge (resp. vertex) connected subgraph, where H_k is the kth harmonic number. Indeed, we obtain a 2-approximation algorithm for the edge-connectivity version, and a $2H_k$-approximation algorithm for the vertex-connectivity version. The former follows by using the algorithm by Khuller and Vishkin [16] for the problem of finding a minimum weight k-edge connected spanning subgraph, while the latter follows by using the $2H_k$-approximation for the problem of finding a k-vertex connected spanning subgraph of minimum weight of Jain, Măndoiu, Vazirani, and Williamson [14]. We can also derive a 2-approximation algorithm for the "rational" version of the more general edge-connectivity problem studied by Jain [13].

The scheme can be adapted for maximizing rational objectives, however the result is not exactly symmetric. What we get in this case is the following. For the same class of problems (given by an increasing property), if we have an α-approximation for *minimizing* a nonnegative linear objective, we obtain a $1/\alpha$-approximation for maximizing a nonnegative rational objective. (This corresponds to applying the scheme above to minimize the inverted fraction.) This asymmetry is somehow expected. Indeed, the maximization of a linear function on domains given by an increasing property is trivial (the ground set is optimum). However it is not obvious how to maximize a rational function on the same domain. For instance, it is trivial to find a set cover of maximum weight (when all weights are nonnegative) but how do we find a set cover of maximum average weight?

The main idea behind our algorithm is to use a transformed cost function which depends on a, b and a certain parameter (which is nothing but a guess of the optimal value), and then search for the parameter that gives a "right" answer. Although this trick is fairly standard in parametric optimization, we want to avoid negative costs, so we need to "truncate" the costs. Another difficulty here is that we are dealing with approximate solutions. Therefore we need to prove that if the parameter is sufficiently close to the optimal value, then the approximate solution is close as well (up to a certain factor). Unfortunately, because of the truncated costs, we can only prove a one-sided result (Lemma 1), which makes the search part of the algorithm harder. Nevertheless there is a way around this issue with essentially no extra computational effort as we show in Theorem 1 and Theorem 2.

It is important to mention that although the algorithm we present in Section 2 is very efficient in most cases (in particular if the coefficients are not too large), it does not run in strongly polynomial time. Thus, in Section 3 we quickly show

how to adapt Megiddo's technique to derive a strongly polynomial-time version of our algorithm.

In what follows, if S is a subset of a set U and w is a function that assigns a number to each element of U, we let $w(S)$ denote the sum of w_e for all e in S. Also, given an increasing binary function f, we will say that $S \subseteq U$ is *feasible* if and only if $f(S) = 1$.

2 Approximating Rational Objectives

Let f be an increasing binary function defined on all subsets of a finite set U. Recall that f is *increasing* if and only if $f(U) = 1$ and $f(B) \leq f(A)$, for all $B \subseteq A \subseteq U$. Also, we will assume that f is *polynomially computable*, i.e., there is a polynomial-time algorithm that, given a subset S of U, computes $f(S)$. We are interested in the following problems:

MINLIN (U, w, f): Given a finite set U, a nonnegative rational w_e for each e in U, and a polynomially computable increasing function $f : 2^U \rightarrow \{0, 1\}$ (usually given implicitly), find a subset S of U such that $f(S) = 1$ and $w(S)$ is minimum.

MINRATIONAL (U, a, b, f): Given a finite set U, nonnegative integers a_0, b_0 and a_e and b_e for each e in U, and a polynomially computable increasing function $f : 2^U \rightarrow \{0, 1\}$ (usually given implicitly), find a subset S of U such that $f(S) = 1$ and $(a_0 + a(S))/(b_0 + b(S))$ is minimum. (To avoid divisions by zero, we assume that $b_0 + b(S) > 0$ for all feasible S.)

For instance, the problem of, given a weighted k-edge-connected graph G, finding a k-edge-connected spanning subgraph of G with minimum average weight is an example of MINRATIONAL (U, a, b, f). The set U in this case is the set of edges of G and a_e is the weight of edge e in G while $b_e = 1$ for each edge e of G. Of course, $a_0 = b_0 = 0$. The function f is such that $f(S) = 1$ if and only if the subgraph of G induced by the edge set S is spanning and k-edge-connected.

We now describe how an α-approximation algorithm for MINLIN (U, w, f) can be turned into an α-approximation algorithm for MINRATIONAL (U, a, b, f). To this end let MINWEIGHT$_\alpha(U, w, f)$ denote an α-approximation algorithm for MINLIN (U, w, f). Note that α may depend on the input. It is easy to see that we can assume $\alpha \leq |U|$. Indeed, consider the following algorithm for MINLIN (U, w, f):

TRIVIAL (U, w, f)
1   sort the elements in U in nondecreasing order of w, obtaining $w_{e_1} \leq \cdots \leq w_{e_{|U|}}$
2   find the smallest index i such that $f(\{e_1, \ldots, e_i\}) = 1$
3   **return** $\{e_1, \ldots, e_i\}$

Clearly, $\sum_{j=1}^{i} w_{e_j} \leq i \, w_{e_i} \leq |U| \, w_{e_i}$. Moreover, it is immediate that TRIVIAL finds a feasible set S minimizing $\max\{w_e : e \in S\}$. Thus, any optimal solution to MINLIN(U, w, f) contains an element of U of weight at least w_{e_i}, which implies that TRIVIAL achieves a ratio of at most $|U|$. In summary, we can always run both MINWEIGHT$_\alpha$ and TRIVIAL and therefore assume that $\alpha \leq |U|$.

2.1   The Algorithm

The core of our transformation is given by the following routine, which we call AUXILIAR_α. Observe that it applies the traditional trick used, for instance, by Megiddo [19] and by Hashizume et al. [11], adapted to avoid negative entries in the derived weight function.

$\text{AUXILIAR}_\alpha \ (U, a, b, f, c)$

```
1    L ← {e : aₑ ≤ c bₑ}
2    for each element e in U do
3        wₑ ← max{0, aₑ − c bₑ}
4    S ← MINWEIGHTα (U, w, f)
5    return S ∪ L.
```

In what follows, AUXILIAR_α is used to find an α-approximate solution to MIN-RATIONAL. Algorithm MINFRAC_α consists of two phases: an "approximate" truncated binary search and an extra step needed to assure the ratio α. After the truncated binary search, either the algorithm found a feasible solution within ratio α or the search interval contains the optimum value. The search interval in the end, scaled by α, is small enough (choice of ϵ) to contain at most one feasible solution. If the best solution found so far (S_t in line 13) is not within ratio α, the second phase finds a better feasible solution that will be within ratio α.

Let ratio(S) denote the ratio $(a_0 + a(S))/(b_0 + b(S))$ for any feasible set S. Below, α might be in fact $\alpha(|U|)$, if α is a function, not simply a constant.

$\text{MINFRAC}_\alpha \ (U, a, b, f)$

```
1     ε ← 1/((b₀ + b(U))²α)                        ▷ First phase
2     i ← 1
3     S₀ ← U
4     left ← 0
5     right ← ratio(U)
6     while right − left > ε do
7         middle ← (left + right)/2
8         Sᵢ ← AUXILIARα (U, a, b, f, middle)
9         if ratio(Sᵢ) ≤ α middle
10            then right ← middle
11            else left ← middle
12        i ← i + 1
13    St ← argmin{ratio(Sⱼ) : 0 ≤ j < i}
14    c′ ← ratio(St)/α                             ▷ Second phase
15    S′ ← AUXILIARα (U, a, b, f, c′)
16    S ← argmin{ratio(St), ratio(S′)}
17    return S.
```

2.2   Analysis of the Running Time

Let us show that the above algorithm is polynomial in the size of its input.

Theorem 1. *Algorithm* $\text{MINFRAC}_\alpha(U, a, b, f)$ *runs in polynomial time. More precisely, it runs in time* $O(\log(a_{\max}) + \log(b_{\max}) + \log|U|)$ *times the running time of* $\text{MINWEIGHT}_\alpha(U, w, f)$, *where* $a_{\max} = \max\{a_e : e \in U \cup \{0\}\}$ *and* $b_{\max} = \max\{b_e : e \in U \cup \{0\}\}$.

Proof. First observe that the number of iterations of the **while** in line 6 is

$$\lceil \log(\mathrm{ratio}(U)/\epsilon)) \rceil = \lceil \log((a_0 + a(U))(b_0 + b(U))\alpha) \rceil$$
$$= O(\log(a_0 + a(U)) + \log(b_0 + b(U)) + \log|U|)$$
$$= O(\log(a_{\max}) + \log(b_{\max}) + \log|U|).$$

Here, the second equality comes from $\alpha \le |U|$. Now, the most time consuming operation within each iteration of the **while** is the call to AUXILIAR$_\alpha$. Clearly AUXILIAR$_\alpha$ runs in polynomial time in the size of its input. Moreover, its running time is exactly that of MINWEIGHT$_\alpha$. Therefore, it is enough to verify that, in each call at line 8 of MINFRAC$_\alpha$, the parameter *middle* has size polynomially bounded by the size of (U, a, b, f). Indeed, in the ith call of AUXILIAR$_\alpha$, we have that $middle = \mathrm{ratio}(U)/2^i$, where $i = O(\log(a_{\max}) + \log(b_{\max}) + \log|U|)$ (as the number of iterations of the **while**). Therefore each AUXILIAR$_\alpha$ call runs in polynomial time in the size of (U, a, b, f). □

Observe that, if we are given a PTAS (resp. FPTAS) for MINLIN, then we have a PTAS (resp. FPTAS) for MINRATIONAL. Unfortunately, if we are given a strongly polynomial algorithm for MINLIN, we only obtain a polynomial algorithm for MIN-RATIONAL this way. But in Section 3 we will show that, under some assumptions, we can get a strongly polynomial algorithm for MINRATIONAL from a strongly polynomial one for MINLIN.

2.3   Analysis of the Approximation Ratio

First observe that as MINWEIGHT$_\alpha$ returns a subset S of U such that $f(S) = 1$, AUXILIAR$_\alpha$ also returns a subset S of U such that $f(S) = 1$ (so $b_0 + b(S) > 0$). Therefore, MINFRAC$_\alpha$ returns a subset S of U such that $f(S) = 1$, i.e., a feasible solution. Now we focus on the approximation ratio. To this end, we need to establish a key lemma.

Proposition 1. *Let $c \ge 0$ and $w_e = \max\{0, a_e - c\,b_e\}$ for all $e \in U$. Consider the set $L = \{e \in U : a_e \le c\,b_e\}$ and define the quantity $D = \sum_{e \in L}(a_e - c\,b_e) = a(L) - c\,b(L)$. Then, if $R \subseteq U$,*

$$w(R) \le a(R) - c\,b(R) - D.$$

Moreover, if $L \subseteq R$, then equality holds.

Proof. We have that,

$$w(R) = a(R \setminus L) - c\,b(R \setminus L)$$
$$= a(R) - a(R \cap L) - c\,b(R) + c\,b(R \cap L)$$
$$= a(R) - c\,b(R) - (a(R \cap L) - c\,b(R \cap L))$$
$$\le a(R) - c\,b(R) - D. \tag{1}$$

The last inequality holds because each term in the sum that defines D is negative or zero. Note also that, if $L \subseteq R$, then equality holds in (1). □

Lemma 1. *Let c^* be the optimal value of* MINRATIONAL(U, a, b, f). *For any* $c \geq c^*$, *if $\hat{S}$ is the output of* AUXILIAR$_\alpha(U, a, b, f, c)$, *then*

$$\text{ratio}(\hat{S}) \;=\; \frac{a_0 + a(\hat{S})}{b_0 + b(\hat{S})} \;\leq\; \alpha c. \tag{2}$$

Moreover, the previous inequality is strict whenever $c > c^*$.

Proof. We want to prove (2), which is equivalent to

$$a(\hat{S}) - \alpha\, c\, b(\hat{S}) \;\leq\; \alpha\, c\, b_0 - a_0.$$

Let $L = \{e \in U : a_e \leq c\, b_e\}$ and $D = a(L) - c\, b(L)$. We start showing that $w(\hat{S}) + \alpha\, D \leq \alpha\, c\, b_0 - a_0$, where $w_e = \max\{0, a_e - c\, b_e\}$ as defined in lines 2–3 of the AUXILIAR$_\alpha$ routine. Indeed, since MINWEIGHT$_\alpha(U, w, f)$ is an α-approximation algorithm for MINLIN(U, w, f), we have that $w(\hat{S}) = w(S) \leq \alpha\, \text{opt}(\text{MINLIN}(U, w, f)) \leq \alpha\, w(S^*)$, where S denotes the set defined in line 4 of the AUXILIAR$_\alpha$ routine and S^* denotes an optimum solution for MINRATIONAL(U, a, b, f). This implies that

$$w(\hat{S}) + \alpha\, D \;\leq\; \alpha\, w(S^*) + \alpha\, D \;=\; \alpha(w(S^*) + D).$$

Also, from Proposition 1 we have that $\alpha(w(S^*) + D) \leq \alpha(a(S^*) - c\, b(S^*))$. By noting that $c \geq c^* = \text{ratio}(S^*) = (a_0 + a(S^*))/(b_0 + b(S^*))$, we can conclude that

$$w(\hat{S}) + \alpha\, D \;\leq\; \alpha(a(S^*) - c\, b(S^*)) \;\leq\; \alpha(c\, b_0 - a_0) \;\leq\; \alpha\, c\, b_0 - a_0.$$

Furthermore, the middle inequality is strict if $c > c^*$.

With the previous inequality in hand, we turn to finish the proof. Note that

$$a(\hat{S}) - \alpha\, c\, b(\hat{S}) = w(\hat{S}) + c\, b(\hat{S}) + D - \alpha\, c\, b(\hat{S}) \tag{3}$$
$$= w(\hat{S}) + D + (\alpha - 1)(-c\, b(\hat{S}))$$
$$\leq w(\hat{S}) + D + (\alpha - 1)D \tag{4}$$
$$= w(\hat{S}) + \alpha\, D$$
$$\leq \alpha\, c\, b_0 - a_0. \tag{5}$$

As $\hat{S}$ contains L, equality (3) holds by Proposition 1. Inequality (4) holds because $D = a(L) - c\, b(L) \geq -c\, b(L) \geq -c\, b(\hat{S})$ (again since $\hat{S}$ contains L). Finally, inequality (5) follows by the previous argument, and it is strict if $c > c^*$. The proof is complete. □

Unfortunately, we do not have any guarantee on $\text{ratio}(\hat{S}) = (a_0 + a(\hat{S}))/(b_0 + b(\hat{S}))$ for $c < c^*$. Nevertheless, MINFRAC$_\alpha$ gets around this and still provides an α-approximation for MINRATIONAL, as we show next.

Theorem 2. *Let S be the output of algorithm* MINFRAC$_\alpha(U, a, b, f)$ *and S^* be an optimal solution to problem* MINRATIONAL(U, a, b, f). *We have that*

$$\text{ratio}(S) \;\leq\; \alpha\, \text{ratio}(S^*).$$

Proof. Let $c^* = \text{ratio}(S^*)$. Suppose that at some iteration, say iteration i, of MINFRAC$_\alpha(U, a, b, f)$ step 10 was executed and at that point we had *middle* $\leq c^*$. Clearly, in this case we are done as this implies that ratio$(S_i) \leq \alpha$ *middle* $\leq \alpha c^*$, and moreover ratio$(S) \leq$ ratio(S_j) for all j. Thus, we may assume that whenever step 10 was executed, we had that *middle* $> c^*$. On the other hand, if at some iteration i step 11 was executed, we had that *middle* $< c^*$. Otherwise, Lemma 1 would imply that AUXILIAR$_\alpha(U, a, b, f, middle)$ returns a set S_i such that ratio$(S_i) \leq \alpha$ *middle*, contradicting the fact that step 11 was executed.

Thus it remains to analyze the case in which, at each iteration, either step 10 is executed and *middle* $> c^*$, or step 11 is executed and *middle* $< c^*$. In this case, at step 13 we have that *left* $\leq c^* \leq$ *right* and *right* — *left* $\leq \epsilon$. Note that this is enough to justify that MINFRAC$_\alpha(U, a, b, f)$ is an $(\alpha + \delta)$-approximation, for some $\delta > 0$. (Indeed, ratio$(S_t) \leq \alpha$ *right* $\leq \alpha(c^* + \epsilon) \leq (\alpha + 1/(b_0 + b(U)))c^*$.) Steps 14–16 are what we need to get rid of the additive term δ.

So, suppose that at step 13 we have ratio$(S_t) > \alpha c^*$. Now, consider k, the last iteration of MINFRAC$_\alpha(U, a, b, f)$ at which step 10 was executed (if step 10 was never executed, we let $k = 0$). It is straightforward to see that

$$\alpha \text{ } left \leq \alpha c^* < \text{ratio}(S_t) \leq \text{ratio}(S_k) \leq \alpha \text{ } right,$$

where *right* denotes the final value of this variable in the execution of the algorithm (or its value after iteration k). Thus, we can conclude that c' defined in step 14 is strictly greater than c^*. But then Lemma 1 implies that the set S' defined in step 15 is such that

$$\text{ratio}(S') < \alpha c' = \alpha \frac{\text{ratio}(S_t)}{\alpha} = \text{ratio}(S_t).$$

Now, observe that, for any two feasible solutions $F, G \subseteq U$ of different values, we have that $|\text{ratio}(F) - \text{ratio}(G)| > 1/(b_0 + b(U))^2$. Thus, in particular

$$\text{ratio}(S') < \text{ratio}(S_t) - \frac{1}{(b_0 + b(U))^2}$$
$$\leq \alpha \text{ } right - \frac{1}{(b_0 + b(U))^2}$$
$$\leq \alpha(right - \epsilon)$$
$$\leq \alpha \text{ } left$$
$$\leq \alpha c^*.$$

This concludes the proof of the theorem.   □

3   Applying Megiddo's Technique

In this section we outline how to turn MINFRAC$_\alpha$ into a strongly polynomial-time algorithm using the approach of Megiddo [19].

Recall that Megiddo showed how to derive a polynomial-time algorithm for MINRATIONAL (without the increasing assumption) from a polynomial-time algorithm for MINLIN. Assuming that comparisons and additions are the only operations the algorithm for $\text{MINLIN}(U, w, f)$ does on w, Megiddo's scheme leads to a strongly polynomial-time algorithm for MINRATIONAL as long as the algorithm for MINLIN is also strongly polynomial. Algorithm MINFRAC_α in Section 2 shows how to get an α-approximation algorithm for MINRATIONAL from an α-approximation algorithm for MINLIN. Nevertheless, in the description given in Section 2, even if $\text{MINWEIGHT}_\alpha(U, w, f)$ runs in strongly polynomial time, $\text{MINFRAC}_\alpha(U, a, b, f)$ will not be strongly polynomial. In what follows, we describe how to get a strongly polynomial-time α-approximation algorithm for MINRATIONAL if, as in Megiddo's, comparisons and additions are the only operations $\text{MINWEIGHT}_\alpha(U, w, f)$ does on w. The idea is that of Megiddo with a few adjustments to accommodate the non-negativity of the w function as well as the approximation goal.

$\text{MEGIDDO\_APPROX}_\alpha\ (U, a, b, f)$

    ▷ **First phase**

1    $n \leftarrow |U|$

2    $c_0 \leftarrow 0$

3    $S_0 \leftarrow U$

4   let $c_1, \ldots, c_{n+1}$ be the ratios in $\{a_e/b_e : e \in U \cup \{0\}\}$ sorted in increasing order

5    $k \leftarrow 0$

6   **for** $j \leftarrow 1$ **to** $n+1$ **do**   ▷ This could be a binary search instead.

7        $S_j \leftarrow \text{AUXILIAR}_\alpha(U, a, b, f, c_j)$

8   **if** $\text{ratio}(S_j) > \alpha\, c_j$

9   **then** $k \leftarrow j$

    ▷ **Second phase**

10    $i \leftarrow n+2$

11    $left \leftarrow c_k$

12    $right \leftarrow c_{k+1}$

13   **for** each element e in U **do**   ▷ Observe that w is linear in $[left .. right]$.

14        $w_e(c) \leftarrow \max\{0, a_e - c\, b_e\}$ for c in $[left .. right]$

15    $finished \leftarrow \text{FALSE}$

16   **while not** $finished$ **do**

17   follow $\text{MINWEIGHT}_\alpha(U, w(c), f)$ for all c in $[left .. right]$ simultaneously, from the start or from the recent point of resumption, to the next comparison

18   **if** there are no more comparisons and $\text{MINWEIGHT}_\alpha(U, w(c), f)$ terminates

19   **then** $finished \leftarrow \text{TRUE}$

20   **else** let $g_1(c)$ and $g_2(c)$ be the functions to be compared over $[left .. right]$

21   **if** there is no unique solution of $g_1(c) = g_2(c)$ in $[left .. right]$

22   **then** resume algorithm $\text{MINWEIGHT}_\alpha(U, w(c), f)$

23   **else** let c' be the unique solution of $g_1(c) = g_2(c)$ in $[left .. right]$

24                        $S_i \leftarrow \text{AUXILIAR}_\alpha(U, a, b, f, c')$

25   **if** $\text{ratio}(S_i) \leq \alpha\, c'$

26   **then** $right \leftarrow c'$

27   **else** $left \leftarrow c'$

28                    $i \leftarrow i + 1$

29    $S \leftarrow \text{argmin}\{\text{ratio}(S_j) : 0 \leq j < i\}$

30   **return** S.

We start with a general description of the algorithm. It consists of two phases. In the first phase, we sort the ratios in $\{a_e/b_e : e \in U \cup \{0\}\}$ in increasing order (consider $a_e/b_e = \infty$ if $b_e = 0$) and let $c_1, \ldots, c_{n+1}$ be the result of this sorting, where $n = |U|$. Let $c_0 = 0$ and S_j be the output of AUXILIAR$_\alpha(U, a, b, f, c_j)$ for $j = 0, \ldots, n+1$. Also, let k be the largest j such that ratio$(S_j) > \alpha c_j$. Note that $k < n+1$. (Indeed, $S_{n+1} = U$ and ratio$(U) \leq c_{n+1}$.) For c in the interval $[c_k \mathinner{\ldotp\ldotp} c_{k+1}]$, function w defined in lines 2–3 of the AUXILIAR$_\alpha$ routine is linear. The idea now is to follow Megiddo's strategy starting from interval $[c_k \mathinner{\ldotp\ldotp} c_{k+1}]$. (Observe that the addition of two linear functions is a linear function, so additions do not affect the algorithm.) Below, we show the resulting algorithm in pseudo-code. The description of the second phase follows the one of algorithm B of Megiddo [19, p. 416].

Observe that we could also have used the first phase of the above algorithm in MINFRAC$_\alpha$. After this first phase, we would apply the "truncated" binary search as before, but starting with the interval $[c_k \mathinner{\ldotp\ldotp} c_{k+1}]$. The worst-case running time of this modification however is the same.

Due to the lack of space, the analysis of the algorithm and its approximation guarantee is left to the full version of the paper

4   Final Remarks

There is a way to convert the scheme proposed in this paper so that it applies to optimizing rational objectives for problems described by a *decreasing* property. This class of problems would for example include most packing problems. The scheme would depend on the existence of an approximation algorithm for the maximization of linear objectives for this class of problems. However, for this class, the scheme of Hashizume et al. [11] already implies the same results. In other words, avoiding negative coefficients in the linear objective function is not necessary for this class of problems: elements with negative weight can be ignored without any loss.

Recently, a related class of problems appeared in the literature. In these problems, the objective is to minimize (or maximize) the sum of rational functions [4, 23]. Maybe one can apply an idea similar to the one proposed here to derive approximation algorithms for this class of problems based on approximation algorithms for their linear counterpart.

References

1. A. Billionnet. Approximation algorithms for fractional knapsack problems. *Operations Research Letters*, 30(5):336–342, 2002.
2. G.R. Bitran and T.L. Magnanti. Duality and sensitivity analysis for fractional programs. *Operations Research*, 24:675–699, 1976.
3. J. Carlson and D. Eppstein. The weighted maximum-mean subtree and other bicriterion subtree problems. In *ACM Computing Research Repository, cs.CG/0503023.* 2005.

4. D.Z. Chen, O. Daescu, Y. Dai, N. Katoh, X. Wu, and J. Xu. Efficient algorithms and implementations for optimizing the sum of linear fractional functions, with applications. *Journal of Combinatorial Optimization*, 9:69–90, 2005.

5. G.B. Dantzig, W. Blattner, and M.R. Rao. Finding a cycle in a graph with minimum cost to time ratio with applications to a ship routing problem. In P. Rosensthiel (ed.), *Theory of Graphs: Int. Symposium*, pages 77–84. Dunod, Paris, 1967.

6. A. Dasdan and R.K. Gupta. Faster maximum and minimum mean cycle algorithms for system-performance analysis. *IEEE Transactions on Computer-Aided Design of Integrated Circuits and Systems*, 17(10):889–899, 1998.

7. A. Dasdan, S.S. Irani, and R.K. Gupta. Efficient algorithms for optimum cycle mean and optimum cost to time ratio problems. In *Proceedings of the 36th ACM/IEEE Conference on Design Automation*, pages 37–42, 1999.

8. W. Dinkelbach. On nonlinear fractional programming. *Management Science*, 13:492–498, 1967.

9. M.R. Garey and D.S. Johnson, *Computers and Intractability: a Guide to the Theory of NP-Completeness*. W.H. Freeman and Co., 1979.

10. P. Gubbala and B. Raghavachari. Finding k-connected subgraphs with minimum average weight. In *Proceedings of the 6th Latin American Theoretical Informatics Symposium (LATIN)*, volume 2976 of *Lecture Notes in Computer Science*, pages 212–221. Springer, Berlin, 2004.

11. S. Hashizume, M. Fukushima, N. Katoh, and T. Ibaraki. Approximation algorithms for combinatorial fractional programming problems. *Mathematical Programming*, 37:255–267, 1987.

12. R. Jagannathan. On some properties of programming problems in parametric form pertaining to fractional programming. *Management Science*, 12:609–615, 1966.

13. K. Jain. A factor 2 approximation algorithm for the generalized Steiner network problem. *Combinatorica*, 21(1):39–60, 2001.

14. K. Jain, I. Măndoiu, V.V. Vazirani, and D.P. Williamson. A primal-dual schema based approximation algorithm for the element connectivity problem. *Journal of Algorithms*, 45(1):1–15, 2002.

15. N. Katoh. A fully polynomial-time approximation scheme for minimum cost-reliability ratio problems. *Discrete Applied Mathematics*, 35(2):143–155, 1992.

16. S. Khuller and U. Vishkin. Biconnectivity approximations and graph carvings. *Journal of the Association for Computing Machinery*, 41(2):214–235, 1994.

17. G. Klau, I. Ljubi, P. Mutzel, U. Pferschy, and R. Weiskircher. The fractional prize collecting Steiner tree problem on trees. In *Proceedings of the 11th European Symposium on Algorithms (ESA 2003)*, volume 2832 of *Lecture Notes in Computer Science*, pages 691–702. Springer, Berlin, 2003.

18. E.L. Lawler. Optimal cycles in doubly weighted directed linear graphs. In P. Rosensthiel (ed.), *Theory of Graphs: Int. Symposium*, pages 209–214. Dunod, Paris, 1967.

19. N. Megiddo. Combinatorial optimization with rational objective functions. *Mathematics of Operations Research*, 4(4):414–424, 1979.

20. J.B. Orlin and R.K. Ahuja. New scaling algorithms for the assignment and minimum mean cycle problems. *Mathematical Programming*, 54(1):41–56, 1992.

21. T. Radzik. Newton's method for fractional combinatorial optimization. In *Proc. 33rd Annual Symposium on Foundations of Computer Science*, pages 659–669, 1992.

22. M. Shigeno, Y. Saruwatari, and T. Matsui. An algorithm for fractional assignment problems. *Discrete Applied Mathematics*, 56(2–3):333–343, 1995.

23. C.C. Skiscim and S.W. Palocsay. The complexity of minimum ratio spanning tree problems. *Journal of Global Optimization*, 30:335–346, 2004.

In-Place Algorithms for Computing (Layers of) Maxima

Henrik Blunck and Jan Vahrenhold

Westfälische Wilhelms-Universität Münster, Institut für Informatik,
48149 Münster, Germany
{blunck, jan}@math.uni-muenster.de

Abstract. We describe space-efficient algorithms for solving problems related to finding maxima among points in two and three dimensions. Our algorithms run in optimal $\mathcal{O}(n \log n)$ time and occupy only constant extra space in addition to the space needed for representing the input.

1 Introduction

Space-efficient solutions for fundamental algorithmic problems such as merging, sorting, and partitioning have been studied over a long period of time; see [11, 12, 14, 21]. The advent of small-scale, handheld computing devices and an increasing interest in utilizing fast but limited-size memory, e.g., caches, recently led to a renaissance of space-efficient computing with a focus on processing geometric data. Brönnimann *et al.* [7] were the first to consider space-efficient geometric algorithms and showed how to optimally compute $2d$-convex hulls using constant extra space. Subsequently, a number of space-efficient geometric algorithms, e.g., for computing $3d$-convex hulls and its relatives, as well as for solving intersection and proximity problems, have been presented [4, 5, 6, 9, 23].

In this paper, we consider the fundamental geometric problems of computing the maxima of point sets in two and three dimensions and of computing the layers of maxima in two dimensions. Given two points p and q, the point p is said to *dominate* the point q iff the coordinates of p are larger than the coordinates of q in all dimensions. A point p is said to be a *maximal point* (or: a *maximum*) of $\mathcal{P}$ iff it is not dominated by any other point in $\mathcal{P}$. The union $\texttt{MAX}(\mathcal{P})$ of all points in $\mathcal{P}$ that are maximal is called the *set of maxima* of $\mathcal{P}$. This notion can be extended in a natural way to compute *layers* of maxima [8]. After $\texttt{MAX}(\mathcal{P})$ has been identified, the computation is repeated for $\mathcal{P} := \mathcal{P} \setminus \texttt{MAX}(\mathcal{P})$, i.e., the next layer of maxima is computed. This process is iterated until $\mathcal{P}$ becomes empty.

Related Work. The problem of finding maxima of a set of n points has a variety of applications in statistics, economics, and operations research (as noted by Preparata and Shamos [20]), and thus was among the first problems having been studied in Computational Geometry: In two and three dimensions, the best known algorithm which has been developed by Kung, Luccio, and Preparata [16] identifies the set of maxima in $\mathcal{O}(n \log n)$ time which is optimal since the problem

L. Arge and R. Freivalds (Eds.): SWAT 2006, LNCS 4059, pp. 363–374, 2006.

exhibits a sorting lower bound [16, 20]. For constant dimensionality $d \geq 4$, their divide-and-conquer approach yields an algorithm with $\mathcal{O}(n \log^{d-2} n)$ running time [1, 16], and Matoušek [17] gave an $\mathcal{O}(n^{2.688})$ algorithm for the case $d = n$. The problem has also been studied for dynamically changing point sets in two dimensions [13] and under assumptions about the distribution of the input points in higher dimensions [2, 10]. Buchsbaum and Goodrich [8] presented an $\mathcal{O}(n \log n)$ algorithm for computing the layers of maxima for point sets in three dimensions. Their approach is based on the plane-sweeping paradigm and relies on dynamic fractional cascading to maintain a point-location structure for dynamically changing two-dimensional layers of maxima.

The maxima problem has been actively investigated in the database community following Börzsönyi, Kossmann, and Stocker's [3] definition of the SQL "skyline" operator. Such an operator producing the set of maxima[1] is needed in queries that, e.g., ask for hotels that are both close to the beach and have low room rates. A number of results have been presented that use spatial indexes to produce the "skyline", e.g., the set of maxima, practically efficient and/or in a progressive way, that is outputting results while the algorithm is running [15, 19, 22]. For none of these approaches non-trivial upper bounds are known.

The Model. The goal of investigating space-efficient algorithms is to design algorithms that use very little extra space in addition to the space used for representing the input. The input is assumed to be stored in an array A of size n, thereby allowing random access. We assume that a constant-sized memory can hold a constant number of words. Each word can hold one pointer, or an $\mathcal{O}(\log n)$ bit integer, and a constant number of words can hold one element of the input array. The extra memory used by an algorithm is measured in terms of the number of extra words; an *in-place* algorithm uses $\mathcal{O}(1)$ extra words of memory. It has been shown that some fundamental geometric problems such as computing 2D convex hulls and closest pairs can be solved *in-place* and in optimal time [4, 5, 7]. More involved problems (range searching, line-segment intersection) can be (currently) solved *in-place* only if one is willing to accept near-optimal running time [6, 23], and 3D convex hulls and related problems seem to require both (poly-)logarithmic extra space and time [6].

Our Contribution. The main issue in designing in-place algorithms is that most powerful algorithmic tools (unbalanced recursion, sweeping, multi-level data structures, fractional cascading) require at least logarithmic extra space, e.g., for the recursion stack or pointer-based structures. This raises the question of whether there exists a time-space tradeoff for geometric algorithms besides range-searching. In this paper, we demonstrate that $\mathcal{O}(1)$ extra space is sufficient to obtain optimal $\mathcal{O}(n \log n)$ algorithms for computing skylines in two and three dimensions and two-dimensional layers of maxima. The solution to the latter

---

[1] Technically speaking, the operator returns the set of *minima*. To unify the presentation, we do not distinguish between these two variants of the same problem.

problem is of particular interest since it is the first optimal in-place algorithm for a geometric problem that is not amenable to a solution based on balanced divide-and-conquer or Graham's scan.

2   Computing the Skyline in $\mathbb{R}^2$ and $\mathbb{R}^3$

A point p from a point set $\mathcal{P}$ is said to be *maximal* if no other point from $\mathcal{P}$ has larger coordinates in *all* dimensions; ties are broken using a standard shearing technique. This definition has been transferred by Kung *et al.* [16] into a two-dimensional plane-sweeping algorithm and into a divide-and-conquer approach for the higher dimensional case. The output of our algorithm consists of a permutation of the input array A and an index k such that k points constituting the set of maxima are stored sorted by decreasing y-coordinates in $A[0, \ldots, k-1]$.

The algorithm for computing the skyline, i.e., the set of maxima, in two dimensions is a straightforward selection algorithm. We discuss it in some more detail to introduce an important algorithmic template, called SORTEDSUBSET-SELECTION(A, ℓ_b, ℓ_e, π) which processes a sorted (sub)array $A[\ell_b, \ldots, \ell_e - 1]$ from left to right. While doing so, the algorithm evaluates a given predicate π for each of the elements and stably moves all elements for which π evaluates to **true** to the front of $A[\ell_b, \ldots, \ell_e - 1]$. This algorithm, presented by Bose *et al.* [4], runs in linear time provided that π can be evaluated in constant time.

To compute the set of maxima in two dimensions, the algorithm of Kung *et al.* [16] sweeps the point set in *decreasing* y-direction keeping track of the bottommost point m of the skyline seen so far. For each point p encountered during the sweep, the algorithm checks whether p is dominated by m. The sweeping direction ensures $p.y \leq m.y$, thus, it is sufficient to check whether also $p.x < m.x$.

The space-efficient implementation of this algorithm thus first presorts A using an optimal $\mathcal{O}(n \log n)$ in-place sorting algorithm, e.g., *heapsort* [11]. The algorithm then runs an instantiation of the linear-time SORTEDSUBSETSELECTION template where the predicate π evaluates to **true** iff the x-coordinate of the current point $A[i]$ is at least as large as the x-coordinate of the point m, i.e, iff $A[i]$ is a maximal point. The algorithm then moves $A[i]$ to the front of the array and updates m to refer to (the new position of) $A[i]$.

Lemma 1. *The skyline, i.e., the set of maxima of a set $\mathcal{P}$ of n points in two dimensions can be computed in-place and in optimal time $\mathcal{O}(n \log n)$. If $\mathcal{P}$ is sorted according to $<_y$, the running time is in $\mathcal{O}(n)$.*

For the case of a three-dimensional input, we implement Kung *et al.*'s [16] divide-and-conquer algorithm using an in-place divide-and-conquer scheme we have proposed earlier [4]; this scheme is based on in-place routines for median-finding, partitioning, and merging. Since we cannot explicitly keep track of the number of maxima in each subproblem, we have to recover them algorithmically during each merging step. The details are given in the full version of this paper.

Theorem 1. *The skyline, i.e., the set of maxima of an n-element point set in three dimensions can be computed in-place and in optimal time $\mathcal{O}(n \log n)$.*

3 Computing the Layers of Maxima in Two Dimensions

An obvious way of computing the layers of maxima is to iteratively compute (and remove) the maximal points of the given point set $\mathcal{P}$ using, e.g., the in-place algorithm described in Section 2. Since a point set may exhibit a linear number of layers, this leads to an $\mathcal{O}(n^2 \log n)$ worst-case running time. In this section, we show that we can simultaneously peel off multiple layers such that the resulting algorithm runs in optimal $\mathcal{O}(n \log n)$ time; its goal is to rearrange the input such that the points are grouped by layers and each layer is sorted by decreasing y-coordinate.

3.1 Computing the Number of Layers of Maxima

As an introductory example of our approach, we extend the algorithm discussed in Section 2 to compute the number of layers of maxima for a given point set. This algorithm builds upon the fact that a layer of points is monotone in both x- and y-direction: a layer $\mathcal{L}_i$ extends vertically to $y = -\infty$ from the point on $\mathcal{L}_i$ that has maximal x-coordinate. This in turn implies that, during the sweep, the x-coordinate of the intersection of $\mathcal{L}_i$ with the sweepline is the x-coordinate of its "tail" point τ_i, i.e., of the last point that has been classified as belonging to $\mathcal{L}_i$ (see Figure 1).

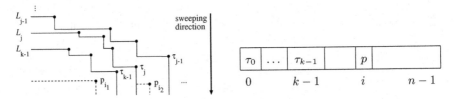

Fig. 1. Classification of a point by binary search (left) and array representation (right)

As usual, we assume that the point set $\mathcal{P}$ to be processed is stored in an array $A[0, \ldots, n-1]$. During the sweep, we ensure that the following invariant holds after having processed a point p:

Invariant (TAILS): Let $k \in \{1, \ldots, n\}$ be the number of layers intersected by the sweepline at $y = p.y$ where p is the point that has just been processed. Then the tail points $\tau_0, \ldots, \tau_{k-1}$ of the layers $\mathcal{L}_0, \ldots, \mathcal{L}_{k-1}$ are stored in decreasing x-order in $A[0, \ldots, k-1]$.

Invariant (TAILS) is certainly true after having processed the first point $p = A[0]$ encountered during the sweep. This point is the y-maximal point of the point set and thus part of the skyline, i.e., of the topmost layer $\mathcal{L}_0$. We thus inductively assume that the invariant holds prior to processing the next point $p := A[i]$. To determine which layer p is part of, we perform a binary search for p w.r.t. the x-coordinate among the points in $A[0, \ldots, k-1]$. If p has a smaller

x-coordinate than $\tau_{k-1} = \mathtt{A}[k-1]$, p is dominated by this point, and thus p is the first point of a new layer $\mathcal{L}_k$ (see p_{i_1} in Figure 1). We then swap $p = \mathtt{A}[i]$ to $\mathtt{A}[k]$ (note that $i \geq k$ trivially holds) and increment k by one. If, on the other hand, p lies right of τ_j but left of τ_{j-1} then p replaces τ_j as the tail point of $\mathcal{L}_j$ (see p_{i_2} in Figure 1), that is, we swap $p = \mathtt{A}[i]$ to $\mathtt{A}[j]$; similarly, p replaces τ_0 if it lies right of τ_0.

The above in-place algorithm maintains Invariant (TAILS) in $\mathcal{O}(\log n)$ time per point processed. Thus after having processed the last point, the index k gives the total number of layers.

Lemma 2. *The number of layers of maxima exhibited by an n-element point set in two dimensions can be compute in-place and in $\mathcal{O}(n \log n)$ time.*

The above algorithm can also be modified to *output*, i.e., to print to a write-only stream, in $\mathcal{O}(n \log n)$ time each point processed together with the number of its containing layer.

3.2 Counting the Number of Points on the Topmost • Layers

The algorithm of Section 3.1 can be modified to *count* the number of points on each of the κ topmost layers. For the simplicity of exposition, we assume that we have access to $\mathcal{O}(\kappa)$ extra space that holds a counter c_i for each layer $\mathcal{L}_i$. In Section 3.4, we will get rid of this assumption, which—in an in-place setting—is prohibitive for non-constant κ.

To compute the number of points on each of the topmost κ layers, we simply increment the counter c_i for layer $\mathcal{L}_i$ whenever we update the value of τ_i. We also stop updating the counter k denoting the number of layers being kept track of at $k = \kappa$. Afterwards, we can determine for each point in $\mathcal{O}(1)$ time whether it lies left of $\tau_{\kappa-1}$ (and thus below $\mathcal{L}_{\kappa-1}$). If so, we simply ignore it, and for all other points, we perform a binary search in $\mathtt{A}[0, \ldots, \kappa - 1]$ as described above.

Lemma 3. *The cardinality c_i of each of the topmost κ layers of $\mathtt{A}[0, \ldots, n-1]$ can be computed in $\mathcal{O}(n \log n)$ time. If the points are presorted, the complexity is in $\mathcal{O}(n + \xi \log \kappa)$ where $\xi = \sum_{i=0}^{\kappa-1} c_i$.*

3.3 Extracting the Topmost • Layers in Sorted Order

As mentioned above, a naive iterative approach to computing all layers of maxima leads to an $\mathcal{O}(n^2 \log n)$ worst-case running time for points sets with a linear number of layers. The algorithm we describe in this section processes several layers at a time to reduce the number of iterations.

Extracting the Points on the Topmost κ Layers. Our algorithm imitates counting-sort, i.e., prior to actually partitioning the points into layers, it first computes the number of points for each of the layers. To illustrate the algorithm, let us assume that we have already peeled off some layers and stored the result in $\mathtt{A}[0, \ldots, \ell_b - 1]$. Inductively, we maintain the following invariant which, prior to the first iteration, can be established by sorting $\mathtt{A}$ and setting $\ell_b := 0$:

Invariant (SORT): The points in $A[0, \ldots, n-1]$ that have not yet been assigned to a layer are stored in $A[\ell_b, \ldots, n-1]$ and are sorted by decreasing y-coordinate.

Let us further assume that the total number of points on the topmost κ layers $\mathcal{L}_0$ through $\mathcal{L}_{\kappa-1}$ of the *remaining* points stored in $A[\ell_b, \ldots, n-1]$ is ξ and that $\ell_b + 2\xi \leq n$. The first step of the algorithm is to stably extract the ξ points on the topmost κ layers and move them to $A[\ell_b, \ldots, \ell_b + \xi - 1]$ while maintaining the sorted y-order in $A[\ell_b + \xi, \ldots, n-1]$:

...	Points on $\mathcal{L}_0$ through $\mathcal{L}_{\kappa-1}$	Points below $\mathcal{L}_{\kappa-1}$
	ℓ_b	$\ell_b + \xi$ $n-1$

This partition is obtained as follows: We run a variant of the algorithm described in Section 3.2 that is combined with SORTEDSUBSETSELECTION. This algorithm maintains the invariant that the tail points τ_0 through $\tau_{\kappa-1}$ are stored in $A[\ell_b, \ldots, \ell_b + \kappa - 1]$ and stably moves all points that are *below* $\mathcal{L}_{\kappa-1}$ to the subarray starting at $A[\ell_b + \kappa]$; thus, it keeps all points below $\mathcal{L}_{\kappa-1}$ in sorted order. At the same time, we also maintain a counter c_i for the size of each layer $\mathcal{L}_i$—as mentioned in Section 3.2, we will later discuss how to do this in-place. The correctness of this algorithm follows from the observation that none of the first κ points in $A[\ell_b, \ldots, \ell_b + \kappa - 1]$ can lie below $\mathcal{L}_{\kappa-1}$ (there have to be at least κ points on κ layers).

The algorithm maintains the invariant that, when processing point $A[j]$, all points that already have been identified as "below $\mathcal{L}_{\kappa-1}$" are stored in decreasing y-order in $A[\ell_b + \kappa, \ldots, i-1]$.

...	$\tau_0 \ldots \tau_{\kappa-1}$	Points below $\mathcal{L}_{\kappa-1}$	Points on $\mathcal{L}_0 \ldots \mathcal{L}_{\kappa-1}$		
	ℓ_b	$\ell_b + \kappa$	i	j	$n-1$

The point $A[j]$ now either is classified as "below $\mathcal{L}_{\kappa-1}$" or replaces the tail τ_h of some layer $\mathcal{L}_h$. In the first situation, $A[j]$ is stably moved directly behind $A[\ell_b + \kappa, \ldots, i-1]$, i.e., it is swapped with $A[i]$ and i is then incremented by one, and in the second situation, $A[j]$ is swapped with τ_h, i.e., with $A[\ell_b + h]$. When we have reached the end of the array, we inductively see that $A[\ell_b + \kappa, \ldots, i-1]$ contains the points below $\mathcal{L}_{\kappa-1}$ in sorted order. Furthermore, by the definition of ξ, we know that the two subarrays $A[\ell_b, \ldots, \ell_b + \kappa - 1]$ (containing the tails) and $A[i, \ldots, n-1]$ (containing the remaining points on the layers $\mathcal{L}_0$ through $\mathcal{L}_{\kappa-1}$) together consist of exactly ξ points. We then swap (in linear time) $A[\ell_b + \kappa, \ldots, i-1]$ (containing the elements below $\mathcal{L}_{\kappa-1}$) and $A[i, \ldots, n-1]$ such that the ξ elements on the layers $\mathcal{L}_0$ through $\mathcal{L}_{\kappa-1}$ (tails and non-tails) are blocked in $A[\ell_b, \ldots, \ell_b + \xi - 1]$. To re-establish the y-order of these ξ points, we sort them in $\mathcal{O}(\xi \log \xi) \subset \mathcal{O}(\xi \log n)$ time, that is, we establish Invariant (SORT) for $A[\ell_b, \ldots, \xi - 1]$. Thus, the overall running time for counting the number of points on the topmost κ layers and for re-establishing Invariant (SORT) is in $\mathcal{O}(n + \xi \log n)$ where $\xi = \sum_{i=0}^{\kappa-1} c_i$.

Sorting the Points by Layer. Using the counters c_i computed during the previous step, we now run a variant of counting sort to extract the layers $\mathcal{L}_0$ through $\mathcal{L}_{\kappa-1}$ in sorted y-order. To do this in-place, we use the subarray $\mathtt{A}[\ell_b+\xi, \ldots, \ell_b+2\xi-1]$ as scratch space that will hold the layers to be constructed (note that we assume $\ell_b + 2\xi \le n$ and that $\xi = \sum_{i=0}^{\kappa-1} c_i$ holds by definition). We traverse the subarray $\mathtt{A}[\ell_b, \ldots, \ell_b + \xi - 1]$ maintaining the tails of all layers in $\mathtt{A}[\ell_b, \ldots, \ell_b + \kappa - 1]$ as before, but whenever we update the tail τ_i of a layer $\mathcal{L}_i$, we swap the old tail to the next available position in the subarray of $\mathtt{A}[\ell_b + \xi, \ldots, \ell_b + 2\xi - 1]$ that is reserved to hold $\mathcal{L}_i$.

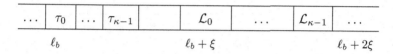

After having constructed a sorted representation of the layers $\mathcal{L}_0$ through $\mathcal{L}_{\kappa-1}$, the two subarrays $\mathtt{A}[\ell_b, \ldots, \ell_b + \xi - 1]$ and $\mathtt{A}[\ell_b + \xi, \ldots, \ell_b + 2\xi - 1]$ are swapped in-place:

...	$\mathcal{L}_0$	...	$\mathcal{L}_{\kappa-1}$	Points below $\mathcal{L}_{\kappa-1}$	...
	ℓ_b			$\ell_b + \xi$	$\ell_b + 2\xi$

To re-establish Invariant (SORT), we finally sort $\mathtt{A}[\ell_b + \xi, \ldots, \ell_b + 2\xi - 1]$ (note that the points $\mathtt{A}[\ell_b + 2\xi, \ldots, n-1]$ have not been touched and thus still are sorted) and update $\ell_b := \ell_b + \xi$. The running time for sorted extraction of the ξ points on the topmost κ layers and for re-establishing Invariant (SORT) for $\mathtt{A}[\ell_b + \xi, \ldots, n-1]$ is in $\mathcal{O}(n + \xi \log n)$.

3.4 Extracting All Layers in Sorted Order

The exposition of the algorithm presented in the previous section was build on two major assumptions: (1) the algorithm had to have access to κ counters and (2) the subarray $\mathtt{A}[\ell_b, \ldots, n-1]$ had to be large enough to accommodate two subarrays of size ξ. In this section, we demonstrate how to maintain both assumptions in an in-place setting.

The first issue to be resolved is how to maintain a non-constant number κ of counters without using $\Theta(\kappa)$ extra space. Each such counter c_i is required to represent values up to n, and thus has to consist of $\log_2 n$ bits. We resort to a standard technique in the design of space-efficient algorithms, namely to encode a single bit by a permutation of two distinct (but comparable) elements q and r: assuming $q < r$, the permutation rq encodes a binary zero, and the permutation qr encodes a binary one [18]. As the elements in our case are two-dimensional points, we use the (lexicographical) y-order for deciding whether two points encode a binary zero or a binary one.[2]

If we reserve a block of $\frac{1}{3}n$ elements, we can encode $\frac{1}{6}n$ bits, i.e., $\frac{1}{6}n/\log_2 n$ counters that may be used to represent values less than n, and this in turn implies that the maximum number of layers for which we can run the algorithms described in Sections 3.2 and 3.3 is bounded by $\kappa = \frac{1}{6}n/\log_2 n$. The analyses at the end of the respective sections gave an $\mathcal{O}(n + \xi \log n)$ bound for each run of the algorithms, and thus we have to make sure that maintaining the counters and the second invariant does not interfere with keeping the overall number of iterations in $\mathcal{O}(\log n)$.

The Case $\bullet_b \bullet \frac{1}{3} \bullet$. If, prior to the current iteration, $\ell_b < \frac{1}{3}n$ holds, we maintain the counters in $\mathtt{A}[\frac{2}{3}n, \dots, n-1]$.

...		Counter representation

$\ell_b \qquad\qquad\qquad \frac{2}{3}n \qquad\qquad n-1$

Counting the Points on the Topmost κ Layers. By Invariant (SORT), $\mathtt{A}[\ell_b, \dots, n-1]$ is sorted by decreasing y-coordinate, so all counters encode the value zero. We set $\kappa := \frac{1}{6}n/\log_2 n$ and run the algorithm for counting the elements on each of the topmost κ layers. We maintain each of the counters c_i in its fixed-size representation by exchanging adjacent elements as needed to implement changing a binary digit, and using the standard analysis for incrementing a binary counter, we observe that all counters can be maintained in $\mathcal{O}(\xi)$ time where $\xi := \sum_{i=0}^{\kappa-1} c_i$. Note that, since the algorithm processes *all* points in $\mathtt{A}[\ell_b, \dots, n-1]$, any point q in $\mathtt{A}[\frac{2}{3}n, \dots, n-1]$ may be swapped to the front of the array since it may become the tail τ_i of some layer $\mathcal{L}_i$. Using a more careful implementation of the approach given in Section 3.2, we can compute all counters and re-establish Invariant (SORT) in $\mathcal{O}(n + \xi \log n)$ time. After we have computed the values of all counters c_i—but prior to re-establishing Invariant (SORT)—we compute the prefix sums of c_0 through $c_{\kappa-1}$, i.e., we replace c_j by $\hat{c}_j := \sum_{i=0}^{j} c_i$. This can be done in-place spending $\mathcal{O}(\log n)$ time per counter, and thus in $\mathcal{O}(n)$ overall time. While doing so, we maintain the maximal index κ' such that $\ell_b + 2\hat{c}_{\kappa'} < \frac{2}{3}n$.

Extracting and Sorting the Points on the Topmost κ Layers. If the index κ' described above exists, we run (a slightly modified implementation of) the algorithm for extracting the $\xi' := \hat{c}_{\kappa'}$ points on the κ' topmost layers as described in Section 3.3. Because of the way κ' was chosen, we can guarantee that the scratch space of size $\hat{c}_{\kappa'}$ needed for the counting-sort-like partitioning does not interfere

---

[2] We point out that a *set* of elements cannot contain duplicates; hence the relative order of two points is unique. Furthermore, the set of maxima of a *multiset* M consists of the same points as the set of maxima of the *set* that is obtained by removing the duplicates from M. Duplicate removal can be done in-place and in $\mathcal{O}(n \log n)$ time by first sorting M according to $<_y$ and then stably selecting exactly one occurrence of each point.

with the space $A[\frac{2}{3}n, \ldots, n-1]$ reserved for representing the counters. Also, by Invariant (SORT), $A[\frac{2}{3}n, \ldots, n-1]$ is guaranteed to be sorted by decreasing y-coordinate. Both conditions imply that, once we have extracted all ξ' points, we can "reset" the counters in $\mathcal{O}(n)$ time (thus re-establishing Invariant (SORT) for $A[\frac{2}{3}n, \ldots, n-1]$) by scanning through $A[\frac{2}{3}n, \ldots, n-1]$ and swapping only adjacent points.

The counters c_i used during the following counting-sort-like partitioning are initialized with $c_0 := 0$ and $c_j := \hat{c}_{j-1}$ for $0 < j \leq \kappa'$; these counters are stored in $A[\frac{2}{3}n, \ldots, n-1]$. Whenever a tail τ_i is updated, the old point p representing τ_i is swapped to $A[\ell_b + \xi' + c_i]$ and then c_i is incremented by one. Decoding and incrementing c_i can be done in $\mathcal{O}(\log n)$ time, and this cost is charged to the point p moved to its containing layer. The total cost for extracting ξ' points is in $\mathcal{O}(n + \xi' \log n)$; this also includes the cost for sorting the scratch space $A[\ell_b + \xi', \ldots, \ell_b + 2\xi' - 1]$ and re-establishing Invariant (SORT) by merging $A[\ell_b + \xi', \ldots, \ell_b + 2\xi' - 1]$ with the (sorted) subarray $A[\ell_b + 2\xi', \ldots, n-1]$ (see Section 3.3).

If $\kappa' < \kappa$, i.e., if we extract some but not all $\kappa = \frac{1}{6}n/\log_2 n$ layers, we additionally run the $\mathcal{O}(n)$ skyline computation algorithm described in Section 2 to extract the points on the next topmost layer, regardless of its size. Similarly, if the index κ' does not exist at all, we extract the topmost layer $\mathcal{L}_0$ using the $\mathcal{O}(n)$ skyline computation algorithm on $A[\ell_b, \ldots, n-1]$—note that in this case the topmost layer $\mathcal{L}_0$ contains $c_0 > \frac{1}{2}\left(\frac{2}{3}n - \ell_b\right) > \frac{1}{2}\left(\frac{2}{3}n - \frac{1}{3}n\right) \in \Theta(n)$ points. In any case, we spend another $\mathcal{O}(n \log n)$ time to re-establish Invariant (SORT) by sorting.

Analysis. Our analysis classifies each iteration according to whether or not all ξ points on the topmost $\kappa = \frac{1}{6}n/\log_2 n$ layers are moved to their final position in the array. If all ξ points are moved, we know that $\xi \geq \frac{1}{6}n/\log_2 n$, and thus only a logarithmic number of such iterations can exist. Also, we can distribute the $\mathcal{O}(n + \xi \log n)$ time spent per iteration such that each iteration gets charged $\mathcal{O}(n)$ time and that each of the ξ points moved to its final position gets charged $\mathcal{O}(\log n)$ time, so the overall cost for all such iterations is in $\mathcal{O}(n \log n)$.

If less than κ layers can be processed in the iteration in question (this also includes the case that κ' does not exist), the $\mathcal{O}(n + \xi \log n)$ cost for counting the ξ points on the topmost κ layers and the $\mathcal{O}(n + \xi' \log n)$ cost for extracting ξ' points on the topmost κ' layers is dominated by the $\mathcal{O}(n \log n)$ cost for the successive skyline computation and re-establishing Invariant (SORT). The definition of κ' guarantees that, after we have performed the skyline computation, we have advanced the index ℓ_b by at least $\frac{1}{2}\left(\frac{2}{3}n - \ell_b\right)$ steps. Combining this with the fact that $\ell_b < \frac{1}{3}n$, we see that there may exist only a constant number of such iterations, and hence their overall cost is in $\mathcal{O}(n \log n)$.

The Case $\bullet_b \geq \frac{1}{3}\bullet$. If, prior to the current iteration, $\ell_b \geq \frac{1}{3}n$ holds, we maintain the counters in $A[0, \ldots, \frac{1}{3}n - 1]$:

Counter representation		
0	$\frac{1}{3}n$ $\quad$ ℓ_b	$n-1$

Note that this subarray contains (part of) the layers that have been computed already. Since maintaining a counter involves swapping some of the elements in $A[0, \ldots, \frac{1}{3}n - 1]$, this disturbs the y-order of (some of) the layers already computed, and we have to make sure that we can reconstruct the layer order. We will discuss this at the end of this section.

Counting the Points on the Topmost κ Layers. The algorithm for counting the points on the topmost κ layer proceeds exactly as described above, i.e., starting with $\kappa := \frac{1}{6}n/\log_2 n$ and updating the κ counters (which now are represented in $A[0, \ldots, \frac{1}{3}n - 1]$). The only difference is that the algorithm's selection process will not touch the space reserved for the counters, and thus, when computing the prefix sums, the algorithm finds the maximal index κ' such that $\ell_b + 2\hat{c}_{\kappa'} < n$ (instead of $\ell_b + 2\hat{c}_{\kappa'} < \frac{2}{3}n$). Then, we can run (a simplified version of) the algorithm we used for the case $\ell_b < \frac{1}{3}n$. Thus we spend $\mathcal{O}(n + \xi \log n)$ time per iteration including the cost for re-establishing Invariant (SORT).

Extracting and Sorting the Points on the Topmost κ Layers. As for the case $\ell_b < \frac{1}{3}n$ we either extract all ξ points on the topmost κ layers in $\mathcal{O}(n + \xi \log n)$ time or extract less than κ layers followed by a skyline computation in $\mathcal{O}(\nu \log \nu)$ time where $\nu := n - \ell_b$. In both cases, the complexity given also includes the cost for re-establishing Invariant (SORT).

Analysis. To estimate the overall running time for the case $\ell_b > \frac{1}{3}n$, we again classify the iterations according to whether or not all ξ points on the topmost $\kappa := \frac{1}{6}n/\log_2 n$ layers can be moved to their final destination. If this is the case, we know that we have moved $\xi \geq \kappa = \frac{1}{6}n/\log_2 n$ points and can charge $\mathcal{O}(\log n)$ time to each point moved and the remaining $\mathcal{O}(n)$ time to the iteration. Moving $\xi \geq \frac{1}{6}n/\log_2 n$ points also implies that the total number of such iterations is bounded by $\mathcal{O}(\log n)$, and hence the *global* cost incurred by assigning $\mathcal{O}(n)$ cost to such an iteration is in $\mathcal{O}(n \log n)$.

If we move $\kappa' < \kappa$ layers, the next step of the algorithm is a skyline computation (or the algorithm terminates), and we analyze these steps together. After we have performed these steps, we know (by the definition of κ') that we have advanced ℓ_b by at least $\frac{1}{2}(n - \ell_b)$. Thus, the next time, this situation occurs, it will occur for a subarray of at most half the size. This geometrically decreasing series implies that the cost for all iterations in which $\kappa' < \kappa$ layers are moved is dominated by the cost of the first such iteration (if any), i.e., it is in $\mathcal{O}(n \log n)$.

Restoring the Layers Stored in $A[0, \ldots, \frac{1}{3}n - 1]$. After the last iteration of the algorithm, we need to restore the y-order of the layers stored in $A[0, \ldots, \frac{1}{3}n - 1]$. The main problem when doing this is that we have no memory of the size of each layer, and thus we cannot simply sort the points. However, we know that each point (having been used for bit encoding) can only be one position off its correct location. Our algorithm for reconstructing the layers exploits this and iterates over pairs of "bit-neighbors" in $A[0, \ldots, \frac{1}{3}n - 1]$ while maintaining the invariant that, when processing $q := A[2i]$ and $r := A[2i + 1]$, all points in $A[0, \ldots, 2i - 1]$

have been restored to their correct order using the algorithm described in the rest of this section.

If one of the two points q and r dominates the other point, the two points cannot be part of the same layer. Thus the point with larger y-coordinate is the last point of the current layer, and the other point is the first point of the next layer—see the left part of Figure 2. The situation that no point dominates the other is further detailed in the right part of Figure 2; in order to bring q and r in their correct order, we need to access the point $p := A[2i-1]$. If $i = 0$, p does not exist, but then q and r are the first two points and thus their y-order gives their correct and final position. If p exists, it is (by the invariant) the right- and bottommost point of its containing layer in $A[0, \ldots, 2i-1]$, and thus no point left of p can belong to the same layer. A careful analysis (omitted due to space constraints) shows that each point of r and q that is right of p belongs to the same layer as p. Also, if both r and q are left of p, we can show that (due to their relative y-order) they have to be part of the same layer. We conclude that either one point dominates the other or that a simple, constant-time test, namely comparing the relative x-order of p, q, and r is sufficient to reconstruct the correct layer order of q and r. Consequently, the algorithm for reconstructing the layer order runs in linear time.

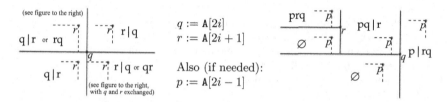

Fig. 2. Restoring the layer order for q and r. A "|" represents a break between layers.

Conclusions. Summing up, the cost for all iterations in which $\ell_b < \frac{1}{3}n$ and for all iterations in which $\ell_b \geq \frac{1}{3}n$ is $\mathcal{O}(n \log n)$. Combining this with the fact that each point gets charged $\mathcal{O}(\log n)$ cost for the iteration in which it is moved to its final location, we obtain our main result:

Theorem 2. *All layers of maxima of an n-element point set in two dimensions can be computed in-place and in optimal time $\mathcal{O}(n \log n)$ such that the points in each layer are sorted by decreasing y-coordinate.*

References

1. J. L. Bentley. Multidimensional divide-and-conquer. *Communications of the ACM*, 23(4):214–229, 1980.
2. J. L. Bentley, K. L. Clarkson, and D. B. Levine. Fast linear expected-time algorithms for computing maxima and convex hulls. *Algorithmica*, 9(2):168–183, 1993.
3. S. Börzsönyi, D. Kossmann, and K. Stocker. The skyline operator. In *Proceedings of the 17th International Conference on Data Engineering*, pages 421–430. 2001.

4. P. Bose, A. Maheshwari, P. Morin, J. Morrison, M. Smid, and J. Vahrenhold. Space-efficient geometric divide-and-conquer algorithms. *Computational Geometry: Theory & Applications*, 2006. To appear, accepted Nov. 2004.

5. H. Brönnimann and T. M.-Y. Chan. Space-efficient algorithms for computing the convex hull of a simple polygonal line in linear time. *Computational Geometry: Theory & Applications*, 34(2):75–82, 2006.

6. H. Brönnimann, T. M.-Y. Chan, and E. Y. Chen. Towards in-place geometric algorithms. In *Proceedings of the 20th Annual Symposium on Computational Geometry*, pages 239–246. 2004.

7. H. Brönnimann, J. Iacono, J. Katajainen, P. Morin, J. Morrison, and G. T. Toussaint. Space-efficient planar convex hull algorithms. *Theoretical Computer Science*, 321(1):25–40, 2004.

8. A. L. Buchsbaum and M. T. Goodrich. Three-dimensional layers of maxima. *Algorithmica*, 39(4):275–286, 2004.

9. E. Y. Chen and T. M.-Y. Chan. A space-efficient algorithm for line segment intersection. In *Proceedings of the 15th Canadian Conference on Computational Geometry*, pages 68–71, 2003.

10. H. K. Dai and X. W. Zhang. Improved linear expected-time algorithms for computing maxima. In *Proceedings of the Latin American Theoretical Informatics Symposium*, Lecture Notes in Computer Science 2976, pages 181–192. Springer, 2004.

11. R. W Floyd. Algorithm 245: Treesort. *Communications of the ACM*, 7(12):701, 1964.

12. V. Geffert, J. Katajainen, and T. Pasanen. Asymptotically efficient in-place merging. *Theoretical Computer Science*, 237(1–2):159–181, 2000.

13. S. Kapoor. Dynamic maintenance of maxima of 2-D point sets. *SIAM Journal on Computing*, 29(6):1858–1877, 2000.

14. J. Katajainen and T. Pasanen. Stable minimum space partitioning in linear time. *BIT*, 32:580–585, 1992.

15. D. Kossmann, F. Ramsak, and S. Rost. Shooting stars in the sky: An online algorithm for skyline queries. In *Proceedings of the 28th International Conference on Very Large Data Bases*, pages 275–286. 2002.

16. H. T. Kung, F. Luccio, and F. P. Preparata. On finding the maxima of a set of vectors. *Journal of the ACM*, 22(4):469–476, 1975.

17. J. Matoušek. Computing dominances in E^n. *Information Processing Letters*, 38(5):277–278, 1991.

18. J. I. Munro. An implicit data structure supporting insertion, deletion, and search in $O(\log^2 n)$ time. *Journal of Computer and System Sciences*, 33(1):66–74, 1986.

19. D. Papadias, Y. Tao, G. Fu, and B. Seeger. Progressive skyline computation in database systems. *ACM Transactions on Database Systems*, 30(1):41–82, 2005.

20. F. P. Preparata and M. I. Shamos. *Computational Geometry. An Introduction*. Springer, 1988.

21. J. S. Salowe and W. L. Steiger. Stable unmerging in linear time and constant space. *Information Processing Letters*, 25(3):285–294, 1987.

22. K.-L. Tan, P.-K. Eng, and B. C. Ooi. Efficient progressive skyline computation. In *Proceedings of the 27th International Conference on Very Large Data Bases*, pages 301–310. 2001.

23. J. Vahrenhold. Line-segment intersection made in-place. In *Proceedings of the 9th International Workshop on Algorithms and Data Structures*, Lecture Notes in Computer Science 3608, pages 146–157. Springer, 2005.

Largest and Smallest Tours and Convex Hulls for Imprecise Points*
(Extended Abstract)

Maarten Löffler and Marc van Kreveld

Institute of Information and Computing Sciences
Utrecht University, the Netherlands
{mloffler, marc}@cs.uu.nl

Abstract. Assume that a set of imprecise points is given, where each point is specified by a region in which the point may lie. We study the problem of computing the smallest and largest possible tours and convex hulls, measured by length, and in the latter case also by area. Generally we assume the imprecision region to be a square, but we discuss the case where it is a segment or circle as well. We give polynomial time algorithms for several variants of this problem, ranging in running time from $O(n)$ to $O(n^{13})$, and prove NP-hardness for some geometric problems on imprecise points.

1   Introduction

In computational geometry, many fundamental problems take a point set as input, on which some computation is done, such as the convex hull, the Voronoi diagram, or a traveling sales route. These problems have been studied for decades. The vast majority of research assumes the locations of the input points to be known exactly. In practice, however, this is often not the case. Coordinates of the points may have been obtained from the real world, using equipment that has some error interval, or they may have been stored as floating points with a limited number of decimals. In real applications, it is important to be able to deal with such imprecise points.

When considering imprecise points, various interesting questions arise. Sometimes it is sufficient to know just a possible solution, which can be achieved by just applying existing algorithms to some point set that is possibly the true point set. More information about the outcome can be obtained by computing a probability distribution over all possibilities, for example using Monte Carlo methods and a probability distribution over the input points. In many applications it is also important to know concrete lower and upper bounds on some measure on the outcome, given concrete bounds on the input: every point is known to be somewhere in a prescribed region.

---

* This research was partially supported by the Netherlands Organisation for Scientific Research (NWO) through the BRICKS project GADGET and through the project GOGO.

L. Arge and R. Freivalds (Eds.): SWAT 2006, LNCS 4059, pp. 375–387, 2006.

Related Work. A lot of research about imprecision in computational geometry is directed at computational imprecision rather than data imprecision. Regarding data imprecision, there is a fair amount of work that uses stochastic or fuzzy models of imprecision. Alternatively, an exact model of imprecision can be used.

Nagai and Tokura [15] compute the union and intersection of all possible convex hulls to obtain bounds on any possible solution. As imprecision regions they use circles and convex polygons, and they give an $O(n \log n)$ time algorithm.

Espilon Geometry is a framework for robust computations on imprecise points. Guibas *et al.* [11] define the notion of *strongly convex* polygons: polygons that are certain to remain convex even if the input points are perturbed within a disc of radius ε. A related concept is that of *tolerance* [1]; see also [12] and [2].

Boissonnat and Lazard [4] study the problem of finding the shortest convex hull of bounded curvature that contains a set of points, and they show that this is equivalent to finding the shortest convex hull of a set of imprecise points modeled as circles that have the specified curvature. They give a polynomial time approximation algorithm.

Goodrich and Snoeyink [10] study a problem where they are given a set of parallel line segments, and must choose a point on each segment such that the resulting point set is in convex position. Given a sequence of k polygons with a total of n vertices, Dror *et al.* [7] study the problem of finding a tour that touches all of them in order that is as short as possible. Higher dimensions are considered in [17].

Fiala *et al.* [9] consider the problem of finding distant representatives in a collection of subsets of a given space. Translated to our setting, they prove that maximizing the smallest distance in a set of n imprecise points, modeled as circles or squares, is NP-hard. Finally, we mention de Berg *et al.* [6] for a problem with data imprecision motivated from computational metrology, Cai and Keil [5] for visibility in an imprecise simple polygon, Sellen *et al.* [18] for precision sensitivity, and Yap [19] for a survey on robustness, which deals with computational imprecision rather than data imprecision.

Problem Definition. All in all there has been little structured research into concrete bounds on the possible outcomes of geometric problems in the presence of data imprecision. When placing a traditional problem that computes some structure on a set of points in this context, two important questions arise:

The first question is what we are given. We model imprecise points by requiring the points to be inside some fixed region, without any assumption on where exactly in their regions the points are, but with absolute certainty that they are not outside their regions. The question then arises what shape these regions should be given. Some natural choices are the square and circular region. The square model for example occurs when points have been stored as floating point numbers, where both the x and y coordinates have an independent uncertainty interval, or with raster to vector conversion. The circular model occurs when the point coordinates have been determined by a scanner or by GPS, for example. Another question is what kind of restrictions we impose on those regions. For example, all points can have the same kind of shape, but are they all of the same size? Do they have the same orientation? Are they allowed to overlap?

The second question is what we actually want to know. Geometric problems usually output some complex structure, not just a number, so a measure on this structure is needed. For example, the convex hull of a set of points can be measured by area or perimeter, or maybe even other measures in some applications. Once a measure has been established, the question is whether you want an upper or a lower bound, or both, on it.

Table 1. Results

goal	measure	model	restrictions	running time
largest	area	line segments	parallel	$O(n^3)$
largest	area	squares	non-intersecting	$O(n^7)$
largest	area	squares	non-intersecting, equal size	$O(n^3)$
largest	area	squares	equal size	$O(n^5)$
largest	perimeter	line segments	parallel	$O(n^5)$
largest	perimeter	squares	non-intersecting	$O(n^{10})$
largest	perimeter	squares	equal size	$O(n^{13})$
smallest	area	line segments	parallel	$O(n \log n)$
smallest	area	squares	-	$O(n^2)$
smallest	perimeter	line segments	parallel	$O(n \log n)$
smallest	perimeter	squares	-	$O(n \log n)$

Results. All these questions together lead to a large class of problems that are all closely related to each other. This paper aims to find out how exactly they are related, which variants are easy and which are hard, and to provide algorithms for the problems that can be solved in polynomial time. Since this type of problem has hardly been studied, we consider the classical planar convex hull problem.

We studied various variants of this problem, and our results are summarized in Table 1. These results are treated in detail in Sections 3, 4 and 5. First, in the next section, some related issues are discussed.

2   Some Results on Spanning Trees and Tours

In this section we briefly discuss the impact of imprecision on another classical geometric problem, the Minimum Spanning Tree. Then we discuss our results on tours. Due to space limitations, we only give the results and very globally, the ideas needed to obtain them. Details can be found in the full paper [14].

Minimum Spanning Tree. To get an idea of how imprecision affects the complexity of geometric problems, consider the Minimum Spanning Tree (MST) problem in an imprecise context. In this case, we have a collection of imprecise points, and we want to determine the MST of, for example, minimal length. This

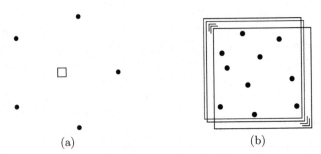

Fig. 1. (a) It is algebraically difficult to find the minimal MST. (b) It is combinatorially difficult to find the minimal MST.

means that we want to choose the points in such a way that the MST of the resulting point set is as small as possible. This problem is both algebraically and combinatorially hard. In Figure 1(a), there are five fixed points and one imprecise one (in the square model, but it could also be a circle). No matter where the point is chosen in this square, the MST of the resulting set will connect all of the fixed points to the imprecise one. Thus the problem reduces to minimizing the sum of the distances from the imprecise point to the fixed points. This is algebraically a hard problem [3]. Furthermore, we can prove NP-hardness of smallest MST by reduction from the Steiner Minimal Tree problem. Given a set of n fixed points P in the plane, we can compute its Steiner Minimal Tree using a solution to the imprecise MST problem as follows. Take P as precise points, and add a set P' of $n-2$ imprecise points whose regions are squares or circles that contain P, see Figure 1(b). The shortest MST of $P \cup P'$ is the Steiner Minimal Tree of P.

Longest Tour. We consider the problem of computing the longest tour that visits a sequence of n axis-parallel squares in a given order. The tour may have self-intersections, see Figure 2(a). We can prove that every vertex of the tour will be at a corner of a square. Given an arbitrary starting square and some vertex v of it, the longest tour up to some vertex w of the i-th square consists of a longest tour to one of the four vertices of the $(i-1)$-st square, and one more segment to w. Hence, the longest tour can be constructed incrementally in $O(1)$ time for each next square. We obtain:

Theorem 1. *Given an ordered set of n arbitrarily sized, axis-aligned squares, the problem of choosing a point in each square such that the perimeter of the resulting polygon is as long as possible can be solved in $O(n)$ time.*

Shortest Tour. Next we study the problem of computing the shortest tour that visits a sequence of n axis-parallel squares in a given order. In this case, vertices of the optimal tour can also be on edges of squares, see Figure 2(b). We can show that the shortest tour can be seen as a combination of two shortest one-dimensional tours, one in the x-projection and one in the y-projection. Therefore we know where the shortest tour changes direction from top to bottom or vice versa, and left to right or vice versa. The shortest tour also satisfies the principle

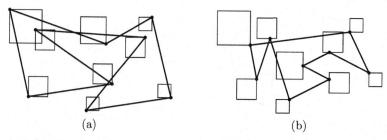

Fig. 2. (a) The longest perimeter solution. (b) The shortest perimeter solution.

of reflection, and therefore we can transform the shortest tour problem to a geodesic shortest path problem in a simple polygon (ignoring some details and complications that are handled in the full paper). We obtain:

Theorem 2. *Given an ordered set of n arbitrarily sized, axis-aligned squares, the problem of choosing a point in each square such that the perimeter of the resulting polygon is as short as possible can be solved in $O(n)$ time.*

Largest or Smallest Area Simple Tour. If we require that the resulting tour has no self-intersections, that is, it is a simple polygon, then we can also minimize or maximize the enclosed area. We can show that this problem is NP-hard. The reduction from planar 3-SAT is in the full paper. It is also NP-hard to determine the longest simple tour, but the proof does not extend to the shortest simple tour. We have:

Theorem 3. *Given an ordered set of n arbitrarily oriented line segments, the problem choosing a point on each segment such that the area of the resulting polygon is as large as possible is NP-hard. The same problem for smallest area and for largest perimeter is also NP-hard.*

3   Largest Convex Hull

We now present our results on the imprecise convex hull problem. This section deals with computing the largest possible convex hull, the smallest convex hull is treated in the next section. We first use the line segment model, where every point can be anywhere on a line segment. This problem does not have much practical use, but it will be extended to the square model later.

Line Segments. The problem we discuss in this section is the following:

Problem 1. *Given a set of parallel line segments, choose a point on each line segment such that the area of the convex hull of the resulting point set is as large as possible (see Figure 3(a)).*

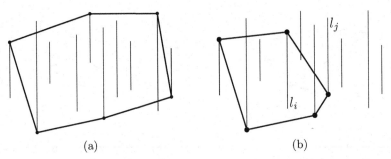

Fig. 3. (a) The largest convex hull for a set of line segments. (b) The polygon p_{ij}.

Observations. First we will show that we can ignore the interiors of the segments in this problem, that is, we only have to consider the endpoints.

Lemma 1. *There is an optimal solution to Problem 1 such that all points are chosen at endpoints of the line segments.*

Algorithm. Let $L = \{l_1, l_2, \ldots, l_n\}$ be a set of n line segments, where l_i lies to the left of l_j if $i < j$. Let l_i^+ denote the upper endpoint of l_i, and l_i^- denote the lower endpoint of l_i. We use a dynamic programming algorithm that runs in $O(n^3)$ time and $O(n^2)$ space. The key element of this algorithm is a polygon which is defined for each pair of line segments. For $i \neq j$, we consider the polygon that starts at l_i^+ and ends at l_j^-, and optimally solves the subproblem to the left of these points, that is, contains only vertices l_k^+ with $k < i$ or l_k^- with $k < j$, but not both for the same k, such that the area of the polygon is maximal, see Figure 3(b). Note that p_{ij} will be convex.

Now, we will show how to compute all p_{ij} using dynamic programming. The solution to the original problem will be either of the form p_{kn} or p_{nk} for some $0 < k < n$, and can thus be computed in linear time once all p_{ij} are known.

When $1 < i < j$, then we can write $p_{ij} = \max_{k<j} \left(p_{ik} + \triangle l_i^+ l_j^- l_k^- \right)$. Of course we maximize over the area of the polygons. In words, we choose one of the lower points to the left of l_j, and add the new point l_j^- to the polygon p_{ik} that optimally solves everything to the left of the chosen point l_k^-. When $1 < j < i$ the expression is symmetric, and $i = 1$ or $j = 1$ is a similar but simpler case. The algorithm runs in $O(n^3)$ time and requires $O(n^2)$ space. We do not need to worry about convexity, because a non-convex solution can never be optimal.

Theorem 4. *Given a set of n arbitrarily sized, parallel line segments, the problem of choosing a point on each segment such that the area of the convex hull of the resulting point set is as large as possible can be solved in $O(n^3)$ time.*

Squares. The problem we discuss in this section is the following:

Problem 2. *Given a set of axis-aligned squares, choose a point in each square such that the area of the convex hull of the resulting point set is as large as possible (see Figure 4(a)).*

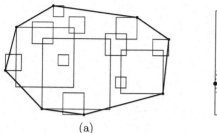
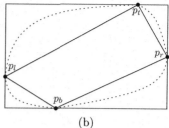

(a)   (b)

Fig. 4. (a) The largest area convex hull for a set of squares. (b) The four extreme points.

Observations. Once again we observe that the points will not have to be chosen in the interior of the squares. In fact we only have to take the corners of the squares into account.

Lemma 2. *There is an optimal solution where all points lie at a corner of their square.*

First we define the four *extreme* points of the convex hull we are trying to compute as the leftmost, topmost, rightmost and bottommost points. These points divide the hull into four chains that connect them. The extreme points and the triangles that surround the four chains are shown in Figure 4(b).

Lemma 3. *All vertices on the top left chain are top left corners of their squares, and similar for the other chains.*

In general it is not easy to find the extreme points. For example, it could be that none of the extreme points in the optimal solution is in one of the extreme squares in the input, see for example Figure 5(a). Here the topmost and bottommost squares are the large ones, and the leftmost and rightmost squares are the medium ones. However, in the optimal solution the extreme points will all be corners of the small squares.

Algorithm for Non-overlapping Squares. When we restrict the problem to non-overlapping squares, we can solve this problem in $O(n^7)$ time. The idea behind the solution is to divide the squares into groups of squares of which we know that only two of their corners are feasible for an optimal solution, and then to reuse the algorithm for Problem 1 on these groups. When the four extreme points are known, we can use this information to solve the problem in $O(n^3)$ time. However, how to find those points still remains a difficult problem, so we try all possible combinations, hence the total of $O(n^7)$.

We call a corner of a square *candidate* if it is in the correct triangle to possibly be part a chain, so for example the top left corner is candidate if it is in the top left triangle, see Figure 4(b). If the squares do not overlap, there can be only two squares that have more than two candidate corners. We ignore these squares (we

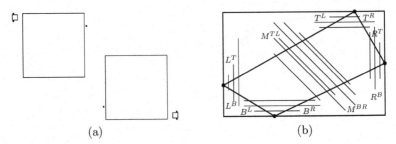

Fig. 5. (a) The four extreme points need not be in the extreme squares. (b) The squares can be divided into five groups of parallel line segments.

just try all possibilities), and note that the rest of the squares all have at most two candidate corners, and can therefore be reduced to line segments. Further note that there are only a limited number of orientations, and those of the same kind are adjacent, as in Figure 5(b). There are six possible kinds of line segments, of which only five may appear at the same time, which implies that we can divide the segments into five groups. The figure is schematic since the segments cannot be extended to non-overlapping squares, but it would require squares of very different sizes to obtain a linear number in each group.

We will now solve the situation of Figure 5(b) in $O(n^3)$ time. Note that any convex hull of a choice of points in this situation must follow these sets of endpoints in the correct order. That is, it starts at the left extreme point, then goes to a number of points of L^B, then to a number of points of B^L, then to the bottom extreme point, and so on. It cannot, for example, go to a point of L^B, then to a point of B^L, and then back to a point of L^B.

The idea of the algorithm is to compute, for each pair of endpoints, the optimal solution connecting them via the lower left side. This can be done by reusing the algorithm for the parallel line segment problem, and distinguishing cases for in which group the two points are. The details can be found in [13].

Theorem 5. *Given a set of n arbitrarily sized, non-overlapping, axis-aligned squares, the problem of choosing a point in each square such that the area of the convex hull of the resulting point set is as large as possible can be solved in $O(n^7)$ time.*

Unit Size Squares. The extra $O(n^4)$ that comes from the fact that it is hard to determine the extreme points, relies on situations where the size of the squares differs greatly, such as in Figure 5(a). When the squares have equal size, we show that there are only constantly many squares that can give the extreme points, thus reducing the running time of the above algorithm to $O(n^3)$.

Lemma 4. *In the largest area convex hull problem for axis-aligned unit squares, an extreme square in the input set gives one of the extreme points of the optimal solution.*

As a consequence of this lemma, the largest convex hull problem for non-overlapping axis-aligned unit squares can be solved in $O(n^3)$ time, since now there are only a constant number of possibilities for the extreme points.

Theorem 6. *Given a set of n equal size, non-overlapping, axis-aligned squares, the problem of choosing a point in each square such that the area of the convex hull of the resulting point set is as large as possible can be solved in $O(n^3)$ time.*

For overlapping squares, the problem remains open. However, for overlapping squares of equal size, we can solve the problem in $O(n^5)$ time, see [13].

4 Smallest Convex Hull

In this section we will investigate the problem of finding the smallest area convex hull of a set of imprecise points. As in the previous section we will first look into the line segment model, and then move on to squares.

Line Segments. The problem we discuss in this section is the following:

Problem 3. *Given a set of parallel line segments, choose a point on each line segment such that the area of the convex hull of the resulting point set is as small as possible.*

Lemma 5. *In the optimal solution, if a line segment defines a vertex of the convex hull, and there are other vertices on the hull strictly on both sides of the supporting line of this segment, then the point on this segment must be chosen at one of the endpoints.*

We denote the leftmost segment by s_l and the rightmost segment by s_r. We define two chains, the *top chain* c_t and the *bottom chain* c_b of the set of segments. The top chain is a polyline connecting the lower endpoint of s_l to the lower endpoint of s_r, and is defined as the upper half of the convex hull of the set of all lower endpoints of the input segments. Symmetrically, the bottom chain is the lower half of the convex hull of the set of all upper endpoints of the input segments, see Figure 6(a). If the top and bottom chains do not intersect, there is a zero

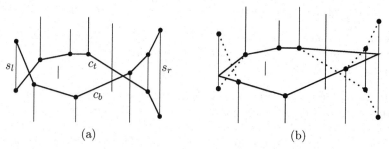

(a)   (b)

Fig. 6. (a) The top chain c_t and bottom chain c_b. (b) The optimal solution.

area solution that can be found in linear time [8]. Therefore, we assume next that they intersect.

For a point p on s_l, there is a *tangent point* $a_l(p)$ on the top chain such that the line through p and $a_l(p)$ does not go through the region below the top chain. When there are more than one such points we choose the one that lies most to the right. Similarly, we define $b_l(p)$ as the tangent point on the bottom chain, and for q on s_r we define two tangent points $a_r(q)$ and $b_r(q)$ on the top and bottom chains. All those tangent points are vertices of the chains.

Lemma 6. *If the points p on s_l and q on s_r are known, the optimal solution is the polygon that consists of p, $a_l(p)$, the piece of the top chain between $a_l(p)$ and $a_r(q)$, $a_r(q)$, q, $b_r(q)$, the piece of the bottom chain between $b_r(q)$ and $b_l(p)$, $b_l(p)$, and back to p, provided that this polygon is convex. If it is not, then p and q will be connected by a straight line above the top chain or below the bottom chain (see Figure 6(b)).*

Algorithm. We will use these observations to construct an efficient algorithm. First we note that the two chains can be computed in $O(n \log n)$ time using conventional convex hull algorithms, and then we show that we can find the optimal solution using the chains in $O(n)$ time, yielding a total of $O(n \log n)$ time.

To find the location of the points p on s_l and q on s_r, we use the fact that they can be found independent of each other.

Lemma 7. *The individual optimal locations for p and q, minimizing the area of p and q respectively add to the intersection of the chains, are the same as the location of p and q in the optimal solution.*

The important point is that, in the optimal solution, p and q will never be connected directly to each other, but always via the chain. The individual solutions can be computed in linear time, after the chains are known. The computation of the chains takes $O(n \log n)$ time.

Theorem 7. *Given a set of n arbitrarily sized, parallel line segments, the problem of choosing a point on each segment such that the area of the convex hull of the resulting point set is as small as possible can be solved in $O(n \log n)$ time.*

Squares. The problem we discuss in this section is the following:

Problem 4. *Given a set of axis-aligned squares, choose a point in each square such that the area of the convex hull of the resulting point set is as small as possible (see Figure 7(a)).*

Lemma 8. *In the optimal solution, only the leftmost, rightmost, topmost, and bottommost vertices of the hull need not be corners of their squares.*

The situation is similar to the line segment case. There are now four *extreme squares* S_l, S_r, S_t and S_b, and for these four squares, the points must lie on the inner edge. We call the points p_l, p_r, p_t and p_b. We now have four chains of corners that could be included in the convex hull, see Figure 7(b). The optimal solution for fixed p_l, p_r, p_t and p_b connects these points to their tangent points on the chains or directly to each other if the result would not be convex.

The critrical difference between the line segment case and the square case, is that the locations of the four extreme points are no longer independent. It can really happen that in the optimal solution two or more of the extreme points are connected by straight line segments, rather than via the chains. This means we need a different approach to solve the problem, and is the reason why this variant cannot be easily solved in $O(n \log n)$. We describe a case distinguishing algorithm that runs in $O(n^2)$ time in [13].

Theorem 8. *Given a set of n arbitrarily sized, possibly overlapping, axis-aligned squares, the problem of choosing a point in each square such that the area of the convex hull of the resulting point set is as small as possible can be solved in $O(n^2)$ time.*

5   Perimeter Versus Area

Until now we have only considered area of the convex hull as the measure to maximize or minimize, but there are other measures that can be used, such as the perimeter. In this section we will briefly consider the relevant differences between the two measures.

One important observation concerns the way the size of a polygon changes when only one point is moving, while the rest remains fixed. The area of the polygon will be a linear function of the moving point, while the perimeter is a hyperbolic function with a minimum. In the case of convex hulls, this only applies as long as the combinatorial structure of the hull does not change. Secondly, note that when we want to maximize the area of a polygon, convexity is automatically achieved. When we want to maximize the perimeter, however, convexity has to be explicitly taken care of. When looking for minimal size, this works the other way around. A minimal perimeter polygon will automatically be convex, while a minimal area polygon is generally not.

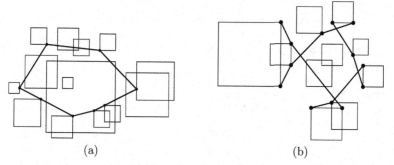

(a)   (b)

Fig. 7. (a) The smallest convex hull for a set of squares. (b) The top left, bottom left, top right, and bottom right chains.

We can adjust all of the above algorithms to the perimeter measure in a more or less straightforward fashion. The time bounds for the largest convex hull indeed become worse, $O(n^3)$ for line segments becomes $O(n^5)$, and $O(n^7)$ for squares becomes $O(n^{10})$. On the other hand, the time bounds for the smallest convex hull become better; all problems considered can be solved in only $O(n \log n)$ time. The details of the changed algorithms can be found in [13].

6   Conclusions

We studied the problem of computing the largest or smallest convex hull of a set of imprecise points; our results are in Table 1.   The problem of finding the smallest convex hull seems to be easier than finding the largest convex hull: the running times are better, and there are fewer restrictions. It also seems that for the largest convex hull the area is easier to maximize than the perimeter, while for the smallest convex hull the perimeter is easier to minimize than the area.

Many problems are open, and there are various directions of research to be pursued. Most notably, what is the status of the problem of finding the largest convex hull when the regions are allowed to intersect? Also, what results can be obtained for the circle model? For the problems that do not have an efficient solution, the study of approximation algorithms is interesting. Thirdly, for many other problems in computational geometry, imprecision in the data and the bounds on the effect on the outcome of an algorithm should be studied.

References

1. M. ABELLANAS, F. HURTADO, AND P. A. RAMOS, Structural tolerance and Delaunay triangulation, *Inf. Proc. Lett.* 71:221–227, 1999.
2. D. BANDYOPADHYAY AND J. SNOEYINK, Almost-Delaunay simplices: nearest neighbour relations for imprecise points, *Proc. 15th ACM-SIAM Sympos. on Discr. Algorithms*, pages 410–419, 2004.
3. C. BAJAJ, The algebraic degree of geometric optimization problems, *Discr. Comput. Geom.*, 3:177–191, 1988.
4. J.-D. BOISSONNAT AND S. LAZARD, Convex hulls of bounded curvature, *Proc. 8th Canad. Conf. on Comput. Geom.*, pages 14–19, 1996.
5. L. CAI AND J. M. KEIL, Computing visibility information in an inaccurate simple polygon, *Internat. J. Comput. Geom. Appl.* 7:515–538, 1997.
6. M. DE BERG, H. MEIJER, M.H. OVERMARS, AND G.T. WILFONG, Computing the angularity tolerance, *Int. J. Comput. Geometry Appl.* 8:467–482, 1998.
7. M. DROR, A. EFRAT, A. LUBIW, AND J. S. B. MITCHELL, Touring a Sequence of Polygons, *Proc. 35th ACM Sympos. Theory Comput.*, 2003
8. H. EDELSBRUNNER, Finding transversals for sets of simple geometric figures, *Theor. Comp. Science* 35:55–69, 1985.
9. J. FIALA, J. KRATOCHVIL, AND A. PROSKUROWSKI, Systems of distant representatives, *Discrete Applied Mathematics* 145:306–316, 2005.
10. M. T. GOODRICH AND J. SNOEYINK, Stabbing parallel segments with a convex polygon, *Computer Vision, Graphics, and Image Processing* 49:152–170, 1990.

11. L. GUIBAS, D. SALESIN, AND J. STOLFI, Constructing strongly convex approximate hulls with inaccurate primitives, *Algorithmica* 9:534–560, 1993.
12. A. A. KHANBAN AND A. EDALAT, Computing Delaunay triangulation with imprecise input data, *Proc. 15th Canad. Conf. on Comput. Geom.*, pages 94–97, 2003.
13. M. LÖFFLER AND M. VAN KREVELD, *Largest and Smallest Convex Hulls for Imprecise Points*, Technical Report UU-CS-2006-019, Department of Computing and Information Science, Utrecht University, 2006.
14. M. LÖFFLER AND M. VAN KREVELD, *Largest and Smallest Tours for Imprecise Points*, manuscript, 2006.
15. T. NAGAI AND N. TOKURA, Tight error bounds of geometric problems on convex objects with imprecise coordinates, *Jap. Conf. Discr. Comp. Geom.*, 252–263, 2000.
16. Y. OSTROVSKY-BERMAN AND L. JOSKOWICZ, Uncertainty Envelopes, *Abstracts 21st European Workshop on Comput. Geom.*, pages 175–178, 2005.
17. V. POLISHCHUK, J. S. B. MITCHELL, Touring Convex Bodies - A Conic Programming Solution, *Proc. 17th Canad. Conf. on Comput. Geom.*, pages 290–293, 2005.
18. J. SELLEN, J. CHOI, AND C.-K. YAP, Precision-sensitive Euclidean shortest paths in 3-space, *SIAM J. Comput.* 29:1577–1595, 2000.
19. C.-K. YAP, Robust geometric computation. In J. E. Goodman and J. O'Rourke, editors. *Handbook of Discrete and Computational Geometry*, Chapman & Hall/CRC, 2004. Chapter 41, pages 927–952.

On Spanners of Geometric Graphs[*]

Joachim Gudmundsson[1] and Michiel Smid[2]

[1] National ICT Australia Ltd., Sydney, Australia
[2] School of Computer Science, Carleton University, Ottawa, Ontario, Canada

Abstract. Given a connected geometric graph G, we consider the problem of constructing a t-spanner of G having the minimum number of edges. We prove that for every t with $1 < t < \frac{1}{4}\log n$, there exists a connected geometric graph G with n vertices, such that every t-spanner of G contains $\Omega(n^{1+1/t})$ edges. This bound almost matches the known upper bound, which states that every connected weighted graph with n vertices contains a t-spanner with $O(tn^{1+2/(t+1)})$ edges. We also prove that the problem of deciding whether a given geometric graph contains a t-spanner with at most K edges is **NP**-hard. Previously, this **NP**-hardness result was only known for non-geometric graphs.

1 Introduction

Let $G = (V, E)$ be a connected undirected graph in which every edge e has a positive weight $\omega(e)$. We define the weight of a path in G to be the sum of the weights of the edges on this path. For any two vertices u and v of G, we denote the weight of a shortest path in G between u and v by $\delta_G(u, v)$. For a given subgraph $G' = (V, E')$ of G (hence, $E' \subseteq E$), we define the *dilation of G' with respect to G* to be the maximum value $\delta_{G'}(u, v)/\delta_G(u, v)$, over all $u, v \in V$ with $u \neq v$. For a given real number $t > 1$, we say that G' is a t-spanner of G, if the dilation of G' with respect to G is at most t.

The following problem has been studied extensively in the literature: Given a connected weighted graph G, and given a real number $t > 1$, does G contain a t-spanner having "few" edges?

Althöfer *et al.* [1] showed that, for every connected weighted graph G with n vertices, and for every real number $t > 3$, there exists a t-spanner of G that consists of $O(n^{1+2/(t-1)})$ edges. This result was improved by Baswana and Sen [2] and Roditty *et al.* [16], who showed that for every integer $t \geq 3$, any connected weighted graph with n vertices contains a t-spanner with $O(tn^{1+2/(t+1)})$ edges.

The following lower bound was proved by Althöfer *et al.* [1]: For every real number $t > 1$, there exists a connected weighted graph G with n vertices, such that every t-spanner of G contains $\Omega(n^{1+4/(3(t+2))})$ edges.

[*] JG was funded by the Australian Government's Backing Australia's Ability initiative, in part through the Australian Research Council. MS was supported by the Natural Sciences and Engineering Research Council of Canada (NSERC). Part of this work was done while MS visited NICTA.

L. Arge and R. Freivalds (Eds.): SWAT 2006, LNCS 4059, pp. 388–399, 2006.
© Springer-Verlag Berlin Heidelberg 2006

We remark that the corresponding problem for unweighted graphs has been considered before by Peleg and Schäffer [15]; see also the book by Peleg [14].

In this paper, we consider the above spanner problem for *geometric graphs*. A graph $G = (S, E)$ is called a geometric graph, if the vertex set S of G is a set of points in $\mathbb{R}^d$, and the weight of every edge $\{u, v\}$ in E is equal to the Euclidean distance $|uv|$ between u and v.

Since the upper bounds in [1, 2, 16] mentioned above are valid for arbitrary connected weighted graphs, they also hold for geometric graphs. The graph constructed in the proof of the lower bound in [1], however, is not a geometric graph. The difficulty is in mapping the vertices to points in the plane, such that the weight of each edge $\{u, v\}$ is exactly equal to the Euclidean distance $|uv|$. In Section 2, we prove the following theorem, which states that the lower bound of Althöfer *et al.* can almost be achieved by geometric graphs:

Theorem 1. *For every sufficiently large integer n, and for every real number t with $1 < t < \frac{1}{4} \log n$, there exists a connected geometric graph G with $2n$ vertices, such that every t-spanner of G contains $\Omega(n^{1+1/t})$ edges.*

The proof of Theorem 1 uses an $n \times n$ connected bipartite graph with $\Omega(kn)$ edges and whose girth is $\Omega(\log n / \log k)$. The probabilistic method has been used to prove the existence of a dense (not necessarily bipartite) graph with high girth; see, for example, Mitzenmacher and Upfal [13]. This existence proof can easily be extended to bipartite graphs. Lazebnik and Ustimenko [12] used algebraic methods to give an explicit construction of a dense bipartite graph with high girth. Chandran [7] used a purely combinatorial approach to construct such a graph, which is, however, not bipartite. In Section 3, we modify Chandran's construction and obtain a simple deterministic algorithm that produces the bipartite graph that is needed to prove Theorem 1.

The spanner problem naturally leads to the following optimization problem: Given a connected weighted graph G with n vertices, and given a real number $t > 1$, compute a t-spanner of G, having the minimum number of edges.

Cai [4] proved that, for any fixed $t \geq 2$, this optimization problem is **NP**-hard for unweighted graphs. Cai and Corneil [5] considered the problem for weighted graphs, and showed it to be **NP**-hard for any fixed $t > 1$. The problem has also been shown to be **NP**-hard for restricted classes of graphs, such as planar graphs (see Brandes and Handke [3]), chordal graphs, and bipartite graphs (see Venkatesan et al. [20]). However, the complexity of the optimization problem has not been considered for geometric graphs. In Section 4, we prove this version of the problem to be **NP**-hard as well. Our proof consists of generalizing the approach of Cai [4]: We show that any Boolean formula φ in 3-conjunctive normal form can be transformed, in polynomial time, to a geometric graph G and an integer K, such that φ is satisfiable if and only if G contains a t-spanner with at most K edges. Again, the main difficulty is in defining G in such a way that its vertices are points in the plane and the weight of each edge $\{u, v\}$ is exactly equal to the Euclidean distance $|uv|$. Recall that the transformation from φ to the pair (G, K) has to be done on a Turing machine. Since Turing machines can only deal with finite strings, we take care that the vertices of G are points in the plane having *rational* coordinates. Thus, the optimization problem for geometric graphs is formally defined as follows, for any fixed rational number $t > 1$:

Problem GEOMMINSPANNER(t):
   Instance: A connected geometric graph $G = (S, E)$, where $S \subseteq \mathbb{Q}^2$, and a positive integer K.
   Question: Does G contain a t-spanner with at most K edges?

In Section 4, we prove the following result:

Theorem 2. *For any fixed rational number $t > 1$, GEOMMINSPANNER(t) is NP-hard.*

1.1 Related Work

The problem of constructing geometric spanners with few edges has been considered for point sets. A graph G', whose vertex set is a set S of points in $\mathbb{R}^d$, is said to be a t-spanner for S, if G' is a t-spanner of the complete geometric graph on S. Salowe [17], Vaidya [19], and Callahan and Kosaraju [6] have shown that, for any set S of n points in $\mathbb{R}^d$, and for any real constant $t > 1$, a t-spanner for S with $O(n)$ edges can be computed in $O(n \log n)$ time. See also the survey papers by Eppstein [8], Smid [18], and Gudmundsson and Knauer [9].

Gudmundsson *et al.* [10] (see also [11]) have shown that if S is a set of n points in $\mathbb{R}^d$, $t > 1$ is a real number, and G is a $(1 + \epsilon)$-spanner for S, then G contains a t-spanner with $O(n)$ edges.

Thus, the problem of constructing sparse spanners of geometric graphs G has been considered for the cases when G is the complete geometric graph or when G itself is a spanner of its vertex set. The problem has not been considered for arbitrary geometric graphs G.

2   A Geometric Graph That Contains Only Dense Spanners

In this section, we will prove Theorem 1. Consider a connected (not necessarily geometric) graph G, in which every edge e has a positive weight $\omega(e)$. Recall that the *girth* of G is the minimum number of edges on any cycle in G. We denote by $\omega(C)$ the weight of any cycle C in G. Thus, $\omega(C)$ is equal to the sum of the weights of the edges on C. We define the *weighted girth* g_ω of G to be the minimum value of $\omega(C)/\omega(e)$, where C is any cycle in G and e is an edge of maximum weight on C. The following lemma relates the dilation of every proper subgraph of G to the weighted girth of G.

Lemma 1. *Let $f = \{u, v\}$ be an arbitrary edge of G, and let G' be the graph obtained by deleting f from G. Then the dilation t of G' with respect to G satisfies $t \geq g_\omega - 1$.*

The next lemma shows that any connected bipartite graph with girth g can be transformed to a connected geometric graph whose weighted girth is $\Omega(g)$.

We say that a graph G is an $n \times n$ bipartite graph, if its vertex set can be partitioned into two sets L and R, each having size n, such that every edge of G is between a vertex in L and a vertex in R.

Lemma 2. *Let G be a connected $n \times n$ bipartite graph with m edges and girth g. Then for every real number ϵ with $0 < \epsilon < 1$, there exists a set S of $2n$ points in the plane and a connected geometric graph with vertex set S that consists of m edges and whose weighted girth is at least $(1 - \epsilon)g$.*

Proof. Let ℓ_1 be the vertical line segment with endpoints $(0,0)$ and $(0, \epsilon/2)$, and let ℓ_2 be the vertical line segment with endpoints $(1 - \epsilon, 0)$ and $(1 - \epsilon, \epsilon/2)$. We embed the graph G in the plane, by mapping the vertices of L to a set S_L of n points on ℓ_1, and mapping the vertices of R to a set S_R of n points on ℓ_2. Let $S = S_L \cup S_R$, and let G' denote the embedded geometric graph. The lemma follows from the fact that the length of each edge of G' is in the interval $[1 - \epsilon, 1]$. $\square$

The previous lemmas imply that we can prove Theorem 1, by constructing a dense bipartite graph whose girth is large. The following lemma states that such a graph exists; the proof will be given in Section 3.

Lemma 3. *Let n and k be positive integers with $n \geq 3k + 4$ and $k \geq 2$. There exists a connected $n \times n$ bipartite graph with kn edges, in which the degrees of all vertices are in $\{k - 1, k, k + 1\}$, and whose girth is at least*

$$\frac{\log(3n/8)}{\log(k + 1)} + 1 = \log_k n - O(1).$$

Consider the bipartite graph of Lemma 3, and denote its girth by g. By Lemma 2, we can transform this graph to a geometric graph G, whose weighted girth is at least $(1 - \epsilon)g$. Then, Lemma 1 implies that every proper subgraph of G has dilation at least $(1 - \epsilon)g - 1$. Thus, we obtain the following result.

Lemma 4. *Let n and k be positive integers with $n \geq 3k + 4$ and $k \geq 2$, and let ϵ be a real number with $0 < \epsilon < 1$. There exists a connected geometric graph G with $2n$ vertices and kn edges, such that for every proper subgraph G' of G, the dilation of G' with respect to G is at least*

$$(1 - \epsilon)\frac{\log(3n/8)}{\log(k + 1)} - \epsilon = (1 - \epsilon)\log_k n - O(1).$$

We are now ready to prove Theorem 1. Let n be a sufficiently large integer, and let t be a real number with $1 < t < \frac{1}{4}\log n$. Let $k = (n/4)^{(1-\epsilon)/(t+\epsilon)} - 1$, where $\epsilon = 2t/\log n$. Let G be the geometric graph in Lemma 4. We claim that this graph has the properties stated in Theorem 1. Indeed, let G' be an arbitrary t-spanner of G. Our choice of k implies that

$$t = (1 - \epsilon)\frac{\log(n/4)}{\log(k + 1)} - \epsilon < (1 - \epsilon)\frac{\log(3n/8)}{\log(k + 1)} - \epsilon.$$

Algorithm. BIPARTITEHIGHGIRTH(n, k)
Input: Integers n and k, such that $n \geq 3k + 4$ and $k \geq 2$.
Output: A connected $n \times n$ bipartite graph G with kn edges and girth at least $\log_k n - O(1)$, such that the degree of each vertex is in $\{k - 1, k, k + 1\}$.

let L and R be two disjoint sets, each having size n;
let $V = L \cup R$;
initialize G to be a Hamiltonian cycle in the complete bipartite graph on $L \cup R$;
for $i = 2n + 1$ to kn
do let M be the set of all vertices in V having minimum degree in G;
  let $P = ((M \cap L) \times R) \cup ((M \cap R) \times L)$;
  let T be the set of all ordered pairs (u, v) in P, such that $\{u, v\}$ is not an edge in G and $deg_G(v) \leq \lceil i/n \rceil$;
  let (u, v) be any pair in T, such that $\delta_G(u, v)$ is maximum;
  add the edge $\{u, v\}$ to G
endfor;
return the graph G

Fig. 1. The algorithm that constructs a dense bipartite graph with high girth

However, if G' is a proper subgraph of G, then Lemma 4 states that t must be at least the quantity on the right-hand side. Thus, G' is equal to G and, therefore, the number of edges of G' is equal to

$$kn = \Omega\left(n^{1+(1-\epsilon)/(t+\epsilon)}\right) = \Omega\left(n^{1+(1-2\epsilon)/t}\right) = \Omega\left(n^{1+1/t}\right).$$

3   Constructing a Dense Bipartite Graph with High Girth

In this section, we prove Lemma 3. Our construction is a modification of a construction due to Chandran [7], who proved the same result for general, i.e., non-bipartite, graphs. In this section, $\delta_G(u, v)$ denotes the minimum number of edges on any path in G between u and v. The algorithm is given in Figure 1.

3.1   Analyzing the Size and the Degree

We number the iterations of the for-loop according to the value of the variable i. In other words, the iterations are numbered $2n + 1, 2n + 2, \ldots, kn$. Iteration j denotes the iteration in which the value of the variable i is equal to j. In this section, we will prove the following lemma.

Lemma 5. *Let d be an integer with $2 \leq d \leq k$. At the moment when iteration dn of the for-loop is completed, the following are true:*

1. *The graph G consists of dn edges.*
2. *The degree in G of every vertex of V is in $\{d - 1, d, d + 1\}$.*
3. *Let X and Z be the sets of vertices of V, whose degrees in G are equal to $d - 1$ and $d + 1$, respectively. Then, $|X| = |Z|$.*

Thus, for $d = k$, this lemma implies the claims in Lemma 3 about the number of edges and the degrees of the vertices. The proof of Lemma 5 is by induction on d. The lemma obviously holds for $d = 2$.

We choose an integer d such that $2 \leq d < k$, and assume that Lemma 5 holds for d. We will prove in Lemmas 6–9 below that the lemma then also holds for $d + 1$. To prove this, we consider iterations $dn + 1, dn + 2, \ldots, (d + 1)n$ of the for-loop. We will refer to this sequence of n iterations as the *current batch*. Observe that during the current batch, the value of $\lceil i/n \rceil$ is equal to $d + 1$.

Lemma 6. *At the end of the current batch, all degrees are at most $d + 2$.*

Proof. Let x be an arbitrary vertex in V. Consider any edge $\{u, v\}$, where $v = x$, that is added to G during the current batch, because the algorithm chooses the pair (u, v) in T. Then, prior to the moment this edge is added, $deg_G(v) \leq d+1$. Therefore, the addition of edges of this type cannot lead to a degree of x that is larger than $d + 2$.

Consider any edge $\{u, v\}$, where $u = x$, that is added to G during the current batch, because the algorithm chooses the pair (u, v) in T. Assume that this addition makes the degree of x to be at least $d + 3$. It follows from the algorithm that, prior to the addition of $\{u, v\}$, x has minimum degree in G. Therefore, the degree of v is at least $d + 2$ at that moment. But this implies that the ordered pair (u, v) is not in the set T. □

Lemma 7. *In each iteration of the current batch, exactly one edge is added.*

Proof. By the induction hypothesis, the graph G consists of dn edges at the beginning of the current batch. During this batch, at most n edges are added to G. It follows that, at any moment during the current batch,

$$\sum_{v \in V} deg_G(v) \leq 2(d + 1)n. \tag{1}$$

Consider one iteration of the current batch, and let G' be the graph G at the start of this iteration. Let u be a vertex of V, whose degree in G' is minimum. We may assume without loss of generality that $u \in L$.

We claim that, at the start of this iteration, there exists a vertex v in R, such that $\{u, v\}$ is not an edge in G' and $deg_{G'}(v) \leq d+1$. Assuming this claim is true, it follows from the algorithm that, during this iteration, the set T is non-empty and, therefore, an edge is added to G'.

It remains to prove the claim. Let d' be the degree of u in G', and let $v_1, v_2, \ldots, v_{d'}$ be all vertices of R that are connected to u by an edge of G'. It follows from the induction hypothesis that $\sum_{j=1}^{d'} deg_{G'}(v_j) \geq d'(d-1)$. Moreover, by (1), we have

$$\sum_{v \in R} deg_{G'}(v) = \frac{1}{2} \sum_{v \in V} deg_{G'}(v) \leq (d + 1)n. \tag{2}$$

Assume that the claim does not hold. Then, we have $deg_{G'}(v) \geq d + 2$ for each $v \in R \setminus \{v_1, v_2, \ldots, v_{d'}\}$. It follows that

$$\sum_{v \in R} deg_{G'}(v) \geq d'(d-1) + (n-d')(d+2). \tag{3}$$

By combining (2) and (3), we obtain $d'(d-1) + (n-d')(d+2) \leq (d+1)n$, which can be rewritten as $n \leq 3d'$. By Lemma 6, we have $d' \leq d + 2 \leq k + 1$, which implies that $n \leq 3k + 3$, contradicting our assumption that $n \geq 3k + 4$. □

Lemma 8. *At the end of the current batch, every vertex has degree at least d.*

Proof. Consider the sets X and Z of vertices of V, whose degrees in G, at the beginning of the current batch, are equal to $d - 1$ and $d + 1$, respectively. Since $|X| = |Z|$, we have $|X| \leq n$. In each iteration of the current batch, one edge $\{u, v\}$, where u has minimum degree in the current graph G, is added to G. The induction hypothesis implies that, after this edge has been added, the degree of u is at least d. Therefore, after the first $|X|$ iterations of the current batch, G does not contain any vertex of degree at most $d - 1$. □

Lemma 9. *Let X', Y', and Z' be the sets of vertices of V, whose degrees in G are equal to d, $d + 1$, and $d + 2$, respectively, at the end of the current batch. Then, $|X'| = |Z'|$.*

Proof. By Lemmas 6–8, we have $|X'| + |Y'| + |Z'| = 2n$ and $d|X'| + (d+1)|Y'| + (d+2)|Z'| = 2(d+1)n$. By multiplying the first equation by $d+1$, and subtracting the result from the second equation, the lemma follows. □

3.2   A Lower Bound on the Girth

Consider the graph G that is returned by algorithm BIPARTITEHIGHGIRTH(n, k), and let g be its girth. In this section, we prove the lower bound on g as stated in Lemma 3. Let C be a cycle in G consisting of g edges, and let $\{u, v\}$ be the last edge of C that is added to G. Let j be the integer such that $\{u, v\}$ is added to G during iteration j of the for-loop. We may assume that $j \geq 2n + 1$, because otherwise, $g = 2n$. Let $d = \lceil j/n \rceil$, and let G_j be the graph G at the start of iteration j. Consider the ordered pair (u, v) in T that corresponds to the edge $\{u, v\}$. We observe that $\delta_{G_j}(u, v) \leq g - 1$. We may assume without loss of generality that $u \in L$. Define $B = \{x \in R : \delta_{G_j}(u, x) \geq g\}$. Let x be an arbitrary element in B. Then $\{u, x\}$ is not an edge in G_j and $\delta_{G_j}(u, x) > \delta_{G_j}(u, v)$. Since the edge $\{u, v\}$ is added to G_j in iteration j, it follows that $(u, x) \notin T$ and, therefore, $deg_{G_j}(x) \geq d + 1$. In fact, by Lemma 5, we have $deg_{G_j}(x) = d + 1$. Hence, $B \subseteq \{x \in R : deg_{G_j}(x) = d + 1\}$. Let G' be the graph G at the end of iteration dn, and define $Z_R = \{x \in R : deg_{G'}(x) = d+1\}$. Since $dn \geq j$, and using Lemma 5, we obtain $B \subseteq Z_R$. Define $X_R = \{x \in R : deg_{G'}(x) = d - 1\}$. Then, as in the proof of Lemma 9, it can be shown that $|X_R| = |Z_R|$, implying that $|Z_R| \leq n/2$. Thus, since $B \subseteq Z_R$, we have $|B| \leq n/2$ and, hence, $|R \setminus B| \geq n/2$.

Since $R \setminus B = \{x \in R : \delta_{G_j}(u, x) \le g - 1\}$, and since, by Lemma 5, the degree of every vertex of G_j is at most $d + 1 \le k + 1$, it follows that

$$|R \setminus B| \le (k+1) + (k+1)^3 + (k+1)^5 + \ldots + (k+1)^{g-1} \le \frac{4}{3}(k+1)^{g-1}.$$

By combining the lower and upper bounds on the size of $R \setminus B$, we obtain the lower bound on g as stated in Lemma 3.

4   The NP-Hardness Proof

We now prove Theorem 2, i.e., the decision problem GEOMMINSPANNER(t) is **NP**-hard. Throughout this section, we fix a rational number $t > 1$. Recall that $3SAT$ is the **NP**-complete problem of deciding whether or not any given Boolean formula in 3-conjunctive normal form is satisfiable. To prove Theorem 2, it suffices to design a polynomial-time reduction from $3SAT$ to GEOMMINSPANNER(t). Note that *time* refers to the number of bit operations made in the reduction. In Section 4.2, we present such a reduction. Our approach is to extend Cai's reduction in [4], which shows that constructing a t-spanner with the minimum number of edges in any unweighted graph is **NP**-hard. First, in Section 4.1, we introduce so-called forced paths, which are paths in a geometric graph G that must be in any t-spanner of G.

4.1   Forced Paths

We fix a rational number $t > 1$ and an even integer k, such that $k \ge 4$ and $k \ge t + 1$. Let $\ell > 0$ be a rational number, and let $x = (x_1, x_2)$ and $y = (y_1, y_2)$ be two points in $\mathbb{Q}^2$. Let μ be a rational number, such that $1/|xy| \le \mu \le 1/|xy| + 1/\ell$, and define the rational number λ as $\lambda = \ell\mu/k$. Let v be the point in $\mathbb{Q}^2$ defined as $v = (\lambda(y_2 - x_2), \lambda(x_1 - y_1))$. Observe that the vector from the origin to v is orthogonal to the line segment joining x and y. For $i = 0, 1, \ldots, k/2$, we define the points a_i and b_i in $\mathbb{Q}^2$ as $a_i = x + iv$ and $b_i = y + iv$. Finally, we define P to be the path consisting of the edges

$$\{a_0, a_1\}, \ldots, \{a_{k/2-1}, a_{k/2}\}, \{a_{k/2}, b_{k/2}\}, \{b_{k/2}, b_{k/2-1}\}, \ldots, \{b_1, b_0\}.$$

We will refer to the path P as the *forced path* of x and y (with respect to ℓ), and denote it by $FP(x, y; \ell)$. Lemma 11 explains this terminology.

Lemma 10. *The length $|P|$ of the forced path $P = FP(x, y; \ell)$ satisfies*

$$\ell \le |P| \le \ell + 2|xy|.$$

Lemma 11. *Let G be a connected geometric graph, whose vertices are points in $\mathbb{Q}^2$, and let x and y be two distinct vertices of G that are not connected by an edge, such that $|xy| \le \ell/(t - 1)$. Assume that G contains the forced path $P = FP(x, y; \ell)$. Also, assume that each vertex of $P \setminus \{x, y\}$ has degree two in G. Then, every t-spanner of G contains the path P.*

Proof. Let G' be an arbitrary t-spanner of G. Let $0 \leq i < k/2$, and assume that the edge $\{a_i, a_{i+1}\}$ of P is not an edge in G'. Then, $\delta_{G'}(a_i, a_{i+1}) > |P| - |a_i a_{i+1}| > (k-1)|a_i a_{i+1}|$. Since $k \geq t+1$, it follows that $\delta_{G'}(a_i, a_{i+1}) > t|a_i a_{i+1}|$, contradicting the fact that G' is a t-spanner of G. By a symmetric argument, all edges $\{b_i, b_{i+1}\}$, with $0 \leq i < k/2$, are contained in G'. Assume that the edge $\{a_{k/2}, b_{k/2}\}$ of P is not an edge in G'. Then, $\delta_{G'}(a_{k/2}, b_{k/2}) > |P| = (\ell\mu+1)|xy| \geq (\ell/|xy| + 1)|xy|$. Since $|xy| \leq \ell/(t-1)$, it follows that $\delta_{G'}(a_{k/2}, b_{k/2}) > t|xy| = t|a_{k/2} b_{k/2}|$, which is again a contradiction. □

Lemma 12. *Assume that $\ell > 0$ is a rational constant. Given the points x and y in $\mathbb{Q}^2$, the path $FP(x, y; \ell)$ can be constructed in time that is polynomial in L, where L is the total number of bits in the binary representations of the numerators and denominators of the coordinates of x and y.*

4.2   The Reduction

We are now ready to give the reduction from $3SAT$ to GEOMMINSPANNER(t). Recall that $t > 1$ is a rational number, and k is an even integer, such that $k \geq 4$ and $k \geq t+1$. We define the rational number ℓ as $\ell = 2(t-1)/3$. We consider t, k, and ℓ to be constants. We need the following lemma, which will be used to obtain points on the unit-circle that have rational coordinates and that are close together.

Lemma 13. *Let $\rho = \min(2/3, \ell/4)$, let C be the circle of radius $\rho/2$ centered at the point $(1, 0)$, let i be an integer, such that $i \geq 4/\rho$, and let $Q(i) \in \mathbb{Q}^2$ be the point $Q(i) = ((i^2 - 1)/(i^2 + 1), 2i/(i^2 + 1))$. Then, $Q(i)$ is on the unit-circle and in the interior of C.*

Let φ be a Boolean formula in 3-conjunctive normal form, with variables $x_1, x_2, \ldots, x_N$, consisting of M clauses $c_1, c_2, \ldots, c_M$. Thus, for each j with $1 \leq j \leq M$, the clause c_j is of the form $c_j = y_1 \vee y_2 \vee y_3$, where each of y_1, y_2, and y_3 is either a variable or the negation of a variable.

Our task is to map φ to an instance of GEOMMINSPANNER(t), i.e., a connected geometric graph G, whose vertex set is a set of points in $\mathbb{Q}^2$, and an integer K, such that φ is satisfiable if and only if G contains a t-spanner having at most K edges.

Let z denote the origin in $\mathbb{R}^2$, and define $i^* = \lceil 4/\rho \rceil$. For each i with $1 \leq i \leq N$, we define the following geometric graph G_i:

1. Let $p_i = Q(i^* + 4i)$, $p'_i = Q(i^* + 4i + 1)$, $q_i = Q(i^* + 4i + 2)$, and $q'_i = Q(i^* + 4i + 3)$.
2. G_i contains the four edges $\{z, p_i\}$, $\{z, p'_i\}$, $\{z, q_i\}$, and $\{z, q'_i\}$.
3. G_i contains the five forced paths $FP(p_i, p'_i; \ell)$, $FP(p_i, q_i; \ell)$, $FP(p_i, q'_i; \ell)$, $FP(p'_i, q_i; \ell)$, and $FP(p'_i, q'_i; \ell)$.

For each clause $c_j = (y_1 \vee y_2 \vee y_3)$, $1 \leq j \leq M$, we define the following geometric graph G^j:

1. Let $r_j = Q(i^* + 4N + 3 + j)$.
2. The graph G^j contains the edge $\{z, r_j\}$.

3. For each m with $1 \leq m \leq 3$, if y_m is equal to the variable, say, x_i, then G^j contains the forced path $FP(r_j, p_i; \ell)$. On the other hand, if y_m is equal to the negation of the variable, say, x_i, then G^j contains the forced path $FP(r_j, p_i'; \ell)$.

We define G to be the union of the graphs G_i ($1 \leq i \leq N$) and the graphs G^j ($1 \leq j \leq M$). Observe that G is a connected geometric graph, whose vertices are points in $\mathbb{Q}^2$. Recall that each forced path consists of $k+1$ edges. The graph G consists of $1 + (5k+4)N + (3k+1)M$ vertices and $(5k+9)N + (3k+4)M$ edges. We define $K = (5k+6)N + (3k+3)M$.

If L denotes the number of bits in the representation of the Boolean formula ψ, then it follows from Lemma 12 that the graph G can be constructed in time that is polynomial in L.

In the rest of this section, we will prove that the Boolean formula φ is satisfiable if and only if the graph G contains a t-spanner with at most K edges. The following lemma follows from Lemmas 10 and 13.

Lemma 14. *The length of each forced path in the graph G is in the interval $[\ell, 3\ell/2]$.*

The next lemma explains our choice for the integer K.

Lemma 15. *Let G' be an arbitrary t-spanner of G. Then, the following two claims are true:*

1. *G' contains at least K edges.*
2. *If G' consists of exactly K edges, then, for each i with $1 \leq i \leq N$, exactly one of the edges $\{z, p_i\}$ and $\{z, p_i'\}$ is in G'.*

Proof. We first observe that, by Lemmas 11 and 13, all forced paths in G are contained in G'. The total number of edges in these forced paths is equal to $(5N + 3M)(k+1) = K - N$. We will prove below that, for each $1 \leq i \leq N$, the graph G' contains at least one of the four edges $\{z, p_i\}$, $\{z, p_i'\}$, $\{z, q_i\}$, and $\{z, q_i'\}$. This will imply that G' contains at least K edges and, thus, prove the first claim.

Let $1 \leq i \leq N$, and assume that none of the edges $\{z, p_i\}$, $\{z, p_i'\}$, $\{z, q_i\}$, and $\{z, q_i'\}$ is contained in G'. Then, any path in G' between z and q_i contains at least one edge of length one and at least two forced paths. Thus, using Lemma 14, we have

$$\delta_{G'}(z, q_i) \geq 1 + 2\ell = 1 + 2 \cdot 2(t-1)/3 > t = t \cdot \delta_G(z, q_i),$$

contradicting the fact that G' is a t-spanner of G.

To prove the second claim, assume that G' consists of exactly K edges. Let $1 \leq i \leq N$. It follows from the argument above that G' contains exactly one of the edges $\{z, p_i\}$, $\{z, p_i'\}$, $\{z, q_i\}$, and $\{z, q_i'\}$. If G' contains $\{z, q_i'\}$, then, by the same argument as above, we must have $\delta_{G'}(z, q_i) > t \cdot \delta_G(z, q_i)$, contradicting our assumption that G' is a t-spanner of G. Similarly, if G' contains $\{z, q_i\}$, then $\delta_{G'}(z, q_i') > t \cdot \delta_G(z, q_i')$, which is also a contradiction. $\square$

Lemma 16. *If G contains a t-spanner with at most K edges, then the Boolean formula φ is satisfiable.*

Proof. Let G' be a t-spanner of G consisting of at most K edges. Then, by Lemma 15, G' contains exactly K edges and, for each $1 \leq i \leq N$, G' contains exactly one of the edges $\{z, p_i\}$ and $\{z, p_i'\}$.

For each $1 \leq i \leq N$, if $\{z, p_i\}$ is an edge of G', then we give the variable x_i the value *true*, otherwise, we give the variable x_i the value *false*. We claim that for this assignment of truth values, the Boolean formula φ evaluates to *true*. To prove this, let $1 \leq j \leq M$, and consider the clause c_j in φ. For ease of notation, let us assume that $c_j = x_1 \vee \overline{x_2} \vee \overline{x_3}$. To prove that c_j evaluates to *true*, we have to show that at least one of the edges $\{z, p_1\}$, $\{z, p_2'\}$, and $\{z, p_3'\}$ is in G'. Assume that neither of these edges is in G'. Observe that $\{z, r_j\}$ is not an edge in G', because otherwise, G' contains more than K edges. Thus, every path in G' between z and r_j contains at least one edge of length one and at least two forced paths. Therefore, we have $\delta_{G'}(z, r_j) \geq 1 + 2\ell > t \cdot \delta_G(z, r_j)$. This contradicts our assumption that G' is a t-spanner of G. □

Lemma 17. *If the Boolean formula φ is satisfiable, then G contains a t-spanner with at most K edges.*

Proof. Assume that φ is satisfiable. We fix an assignment of truth values for the variables $x_1, x_2, \ldots, x_N$ for which φ evaluates to *true*. Define the following subgraph G' of G: First, G' contains all forced paths in G. Second, for each $1 \leq i \leq N$, if $x_i = true$, then G' contains the edge $\{z, p_i\}$, otherwise, G' contains the edge $\{z, p_i'\}$. The graph G' contains exactly K edges. To show that G' is a t-spanner of G, it suffices to prove the following claim: For each edge $\{a, b\}$ of G that is not in G', we have $\delta_{G'}(a, b) \leq t|ab|$.

Let $1 \leq i \leq N$. We may assume without loss of generality that $\{z, p_i'\}$ is an edge in G'. Consider the edge $\{z, p_i\}$ of G, which is not an edge in G'. The edge $\{z, p_i'\}$ and the forced path $FP(p_i, p_i'; \ell)$ form a path in G' between z and p_i. Thus, using Lemma 14, we have $\delta_{G'}(z, p_i) \leq 1 + 3\ell/2 = t = t|zp_i|$. In a similar way, it can be shown that $\delta_{G'}(z, q_i) \leq t = t|zq_i|$ and $\delta_{G'}(z, q_i') \leq t = t|zq_i'|$.

Let $1 \leq j \leq M$. Write the clause c_j as $c_j = y_1 \vee y_2 \vee y_3$, and consider the edge $\{z, r_j\}$ of G, which is not an edge in G'. Since c_j evaluates to *true*, at least one of the literals in c_j is true. We may assume without loss of generality that y_1 is *true*. If $y_1 = x_i$, for some i, then G' contains the edge $\{z, p_i\}$ and the forced path $FP(r_j, p_i; \ell)$. It follows that $\delta_{G'}(z, r_j) \leq 1 + 3\ell/2 = t = t|zr_j|$. On the other hand, if $y_1 = \overline{x_i}$, for some i, then G' contains the edge $\{z, p_i'\}$ and the forced path $FP(r_j, p_i'; \ell)$. Thus, in this case, we have $\delta_{G'}(z, r_j) \leq 1 + 3\ell/2 = t = t|zr_j|$. Hence, we have shown that G' is a t-spanner of G. □

References

1. I. Althöfer, G. Das, D. P. Dobkin, D. Joseph, and J. Soares. On sparse spanners of weighted graphs. *Discrete & Computational Geometry*, 9:81–100, 1993.
2. S. Baswana and S. Sen. A simple linear time algorithm for computing a $(2k - 1)$-spanner of $O(n^{1+1/k})$ size in weighted graphs. In *Proceedings of the 30th International Colloquium on Automata, Languages and Programming*, volume 2719 of *Lecture Notes in Computer Science*, pages 384–396, Berlin, 2003. Springer-Verlag.

3. U. Brandes and D. Handke. NP-completeness results for minimum planar spanners. *Discrete Mathematics and Theoretical Computer Science*, 3:1–10, 1998.

4. L. Cai. NP-completeness of minimum spanner problems. *Discrete Applied Mathematics*, 48:187–194, 1994.

5. L. Cai and D. Corneil. Tree spanners. *SIAM Journal on Discrete Mathematics*, 8:359–387, 1995.

6. P. B. Callahan and S. R. Kosaraju. Faster algorithms for some geometric graph problems in higher dimensions. In *Proceedings of the 4th ACM-SIAM Symposium on Discrete Algorithms*, pages 291–300, 1993.

7. L. S. Chandran. A high girth graph construction. *SIAM Journal on Discrete Mathematics*, 16:366–370, 2003.

8. D. Eppstein. Spanning trees and spanners. In J.-R. Sack and J. Urrutia, editors, *Handbook of Computational Geometry*, pages 425–461. Elsevier Science, Amsterdam, 2000.

9. J. Gudmundsson and C. Knauer. Dilation and detours in geometric networks. In T. F. Gonzalez, editor, *Handbook on Approximation Algorithms and Metaheuristics*. Chapman & Hall/CRC, Boca Raton, 2006.

10. J. Gudmundsson, C. Levcopoulos, G. Narasimhan, and M. Smid. Approximate distance oracles for geometric graphs. In *Proceedings of the 13th ACM-SIAM Symposium on Discrete Algorithms*, pages 828–837, 2002.

11. J. Gudmundsson, G. Narasimhan, and M. Smid. Fast pruning of geometric spanners. In *Proceedings of the 22nd Symposium on Theoretical Aspects of Computer Science*, volume 3404 of *Lecture Notes in Computer Science*, pages 508–520, Berlin, 2005. Springer-Verlag.

12. F. Lazebnik and V.A. Ustimenko. Explicit construction of graphs with an arbitrary large girth and of large size. *Discrete Applied Mathematics*, 60:275–284, 1995.

13. M. Mitzenmacher and E. Upfal. *Probability and Computing*. Cambridge University Press, Cambridge, UK, 2005.

14. D. Peleg. *Distributed Computing: A Locality-Sensitive Approach*. Monographs on Discrete Mathematics and Applications. Society for Industrial and Applied Mathematics, Philadelphia, 2000.

15. D. Peleg and A. A. Schäffer. Graph spanners. *Journal of Graph Theory*, 13:99–116, 1989.

16. L. Roditty, M. Thorup, and U. Zwick. Deterministic constructions of approximate distance oracles and spanners. In *Proceedings of the 32nd International Colloquium on Automata, Languages and Programming*, volume 3580 of *Lecture Notes in Computer Science*, pages 261–272, Berlin, 2005. Springer-Verlag.

17. J. S. Salowe. Constructing multidimensional spanner graphs. *International Journal of Computational Geometry & Applications*, 1:99–107, 1991.

18. M. Smid. Closest-point problems in computational geometry. In J.-R. Sack and J. Urrutia, editors, *Handbook of Computational Geometry*, pages 877–935. Elsevier Science, Amsterdam, 2000.

19. P. M. Vaidya. A sparse graph almost as good as the complete graph on points in K dimensions. *Discrete & Computational Geometry*, 6:369–381, 1991.

20. G. Venkatesan, U. Rotics, M.S. Madanlal, J.A. Makowsky, and C. Pandu Rangan. Restrictions of minimum spanner problems. *Information and Computation*, 136:143–164, 1997.

The Weighted Maximum-Mean Subtree and Other Bicriterion Subtree Problems

Josiah Carlson and David Eppstein

Computer Science Department
University of California, Irvine CA 92697, USA
{jcarlson, eppstein}@uci.edu

Abstract. We consider problems where we are given a rooted tree as input, and must find a subtree with the same root, optimizing some objective function of the nodes in the subtree. When the objective is the sum of linear function weights of a parameter, we show how to list all optima for all parameter values in $O(n \log n)$ time. This can be used to solve many bicriterion optimizations problems in which each node has two values x_i and y_i associated with it, and the objective function is a bivariate function $f(\sum x_i, \sum y_i)$ of the sums of these two values. When f is the ratio of the two sums, we have the Weighted Maximum-Mean Subtree Problem, or equivalently the Fractional Prize-Collecting Steiner Tree Problem on Trees; we provide a linear time algorithm when all values are positive, improving a previous $O(n \log n)$ solution, and prove NP-completeness when certain negative values are allowed.

1 Introduction

Suppose we are given a rooted tree, in which each node i has two quantities x_i and y_i associated with it, and a bivariate objective function f. For any subtree S, let $X_S = \sum_{i \in S} x_i$ and $Y_S = \sum_{i \in S} y_i$. We wish to find the subtree S, having the same root as the input tree, that maximizes $f(X_S, Y_S)$. (We note that finding the tree minimizing or maximizing $\sum x_i$ is a simple variant of the Open Pit Mining problem on DAGs [2], and can be easily solved in $O(n)$ time [7].)

For instance, if $f(X, Y) = X/Y$, we can interpret x_i as the *profit* of a node, and y_i as the *cost* of a node; the optimal subtree is the one that maximizes the return on investment. This problem, which we call the Weighted Maximum-Mean Subtree Problem (WMMSTP), can also be viewed as a special case of the Fractional Prize-Collecting Steiner Tree Problem (FPCSTP); in the FPCSTP, one is given a graph, with costs on the edges and profits on the vertices, a starting vertex v_0, and a starting cost c_0, and must find a tree rooted at v_0 that maximizes the total profit divided by the total cost. It is easy to see that the WMMSTP is equivalent to a special case of the FPCSTP in which the input is a rooted tree, v_0 is the root of the tree, c_0 is the cost of the root node, and each additional tree node has a cost in the WMMSTP that is equal to the cost in the FPCSTP of the edge connecting the node to its parent. For this special case of

L. Arge and R. Freivalds (Eds.): SWAT 2006, LNCS 4059, pp. 400–410, 2006.

FPCSTP on Trees (or equivalently WMMSTP), when all costs are positive, an $O(n \log n)$ time algorithm is known [7].

The Weighted Maximum-Mean Subtree Problem (and equivalently the FPC-STT problem) can be applied to maximizing return on investment for adding services to a preexisting tree structured utility networks. An example of this lies within the development of DSL services over preexisting telephone networks, as has been in construction in recent years. The costs associated with adding such a service involve additional equipment (repeaters, hubs, switches, filters, and line upgrades) placed along the standard wired telephone services, and profits are gained from providing such a service to homes and businesses connected to the upgraded network. In running the algorithm we present, a telephone company that desires to upgrade their lines would discover where they should offer such services to maximize the percent return on their service upgrade investment. More generally this problem applies to situations where return on investment is optimized among hierarchical sets of business opportunities.

More generally let $g(X,Y) = X + \lambda Y$, and consider the sequence of subtrees that maximize $g(X_S, Y_S)$ as the parameter λ varies from $-\infty$ to $+\infty$. We call the problem of computing this sequence the Parametric Linear Maximum Subtree Problem. This sequence of parametric optimal subtrees can be viewed as forming the upper convex hull of the planar point set formed by taking one point (X_S, Y_S) for each possible subtree S of the input tree, and the lower convex hull can be formed similarly as the sequence of trees minimizing $g(X_S, Y_S)$ as λ varies.

As seen for related bicriterion spanning tree problems [1,5,6], many bicriterion optimal subtree problems can be solved by this parametric approach. If $f(X,Y)$ is any convex or quasiconvex function, its maximum over the set of points (X_S, Y_S) is achieved at a vertex of the convex hull, so by computing and testing all parametric optima we can be certain to find the subtree that maximizes a convex or quasiconvex $f(X,Y)$ or equivalently that minimizes a concave or quasiconcave $f(X,Y)$. The following problems can be solved with this parametric formulation:

- The Weighted Maximum-Mean Subtree Problem can be viewed as a special case of this formulation where $f(X,Y) = X/Y$; this function is convex on the upper halfplane (positive costs) but not on the whole plane (where costs may be negative). We supply a more efficient algorithm for this problem when weights are positive, but the parametric approach succeeds more generally when weights may be negative but all subtrees containing the tree root have positive total weight.
- Suppose node i may fail with probability p_i and has a cost x_i, and let $y_i = -\ln(1 - p_i)$. Then the reliability probability of the overall subtree is e^Y, and the subtree that minimizes the ratio of cost to reliability can be found by minimizing $f(X,Y) = Xe^Y$.
- If nodes have weights that are unknown random variables with known mean and variance, the stochastic programming problem of finding a subtree with high probability of having low weight can be expressed as minimizing

$f(X,Y) = X + \sqrt{Y}$, and the problem of finding a subtree minimizing the variance in weight can be expressed as minimizing $f(X,Y) = X - Y^2$.

As we will show, this parametric approach leads to efficient algorithms for these and many other bicriterion optimal subtree problems.

Our algorithms assume that the values x_i and y_i associated with tree nodes are real-valued, and that any single arithmetic operation or comparison on two real values can be performed in constant time.

2   New Results

We provide the following results:

- A linear time algorithm for solving the Weighted Maximum-Mean Subtree Problem with positive weights, improving a previous $O(n \log n)$ time solution [7].
- A proof that the Weighted Maximum-Mean Subtree Problem is NP-complete when the weights are allowed to be negative.
- An optimal $O(n \log n)$ time algorithm for listing all solutions, in order by parameter value, to the Parametric Linear Maximum Subtree Problem.
- An $O(n \log n)$ time algorithm for solving any bicriterion optimal subtree problem that maximizes a convex function $f(X,Y)$ or minimizes a concave function $f(X,Y)$ of the sums X and Y of the two node values.

3   The Weighted Maximum-Mean Subtree Problem

We are given a rooted tree of nodes such that each node i has a real valued *profit* x_i, and we produce a subtree that maximizes the average profit of the remaining nodes. By *pruning a node* we mean removing it and all of its descendants from the input; our task then becomes finding an optimal set of nodes to prune.

A generalization of this problem gives each node a positive real valued *cost* y_i; the original problem can be viewed as assigning each node a unit cost. The overall average of a tree is the sum of the profits divided by the sum of the costs, including only unpruned nodes. In this section we show the problem to be NP-complete when the costs can be negative, and present an algorithm that solves the generalization with both profits and positive costs per node in time $O(n)$.

We assume that our input consists of a rooted tree T, whose nodes each have positive or negative real valued profits and positive real valued costs. The output should be a pruned subtree $P(T)$ and an average profit

$$OPTAVG = AVG(P(T)) = \sum_{x \in U} profit(x) / \sum_{x \in U} cost(x),$$

where U denotes the set of unpruned nodes in $P(T)$, and where this average profit is at least as large as that of any other subtree of T having the same root.

```
def HasAverageAtLeast(tree, cutoff):
    tree.subprofit = tree.profit
    tree.subcost = tree.cost

    for child in tree.children:
        if HasAverageAtLeast(child, cutoff):
            tree.subprofit += child.subprofit
            tree.subcost += child.subcost

    unpruned = tree.subprofit/tree.subcost >= cutoff
    .tree.pruned = not unpruned
    return unpruned
```

Listing 1: Testing whether $OPTAVG$ is at least a given cutoff.

Theorem 1. *The Weighted Maximum-Mean Subtree Problem with negative cost nodes is NP-complete.*

Proof. Given an instance of Integer Subset-Sum with a set S of values, and a desired total U, we create a rooted tree with a root node and $|S|$ leaf nodes hanging from the root. Set the root's profit to one, the root's cost to U, and all leaf profits to zero. Assign each leaf's cost as the negation of one of the values from S. The maximum mean for such a tree is ∞, if and only if the Subset Sum instance has a subset of values summing to U, where in this case the optimal subtree includes only those leaf nodes whose costs sum to U.   ∎

By slightly adjusting the root cost in this reduction we can avoid the issue of division by zero while still preserving the computational complexity of the problem. Due to this result, in the rest of the section we restrict costs to being strictly positive. However, our algorithm correctly finds the maximum mean subtree even when profits are allowed to be negative.

We first define the algorithm provided in Listing 1 which tells us whether or not some tree has a pruning with average greater than or equal to some provided cutoff. Essentially, the algorithm traverses the tree bottom-up, pruning any node when the average of it and its unpruned descendants falls below the cutoff.

Lemma 1. *Suppose that there exists a tree T with average value at least cutoff. Then the tree U that HasAverageAtLeast forms by pruning the input tree also has average value at least cutoff.*

Proof. T and U may differ, by the inclusion of some subtrees and the exclusion of others. For each subtree s that is included in U and excluded from T, s must have average value at least cutoff (otherwise, *HasAverageAtLeast* would have pruned it) so combining its value from that of T can not bring the average below cutoff. For each subtree s that is excluded from U and included in T, s must have average value below cutoff (by an inductive application of the lemma to the subtree rooted at the root of s) so removing its value from that of T can only increase the overall value.   ∎

Corollary 1. *HasAverageAtLeast returns True if and only if there exists a tree with average value at least cutoff.*

A very similar subroutine, which we call *HasAverageGreaterThan*, replaces the greater-or-equal test in the assignment to *unpruned* by a strict inequality. A suitably modified version of Lemma 1 and its proof would also prove its correctness. Note that *tree.subprofit*, *tree.subcost*, and *tree.pruned* variables provided above are implementation details.

Klau *et al.* [7] provide three algorithms which solve the Maximum-Mean Subtree problem. The first is a binary search, based on the linear time decision algorithms presented above, which runs in $O(nk)$ time, where k is the desired precision of the answer in bits, and n is the number of nodes in the input tree. A second algorithm, which they present as a form of Newton's Method, runs in worst-case $O(n^2)$. Their third algorithm, which they present as their main result, is based on Megiddo's Parametric Search and runs in $O(n \log n)$ time.

We briefly describe this third algorithm, as our linear time algorithm is closely related to it. It performs a sequence of iterations, each of which performs a binary search among the profit/cost ratios of the remaining tree nodes to determine their relation to the optimal solution value. Once all of these values are known, their algorithm uses this information to reduce the input tree to a smaller tree with the same final solution value, continuing recursively on that tree. Pruning and merging steps are introduced which further reduce the remaining tree nodes.

The basic difference between Klau *et al.*'s third algorithm and ours, is that where Klau *et al.* binary search from among a set of node profit/cost ratios to constrain the range of solution values, we choose the median from the set and perform a single call to the decision algorithm to constrain our range. We prove that our method is sufficient to reduce a potential function related to the size of the tree by a constant fraction every pass. When the tree has been reduced to a single node, we are done. Because each pass only calls the decision algorithm once, the time per pass is reduced from $O(n \log n)$ to $O(n)$, and the running time of our algorithm reduces to $O(n)$ via geometric sum. An outline for our algorithm with pseudocode for *merge* and *prune* are provided in Listing 2. Our provided algorithm returns the optimal average, but by calling *HasAverageAtLeast* on the original input tree with the optimal average and obeying the pruning decisions it makes, we can produce $P(T)$.

Lemma 2. *The merge decisions made in the algorithm are correct.*

Proof. The only way merge decisions could be incorrect would be if the optimal pruning cuts between x and y (x being the parent of y); we show that cannot happen. First suppose x has value below the low bound. If a pruning were made between x and y, the remaining subtree rooted at x would consist of x itself, which has a low value, so pruning x could only improve the tree. Suppose y is above the high bound, and suppose that a tree includes x but does not include y. Then y could be included with no other nodes, increasing the average, so the tree excluding y could not be optimal. ∎

Our Algorithm:

1. Set *low* and *high* to be outside the range of all node values.
2. While the root of the tree has children, repeat the following steps:
   (a) Find the set of tree nodes whose values are within the range between *low* and *high,* call them in-range nodes.
   (b) Reduce the range by applying the decision algorithm to the median value of these tree nodes.
   (c) For each node in a post-order traversal of the tree:
       i. Prune any leaf node whose value is below *low,* calling them low nodes.
       ii. Merge any node whose value is above *high* with its parent, calling them high nodes.
       iii. Merge with its child any node that has a single child and that has value below *low,* also calling them low nodes.

To merge a child *ch* with its parent *pa:*

1. Remove *ch* from *pa*'s list of children.
2. Merge *ch*'s list of children with *pa*'s list of children.
3. Increment the profit of *pa* with the profit of *ch.*
4. Increment the cost of *pa* with the cost of *ch.*

To prune a child *ch* from its parent *pa:*

1. Remove *ch* from *pa*'s list of chilren.

Listing 2: Outline for our algorithm with *merge* and *prune* subprocedures.

Lemma 3. *If there are m nodes remaining in the tree after any iteration of the algorithm, then at least $m/2$ of these nodes are in range $(low, high)$.*

Proof. Let T be a tree in which no further cutting or merging steps can be performed, which minimizes the fraction of nodes in range $(low, high)$. The root may be low (with more than 1 child), in range, or high. All leaves must be in range. All internal nodes with 1 child must be in range. All remaining nodes must have at least two children, and must be low or in range. Because there are at least $m/2$ nodes with 0 or 1 children, and all such nodes are in range, then at least $m/2$ of the nodes must be in range. ∎

Theorem 2. *Our algorithm solves the Weighted Maximum Mean Subtree Problem for inputs with positive node costs in time $O(n)$.*

Proof. Let Φ be the number of tree nodes plus the number of nodes in range $(low, high)$. Initially Φ is $2n$; it is reduced by each step where we cut nodes or shrink the range, and reduced or unchanged by each step where we merge nodes. By Lemma 3, in each iteration of the algorithm there are at least $\Phi/3$ nodes in range, half of which become low or high by the range-shrinking step of the iteration, so Φ is reduced by a factor of $5/6$ or better per iteration. The time per iteration is $O(\Phi)$, so the total time adds in a geometric series to $O(n)$. ∎

4   The Parametric Linear Maximum Subtree Problem

We now consider the Parametric Linear Maximum Subtree Problem. As discussed in the introduction, the solution can be used to solve many bicriterion optimal subtree problems.

We are given as input a tree T, each node i of T has two values a_i and b_i, which defines the weight of the node as a linear function $a_i \lambda + b_i$ with slope a_i and y-intercept b_i. We wish to produce as output a function F, which describes the weight of the maximum weight subtree of T for each parameter value $0 \leq \lambda < \infty$. We note that F is the pointwise maximum of a set of linear functions (one per subtree of T); therefore, F is convex and piecewise linear, and can be described by the breakpoints, slopes, and y-intercepts of each linear segment of F.

We do not output the sequence of optimal subtrees explicitly; the output size would dwarf the time complexity of our algorithms. When using the parametric problem to solve bicriterion optimization problems, all we need is the function F, as the value of the optimal tree can be determined from the slope and y-intercept of one of the pieces of F; once that piece is found, the optimal tree itself can be determined by fixing λ to a value within the range over which that piece determines the value of F, and solving a maximum weight subtree problem for that fixed value of λ. If the sequence of trees is needed, it can be represented concisely by a sequence of $O(n)$ prune and unprune events on edges of T.

For any node i, let $F_i(\lambda)$ denote the output for the Parametric Linear Maximum Subtree Problem restricted to the subtree of T rooted at i; then (if i is not itself the root of T) we will prune node i and its descendents for exactly those values of λ for which $F_i(\lambda) < 0$.

Lemma 4. *The function F has at most $2n$ breakpoints.*

Proof. Each breakpoint in F occurs when some node i becomes or ceases being pruned, when $F_i(\lambda) = 0$. Since each F_i is convex, $F_i(\lambda) = 0$ can only occur for two values of λ per node i, and the node contributes at most two breakpoints to F.   ∎

Let $F_i(\lambda) = \max(0, G_i(\lambda))$. G_i measures the contribution of node i and its descendants of the tree rooted at i's parent; it is negative or zero when the node is pruned, and otherwise sums the weights of the unpruned descendants of i. We compute F_i and G_i recursively with the formula

$$G_i(\lambda) = a_i \lambda + b_i + \sum_{c \in children(i)} F_c(\lambda).$$

We have seen above a formula with which we can compute the desired function F, by computing similar functions bottom-up through the input tree T. To show how to implement this formula efficiently, we adapt an algorithm of Shah and Farach-Colton [8] which uses a similar computation of piecewise linear functions to solve tree partitioning problems.

The execution of our algorithm is as follows. During a post order traversal of the tree, we generate a piecewise linear function for each node i by adding the

1. For tree node j in *postorderTraversal*(T):
   (a) Set $fcn = create(0, j.a, j.b)$
   (b) For tree node k in $j.children$:
       i. Set $fcn = add(fcn, k.fcn)$
   (c) Set $j.fcn = trim(fcn)$
2. Return $T.fcn$

Listing 3: Outline for algorithm solving the Parametric Linear Maximum Subtree Problem.

function defined by a_i and b_i with the functions defined by the subtrees rooted at the i's children. We then set the function for node i to be the maximum of this sum and zero, handling the case if or when the value of node i drops below zero and must be pruned.

As in the algorithm of Shah and Farach-Colton [8], we manipulate objects representing piecewise linear functions, with operations that create new functions, add two functions, and take the maximum of one such function with the constant zero function. A detailed API for these objects is provided in Listing 4, and an outline of our overall algorithm that uses this API to compute F_i and G_i can be seen in Listing 3.

In more detail, we represent each piecewise linear function as an AVL tree, sorted by keys which are the x-coordinates of the left endpoints of each linear segment of the function. Each tree node in stores values $deltaA$ and $deltaB$ which represent the change in slope and intercept of the piecewise linear function at that breakpoint. We further store in each node the total sums $daTotal$ and $dbTotal$ of all values contained in the subtree rooted at that node, maintained during rotation in a fashion identical to that of Order Statistic Trees as described by Cormen *et al.* [4], extended to support tree merging [3].

To add two piecewise linear functions, as described in Listing 5, we merge the two AVL trees representing the two functions, then recompute the sums $daTotal$ and $dbTotal$ at all ancestors of nodes changed by the merge. In this way, adding a tree with n_1 breakpoints to a tree with n_2 breakpoints can be performed in time

$$O\left(n_1 \log\left(\frac{n_1 + n_2}{n_1}\right) + n_2 \log\left(\frac{n_1 + n_2}{n_2}\right)\right).$$

To take the maximum of a convex piecewise linear function f with the zero function, we use the binary search tree structure to search for the values of λ for which $f(\lambda) = 0$, add breakpoints at these two points, and remove all breakpoints between these points, as described in Listing 6. This operation takes $O(\log n)$ amortized time per call.

Theorem 3. *The algorithm described above solves the Parametric Linear Maximum Subtree Problem in time $O(n \log n)$.*

Proof. The time is dominated by add and trim operations used to maintain piecewise linear functions. There is one trim call per node, taking $O(\log n)$ time

1. $create(z, a, b)$ - creates a function starting at x-offset z, with slope a and y-intercept b.
2. $add(f, g)$ - adds function g to the function f, merging their contents using the Brown and Tarjan algorithm [3] with our total updating procedure given in Listing 5 . In this process, the larger of f or g being modified in place and returned.
3. $delete(f, i)$ - deletes the node i from the function tree defined by f using the standard AVL tree deletion methods.
4. $functionAt(f, z)$ - discovers the slope a and y-intercept b at x-offset z of the function f, using an algorithm similar to order discovery in Order Statistic Trees as described by Cormen et al. [4], not provided here.
5. $trim(f)$ - trims the function f such that for all $x \geq 0$, $f(x) \geq 0$, outlined in Listing 6.

Listing 4: API for data structure representing piecewise linear functions, used by our algorithm.

When adding two functions F and G:

1. Keep an auxillary list of all nodes that have been inserted or changed during the merge.
2. Generate a secondary tree that contains those nodes from the auxillary list and their ancestors, with structure identical to the current tree with links back to the function tree nodes.
3. For each node j in the postorder traversal of the secondary tree:
   (a) $i = j.link$
   (b) $i.daTotal = i.left.daTotal + i.right.daTotal + i.deltaA$
   (c) $i.dbTotal = i.left.dbTotal + i.right.dbTotal + i.deltaB$

Listing 5: Details of data structure operations for adding two piecewise linear functions.

to discover the range of breakpoints that must be removed, giving us $O(n \log n)$ total time. If we charge a breakpoint's possible removal to its insertion, removal is ammortized free within each trim call.

By Lemma 4 there are $O(n)$ breakpoints active during algorithm execution. If a breakpoint is active within a sequence of piecewise linear functions of size n_0, $n_1, n_2, \ldots$ then the amount it contributes to the time bounds for adding these functions is $\log(n_0), \log(n_1/n_0), \log(n_2/n_1), \ldots$ These times add to $O(\log n)$, so the total time for adding piecewise linear functions is $O(n \log n)$. ∎

Corollary 2. *We can solve any bicriterion optimal subtree problem in which we attempt to maximize a convex function $f(X, Y)$ or minimize a concave function $f(X, Y)$, where X and Y are respectively the sums of values x_i and y_i associated with each tree node in the subtree, in time $O(n \log n)$.*

Proof. Let $a_i = x_i$ and $b_i = y_i$. Then any subtree S with sums X_S and Y_S (as defined in the introduction) corresponds to a line $X_S \lambda + Y_S = 0$, and the upper convex hull of the points (X_S, Y_S) for the set of all subtrees S is projectively dual

1. Follow the slopes of the function downwards via a binary search to discover a node that is, or is adjacent to, the minimum of the function.
2. Check successor and predecessor nodes for the true function minimum.
3. Given $a, b = functionAt(f, minnode.x)$, if $minnode.x * a + b \geq 0$, return.
4. Discover left crossing point $lcx > 0$ where the function crosses the x-axis if one exists, via binary search, else $lcx = 0$.
5. Discover right crossing point rcx where the function crosses the x-axis if one exists, via binary search, else $rcx = \infty$.
6. If $lcx == 0$ and $rcx == \infty$, then $f = create(0, 0, 0)$, return.
7. $saveda, savedb = functionAt(f, rcx)$
8. For each node where $lcx \leq node.x \leq rcx$
   (a) $f = add(f, create(node.x, -node.a, -node.b))$
   (b) $delete(f, node)$
9. $a, b = functionAt(f, lcx)$
10. $f = add(f, create(lcx, -a, -b))$
11. If $rcx < \infty$
   (a) $a, b = functionAt(f, rcx)$
   (b) $f = add(f, create(rcx, saveda - a, savedb - b))$

Listing 6: Details of data structure operations for taking the maximum of a piecewise linear function with the constant zero function.

to the upper envelope of all such lines, which is our Parametric Linear Maximum Subtree Problem. That is, if we solve the Parametric Linear Maximum Subtree Problem, and an equivalent Parametric Linear Minimum Subtree Problem, the slopes and y-intercepts of the segments of the output functions from these problems give us the points (X_S, Y_S) belonging to the convex hull of the set of such points for all trees. The optimal pair (X, Y) can be found by testing these $O(n)$ points and choosing the best of them. The optimal tree itself can be then found by letting λ be a parameter value contained in the function segment with slope X and y-intercept Y, and finding the maximum weight subtree for the weights given by that value of λ. ∎

To see that our $O(n \log n)$ time bound for the Parametric Linear Maximum Subtree problem is optimal, consider the following simple reduction from sorting. Given n values x_i to sort, create a tree with a root that is the parent of n leaves, one leaf having $a_i = 1$ and $b_i = x_i$. Then, for each leaf i, the function G_i will have a breakpoint exactly at x-coordinate x_i, so the sequence of breakpoints of the output F (or equivalently the sequence of prune and unprune operations generating the sequence of optimal subtrees) is exactly the sorted sequence of the input values.

However, this lower bound does not apply to the bicriterion optimal subtree problems that we solve by our parametric approach, and we have seen that for one such problem (the Weighted Maximum-Mean Subtree Problem) a faster linear time algorithm is possible. It would be interesting to determine for what other bicriterion optimal subtree problems such speedups are possible.

References

1. P. K. Agarwal, D. Eppstein, L. J. Guibas, and M. R. Henzinger. Parametric and kinetic minimum spanning trees. In *Proc. 39th Symp. Foundations of Computer Science*, pages 596–605. IEEE, November 1998.
2. R. Ahuja, T. Magnanti, and J. Orlin. *Network Flows*. Prentice Hall, 1993.
3. M. Brown and R. Tarjan. A fast merging algorithm. *J. ACM*, 26(2):211–226, 1979.
4. T. Cormen, C. Leiserson, R. Rivest, and C. Stein. *Introduction to Algorithms*. MIT Press, 2001.
5. N. Katoh. Parametric combinatorial optimization problems and applications. *J. Inst. Electronics, Information and Communication Engineers*, 74:949–956, 1991.
6. N. Katoh. Bicriteria network optimization problems. *IEICE Trans. Fundamentals of Electronics, Communications and Computer Sciences*, E75-A:321–329, 1992.
7. G. Klau, I. Ljubi, P. Mutzel, U. Pferschy, and R. Weiskircher. The Fractional Prize-Collecting Steiner Tree Problem on Trees. In *Proc. 11th Eur. Symp. Algorithms (ESA 2003)*, number 2832 in Lecture Notes in Computer Science, pages 691–702, 2003.
8. R. Shah and M. Farach-Colton. Undiscretized dynamic programming: faster algorithms for facility location and related problems on trees. In *Proc. 13th ACM-SIAM Symp. Discrete Algorithms (SODA 2002)*, pages 108–115, 2002.

Linear-Time Algorithms for Tree Root Problems

Maw-Shang Chang[1,*], Ming-Tat Ko[2], and Hsueh-I Lu[3,**]

[1] Department of Computer Science and Information Engineering,
National Chung Cheng University, Ming-Shiun, Chiayi 621, Taiwan
mschang@cs.ccu.edu.tw
[2] Institute of Information Science,
Academia Sinica, Taipei 115, Taiwan
mtko@iis.sinica.edu.tw
[3] Department of Computer Science and Information Engineering,
National Taiwan University, Taipei 106, Taiwan
hil@csie.ntu.edu.tw
http://www.csie.ntu.edu.tw/~hil

Abstract. Let T be a tree on a set V of nodes. The *p-th power* T^p of T is the graph on V such that any two nodes u and w of V are adjacent in T^p if and only if the distance of u and w in T is at most p. Given an n-node m-edge graph G and a positive integer p, the *p-th tree root problem* asks for a tree T, if any, such that $G = T^p$. Given a graph G, the *tree root problem* asks for a positive integer p and a tree T, if any, such that $G = T^p$. Kearney and Corneil gave the best previously known algorithms for both problems. Their algorithm for the former (respectively, latter) problem runs in $O(n^3)$ (respectively, $O(n^4)$) time. In this paper, we give $O(n + m)$-time algorithms for both problems.

1 Introduction

Let H be a graph on a set V of nodes. The *p-th power* H^p of H is the graph on V such that any two nodes u and w of V are adjacent in H^p if and only if the distance of u and w in H is at most p. If $G = H^p$, then we say that graph H is a *p-th root* of graph G or, equivalently, G is the *p-th power* of H. Determining whether the input graph is a power of some other graph is the *graph root problem*. Graph roots and powers have been extensively studied in the literature. Motwani and Sudan [12] proved the NP-completeness of recognizing squares of graphs. Lau [9] showed that squares of bipartite graphs can be recognized in polynomial time and proved the NP-completeness of recognizing cubes of bipartite graphs. Lau and Corneil [10] also studied the tractability of recognizing powers of proper interval, split, and chordal graphs. Lin and Skiena [11] gave a linear-time algorithm to find square roots of planar graphs.

If $G = T^p$ for some tree T and number p, we call such a tree T a *p-th tree root* of G. Given a graph G and a positive integer p, the *p-th tree root problem*

---

* The author thanks the Institute of Information Science of Academia Sinica of Taiwan
for their hospitality and support where part of this research took place.
** Corresponding author.

L. Arge and R. Freivalds (Eds.): SWAT 2006, LNCS 4059, pp. 411–422, 2006.

asks for a tree T, if any, with $G = T^p$. Given a graph G, the *tree root problem* asks for a tree T and a number p, if any, with $G = T^p$. Ross and Harary [14] characterized squares of trees and showed that square tree roots, when they exist, are unique up to isomorphism. Lin and Skiena [11] gave a linear-time algorithm to recognize squares of trees. Kearney and Corneil [7] gave the best previously known algorithms for the p-th tree root problem and the tree root problem. Their algorithm for the p-th tree root problem runs in $O(n^3)$ time for any n-node graph, leading to an $O(n^4)$-time algorithm for the tree root problem. Gupta and Singh [4] gave a characterization of graphs which are the p-th powers of trees and proposed a heuristic algorithm to construct a p-th tree root. The running time of their algorithm is $O(n^3)$, but they did not prove the correctness of their algorithm. It was unknown whether the p-th tree root problem can be solved in $o(n^3)$ time [7,9].

In this paper we improve Kearney and Corneil's result [7] by giving linear time algorithms, in the size of the input graph, for the tree root problem as well as the p-th tree root problem for any given p. For the p-th tree root problem, our linear-time algorithm processes the input graph in two different but analogous ways, depending on whether p is even. If p is even, our algorithm is based upon a new observation that the p-th power T^p of a tree T is a chordal graph admitting a unique clique tree which is isomorphic to the $\frac{p}{2}$-centroid $T(\frac{p}{2})$ of T. Since it takes linear time to compute a clique tree for a chordal graph, $T(\frac{p}{2})$ can thus be obtained efficiently. To determine the remaining topology of T, we resort to a linear-time computable *clique-position function* for all nodes. We also prove that the existence of clique-position function is a new characterization for graphs admitting tree roots. To be more specific, there is a one-to-one mapping between the nodes of $T(\frac{p}{2})$ and the maximal cliques of T^p. The clique-position for a node u of T can be used to determine the distance between u and the node w in $T(\frac{p}{2})$ that is closest to u in T. Having determined the clique positions of all nodes, we grow the remaining tree topology from the outermost layer to the innermost layer. As for the case that p is odd, our algorithm works in an analogous way. In particular, the minimal node separators of T^p plays the role of maximal cliques of T^p for the case that p is even. The separator tree of T^p is unique and isomorphic to the $\frac{p+1}{2}$-centroid $T(\frac{p+1}{2})$ of T. The remaining topology of T can be determined by a linear-time computable *separator-position function* of T^p.

Our linear-time algorithm for the p-th tree root problem immediately yields a quadratic-time algorithm for the tree root problem. Deriving from (1) the diameters of the clique tree and separator tree of the input graph and (2) the clique positions and separator positions of nodes in the input graph, we also show a linear-time algorithm for finding the minimum p, if any, such that the input graph admits a p-th tree root. As a result, we have a linear-time algorithm for the tree root problem.

Due to space limit, the case for odd p is omitted in this extended abstract. The rest of the paper is organized as follows. Section 2 gives the preliminaries. Section 3 gives our linear-time algorithm for the p-th tree root problem for the case that p is even. Section 4 gives our linear-time algorithm for finding the smallest even number p such that the input graph admits a p-th tree root.

2   Preliminaries

For any set S, let $|S|$ denote the cardinality of S. All graphs in this paper are undirected, simple, and have no self-loops. Let $V(G)$ (respectively, $E(G)$) consist of the nodes (respectively, edges) of graph G. For any subset U of $V(G)$, let $G[U]$ denote the subgraph of G induced by U. A node is *dominating* in G if it is adjacent to all other nodes in G. Let $Dom(G)$ consist of the dominating nodes of G, which can be computed from G in linear time.

2.1   Notation for Trees

Let T be a tree. Let $Path_T(u, w)$ denote the path of T between u and w. Let $d_T(u, w)$ denote the distance of nodes u and w in T. Define $d_T(u) = \max_{w \in V(T)} d_T(u, w)$. The *diameter* d_T of a tree T is $\max_{u \in V(T)} d_T(u)$. Let $\Gamma_T(u, i)$ consist of the nodes w with $d_T(u, w) \le i$.

Define the *i-centroid* $T(i)$ of T as follows. Let $T(0) = T$. For each i with $1 \le i \le \frac{d_T}{2}$, let $T(i)$ be the tree obtained by deleting the leaf nodes of $T(i-1)$. If $i > \frac{d_T}{2}$, then $T(i)$ is the empty graph. The *centroid* $Cent(T)$ of T is $T(\lfloor \frac{d_T}{2} \rfloor)$, which is either a single node or a single edge. The centroid of T can be computed from T in $O(|V(T)|)$ time.

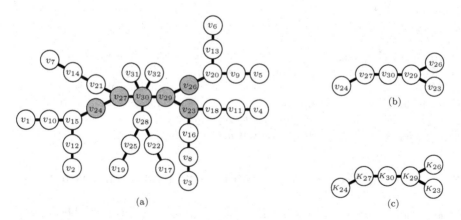

Fig. 1. (a) A tree T. (b) The 3-centroid $T(3)$ of T. (c) The clique tree of T^6, where $K_i = \Gamma_T(v_i, 3)$.

We use the tree T shown in Figure 1(a), which also appeared in [7], to illustrate the aforementioned notation: $\Gamma_T(v_{10}, 3) = \{v_1, v_2, v_{10}, v_{12}, v_{15}, v_{24}, v_{27}\}$. The 3-centroid $T(3)$ of T, as shown in Figure 1(b), is the subtree of T induced by $\{v_{24}, v_{27}, v_{30}, v_{29}, v_{26}, v_{23}\}$,

2.2   Chordal Graphs

A graph G is *chordal* if it contains no induced subgraph which is a cycle of size greater than three. Chordal graphs, which can be recognized in linear time [15, 13], have been extensively studied in the literature. It is well known that any tree power is chordal (see, e.g., [7]).

A subset S of $V(G)$ is a *separator* of a connected graph G if $G[V(G) \setminus S]$ has at least two connected components. A separator S of G is *minimal* if any proper subset of S is not a separator of G. A separator S of G is a (u, w)-*separator* of G if nodes u and w are in different connected components of $G[V(G) \setminus S]$. A (u, w)-separator S is *minimal* if any proper subset of S is not a (u, w)-separator of G. A *minimal node separator* is a minimal (u, w)-separator for some nodes u and w. Note that a minimal node separator is not necessarily a minimal separator, as the minimal (u, w)-separator may contain the minimal (x, y)-separator for some other nodes x and y. Graph G is chordal if and only if every minimal node separator of G induces a clique in G [2].

Let G be a chordal graph. Let $\mathcal{K}_G$ consist of the maximal cliques of G. For each node u of G, let $\mathcal{K}_G(u)$ consist of the maximal cliques of G containing u. A *clique tree* of a chordal graph G is a tree $\mathcal{T}$ with $V(\mathcal{T}) = \mathcal{K}_G$ such that each $\mathcal{K}_G(u)$ with $u \in V(G)$ induces a subtree of $\mathcal{T}$. Gavril [3] and Buneman [1] ensured that graph G is chordal if and only if G has a clique tree. Moreover, a clique tree of any chordal graph G can be computed in $O(|V(G)| + |E(G)|)$ time [6]. A chordal graph may have more than one clique tree [5]. A chordal graph is *uniquely representable* [8] if it admits a unique clique tree.

Lemma 1. *Let $\mathcal{T}$ be a clique tree of a chordal graph G. Then, S is a minimal node separator of G if and only if $S = K_1 \cap K_2$ for some edge (K_1, K_2) of $\mathcal{T}$.*

Proof. For each edge e of $\mathcal{T}$, let S_e consist of the nodes u of G such that e belongs to the subtree of $\mathcal{T}$ induced by $\mathcal{K}_G(u)$. That is, $S_e = \{u \in V(G) \mid e \in E(\mathcal{T}[\mathcal{K}_G(u)])\}$. Ho and Lee [5] ensured that S is a minimal node separator of G if and only if $S = S_e$ for some $e \in E(\mathcal{T})$. Therefore, it remains to ensure $S_{(K_1, K_2)} = K_1 \cap K_2$ by verifying that (K_1, K_2) is an edge of $\mathcal{T}[\mathcal{K}_G(u)]$ if and only if $\{K_1, K_2\} \subseteq \mathcal{K}_G(u)$ if and only if $u \in K_1 \cap K_2$. □

3   The •-th Tree Root Problem for any Given Even •

This section assumes that the given positive number $p \leq n$ is even. Let $h = \frac{p}{2}$.

3.1   Unique Representability

This subsection shows that if T is a tree, then T^p has a unique clique tree, which has to be isomorphic to $T(h)$.

Lemma 2. *For any tree T, we have that $G = T^p$ if and only if $\mathcal{K}_G = \{\Gamma_T(u, h) \mid u \in V(T(h))\}$.*

Proof. Observe that nodes u and w are adjacent in G if and only if there exists a maximal clique K of G that contains both u and w. Therefore, for any graphs G and H, we have that $G = H$ if and only if $\mathcal{K}_G = \mathcal{K}_H$. Gupta and Singh [4] proved that K is a maximal clique of T^p if and only if there exists a node u of $T(h)$ with $\Gamma_T(u, h) = K$. The lemma is proved. $\qquad \square$

Theorem 1. *If T is a tree, then T^p has a unique clique tree. Moreover, the clique tree of T^p is isomorphic to $T(h)$.*

Proof. Let $\mathcal{T}$ be the tree with

$$V(\mathcal{T}) = \{\Gamma_T(u, h) \mid u \in V(T(h))\};$$
$$E(\mathcal{T}) = \{(\Gamma_T(u, h), \Gamma_T(w, h)) \mid (u, w) \in E(T(h))\}.$$

It is not hard to verify that $T(h)$ and $\mathcal{T}$ are isomorphic. We first show that $\mathcal{T}$ is a clique tree of T^p. Observe that by Lemma 2, we have

$$\mathcal{K}_{T^p}(u) = \{\Gamma_T(w, h) \mid w \in V(T(h)), d_T(u, w) \leq h\}$$
$$= \{\Gamma_T(w, h) \mid w \in V(T(h)) \cap \Gamma_T(u, h)\}.$$

Since $T[V(T(h)) \cap \Gamma_T(u, h)]$ is a subtree of $T(h)$, $\mathcal{T}[\mathcal{K}_{T^p}(u)]$ is a subtree of $\mathcal{T}$. Therefore, $\mathcal{T}$ is a clique tree of T^p.

To show that T^p has no other clique tree, we resort to an observation of Kumar and Madhavan [8] stating that a chordal graph G is uniquely representable if (and only if) every minimal node separator of G is contained in exactly two maximal cliques of G. By Lemmas 1 and 2, each minimal node separator of T^p has the form $\Gamma_T(u, h) \cap \Gamma_T(w, h)$ for some nodes u and w adjacent in $T(h)$. Observe that $\Gamma_T(u, h) \cap \Gamma_T(w, h) \not\subseteq \Gamma_T(x, h)$ for any node x of $T(h)$ other than u and w. The theorem is proved. $\qquad \square$

3.2   Clique-Position Function

Let G be a uniquely representable chordal graph. Let $\mathcal{T}$ be the clique tree of G. We say that (K, i) is a *clique position* of u in G if $\mathcal{K}_G(u) = \Gamma_{\mathcal{T}}(K, i)$, i.e., the maximal cliques containing u are exactly those nodes in the clique tree $\mathcal{T}$ that are at a distance up to i from K. For notational brevity, we also write $u \in \Pi_G(K, i)$ to signify that (K, i) is a clique position of u in G, where the subscript G may be omitted when it is clear from the context. Let us use Figure 1 to explain this crucial concept of the paper. For each index i with $v_i \in V(T(3))$, let $K_i = \Gamma_T(v_i, 3)$. By Lemma 2, we know that these K_i are the maximal cliques of T^6. Observe that

$$\mathcal{K}_G(v_{14}) = \{K_{24}, K_{27}, K_{30}\} = \Gamma_{\mathcal{T}}(K_{27}, 1) = \Gamma_{\mathcal{T}}(K_{24}, 2).$$

Therefore, both $(K_{27}, 1)$ and $(K_{24}, 2)$ are clique positions of v_{14} in T^6. One can also verify that $v_{15} \in \Pi_{T^6}(K_{27}, 1) \cap \Pi_{T^6}(K_{24}, 2)$.

A *clique-position function of G with respect to p* is a function $\Phi : V(G) \to \mathcal{K}_G \times \{0, 1, \ldots, h\}$ that satisfies the following conditions.

*Condition C*1: For each $u \in V(G)$, $\Phi(u)$ is a clique position of u in G.

*Condition C*2: For each $K \in \mathcal{K}_G$, there exists a unique node u of G with $\Phi(u) = (K, h)$.

*Condition C*3: If $\Phi(u) = (K, i)$ for some $K \in \mathcal{K}_G$ and $i < h$, then there is a node w of G with $\Phi(w) = (K, i + 1)$.

Given a chordal graph T^p, an integer p, and the clique tree $\mathcal{T}$ of T^p, the rest of the subsection shows how to compute a clique-position function of T^p with respect to p. Note that our algorithm does not know T.

For each node $u \in V(T) \setminus V(T(h))$, define $\ell(u)$ to be the largest index i with $i \leq h - 1$ such that u belongs to $\Pi(K, i)$ for some maximal clique K of T^p. If $u \in V(T) \setminus V(T(h))$, then $u \in \Pi(\kappa(w), h - d_T(u, w))$, where w is the node of $T(h)$ that is closest to u in T. Therefore, $\ell(u)$ is well defined for each node $u \in V(T) \setminus V(T(h))$. To simplify the description of our algorithm, each node u of T^p is initially white, signifying that $\Phi(u)$ is still undefined. If $\Phi(u)$ is defined but may be changed later, then u is gray. If $\Phi(u)$ is defined and will not be changed later, then u is black. Our algorithm is as shown in Algorithm 1, whose correctness is ensured by the following lemma.

Input: A positive even number p, and a graph T^p and its clique tree $\mathcal{T}$.
Output: A clique-position function Φ of T^p with respect to p.

1: For each node u of T^p, compute the clique positions of u in T^p and let u be white.
2: For each $K \in \mathcal{K}_{T^p}$, choose a white node $u \in \Pi(K, h)$, let $\Phi(u) = (K, h)$, and let u be black.
3: Let K^* be a maximal clique of T^p in $V(Cent(\mathcal{T}))$.
4: For each white node $u \in Dom(T^p)$, let $\Phi(u) = (K^*, h - 1)$ and let u be gray.
5: For each white node $u \in V(T^p) \setminus Dom(T^p)$, compute $\ell(u)$.
6: **while** there are still white nodes in $V(T^p) \setminus Dom(T^p)$ **do**
7:     Let u be a white node in $V(T^p) \setminus Dom(T^p)$ with the smallest $\ell(u)$.
8:     Let $\kappa(u)$ be a maximal clique of T^p with $u \in \Pi(\kappa(u), \ell(u))$.
9:     Let $\Phi(u) = (\kappa(u), \ell(u))$ and let u be black.
10:     For each white node $w \in \Pi(\kappa(u), \ell(u))$, let $\Phi(w) = (\kappa(u), \ell(u))$ and let w be gray.
11:     **for** $j = \ell(u) + 1$ to $h - 1$ **do**
12:         Choose a non-black node $w \in \Pi(\kappa(u), j)$. Let $\Phi(w) = (\kappa(u), j)$ and let w be black.
13:         For each white node $w \in \Pi(\kappa(u), j)$, let $\Phi(w) = (\kappa(u), j)$ and let w be gray.
14:     **end for**
15: **end while**

Algorithm 1: Computing a clique-position function of T^p with respect to p

Lemma 3. *Algorithm 1 correctly computes a clique-position function of T^p with respect to p.*

Proof. Our proof is based upon the facts that T is one of the p-th tree roots of T^p and the clique tree $\mathcal{T}$ of T^p is isomorphic to $T(h)$. As we will see, the challenge of the proof lies in showing that the algorithm does not abort at Steps 2 and 12. Let us first assume that the algorithm does not abort at Steps 2 and 12, and show that the function Φ computed by the algorithm has to be a clique-position function of T^p.

- Condition C1. One can verify that the algorithm assigns $\Phi(u) = (K, i)$ only if $u \in \Pi(K, i)$. It is also easy to see that $\Phi(u)$ is defined for every node of T, i.e., no white nodes left, at the end of the algorithm. Thus, Condition C1 holds for the function Φ computed by the algorithm.
- Condition C2. Since the algorithm does not abort at Step 2, the algorithm successfully assigns clique positions for $|\mathcal{K}_{T^p}|$ nodes at the end of Step 2. Since the rest of the algorithm never assigns (K, h) to any $\Phi(u)$, Condition C2 holds for the function Φ computed by the algorithm.
- Condition C3. By Condition C2 of Φ, we know that Condition C3 holds for each node processed at Step 4. For each iteration of the while-loop (Steps 6–15), we first let $\Phi(u) = (\kappa(u), \ell(u))$ and turn u into black. Then, in the for-loop (Steps 11–14), for each index j with $\ell(u) < j < h$, the algorithm assigns $\Phi(w) = (\kappa(u), j)$ for some non-black node w and turns w into black. Therefore, as long as the algorithm does not abort at Step 12, one can see that Condition C3 holds for each node not processed at Step 4.

We then show that the algorithm does not abort at Step 2. Observe that each node u of $T(h)$ belongs to $\Pi(\Gamma_T(u, h), h)$. By Lemma 2, for each maximal clique K of T^p, the number of maximal cliques K' of T^p with $\Gamma_T(K', h) = \Gamma_T(K, h)$ is no more than $|\Pi(K, h)|$. Therefore, Step 2 successfully assigns clique positions for $|\mathcal{K}_{T^p}|$ nodes.

It remains to prove that the algorithm does not abort at Step 12. We first show that if an iteration of the while-loop enters the for-loop, then the node u of the iteration has a unique clique position in T^p. We can focus only on the case with $1 \leq \ell(u) \leq h-2$, because (1) if $\ell(u) = 0$, then u has a unique clique position in T^p, and (2) if $\ell(u) \geq h-1$, then the algorithm does not enter the for-loop. We also observe that $\kappa(u)$ cannot be a leaf of T. The reason is that if $\kappa(u)$ is a leaf of T, then T has at least one leaf whose unique clique position in T^p is $(\kappa(u), 0)$. Since the while-loop processes nodes u in non-decreasing order of $\ell(u)$, $\ell(u) \geq 1$ implies that u cannot be white at the beginning of the iteration of the while-loop. Let v be the node of $T(h)$ such that $\kappa(u) = \Gamma_T(v, h)$. Since $\kappa(u)$ is not a leaf of T, node v is not a leaf of $T(h)$. See Figure 2 for an illustration for the proof. Let $S = \Gamma_{T(h)}(v, \ell(u))$. Since $\ell(u) \geq 1$ and u is not a dominating node of T^p, there is an edge (x, y) of $T(h)$ such that $y \in S - \{v\}$ and $x \notin S$. Since v is not a leaf of $T(h)$, there has to be a neighbor w of v in $T(h)$ such that $Path_{T(h)}(w, y)$ contains v. Since $\ell(u) \leq h - 2$, we know $\Gamma_{T(h)}(w, \ell(u) + 1) \neq S$. There has to be an edge (x', y') of $T(h)$ such that $y' \in S$, $x' \notin S$, and $Path_{T(h)}(x', x)$ contains y, y', w, and v. By the existence of edges (x, y) and (x', y') of $T(h)$, we know that $S = \Gamma_{T(h)}(z, j)$ implies $z = v$ and $j = \ell(u)$. Thus, $(\kappa(u), \ell(u))$ is the unique clique position of u in T^p.

Let u_i be the node u for the i-th iteration of the while-loop that enters the for-loop. We already showed that $(\kappa(u_i), \ell(u_i))$ is the unique clique position of u_i

in T^p. Let v_i be the node with $\kappa(u_i) = \Gamma_T(v_i, h)$. By definition of our algorithm, all the maximal cliques $\kappa(u_i)$ have to be distinct. Thus, all paths $Path_T(u_i, v_i)$ are disjoint. Therefore, the number of indices i such that $Path_T(u_i, v_i)$ contains nodes in any $\Pi(\kappa, \ell)$ is no more than the cardinality of $\Pi(\kappa, \ell)$. It follows that our algorithm does not abort at Step 12.   □

3.3   A New Characterization for Tree Powers

Theorem 2. *A graph G has a p-th tree root if and only if G is a uniquely representable chordal graph admitting a clique-position function with respect to p.*

Proof. Lemma 3 ensures the only-if direction. The rest of the proof shows the other direction. Let T be the clique tree of G. Let Φ be a clique-position function of G with respect to p. We construct a tree T with $V(T) = V(G)$ as follows.

- Let S consist of the nodes u of G with $\Phi(u) = (K, h)$ for some maximal clique K of G. By Condition C2 of Φ, we know that Φ provides a one-to-one mapping between S and $\mathcal{K}_G$. Let $T[S]$ be the tree isomorphic to T via this isomorphism.
- As for each node u of G not in S, we know that $\Phi(u) = (K, i)$ for some maximal clique K of G and some index i with $0 \leq i < h$. We simply add an edge between u and an arbitrary node w with $\Phi(w) = (K, i + 1)$. By Condition C3 of Φ, such a node w always exists.

One can see that the resulting T is a tree. It is also clear that the path of T attached to $T[S]$ has length no more than h.

We first prove $T(h) = T[S]$ by showing that each leaf of $T[S]$ is attached by a length-h path in T. By definition of T, it suffices to ensure that for each leaf K of T, there exists a node v of G with $\Phi(v) = (K, 0)$: Let K' be the maximal clique of G with $(K, K') \in E(T)$. By the maximality of K, there exists a node u in $K \setminus K'$. By definition of clique tree, $\mathcal{K}_G(u)$ induces a subtree of T. Since K is a leaf of T, it follows that $\mathcal{K}_G(u) = \{K\}$, i.e., $(K, 0)$ is the unique clique position of u in G. By Condition C1 of Φ, we have $\Phi(u) = (K, 0)$.

Next we show that $\Phi(v) = (K, h)$ implies $\Gamma_T(v, h) = K$.

- To show $K \subseteq \Gamma_T(v, h)$, let u be a node in K, where $\Phi(u) = (K', i)$. Observe that $d_T(K, K') \leq i$. Let w be the node of S with $\Phi(w) = (K', h)$. Since $T[S]$ is isomorphic to T, we know $d_T(v, w) \leq i$. By Condition C1 of Φ, $u \in K'$. By the definition of T, $d_T(u, w) = h - i$. It follows that $d_T(v, u) = d_T(v, w) + d_T(w, u) \leq h$. Thus, $u \in \Gamma_T(v, h)$.

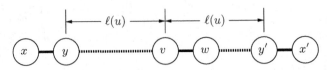

Fig. 2. An illustration for showing that u has only one clique position in T^p

- To show $\Gamma_T(v, h) \subseteq K$, let u be a node in $\Gamma_T(v, h)$, where $\Phi(u) = (K', i)$. Let w be the node with $\Phi(w) = (K', h)$. By the definition of T, $d_T(u, v) = d_T(u, w) + d_T(w, v) = h - i + d_T(w, v)$. Since $d_T(u, v) \leq h$, we have $d_T(w, v) \leq i$. Since $T[S]$ is isomorphic to T, we have that $d_T(K, K') \leq i$. By Condition C1 of Φ, $u \in \Pi(K', i)$. Hence $\mathcal{K}_G(u) = \Gamma_T(K', i)$. Since $d_T(K, K') \leq i$, we have $u \in K$.

Since $\Phi(v) = (K, h)$ implies $\Gamma_T(v, h) = K$, we have that $\mathcal{K}_G = \{\Gamma_T(v, h) \mid v \in S\}$ by Condition C2 of Φ. By $T[S] = T(h)$, we know that $\mathcal{K}_G = \{\Gamma_T(v, h) \mid v \in V(T(h))\}$. By (the "if" direction of) Lemma 2, $G = T^p$. $\qquad\square$

3.4   A Linear-Time Algorithm

Theorem 3. *The p-th tree root problem for any n-node m-edge graph G and any even number p can be solved in $O(n + m)$ time.*

Proof.   The constructive proof for the "if" direction of Theorem 2 can be implemented to run in $O(n)$ time. Since chordal graphs can be recognized in linear time [13, 15] and a clique tree T of the input chordal graph G can be obtained in linear time [6], the remaining task for solving the p-th tree root problem for G in $O(n + m)$ time is to compute a clique-position function, if any, of G with respect to p. Although Algorithm 1 is designed for computing a clique-position function for T^p, we can still run it on any chordal graph G. If the execution of the algorithm aborts at Step 2 or 12, Lemma 3 ensures that G does not admit any p-th tree root. If the algorithm does not abort at Step 12 and successfully outputs a function Φ, it takes $O(n)$ time to determine whether Φ is indeed a clique-position function of G, thereby figuring out whether G admits a p-th tree root.

It suffices to show that Step 1 of Algorithm 1, i.e., determining the clique positions of all nodes in G, can be implemented to run in $O(n+m)$ time. Observe that $\sum_{K \in \mathcal{K}_G} |V(K)| = O(n + m)$. All clique positions of all nodes in G can be computed in time linear in the size of G as follows. For each node $u \in V(G) \setminus Dom(G)$, let X_u consist of the nodes X of $\mathcal{K}_G(u)$ such that the degree of X in $T[\mathcal{K}_G(u)]$ is less than the degree of X in T. Observe that $u \in V(G) \setminus Dom(G)$ implies that X_u is non-empty. The sets $\mathcal{K}_G(u)$ and X_u for all nodes $u \in V(G) \setminus Dom(G)$ can be computed in $O(n+m)$ time. It remains to show how to solve the following problem.

Let τ be a tree, some of whose leaves are marked. A node u of τ is a *center* of τ if u has the same distance δ to all marked leaves of τ and the distance between u and any node of τ is no more than δ. Under the assumption that τ has a center, the problem is to identify all centers of τ in $O(V(\tau))$ time.

Let y be an arbitrary marked leaf of τ. If $|V(Cent(\tau))| = 1$, then let x be the single node in $Cent(\tau)$. If $|V(Cent(\tau))| = 2$, then let x be the node of $Cent(\tau)$ whose distance to y in τ is larger. If τ has centers, then x has to be one of them. Let z be a node of $V(\tau) \setminus \{x\}$. Let Z be the connected component of $\tau - x$ that contains z. One can verify that Z contains a marked leaf of τ if and only if z is not a center of τ. Therefore, the above problem can be solved in $O(|V(\tau)|)$ time. The theorem is proved.   □

4   The •-th Tree Root Problem for Unknown Even •

Theorem 4. *Let G be an n-node m-edge graph. Then, it takes $O(n + m)$ time to determine the smallest even number p, if any, such that G admits a p-th tree root.*

Proof. Since tree powers are chordal and chordal graphs can be recognized in linear time, we may assume without loss of generality that G is chordal, thereby admitting a clique tree $\mathcal{T}$. Let K^* be a node in $Cent(\mathcal{T})$. Let J consist of the even numbers j such that $\Pi(K^*, j/2)$ is non-empty. Let

$$p^* = \begin{cases} \max J & \text{if } |Dom(G)| = 0; \\ d_{\mathcal{T}} + |Dom(G)| - 1 & \text{if } 1 \leq |Dom(G)| \leq 2; \\ 2\lceil d_{\mathcal{T}}/2 \rceil + 2 & \text{if } |Dom(G)| \geq 3. \end{cases}$$

Let P consist of the positive numbers p such that the input graph admits a p-tree root. We first show that if P is non-empty, then $p^* = \min P$. For brevity of proof, we regard $\mathcal{T}$ as being rooted at K^*: For any maximal clique K of G other than K^*, the *parent* of K in $\mathcal{T}$, denote $\pi(K)$, is the maximal clique K' of G such that (K, K') is an edge of $Path_{\mathcal{T}}(K, K^*)$. We say that K is a *child* of $\pi(K)$ in $\mathcal{T}$.

Case 1: $|Dom(G)| = 0$. By Condition C2 of any clique-position function of G with respect to p, we have $P \subseteq J$. We show $P = \{\max J\}$ by proving that that $j \leq p$ holds for any even numbers $j \in J$ and $p \in P$. Assume for a contradiction that $j > p$ holds for some even numbers $j \in J$ and $p \in P$. Let T be a p-th tree root of G. Let $v(\cdot)$ be an isomorphism between T and $T(p/2)$. That is, $v(K)$ is the node in $V(T(p/2))$ such that $\Gamma_T(v(K), p/2) = K$.

Since $Dom(G) = \emptyset$, we know $d_T > 2p$. Therefore, $d_{\mathcal{T}} = d_{T(p/2)} > p$. It follows from $d_{\mathcal{T}} > p$ and $j > p$ that

$$\Gamma_{\mathcal{T}}(K^*, p/2) \subsetneq \Gamma_{\mathcal{T}}(K^*, j/2).$$

Let w be a node in $\Pi(K^*, j/2)$, implying $\mathcal{K}_G(w) = \Gamma_{\mathcal{T}}(K^*, j/2)$. Let $u^* = v(K^*)$. We know $u^* \in \Pi(K^*, p/2)$, implying $\mathcal{K}_G(u^*) = \Gamma_{\mathcal{T}}(K^*, p/2)$. It follows that

$$\mathcal{K}_G(u^*) \subsetneq \mathcal{K}_G(w),$$

thereby $u^* \neq w$. Since K^* is a centroid of $\mathcal{T}$, u^* is also a centroid of $T(p/2)$. By $d_{T(p/2)} > p$ and $u^* \neq w$, there is a node u in $V(T(p/2))$ such that $d_T(u, u^*) = p/2$ and $d_T(u, w) > p/2$. Let $K = \Gamma_T(u, h)$. We have $u^* \in V(K)$ and $w \notin V(K)$. Thus, K is a maximal clique in $\mathcal{K}_G(u^*) \backslash \mathcal{K}_G(w)$, contradicting $\mathcal{K}_G(u^*) \subsetneq \mathcal{K}_G(w)$.

Case 2: $1 \leq |Dom(G)| \leq 2$. We show that P consists of the single number $d_T - |Dom(G)| + 1$ as follows. Let p be a number in P. Let T be a p-th tree root of G. By $1 \leq |Dom(G)| \leq 2$, it is not hard to see that $Dom(G)$ consists of the centroids of T. If $|Dom(G)| = 1$, then $d_T = 2p$. It follows that $d_{T(p/2)} = d_T = p$. If $|Dom(G)| = 2$, then $d_T = 2p - 1$. It follows that $d_{T(p/2)} = d_T = p - 1$. For either case, we have $p = d_T + |Dom(G)| - 1$. Since G has exactly one clique tree T, the diameter of T is fixed. We have $|P| = 1$.

Case 3: $|Dom(G)| \geq 3$. We show that $\min P = 2\lceil d_T/2 \rceil + 2$. Let p be an index in P. Let T be a p-th tree root of G. By $|Dom(G)| \geq 3$, one can verify that $d_T \leq 2p - 2$. Therefore, $d_T = d_{T(p/2)} \leq p - 2$, implying that $\min P \geq d_T + 2$. Observe that $2\lceil d_T/2 \rceil + 2$ is the smallest even number that is no less than $d_T + 2$. We prove the equality by showing that if $p \geq d_T + 4$, then G also admits a $(p-2)$-nd tree root.

The rest of the proof assumes $d_T \leq p - 4$, which directly implies that

$$\Gamma_T(K^*, p/2 - 2) = \mathcal{K}_G. \tag{1}$$

For any maximal clique K of G other than K^*, observe that $d_T \leq p - 4$ also implies

$$\Pi(K, p/2) \subseteq \Pi(\pi(K), p/2 - 1); \tag{2}$$
$$\Pi(K, p/2 - 1) \subseteq \Pi(\pi(K), p/2 - 2). \tag{3}$$

Let Φ be a clique-position function of G with respect to p. Let $\Phi_1(u)$ (respectively, $\Phi_2(u)$) denote the first (respectively, second) component of $\Phi(u)$. Let Φ' be obtained from Φ by the following steps.

1. For each node u with $\Phi_2(u) \leq p/2 - 2$, we simply let $\Phi'(u) = \Phi(u)$.
2. For the node u with $\Phi(u) = (K^*, p/2)$, i.e., $u = v(K^*)$, let $\Phi'(u) = (K^*, p/2 - 1)$. For each node u of G with $\Phi(u) = (K^*, p/2 - 1)$, let $\Phi'(u) = (K^*, p/2 - 2)$.
3. For each maximal clique $K \neq K^*$ of G that is not a leaf of T, we do the following:
   (a) Choose an arbitrary child K_1 of K in T and let $\Phi'(v(K_1)) = (K, p/2 - 1)$.
   (b) For each child K_2 of K other than K_1, let $\Phi'(v(K_2)) = (\pi(K), p/2 - 2)$.
   (c) For each node u of G with $\Phi(u) = (K, p/2 - 1)$, let $\Phi'(u) = (\pi(K), p/2 - 2)$.
4. For each maximal clique $K \neq K^*$ of G that is a leaf of T, we do the following:
   (a) Choose an arbitrary node u with $\Phi(u) = (K, p/2 - 1)$ and let $\Phi'(u) = (K, p/2 - 1)$.
   (b) For each node $w \neq u$ with $\Phi(w) = (K, p/2 - 1)$, let $\Phi'(w) = (\pi(K), p/2 - 2)$.
5. Let u be the node with $\Phi(u) = (K^*, p/2)$. Let $\Phi'(u) = (K^*, p/2 - 1)$. For each node w with $\Phi(w) = (K, p/2)$ and $\pi(K) = K^*$, let $\Phi'(w) = (K^*, p/2 - 2)$.

We show that Φ' is a clique-position function of G with respect to $p - 2$.

- Condition C1: One can verify that $\Phi'(u)$ is well defined for each node u of G. Moreover, by Condition C1 of Φ and Equations (1), (2), and (3), one can also verify that $\Phi'(u)$ is a clique position of u in G.
- Condition C2: By Condition C2 of Φ and Steps 2, 3(a), 4(a), and 5, one can see that for each maximal clique K of G, there exists a unique node u of G with $\Phi'(u) = (K, p/2 - 1)$.
- Condition C3: Condition C2 of Φ' ensures that if u is a node with $\Phi'_2(u) = p/2 - 2$, then there is a node w with $\Phi'_1(w) = \Phi'_1(u)$ and $\Phi'_2(w) = \Phi'_2(u) + 1$. Condition C3 of Φ and Step 1 ensures that if u is a node with $\Phi'_2(u) < p/2 - 2$, then there is a node w with $\Phi'_1(w) = \Phi'_1(u)$ and $\Phi'_2(w) = \Phi'_2(u) + 1$.

Observe that whether P is empty or not, p^* can always be computed from G in linear time. (This includes the case that $J = Dom(G) = \emptyset$, i.e., p^* is undefined.) Thus, the above proof reduces the tree root problem G to the p^*-th tree root problem for G, which by Theorem 3 can be solved in linear time.   □

References

1. P. Buneman. Characterization of rigid circuit graphs. *Discrete Mathematics*, 9:205–212, 1974.
2. G. A. Dirac. On rigid circuit graphs. *Abhandlungen aus dem Mathematischen Seminar der Universität Hamburg*, 25:71–76, 1961.
3. F. Gavril. The intersection graphs of subtrees in trees are exactly chordal graphs. *Journal of Combinatorial Theory, Series B*, 16:47–56, 1974.
4. S. K. Gupta and A. Singh. On tree roots of graphs. *International Journal of Computer Mathematics*, 73:157–166, 1999.
5. C.-W. Ho and R. C. T. Lee. Counting clique trees and computing perfect elimination schemes in parallel. *Information Processing Letters*, 31:61–68, 1989.
6. W.-L. Hsu and T.-H. Ma. Fast and simple algorithms for recognizing chordal comparability graphs and interval graphs. *SIAM Journal on Computing*, 28:1004–1020, 1999.
7. P. E. Kearney and D. G. Corneil. Tree powers. *Journal of Algorithms*, 29:111–131, 1998.
8. P. S. Kumar and C. E. V. Madhavan. Clique tree generalization and new subclasses of chordal graphs. *Discrete Applied Mathematics*, 117:109–131, 2002.
9. L. C. Lau. Bipartite roots of graphs. In *Proceedings of the Fifteenth Annual ACM-SIAM Symposium on Discrete Algorithms*, pages 952–961, 2004.
10. L. C. Lau and D. G. Corneil. Recognizing powers of proper interval, split, and chordal graphs. *SIAM Journal on Computing*, 18(1):83–102, 2004.
11. Y. L. Lin and S. Skiena. Algorithms for square roots of graphs. *SIAM Journal on Discrete Mathematics*, 8:99–118, 1995.
12. R. Motwani and M. Sudan. Computing roots of graphs is hard. *Discrete Applied Mathematics*, 54:81–88, 1994.
13. D. Rose, R. Tarjan, and G. Lueker. Algorithmic aspects of vertex elimination of graph. *SIAM Journal on Computing*, 5(2):266–283, 1976.
14. I. C. Ross and F. Harary. The squares of a tree. *Bell System Technical Journal*, 39:641–647, 1960.
15. R. Tarjan and M. Yannakakis. Simple linear time algorithms to test chordality of graphs, test acyclicity of hypergraphs and selectively reduce acyclic hypergraphs. *SIAM Journal on Computing*, 13(3):566–576, 1984.

Generalized Powers of Graphs and Their Algorithmic Use

Andreas Brandstädt[1], Feodor F. Dragan[2], Yang Xiang[2], and Chenyu Yan[2]

[1] Universität Rostock, FB Informatik, Albert–Einstein–Str. 21,
D–18051 Rostock, Germany
ab@informatik.uni-rostock.de

[2] Department of Computer Science, Kent State University, Kent, OH 44242, USA
dragan, yxiang, cyan@cs.kent.edu

Abstract. Motivated by the frequency assignment problem in heterogeneous multihop radio networks, where different radio stations may have different transmission ranges, we introduce two new types of coloring of graphs, which generalize the well-known Distance-k-Coloring. Let $G = (V, E)$ be a graph modeling a radio network, and assume that each vertex v of G has its own transmission radius $r(v)$, a nonnegative integer. We define r-coloring (r^+-coloring) of G as an assignment $\Phi : V \mapsto \{0, 1, 2, \ldots\}$ of colors to vertices such that $\Phi(u) = \Phi(v)$ implies $d_G(u, v) > r(v) + r(u)$ ($d_G(u, v) > r(v) + r(u) + 1$, respectively). The r-Coloring problem (the r^+-Coloring problem) asks for a given graph G and a radius-function $r : V \mapsto N \cup \{0\}$, to find an r-coloring (an r^+-coloring, respectively) of G with minimum number of colors. Using a new notion of generalized powers of graphs, we investigate the complexity of the r-Coloring and r^+-Coloring problems on several families of graphs.

1 Introduction

The Frequency Assignment Problem (FAP) in multihop radio networks is the problem of assigning frequencies to transmitters exploiting frequency reuse while keeping signal interference to acceptable levels. The FAP is usually modeled by variations of the graph coloring problem. The $L(\delta_1, \delta_2, \ldots, \delta_k)$-coloring of a graph $G = (V, E)$, where δ_is are positive integers, is an assignment function $\Phi : V \mapsto N \cup \{0\}$ such that $|\Phi(u) - \Phi(v)| \geq \delta_i$ when the distance between u and v in G is equal to i ($i \in \{1, 2, \ldots, k\}$). The aim is to minimize the range of the frequencies used, i.e., we search for the minimum λ such that G admits a $L(\delta_1, \delta_2, \ldots, \delta_k)$-coloring with frequencies between 0 and λ. Let us denote that minimum by $\lambda_{\delta_1, \delta_2, \ldots, \delta_k}(G)$. Unfortunately, already a restricted version of this problem, the $L(2, 1)$-coloring problem (called also the *Radiocoloring* problem), is NP-complete even for planar graphs, bipartite graphs, chordal graphs and split graphs [5, 21], the classes of graphs where the ordinary graph coloring problem is easily (polynomial time) solvable. Polynomial time algorithms for optimal $L(2, 1)$-coloring are known only for trees [11, 24], cographs [11], k-almost trees [19] and for very regular graphs such as triangular grids, rectangular grids and hexagonal grids (see [4, 10] and papers cited therein).

L. Arge and R. Freivalds (Eds.): SWAT 2006, LNCS 4059, pp. 423–434, 2006.
© Springer-Verlag Berlin Heidelberg 2006

Another variation of FAP considers k-powers, G^k ($k = 1, 2, 3, \ldots$), of a given graph G. The kth power G^k of $G = (V, E)$ has the same vertex set V but two vertices v and u are adjacent in G^k if and only if their distance in G is at most k. The problem is to color the kth power of G with minimum number of colors, denoted by $\chi(G^k)$. This problem is extensively studied in literature and often called *Distance-k-Coloring*. Again, the problem is NP-complete even for chordal graphs [1] and for planar graphs and $k = 2$ [31]. Moreover, it is computationally hard to approximately color the even powers of n-vertex chordal graphs within an $n^{\frac{1}{2}-\epsilon}$ factor, for any $\epsilon > 0$ [1]. Exact polynomial time algorithms are known only for some special graph classes: for graphs with bounded tree-width [24], for graphs with bounded clique-width [33], and for interval, strongly chordal, doubly chordal, trapezoid, d-trapezoid, and cocomparability graphs as those families of graphs are closed under taking powers [7, 14, 15, 16, 20, 32] and the ordinary coloring problem on them is polynomial time solvable [23, 26, 30]. Approximation algorithms for coloring powers of chordal graphs and squares of planar graphs are presented in [2, 28].

The $L(\delta_1, \delta_2, \ldots, \delta_k)$-coloring problem and the Distance-k-Coloring problem are related. Clearly, $\chi(G^k) - 1 = \lambda_{\underbrace{1, 1, \ldots, 1}_{k}}(G) \le \lambda_{\delta_1, \delta_2, \ldots, \delta_k}(G)$ and, since from

a valid coloring of G^k we can always get a valid $L(\delta_1, \delta_2, \ldots, \delta_k)$-coloring of G by multiplying by $t := max_{1 \le i \le k}\{\delta_i\}$ the assigned color of each vertex, we have also $\lambda_{\delta_1, \delta_2, \ldots, \delta_k}(G) \le t(\chi(G^k) - 1)$. Hence, an algorithm solving the Distance-k-Coloring problem for a class of graphs also provides a ($max_{1 \le i \le k}\{\delta_i\}$)-approximation for the $L(\delta_1, \delta_2, \ldots, \delta_k)$-coloring problem. So, it is natural to investigate graph classes for which powers G^k are easy to color.

In this paper, we extend the second variant of the Frequency Assignment Problem to the so-called heterogeneous multihop radio networks where different radio stations may have different transmission ranges. In this model, two radio stations x and y must not receive the same frequency if there is a third radio station z which is within the transmission ranges of both x and y (to avoid collisions at z). In a more restricted model, we may forbid even two radio stations to have the same frequency if their transmission areas are very close. More formally, let $G = (V, E)$ be a graph modeling a radio network, and assume that each vertex v of G has its own transmission radius $r(v)$, a non-negative integer. We define *r-coloring* of G as an assignment $\Phi : V \mapsto \{0, 1, 2, \ldots\}$ of colors to vertices such that $\Phi(u) = \Phi(v)$ implies $d_G(u, v) > r(v) + r(u)$, and *r+-coloring* of G as an assignment $\Phi : V \mapsto \{0, 1, 2, \ldots\}$ of colors to vertices such that $\Phi(u) = \Phi(v)$ implies $d_G(u, v) > r(v) + r(u) + 1$. Here, $d_G(u, v)$ is the shortest path distance between u and v in G. The *r-Coloring problem* (the *r+-Coloring problem*) asks for a given graph G and a radius-function $r : V \mapsto N \cup \{0\}$, to find an r-coloring (an r^+-coloring, respectively) of G with minimum number of colors. Clearly, if $r(v) = l$ (l is a fixed integer) for each $v \in V$, then r-coloring is just an ordinary coloring of G^{2l} and r^+-coloring is just an ordinary coloring of G^{2l+1}. Hence, the r-Coloring and r^+-Coloring problems generalize the Distance-k-Coloring.

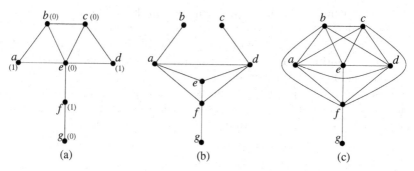

Fig. 1. A graph G with a radius-function $r : V \mapsto N \cup \{0\}$ (a), and the corresponding graphs $L(\mathcal{D}(G,r))$ (b) and $\Gamma(\mathcal{D}(G,r))$ (c).

For a graph $G = (V, E)$ with a radius-function $r : V \mapsto N \cup \{0\}$, we can define two new graphs $\Gamma(\mathcal{D}(G,r))$ and $L(\mathcal{D}(G,r))$ (and call them *generalized powers* of G) as follows. Both $\Gamma(\mathcal{D}(G,r))$ and $L(\mathcal{D}(G,r))$ have the same vertex set V as G has, and vertices $u, v \in V$ form an edge in $\Gamma(\mathcal{D}(G,r))$ (in $L(\mathcal{D}(G,r))$) if and only if $d_G(u, v) \leq r(v) + r(u) + 1$ ($d_G(u, v) \leq r(v) + r(u)$, respectively). Figure 1 shows a graph G with a radius-function $r : V \mapsto N \cup \{0\}$, and the corresponding graphs $\Gamma(\mathcal{D}(G,r))$ and $L(\mathcal{D}(G,r))$. It is easy to see that an r-coloring of G is nothing else than an ordinary coloring of $L(\mathcal{D}(G,r))$ and an r^+-coloring of G is nothing else than an ordinary coloring of $\Gamma(\mathcal{D}(G,r))$.

In this paper, we investigate the r-Coloring and r^+-Coloring problems on special graph classes. We are interested in determining large families of graphs G for which the graphs $L(\mathcal{D}(G,r))$ and/or $\Gamma(\mathcal{D}(G,r))$ have enough structure to exploit algorithmically and to solve the r-Coloring and/or r^+-Coloring problems on G efficiently. Among other results we show that

- if G is a chordal (interval, circular-arc, cocomparability, weakly chordal) graph, then for any radius-function $r : V \mapsto N \cup \{0\}$, the graph $\Gamma(\mathcal{D}(G,r))$ is chordal (resp., interval, circular-arc, cocomparability, weakly chordal);
- if G is a chordal graph with chordal square G^2 (a so called power-chordal graph), then for any radius-function $r : V \mapsto N \cup \{0\}$, the graphs $\Gamma(\mathcal{D}(G,r))$ and $L(\mathcal{D}(G,r))$ are chordal;
- if G is a weakly chordal graph with weakly chordal square G^2, then for any radius-function $r : V \mapsto N \cup \{0\}$, the graphs $\Gamma(\mathcal{D}(G,r))$ and $L(\mathcal{D}(G,r))$ are weakly chordal;
- if G is a distance-hereditary graph, then for any radius-function $r : V \mapsto N \cup \{0\}$, the graph $\Gamma(\mathcal{D}(G,r))$ is weakly chordal and the graph $L(\mathcal{D}(G,r))$ is chordal;
- if G is an AT-free graph, then for any radius-function $r : V \mapsto N$, the graphs $\Gamma(\mathcal{D}(G,r))$ and $L(\mathcal{D}(G,r))$ are cocomparability graphs (note that $r(v) = 0$ is not allowed here);
- if G is a cocomparability (interval, circular-arc) graph, then for any radius-function $r : V \mapsto N$, the graph $L(\mathcal{D}(G,r))$ is cocomparability (resp., interval, circular-arc).

Since the (ordinary) coloring problem on chordal graphs, weakly chordal graphs, interval graphs and cocomparability graphs are polynomial time solvable (see [23, 26, 30]), we immediately obtain polynomial time solvability of the corresponding r-Coloring and/or r^+-Coloring problems on those graphs. Unfortunately, coloring of circular-arc graphs is an NP-complete problem [22]. However, one can use the circular-arc graph coloring approximation algorithm of [27] with a performance ratio of $3/2$ to get an approximate solution for the corresponding r-Coloring and/or r^+-Coloring problems on circular-arc graphs.

2   Notations and Preliminaries

Let $G = (V, E)$ be a finite, undirected, connected and simple (i.e. without loops and multiple edges) graph. For two vertices $x, y \in V$, the *distance* $d_G(x, y)$ is the length (i.e. number of edges) of a shortest path connecting x and y. By $N_G(v) = \{u : uv \in E\}$ and $N_G[v] = N_G(v) \cup \{v\}$ we denote the *open neighborhood* and the *closed neighborhood* of v, respectively. If no confusion can arise we will omit the index G. Let $\mathcal{N}(G) = \{N[v] : v \in V\}$ be the *family of closed neighborhoods* of G. The *disk* centered at v with radius k is the set of all vertices having distance at most k to v: $N^k[v] = \{u : u \in V \text{ and } d(u, v) \leq k\}$. Denote by $\mathcal{D}(G) = \{N^r[v] : v \in V, r \text{ a non-negative integer}\}$ the *family of all disks* of G and by $\mathcal{N}^k(G) = \{N^k[v] : v \in V\}$, where k is a fixed non-negative integer, the *family of all disks of radius* k of G. The *kth power* of a graph $G = (V, E)$ is the graph $G^k = (V, U)$, where two vertices $x, y \in V$ are adjacent in G^k if and only if $d_G(x, y) \leq k$.

For a graph G, consider a family $\mathcal{S} = \{S_1, \ldots, S_l\}$ of subsets of V, i.e., $S_i \subseteq V$, $i = 1, \ldots, l$. The *intersection graph* $L(\mathcal{S})$ of $\mathcal{S}$ is defined as follows. The sets from $\mathcal{S}$ are the vertices of $L(\mathcal{S})$ and two vertices of $L(\mathcal{S})$ are joined by an edge if and only if the corresponding sets intersect. The *visibility graph* $\Gamma(\mathcal{S})$ of $\mathcal{S}$ is defined as follows. The sets from $\mathcal{S}$ are the vertices of $\Gamma(\mathcal{S})$ and two vertices of $\Gamma(\mathcal{S})$ are joined by an edge if and only if the corresponding sets are visible to each other. We say that sets S_i and S_j are *visible to each other* if $S_i \cap S_j \neq \emptyset$ or there is an edge of G with one end in S_i and the other end in S_j.

It is easy to see, from the definitions, that

- two disks $N^p[v]$ and $N^q[u]$ of G are intersecting if and only if $d_G(u, v) \leq p+q$ and are visible to each other if and only if $d_G(u, v) \leq p + q + 1$,
- $L(\mathcal{N}^k(G))$ is isomorphic to G^{2k} ($k \geq 1$), i.e., $G^{2k} \simeq L(\mathcal{N}^k(G))$,
- $\Gamma(\mathcal{N}^k(G))$ is isomorphic to G^{2k+1} ($k \geq 0$), i.e., $G^{2k+1} \simeq \Gamma(\mathcal{N}^k(G))$.

Definitions of graph classes considered are given in appropriate sections.

3   c-Chordal Graphs

In this section we consider the generalized powers of c-chordal graphs. A graph is called *chordal* if it has no induced cycles of size greater than 3 and is called *c-chordal* if it has no induced cycles of size greater than c ($c \geq 3$).

Let a maximal induced cycle of G be an induced cycle of G with maximum number of edges. Denote by $l(G)$ the number of edges of a maximal induced cycle of G. The parameter $l(G)$ of a graph G is often called the *chordality* of G. Clearly, the chordal graphs are exactly the graphs of chordality 3 and c-chordal graphs are exactly the graphs of chordality c. In [7], an important lemma is proven which connects the chordality of $L(\mathcal{D}(G))$ with the chordality of G^2.

Lemma 1. *[7] For any graph G, $l(L(\mathcal{D}(G))) = l(G^2)$.*

One can prove a similar result for graphs G and $\Gamma(\mathcal{D}(G))$ (proof is omitted).

Lemma 2. *For any graph G with $l(G) \geq 3$, $l(\Gamma(\mathcal{D}(G))) = l(G)$.*

From Lemma 1 and Lemma 2 we conclude.

Theorem 1. *For a graph G,*

1) *$\Gamma(\mathcal{D}(G))$ is c-chordal if and only if G is c-chordal,*
2) *$L(\mathcal{D}(G))$ is c-chordal if and only if G^2 is c-chordal.*

Let now $G = (V, E)$ be a graph and $r : V \mapsto N \cup \{0\}$ be a non–negative integer–valued radius-function defined on V. For a graph G with a radius-function $r : V \mapsto N \cup \{0\}$, define a subfamily $\mathcal{D}(G, r)$ of the family of all disks $\mathcal{D}(G)$ of G as follows: $\mathcal{D}(G, r) = \{N^{r(v)}[v] : v \in V\}$.

Clearly, graphs $L(\mathcal{D}(G, r))$ and $\Gamma(\mathcal{D}(G, r))$ are induced subgraphs of the graphs $L(\mathcal{D}(G)))$ and $\Gamma(\mathcal{D}(G))$, respectively. Furthermore, the graph $L(\mathcal{D}(G, r))$ can be viewed (by identifying every disk with its center) as a graph on the vertex set V, where two vertices $u, v \in V$ are adjacent in $L(\mathcal{D}(G, r))$ if and only if $d_G(u, v) \leq r(u) + r(v)$. Similarly, the graph $\Gamma(\mathcal{D}(G, r))$ can be viewed as a graph on the vertex set V, where two vertices $u, v \in V$ are adjacent in $\Gamma(\mathcal{D}(G, r))$ if and only if $d_G(u, v) \leq r(u) + r(v) + 1$. Thus, graphs $\Gamma(\mathcal{D}(G, r))$ and $L(\mathcal{D}(G, r))$ are generalizations of odd and even, respectively, powers of G.

Since induced subgraphs of c-chordal graphs are c-chordal, we can state the following corollaries from Lemma 1 and Lemma 2.

Corollary 1. *For any c-chordal graph G and any radius-function $r : V \mapsto N \cup \{0\}$ defined on the vertex set of G, the graph $\Gamma(\mathcal{D}(G, r))$ is c-chordal. In particular, odd powers G^{2k+1} $(k = 1, 2, \ldots)$ of a c-chordal graph G are c-chordal.*

Corollary 2. *Let G be a graph having c-chordal square G^2. Then, for any radius-function $r : V \mapsto N \cup \{0\}$ defined on the vertex set of G, the graph $L(\mathcal{D}(G, r))$ is c-chordal. In particular, if the square G^2 of some graph G is c-chordal, then all even powers G^{2k} $(k = 1, 2, \ldots)$ of G are c-chordal.*

Note that, Corollary 1 generalizes the known fact that *odd powers of chordal graphs are chordal* [1, 18, 29]. Corollary 2 generalizes the known fact that *even powers of square-chordal graphs are chordal* [7, 18]. Here, a graph G is *square-chordal* if its square G^2 is a chordal graph. Note also that the class of square-chordal graphs comprises such known families of graphs as trees, interval graphs, directed path graphs, strongly chordal graphs, doubly chordal graphs, distance-hereditary graphs, dually chordal graphs, homogeneous graphs, homogeneously orderable graphs and others (see, for example, [7, 8, 9] and papers cited therein). But, it still remains an open question to give a complete characterization of the whole family of square-chordal graphs. As it was shown in [29], squares of chordal graphs are not necessarily chordal; in fact, a square of a chordal graph can have arbitrarily large chordality [29].

In [7], we defined *power-chordal graphs* as graphs G all powers G^k ($k \geq 1$) of which are chordal (or, equivalently, if both G and G^2 are chordal). For this family of graphs we have.

Corollary 3. *For a power-chordal graph G, graphs $\Gamma(\mathcal{D}(G, r))$, $L(\mathcal{D}(G, r))$ are chordal for any radius-function $r : V \mapsto N \cup \{0\}$ defined on the vertex set of G.*

Notice that the class of power-chordal graphs comprises such known families of graphs as trees, interval graphs, directed path graphs, strongly chordal graphs and doubly chordal graphs [7].

4   Weakly Chordal Graphs

In this section we consider the generalized powers of weakly chordal graphs. In what follows, the complement of a graph G is denoted by $\overline{G}$, C_k is an induced cycle on k vertices and $\overline{C_k}$ is the complement of C_k. A graph G is *weakly chordal* if both G and $\overline{G}$ are 4-chordal, i.e., G has neither C_k nor $\overline{C_k}$, $k > 4$, as an induced subgraph.

Lemma 3. *Let G be a graph such that $\overline{L(\mathcal{N}(G))} \simeq \overline{G^2}$ is a 4-chordal graph and $L(\mathcal{D}(G))$ has no induced subgraphs isomorphic to C_5. Then, $\overline{L(\mathcal{D}(G))}$ is a 4-chordal graph, too.*

Proof. Assume that the graph $\overline{L(\mathcal{D}(G))}$ is not 4-chordal. Then, there must exist an induced cycle C_{k+1} in $\overline{L(\mathcal{D}(G))}$ such that $k + 1 > 4$. In fact, $k + 1$ is larger than 5 since $\overline{L(\mathcal{D}(G))}$ cannot have an induced subgraph isomorphic to C_5 (notice that an induced cycle on 5 vertices is self-complementary). We will assume that k is minimal, i.e., any cycle of $\overline{L(\mathcal{D}(G))}$ of length t ($4 < t \leq k$) has a chord. Let C_{k+1} be formed by disks $N^{r_0}[x_0], N^{r_1}[x_1], \ldots, N^{r_{k-1}}[x_{k-1}], N^{r_k}[x_k]$ of G, i.e., in G we have $N^{r_i}[x_i] \cap N^{r_j}[x_j] = \emptyset$ if and only if $i = j \pm 1 (mod(k + 1))$. Among all such induced cycles C_{k+1} of $\overline{L(\mathcal{D}(G))}$ we will choose one with minimum sum $\sigma := r_0 + r_1 + \cdots + r_k$. Clearly, $r_i > 0$ for each i. We will show that $r_0 = r_1 = \cdots = r_k = 1$ holds.

Assume, without loss of generality, that $r_0 > 1$. Consider a neighbor y of x_0 on a shortest path of G from x_0 to x_2 and a neighbor z of x_0 on a shortest path of G from x_0 to x_3. Since $L(\mathcal{D}(G))$ has no induced subgraphs isomorphic to C_5, a cycle of $L(\mathcal{D}(G))$ formed by disks $N^{r_0-1}[y], N^{r_0-1}[z], N^{r_2}[x_2], N^{r_3}[x_3], N^{r_k}[x_k]$ cannot be induced. Therefore, as $N^{r_0}[x_0] \cap N^{r_k}[x_k] = \emptyset$, we must have $N^{r_0-1}[y] \cap N^{r_3}[x_3] \neq \emptyset$ or $N^{r_0-1}[z] \cap N^{r_2}[x_2] \neq \emptyset$. Let, without loss of generality, $N^{r_0-1}[y] \cap N^{r_3}[x_3] \neq \emptyset$.

Thus, there is a neighbor y of x_0 such that $N^{r_0-1}[y] \cap N^{r_2}[x_2] \neq \emptyset$ and $N^{r_0-1}[y] \cap N^{r_3}[x_3] \neq \emptyset$. Next we claim that $N^{r_0-1}[y] \cap N^{r_i}[x_i] \neq \emptyset$ for every $i = 4, \ldots, k-1$. Let i_0 $(3 < i_0 < k)$ be the minimal index such that $N^{r_0-1}[y] \cap N^{r_{i_0}}[x_{i_0}] = \emptyset$. Then, it is easy to see that the disks $N^{r_0-1}[y], N^{r_1}[x_1], N^{r_2}[x_2],$ $N^{r_3}[x_3], \ldots, N^{r_{i_0}}[x_{i_0}]$ form an induced cycle in $\overline{L(\mathcal{D}(G))}$ of length $t := i_0 + 1$ with $4 < t \leq k$, and a contradiction with the choice of k arises.

Now, disks $N^{r_0-1}[y]$ and $N^{r_i}[x_i]$ intersect if and only if $i \in \{2, 3, \ldots, k -2, k-1\}$, i.e., $i \neq 1, k$. But, then the disks $N^{r_0-1}[y], N^{r_1}[x_1], N^{r_2}[x_2], N^{r_3}$ $[x_3], \ldots, N^{r_{k-1}}[x_{k-1}], N^{r_k}[x_k]$ of G induce a cycle in $L(\mathcal{D}(G))$ of length $k+1$ and, since the sum of radiuses of these disks is $\sigma - 1$, a contradiction with the minimality of $\sigma := r_0 + r_1 + \cdots + r_k$ occurs.

Consequently, $r_0 = r_1 = \cdots = r_k = 1$ must hold, implying that C_{k+1} with $k + 1 > 4$ is also an induced cycle of $\overline{L(\mathcal{N}(G))}$. The latter contradicts now with $\overline{L(\mathcal{N}(G))} \simeq \overline{G^2}$ being a 4-chordal graph. Obtained contradictions prove the lemma. □

Combining Lemma 1 with Lemma 3 we obtain the following results. Notice that induced subgraphs of weakly chordal graphs are weakly chordal, too.

Theorem 2. *For a graph G, $L(\mathcal{D}(G))$ is weakly chordal if and only if G^2 is weakly chordal.*

Proof. By Lemma 1, $L(\mathcal{D}(G))$ is 4-chordal if and only if G^2 is so. Assuming now that $L(\mathcal{D}(G))$ is 4-chordal, by Lemma 3, $\overline{L(\mathcal{D}(G))}$ is 4-chordal if and only if $\overline{G^2}$ is 4-chordal (notice that G^2 is an induced subgraph of $L(\mathcal{D}(G))$). Hence, $L(\mathcal{D}(G))$ is weakly chordal if and only if G^2 is weakly chordal. □

Corollary 4. *Let G be a graph having weakly chordal square G^2. Then, for any radius-function $r : V \mapsto N \cup \{0\}$ defined on the vertex set of G, the graph $L(\mathcal{D}(G, r))$ is weakly chordal. In particular, if the square G^2 of some graph G is weakly chordal, then all even powers G^{2k} $(k = 1, 2, \ldots)$ of G are weakly chordal.*

Lemma 4. *Let G be the complement of a 4-chordal graph and $\Gamma(\mathcal{D}(G))$ has no induced subgraphs isomorphic to C_5, C_6. Then, $\overline{\Gamma(\mathcal{D}(G))}$ is a 4-chordal graph.*

Proof of this lemma is omitted. Combining Lemma 2 with Lemma 4 we obtain the following results.

Theorem 3. *For a graph G, $\Gamma(\mathcal{D}(G))$ is weakly chordal if and only if G is weakly chordal.*

Corollary 5. *For any weakly chordal graph G and any radius-function $r : V \mapsto N \cup \{0\}$ defined on the vertex set of G, the graph $\Gamma(\mathcal{D}(G, r))$ is weakly chordal. In particular, odd powers G^{2k+1} $(k = 1, 2, \ldots)$ of a weakly chordal graph G are weakly chordal.*

Recall (see, e.g., [9]) that a graph G is *distance-hereditary* if each induced path of G is shortest. It is known that any distance-hereditary graph G is weakly chordal and its square G^2 is even chordal [3, 8]. Hence the following result holds.

Corollary 6. *For any distance-hereditary graph G and any radius-function r : $V \mapsto N \cup \{0\}$ defined on the vertex set of G, the graph $\Gamma(\mathcal{D}(G, r))$ is weakly chordal and the graph $L(\mathcal{D}(G, r))$ is chordal. In particular [3, 8], odd powers G^{2k+1} $(k = 1, 2, \ldots)$ of a distance-hereditary graph G are weakly chordal, while its even powers G^{2k} $(k = 1, 2, \ldots)$ are chordal.*

5   AT-Free Graphs and Cocomparability Graphs

In this section we consider the generalized powers of AT-free graphs, cocomparability graphs and interval graphs.

In a graph G, an *asteroidal triple* is a triple of vertices such that between any two there is a path of G that avoids the neighbourhood of the third. A graph G is *asteroidal triple-free (AT-free)* if it has no asteroidal triples. It is shown in [12] that any AT-free graph $G = (V, E)$ admits a so-called *strong 2-cocomparability ordering*, i.e., an ordering $\sigma := [v_1, v_2, \ldots, v_n]$ of vertices of G such that for any three vertices x, y, z, if $x \prec y \prec z$ (x precedes y and y precedes z in the ordering) and $d_G(x, z) \leq 2$ then $d_G(x, y) = 1$ or $d_G(y, z) \leq 2$ must hold. Moreover, such an ordering of vertices of an AT-free graph $G = (V, E)$ can be found in time $O(|V| + |E|)$ [13].

Our next lemma shows that a strong 2-cocomparability ordering of an AT-free graph satisfies a useful distance property (proof is omitted).

Lemma 5. *Let G be an AT-free graph and σ be a strong 2-cocomparability ordering of vertices of G. If $x \prec y \prec z$ and $d_G(y, z) > 2$, then $d_G(x, y) + d_G(y, z) \leq d_G(x, z) + 3$.*

It is well known (see, e.g., [16]) that a graph is cocomparability if and only if it admits a *cocomparability ordering*, i.e., an ordering $\sigma := [v_1, v_2, \ldots, v_n]$ of its vertices such that if $x \prec y \prec z$ in σ and $xz \in E(G)$ then $xy \in E(G)$ or $yz \in E(G)$ must hold. Lemma 5 is essential to proving the following result.

Theorem 4. *Let $G = (V, E)$ be an AT-free graph and $r : V \mapsto N$ be a radius-function defined on V. Then, both $L(\mathcal{D}(G, r))$ and $\Gamma(\mathcal{D}(G, r))$ are cocomparability graphs.*

Proof. Using Lemma 5, we will show that any strong 2-cocomparability ordering σ of vertices of G gives a cocomparability ordering for both $L(\mathcal{D}(G, r))$ and $\Gamma(\mathcal{D}(G, r))$. In what follows, we will identify a vertex $N^{r(v)}[v]$ of $L(\mathcal{D}(G, r))$ (and $\Gamma(\mathcal{D}(G, r))$) with v.

Assume, by way of contradiction, that there exist three vertices x, y, z in $L(\mathcal{D}(G, r))$ such that $x \prec y \prec z$ in σ, $xz \in E(L(\mathcal{D}(G, r)))$ but neither xy nor yz is in $E(L(\mathcal{D}(G, r)))$. We know that two vertices $u, v \in V$ are adjacent in $L(\mathcal{D}(G, r))$ if and only if $d_G(u, v) \leq r(u) + r(v)$. Hence, we have $d(x, y) \geq r(x) + r(y) + 1$, $d(y, z) \geq r(y) + r(z) + 1$ and $d(x, z) \leq r(x) + r(z)$, i.e., $d(x, y) + d(y, z) \geq d(x, z) + 2r(y) + 2$. Since, by the theorem assumption, $r(v) \geq 1$ for any $v \in V$, we obtain $d(x, y) + d(y, z) \geq d(x, z) + 4$, which is in a contradiction with Lemma 5 (note that $d(y, z) \geq r(y) + r(z) + 1 \geq 3 > 2$). Thus, a strong 2-cocomparability ordering σ of vertices of G must be a cocomparability ordering for $L(\mathcal{D}(G, r))$, i.e., $L(\mathcal{D}(G, r))$ is a cocomparability graph.

Assume now, by way of contradiction, that there exist three vertices x, y, z in $\Gamma(\mathcal{D}(G, r))$ such that $x \prec y \prec z$ in σ, $xz \in E(\Gamma(\mathcal{D}(G, r)))$ but neither xy nor yz is in $E(\Gamma(\mathcal{D}(G, r)))$. We know that two vertices $u, v \in V$ are adjacent in $\Gamma(\mathcal{D}(G, r))$ if and only if $d_G(u, v) \leq r(u) + r(v) + 1$. Hence, we have $d(x, y) \geq r(x) + r(y) + 2$, $d(y, z) \geq r(y) + r(z) + 2$ and $d(x, z) \leq r(x) + r(z) + 1$, i.e., $d(x, y) + d(y, z) \geq d(x, z) + 2r(y) + 3$. Since, by the theorem assumption, $r(v) \geq 1$ for any $v \subset V$, we obtain $d(x, y) + d(y, z) \geq d(x, z) + 5$, which is in a contradiction with Lemma 5 (note that $d(y, z) \geq r(y) + r(z) + 2 \geq 4 > 2$). Thus, a strong 2-cocomparability ordering σ of vertices of G must be a cocomparability ordering for $\Gamma(\mathcal{D}(G, r))$, i.e., $\Gamma(\mathcal{D}(G, r))$ is a cocomparability graph, too. □

Notice that Theorem 4 is not true if we allow $r(v) = 0$ for some vertices v of G. An induced cycle C_5 on five vertices is an AT-free graph, however $\Gamma(\mathcal{D}(C_5, r))$, where $r(v) = 0$ for each vertex v of C_5, is not a cocomparability graph (since $\Gamma(\mathcal{D}(C_5, r)) \simeq C_5$ and C_5 is not a cocomparability graph). The graph G shown in Figure 1 is an AT-free graph (even an interval graph), however the graph $L(\mathcal{D}(G, r))$ shown in that figure contains an asteroidal triple b, c, g.

As a corollary, we obtain the following result known from [12].

Corollary 7. *[12] If G is an AT-free graph, then G^k is a cocomparability graph for any $k \geq 2$.*

Recall (see, e.g., [9]) that any cocomparability graph is AT-free. Therefore, Theorem 4 holds for any cocomparability graph, too. However, for the class of cocomparability graphs a slightly stronger result can be proven. In [16], it was shown that if G is a cocomparability graph and σ is its cocomparability ordering, then $x \prec y \prec z$ implies $d_G(x, y) + d_G(y, z) \leq d_G(x, z) + 2$. Using this stronger version of Lemma 5, similar to the proof of Theorem 4, one can prove the following.

Theorem 5. *Let $G = (V, E)$ be a cocomparability graph. Then, for any radius-function $r : V \mapsto N$, $L(\mathcal{D}(G, r))$ is a cocomparability graph, and for any radius-function $r : V \mapsto N \cup \{0\}$, $\Gamma(\mathcal{D}(G, r))$ is a cocomparability graph.*

Corollary 8. *[16] All powers G^k ($k \geq 1$) of a cocomparability graph G are cocomparability, too.*

Note that the class of cocomparability graphs contains such known families of graphs as interval graphs, permutation graphs, trapezoid graphs and m-trapezoid graphs. Hence, the graphs $L(\mathcal{D}(G, r))$ and $\Gamma(\mathcal{D}(G, r))$ for a graph G from those families are cocomparability, too. For interval graphs the result can be further strengthened. An *interval* graph is the intersection graph of intervals of a line.

Theorem 6. *Let $G = (V, E)$ be an interval graph. Then, for any radius-function $r : V \mapsto N$, $L(\mathcal{D}(G, r))$ is an interval graph, and for any radius-function $r : V \mapsto N \cup \{0\}$, $\Gamma(\mathcal{D}(G, r))$ is an interval graph.*

Corollary 9. *[14] All powers G^k ($k \geq 1$) of an interval graph G are interval.*

Due to space limitation, we omit results on circular-arc graphs. In the full version of this paper, we show that for any circular-arc graph G and any radius-function $r : V \mapsto N$, both graphs $L(\mathcal{D}(G, r))$ and $\Gamma(\mathcal{D}(G, r))$ are circular-arc.

6   Algorithmic Use of the Generalized Powers of Graphs

Based on the results obtained in the previous sections and known results on ordinary coloring, we deduce the following complexity results for the r-Coloring and r^+-Coloring Problems on the graph families considered in this paper (see Table 1).

Table 1. Complexity results for the r-Coloring and r^+-Coloring Problems on the graph families considered in this paper. $^{(*)}$ marking means that $r(v) = 0$ is not allowed. Here we define the *power-4-chordal* (the *power-weakly-chordal*) graphs as graphs G for which both G and G^2 (equivalently, all powers of G) are 4-chordal (respectively, weakly chordal) graphs. If for a graph G, only G^2 is 4-chordal (is weakly chordal), then we say that G is a *square-4-chordal* (respectively, a *square-weakly-chordal*) graph.

Graph class	Complexity of the r-Coloring problem	Complexity of the r^+-Coloring problem
chordal	hard to approximate	$O(nm)$
4-chordal	hard to approximate	P
weakly chordal	hard to approximate	$O(n^3)$
square-chordal	$O(nm)$	hard to approximate
square-4-chordal	P	hard to approximate
square-weakly-chordal	$O(n^3)$	hard to approximate
power-chordal	$O(nm)$	$O(nm)$
power-4-chordal	P	P
power-weakly-chordal	$O(n^3)$	$O(n^3)$
distance-hereditary	$O(n^2)$	$O(n^3)$
AT-free	$^{(*)}$ $O(n^3)$	$^{(*)}$ $O(n^3)$
cocomparability	$^{(*)}$ $O(n^3)$	$O(n^3)$
interval	$^{(*)}$ $O(n^2)$	$O(n^2)$
circular-arc	$^{(*)}$? and 3/2-approximation	NPc and 3/2-approximation

For a given graph G with n vertices and m edges, we first find the distance matrix of G, then construct the graphs $L(\mathcal{D}(G, r))$ and $\Gamma(\mathcal{D}(G, r))$ using that matrix in $O(n^2)$ time, and finally color $L(\mathcal{D}(G, r))$ and/or $\Gamma(\mathcal{D}(G, r))$ using some known algorithm (depending on what graph family the graph G is from). Note that graphs $L(\mathcal{D}(G, r))$ and $\Gamma(\mathcal{D}(G, r))$ may have now $O(n^2)$ edges. To compute the distance matrix, for distance-hereditary graphs and interval graphs we can use $O(n^2)$ time algorithms presented in [17], for other graph families we use general $O(nm)$ time algorithm. To color chordal graphs (as well as interval graphs), we can use a linear (in the size of the constructed graph) time algorithm from [23]. To color weakly chordal graphs, we use an algorithm from [26] which will color $L(\mathcal{D}(G, r))$ and/or $\Gamma(\mathcal{D}(G, r))$ in $O(n^3)$ time. To color 4-chordal graphs, we can use general polynomial time coloring algorithm designed in [25] for all perfect graphs (4-chordal graphs are perfect). To color cocomparability graphs, we can use $O(n^3)$ time algorithm from [30]. According to [1], it is computationally hard to approximately color the even powers of n-vertex chordal graphs within an $n^{\frac{1}{2}-\epsilon}$ factor, for any $\epsilon > 0$. Consequently, it is computationally hard to approximately r-color any chordal (hence, any weakly chordal, any 4-chordal) graph within the same $n^{\frac{1}{2}-\epsilon}$ factor. According to [6], to (approximately) color a dually chordal graph is as hard as to (approximately) color any graph Since dually chordal graphs are square-chordal, it is computationally hard to approximately r^+-color any square-chordal (hence, any square-weakly-chordal, any square-4-chordal) graph. We know [22] that coloring of circular-arc graphs is an NP-complete problem. Since $\Gamma(\mathcal{D}(G, r)) \simeq G$ when $r(v) = 0$ for every $v \in V$, the general r^+-Coloring problem is also NP-complete on circular-arc graphs. However, we do not know the complexity of the r-Coloring problem on circular-arc graphs G as the graphs $L(\mathcal{D}(G, r))$ may represent only a subclass of circular-arc graphs. One can use the circular-arc graph coloring approximation algorithm of [27] with a performance ratio of $3/2$ to get an approximate solution for the corresponding r-Coloring and/or r^+-Coloring problems on circular-arc graphs with the same performance ratio.

In the full version of this paper, other applications of the generalized powers of graphs (e.g., to r-packing, q-dispersion, k-domination, p-centers, r-clustering, etc.) are discussed.

References

1. G. Agnarsson, R. Greenlaw and M.M. Halldorsson, On powers of chordal graphs and their colorings, *Congressus Numerantium*, 144 (2000), 41-65.
2. G. Agnarsson and M.M. Halldorsson, Coloring powers of planar graphs, *SIAM J. Disc. Math.* 16 (2003), 651-662.
3. H.J. Bandelt, A. Henkmann and F. Nicolai, Powers of distance-hereditary graphs, *Discrete Math.* 145 (1995), 37-60.
4. A.A. Bertossi, M.C. Pinotti and R.B. Tan, Channel Assignment with Separation for Interference Avoidance in Wireless Networks, *IEEE Trans. Parallel Distrib. Syst.* 14 (2003), 222-235.
5. H.L. Bodlaender, T. Kloks, R.B. Tan J.van Leeuwen, λ-Coloring of Graphs, STACS 2000, LNCS 1770, 395–406.
6. A. Brandstädt, V. Chepoi and F. Dragan, The algorithmic use of hypertree structure and maximum neighborhood orderings, *Disc. Appl. Math.* 82 (1998), 43-77.
7. A. Brandstädt, F.F. Dragan, V. Chepoi and V.I. Voloshin, Dually Chordal Graphs, *SIAM J. Discrete Math.* 11 (1998), 437-455.
8. A. Brandstädt, F.F. Dragan and F. Nicolai, Homogeneously orderable graphs, *Theoretical Computer Science* 172 (1997), 209-232.
9. A. Brandstädt, V.B. Le and J. Spinrad, Graph Classes: A Survey, *SIAM Monographs on Discrete Math. Appl.*, (SIAM, Philadelphia, 1999).

10. T. Calamoneri and R. Petreschi, On the Radiocoloring Problem, IWDC 2002, LNCS 2571, 118-127.

11. G.J. Chang and D. Kuo, The $L(2,1)$-labeling problem on graphs, *SIAM J. Disc. Math.* 9 (1996), 309-316.

12. J.M. Chang, C.W. Ho and M.T. Ko, Powers of Asteroidal Triple-free Graphs with Applications, *Ars Combinatoria* 67 (2003), 161-173.

13. J.M. Chang, C.W. Ho and M.T. Ko, LexBFS-ordering in Asteroidal Triple-Free Graphs, ISAAC 1999, LNCS 1741, 163-172.

14. M. Chen and G.J. Chang, Families of Graphs Closed Under Taking Powers, *Graphs and Combinatorics* 17 (2001), 207-212.

15. E. Dahlhaus, P. Duchet, On strongly chordal graphs, *Ars Comb.* 24B (1987), 23–30.

16. P. Damaschke, Distances in cocomparability graphs and their powers, *Discrete Applied Mathematics* 35 (1992), 67-72.

17. F.F. Dragan, Estimating All Pairs Shortest Paths in Restricted Graph Families: A Unified Approach, *Journal of Algorithms* 57 (2005), 1-21.

18. P. Duchet, Classical perfect graphs, *Ann. Discrete Math.* 21 (1984), 67-96.

19. J. Fiala, T. Kloks and J. Kratochvíl, Fixed-parameter complexity of λ-labelings, WG 1999, LNCS 1665, 350-363.

20. C. Flotow, On Powers of Circular Arc Graphs and Proper Circular Arc Graphs, *Discrete Applied Mathematics* 69 (1996), 199-207.

21. D. Fotakis, S.E. Nikoletseas, V.G. Papadopoulou and P.G. Spirakis, NP-Completeness Results and Efficient Approximations for Radiocoloring in Planar Graphs, MFCS 2000, LNCS 1893, 363–372.

22. M.R. Garey, D.S. Johnson, G.L. Miller and C.H. Papadimitriou, The complexity of coloring circular arcs and chords, *SIAM J. Alg. Disc. Meth.* 1 (1980), 216-227.

23. F. Gavril, Algorithms for min. coloring, max. clique, min. covering by cliques and max. independent set of a chordal graph, *SIAM J. Comput.* 1 (1972), 180-187.

24. J.R. Griggs and R.K. Yeh, Labeling graphs with a condition at distance 2, *SIAM J. Discrete Math.* 5 (1992), 586–595.

25. M. Grötschel, L. Lovasz and A. Schrijver, Polynomial algorithms for perfect graphs, In *Topics on perfect graphs*, volume 88, pages 325–356.

26. R. Hayward, J. Spinrad R. Sritharan, Weakly chordal graph algorithms via handles, SODA 2000, 42-49.

27. I.A. Karapetian, On Coloring of Arc Graphs, *Akademiia nauk Armianskoi SSR Doklady*, 70 (1980), 306-311.

28. D. Král', Coloring Powers of Chordal Graphs, *SIAM Journal on Discrete Mathematics*, 18 (2004), 451-461.

29. R. Laskar and D. Shier, Powers and centers of chordal graphs, *Discrete Appl. Math.* 6 (1983), 139-147.

30. R.H. Möhring, Algorithmic aspects of comparability graphs and interval graphs, *Graphs and Order*, (I. Rival, ed.,) (1985) 41–102.

31. S. Ramanathan and E.L. Lloyd, Scheduling Algorithms for Multihop Radio Networks, *IEEE/ACM Transactions on Networking*, 1 (1993), 166-172.

32. A. Raychaudhuri, On powers of strongly chordal and circular arc graphs, *Ars Combin.* 34 (1992), 147-160.

33. I. Todinca, Coloring Powers of Graphs of Bounded Clique-Width, WG 2003, LNCS 2880, 370-382.

Author Index

Lecture Notes in Computer Science

For information about Vols. 1–3964

please contact your bookseller or Springer

Vol. 4007: C. Àlvarez, M. Serna (Eds.), Experimental Algorithms. XI, 329 pages. 2006.

Vol. 4006: L.M. Pinho, M. González Harbour (Eds.), Reliable Software Technologies – Ada-Europe 2006. XII, 241 pages. 2006.

Vol. 4005: G. Lugosi, H.U. Simon (Eds.), Learning Theory. XI, 656 pages. 2006. (Sublibrary LNAI).

Vol. 4004: S. Vaudenay (Ed.), Advances in Cryptology - EUROCRYPT 2006. XIV, 613 pages. 2006.

Vol. 4003: Y. Koucheryavy, J. Harju, V.B. Iversen (Eds.), Next Generation Teletraffic and Wired/Wireless Advanced Networking. XVI, 582 pages. 2006.

Vol. 4001: E. Dubois, K. Pohl (Eds.), Advanced Information Systems Engineering. XVI, 560 pages. 2006.

Vol. 3999: C. Kop, G. Fliedl, H.C. Mayr, E. Métais (Eds.), Natural Language Processing and Information Systems. XIII, 227 pages. 2006.

Vol. 3998: T. Calamoneri, I. Finocchi, G.F. Italiano (Eds.), Algorithms and Complexity. XII, 394 pages. 2006.

Vol. 3997: W. Grieskamp, C. Weise (Eds.), Formal Approaches to Software Testing. XII, 219 pages. 2006.

Vol. 3996: A. Keller, J.-P. Martin-Flatin (Eds.), Self-Managed Networks, Systems, and Services. X, 185 pages. 2006.

Vol. 3995: G. Müller (Ed.), Emerging Trends in Information and Communication Security. XX, 524 pages. 2006.

Vol. 3994: V.N. Alexandrov, G.D. van Albada, P.M.A. Sloot, J. Dongarra (Eds.), Computational Science – ICCS 2006, Part IV. XXXV, 1096 pages. 2006.

Vol. 3993: V.N. Alexandrov, G.D. van Albada, P.M.A. Sloot, J. Dongarra (Eds.), Computational Science – ICCS 2006, Part III. XXXVI, 1136 pages. 2006.

Vol. 3992: V.N. Alexandrov, G.D. van Albada, P.M.A. Sloot, J. Dongarra (Eds.), Computational Science – ICCS 2006, Part II. XXXV, 1122 pages. 2006.

Vol. 3991: V.N. Alexandrov, G.D. van Albada, P.M.A. Sloot, J. Dongarra (Eds.), Computational Science – ICCS 2006, Part I. LXXXI, 1096 pages. 2006.

Vol. 3990: J. C. Beck, B.M. Smith (Eds.), Integration of AI and OR Techniques in Constraint Programming for Combinatorial Optimization Problems. X, 301 pages. 2006.

Vol. 3989: J. Zhou, M. Yung, F. Bao, Applied Cryptography and Network Security. XIV, 488 pages. 2006.

Vol. 3988: A. Beckmann, U. Berger, B. Löwe, J.V. Tucker (Eds.), Logical Apporaches to Computational Barriers. XV, 608 pages. 2006.

Vol. 3987: M. Hazas, J. Krumm, T. Strang (Eds.), Location- and Context-Awareness. X, 289 pages. 2006.

Vol. 3986: K. Stølen, W.H. Winsborough, F. Martinelli, F. Massacci (Eds.), Trust Management. XIV, 474 pages. 2006.

Vol. 3984: M. Gavrilova, O. Gervasi, V. Kumar, C.J. K. Tan, D. Taniar, A. Laganà, Y. Mun, H. Choo (Eds.), Computational Science and Its Applications - ICCSA 2006, Part V. XXV, 1045 pages. 2006.

Vol. 3983: M. Gavrilova, O. Gervasi, V. Kumar, C.J. K. Tan, D. Taniar, A. Laganà, Y. Mun, H. Choo (Eds.), Computational Science and Its Applications - ICCSA 2006, Part IV. XXVI, 1191 pages. 2006.

Vol. 3982: M. Gavrilova, O. Gervasi, V. Kumar, C.J. K. Tan, D. Taniar, A. Laganà, Y. Mun, H. Choo (Eds.), Computational Science and Its Applications - ICCSA 2006, Part III. XXV, 1243 pages. 2006.

Vol. 3981: M. Gavrilova, O. Gervasi, V. Kumar, C.J. K. Tan, D. Taniar, A. Laganà, Y. Mun, H. Choo (Eds.), Computational Science and Its Applications - ICCSA 2006, Part II. XXVI, 1255 pages. 2006.

Vol. 3980: M. Gavrilova, O. Gervasi, V. Kumar, C.J. K. Tan, D. Taniar, A. Laganà, Y. Mun, H. Choo (Eds.), Computational Science and Its Applications - ICCSA 2006, Part I. LXXV, 1199 pages. 2006.

Vol. 3979: T.S. Huang, N. Sebe, M.S. Lew, V. Pavlović, M. Kölsch, A. Galata, B. Kisačanin (Eds.), Computer Vision in Human-Computer Interaction. XII, 121 pages. 2006.

Vol. 3978: B. Hnich, M. Carlsson, F. Fages, F. Rossi (Eds.), Recent Advances in Constraints. VIII, 179 pages. 2006. (Sublibrary LNAI).

Vol. 3977: N. Fuhr, M. Lalmas, S. Malik, G. Kazai (Eds.), Advances in XML Information Retrieval and Evaluation. XII, 556 pages. 2006.

Vol. 3976: F. Boavida, T. Plagemann, B. Stiller, C. Westphal, E. Monteiro (Eds.), NETWORKING 2006. Networking Technologies, Services, and Protocols; Performance of Computer and Communication Networks; Mobile and Wireless Communications Systems. XXVI, 1276 pages. 2006.

Vol. 3975: S. Mehrotra, D.D. Zeng, H. Chen, B. Thuraisingham, F.-Y. Wang (Eds.), Intelligence and Security Informatics. XXII, 772 pages. 2006.

Vol. 3973: J. Wang, Z. Yi, J.M. Zurada, B.-L. Lu, H. Yin (Eds.), Advances in Neural Networks - ISNN 2006, Part III. XXIX, 1402 pages. 2006.

Vol. 3972: J. Wang, Z. Yi, J.M. Zurada, B.-L. Lu, H. Yin (Eds.), Advances in Neural Networks - ISNN 2006, Part II. XXVII, 1444 pages. 2006.

Vol. 3971: J. Wang, Z. Yi, J.M. Zurada, B.-L. Lu, H. Yin (Eds.), Advances in Neural Networks - ISNN 2006, Part I. LXVII, 1442 pages. 2006.

Vol. 3970: T. Braun, G. Carle, S. Fahmy, Y. Koucheryavy (Eds.), Wired/Wireless Internet Communications. XIV, 350 pages. 2006.

Vol. 3969: Ø. Ytrehus (Ed.), Coding and Cryptography. XI, 443 pages. 2006.

Vol. 3968: K.P. Fishkin, B. Schiele, P. Nixon, A. Quigley (Eds.), Pervasive Computing. XV, 402 pages. 2006.

Vol. 3967: D. Grigoriev, J. Harrison, E.A. Hirsch (Eds.), Computer Science – Theory and Applications. XVI, 684 pages. 2006.

Vol. 3966: Q. Wang, D. Pfahl, D.M. Raffo, P. Wernick (Eds.), Software Process Change. XIV, 356 pages. 2006.

Vol. 3965: M. Bernardo, A. Cimatti (Eds.), Formal Methods for Hardware Verification. VII, 243 pages. 2006.